W9-ARZ-140

The History of Modern Physics

1800 – 1950

Volume 11

The History of Modern Physics, 1800–1950

The History of Modern Physics, 1800–1950

TITLES IN SERIES

VOLUME 1
Alsos
by Samuel A. Goudsmit

VOLUME 2
Project Y: The Los Alamos Story
Part I: Toward Trinity
by David Hawkins
Part II: Beyond Trinity
by Edith C. Truslow and Ralph Carlisle Smith

VOLUME 3
American Physics in Transition: A History of Conceptual Change in the Late Nineteenth Century
by Albert E. Moyer

VOLUME 4
The Question of the Atom: From the Karlsruhe Congress to the First Solvay Conference, 1860–1911
by Mary Jo Nye

VOLUME 5
Physics For a New Century: Papers Presented at the 1904 St. Louis Congress

VOLUME 6
Basic Bethe:
Seminal Articles on Nuclear Physics, 1936–1937
by Hans A. Bethe, Robert F. Bacher, and M. Stanley Livingston

VOLUME 7
History of the Theories of Aether and Electricity: Volumes I and II
by Sir Edmund Whittaker

The History of Modern Physics, 1800–1950

VOLUME 8
Radar in World War II
by Henry Guerlac

VOLUME 9
Atoms in the Family: My Life with Enrico Fermi
by Laura Fermi

VOLUME 10
Quantum Physics in America: The Years Through 1935
by Katherine R. Sopka

VOLUME 11
The Theory of Heat Radiation
by Max Planck

INTRODUCTORY NOTE

The Tomash/American Institute of Physics series in the History of Modern Physics offers the opportunity to follow the evolution of physics from its classical period in the nineteenth century when it emerged as a distinct discipline, through the early decades of the twentieth century when its modern roots were established, into the middle years of this century when physicists continued to develop extraordinary theories and techniques. The one hundred and fifty years covered by the series, 1800 to 1950, were crucial to all mankind not only because profound evolutionary advances occurred but also because some of these led to such applications as the release of nuclear energy. Our primary intent has been to choose a collection of historically important literature which would make this most significant period readily accessible.

We believe that the history of physics is more than just the narrative of the development of theoretical concepts and experimental results: it is also about the physicists individually and as a group—how they pursued their separate tasks, their means of support and avenues of communication, and how they interacted with other elements of their contemporary society. To express these interwoven themes we have identified and selected four types of works: reprints of "classics" no longer readily available; original monographs and works of primary scholarship, some previously only privately circulated, which warrant wider distribution; anthologies of important articles here collected in one place; and dissertations, recently written, revised, and enhanced. Each book is prefaced by an introductory essay written by an acknowledged scholar, which, by placing the material in its historical context, makes the volume more valuable as a reference work.

The books in the series are all noteworthy additions to the literature of the history of physics. They have been selected for their merit, distinction, and uniqueness. We believe that they will be of interest not only to the advanced scholar in the history of physics, but to a much broader, less specialized group of readers who may wish to understand a science that has become a central force in society and an integral part of our twentieth-century culture. Taken in its entirety, the series will bring to the reader a comprehensive picture of this major discipline not readily achieved in any one work. Taken individually, the works selected will surely be enjoyed and valued in themselves.

The History of Modern Physics 1800 – 1950 *Volume 11*

The Theory of Heat Radiation

MAX PLANCK

Introduction by
ALLAN A. NEEDELL

Translation by
MORTON MASIUS

Tomash Publishers

American Institute of Physics

Library of Congress Cataloging-in-Publication Data

Planck, Max, 1858–1947.
The theory of heat radiation.
(The History of modern physics, 1800–1950; v. 11)
Text in English and German.
Reprint (1st work). Originally published: Philadelphia: P. Blakiston, c1914.
Reprint (2nd work). Originally published: Leipzig: J. A. Barth, 1906.
Includes bibliographical references.
ISBN 0-88318-597-0
1. Heat—Radiation and absorption. 2. Electromagnetic waves.
I. Title. II. Series
QC331.P73 1989
536′.33 88–34240

CONTENTS

Introduction

Allan A. Needell
Smithsonian Institution

Planck's Career to 1900

Max Planck was born in 1858 in Kiel, an important commercial port city in the then Danish-controlled duchy of Holstein. Son of a professor of jurisprudence, Planck was raised in a comfortable German-speaking household. Planck's youth corresponded with a tumultuous period in Central European history—the period that ultimately produced a politically unified Germany dominated by Prussia. As he matured, Planck came to identify his own life and values closely with Germany and its culture. The ideal of unity became central to his emerging world view.[1]

Planck's study of physics began in 1874 in Munich where his family had moved seven years earlier. His primary interest was in clarifying and building upon the theoretical foundations of his subject.[2] In 1877 Planck spent two semesters in Berlin, where he attended lectures by Helmholtz and Kirchhoff. After he returned to Munich, Planck prepared his doctoral dissertation and was awarded the Ph.D. in 1879. He then served as a *Privatdocent* at Munich until 1885.[3]

Planck's special interest in the foundations of physics was given an important institutional boost by his being called, in the spring of 1885, to Kiel as an associate (extraordinary) professor of theoretical physics. It was a position that had been declined by Heinrich Hertz, who had been a *Privatdocent* at Kiel but who preferred a more standard emphasis on experimental work. Planck, in contrast, was quite satisfied with the emphasis on theory. He remained at Kiel until 1889 when he was called to Berlin to a similar professorship and to direct a newly established institute for theoretical physics.[4]

The institute was modest by all standards. Basically, it amounted to a library and a research assistant. Planck's major responsibility, in addition to research, was to present a multisemester cycle of lectures.[5]

Planck's career blossomed in Berlin. In 1892 he was promoted to the position of full (ordinary) professor; in 1894 he was admitted to membership in the Prussian Academy of Sciences.[6] He used his increasingly influential position as an important platform to articulate a basic, conservative research philosophy. His program emphasized the universality of the most fundamental of the known laws of physics and the need to explore the full implications of those basic laws. His goal was, in part, to provide a check on the excesses of several more specialized and speculative theoretical approaches then current in both continental and English physical thinking.[7]

From his student days onward Planck believed the greatest unifying principles of physics were those of thermodynamics—the principle of the conservation of energy and the principle of the increase of entropy.[8] Thermodynamics became his specialty.

In 1895, Planck began to work on the thermal behavior of ideal oscillating electric dipoles as an extension of his primary interest. In 1900 that effort led him to a new mathematical expression for the energy distribution of blackbody radiation. Shortly thereafter, in attempting to provide a derivation for the new blackbody law, Planck introduced the quantum constant h into his mathematical equations—an act that by most accounts marks the beginning of the twentieth-century revolution in physics.

The Lectures on the Theory of Heat Radiation

In ways that have been duly emphasized in the historical writings of Martin Klein and Thomas Kuhn, the original quantum papers of 1900 and 1901 were difficult for Planck's contemporaries to follow.[9] In contrast, Planck's 1906 *Vorlesungen über die Theorie der Wärmestrahlung* (*Lectures on the Theory of Heat Radiation*) provided the physics community with a comprehensive account of his theory of blackbody radiation and the role of the quantum hypothesis within it. Although the theory it contained left many essential questions unanswered (as Planck freely admitted), the 1906 *Wärmestrahlung* is nonetheless a superb example of a genre, important at the turn of the century, in which leading physicists set forth their personal visions of how the corpus of physical knowledge in a given area should best be understood. Such texts served various purposes: from establishing a unified curriculum for the training of students, to providing a means of staking out intellectual property rights for systematic work not directly associated with new experimental results.[10]

The *Wärmestrahlung* (1906), which is reprinted in the original German as the second part of this volume, was a careful recapitulation of the steps Planck had taken to derive the new radiation law. Planck organized it to make the major points clear and to serve as a text for learning the possibilities and limitations inherent in combining electromagnetic and thermodynamic reasoning in the analysis of a complex physical phenomenon. Section 1 of the text is devoted to basic facts and definitions, including the definition of blackbody radiation. Section 2 presents general results obtainable from Maxwellian electrodynamics and the two laws of thermodynamics. Section 3 presents the details (developed in the years 1895 to 1899) of Planck's theory of ideal electromagnetic oscillators and the interactions of such oscillators with surrounding radiation. In Section 4 Planck turns to the subject of entropy, which he continued to think was centrally important. It is in this section that he introduces the quantum hypothesis. And finally, Planck provides a tentative chapter describing the way in which the systems described in the previous sections could be thought of as demonstrating irreversibility.

In 1913 Planck published a revised edition of the *Wärmestrahlung*. In the meantime, his continuing work and questions raised by scientists like Albert Einstein and H. A. Lorentz had led him to substantially reformulate his own theory. And by that time there was considerable interest in Germany and abroad in a range of questions associated with the quantum hypothesis.

Morton Masius, who received his Ph.D. from Leipzig in 1908 and had since moved to the United States (first as a postdoctoral fellow at Harvard and then as professor at the Worcester Polytechnic Institute), obtained Planck's permission to translate the new version of the *Wärmestrahlung* into English.[11] Published in 1914, Masius's translation is quite literal. As he claims in the translator's preface, the only changes were the addition of a few explanatory footnotes and an appendix consisting of a selected bibliography of "the most important papers on the subjects treated of in this book and others closely related to them." The 1914 translation is reprinted as the first part of this volume.

Planck's work and his text were quite influential in America. As Gerald Holton has documented, an important path was through Robert A. Millikan, the second American scientist to win the Nobel prize, and Edwin C. Kemble, who was "arguably the first American to use quantum-theoretical considerations to predict results that were not yet in the domain of experimental knowledge,..." and who was also the teacher of "a considerable fraction of the new generation of American theoretical physicists."[12]

The version of the *Wärmestrahlung* translated by Masius, in addition to its importance in disseminating quantum ideas, also represents a critical step in the evolution of Planck's own ideas concerning quanta. Historians of physics have described the reformulations of Planck's theory of 1911–1913 as providing progressively weaker versions of the more radical theory of the period 1900–1901.[13] More recently, the revisions of Planck's radiation theory have been seen as "... the first attempts [by Planck] to incorporate discontinuity and develop a non-classical physics." In this view, the earlier theory is seen as essentially classical, containing no postulation of physical discontinuities whatever.[14]

Such widely different interpretations recommended to the editors of the series that the actual texts, now out of print, be made readily available to those who seek to study and understand the history of quantum physics. As an introduction, I will summarize (with selective references to the historical literature) the most widely accepted historical accounts of Planck's contributions to the quantum theory. But I will also depart in several ways from those accounts. Most significantly, I will consciously deemphasize the questions of whether and when Planck first introduced discontinuities into the basic laws of physics. Instead I will place more emphasis on Planck's avoidance in principle of such physical assumptions and on the consistency of Planck's research program through 1914. It is, after all, as products of Planck's thinking during the fifteen or so years following his turn-of-the-century quantum papers that these reprinted editions of Planck's *Wärmestrahlung* are to be understood and appreciated.

It is my view that Planck's personal commitment to the ideal of universal principles guiding mankind in a complex world of confusing detail was extremely powerful. I see that commitment as convincing Planck that the basic laws used to describe small-scale physical processes were fundamentally incomplete. That was before 1900. Such recognized incompleteness is what left room in Planck's thinking for a new universal constant of nature, the interpretation of which was to be sought outside of the existing framework of physical theory. In the years prior to 1914 Planck doggedly sought to apply his own special formulation of the second law of thermodynamics in an effort to clearly define the proper realm of quantum processes and to sharply demarcate that realm from the realms where mechanics and electrodynamic theory had proven successful.

Planck's views were idiosyncratic; at the same time they were expressions of deep cultural values and they reflect important aspects of contemporary scientific thinking. How Planck's own work was transformed from its own local context of creation into the modern quantum theory, given the varying commitments and presumptions of other physicists, remains an important subject for historians to ponder.

The Second Law and Irreversible Radiation Processes

In his doctoral dissertation (1879),[15] Planck identified the second law of thermodynamics with the experimental fact that the processes observed in nature could never be completely undone. Following Clausius, Planck viewed the second law as implying that the return of a system to its initial condition, after a given natural process, could not be accomplished without the simultaneous creation of a change in the surroundings of that system. The free expansion of a gas, for example, left the universe permanently altered. The return of the gas to its original volume and temperature could not be accomplished without some lasting effect (such as, in that case, the lowering of a weight and the heating up of some part of the environment).

For Planck, the essence of the second law was conveyed by the statement that there existed an unambiguous ordering of all the possible observable states of any natural system. Mathematically, such an ordering suggested that a state function could be established whereby the numerical value associated with each state would express quantitatively what Planck described in 1887 as nature's apparent "preference" for some states over others.[16]

The entropy function developed by Clausius (1865) for mechanical systems served just that purpose. Planck believed that it should be possible to associate a specific entropy value with every observable state of any physical system capable of exchanging energy with its environment or with other systems. He believed that this entropy function should be afforded the same fundamental status as energy; and he believed that Clausius's principle of the increase of entropy should stand beside Helmholtz's 1847 principle of energy conservation as an exemplar of nature's absolute lawfulness.[17]

Planck's work to 1895 developed these ideas in many different contexts. He published papers extending the range of thermodynamics from simple physical systems characterized by macroscopic observables like volume, temperature, and pressure, to systems involving, for example, electrochemical processes and complex chemical and phase transitions.[18]

When Planck took up the problem of irreversible radiation processes in 1895 he continued his emphasis on irreversibility; but he also significantly expanded the goals of his research.[19] Planck now hoped to do more than describe irreversibility in precise mathematical terms. In addition, he hoped to lay bare in the laws of either electrodynamics or the mechanics of continuous media, the source of the absolute validity he attributed to the second law of thermodynamics. To his way of thinking, the ability to account for irreversible behavior would amply justify the further pursuit of one or another versions of the rapidly advancing electrodynamic theory, or of theories involving waves in a mechanical continuum.[20] Planck had recently become convinced that irreversibility could never be explained solely on the basis of the kinds of mechanical assumptions customarily made in the kinetic theory of gases.[21] If some other theory promised to succeed where the kinetic theory had failed, Planck believed there would be good reason to suspect that other important results would follow.

Planck's investigations over the next several years focussed upon ideal resonators; that is, simple sinusoidally varying electric dipoles. He believed that such resonators were the simplest possible models for localized centers of the emission and absorption of electromagnetic radiation. They were in no way meant to be specific models of atoms or molecules. To the contrary, Planck's resonators were meant to be abstract and general enough to indicate whether electrodynamics alone could be the universal source of irreversible behavior.[22]

Planck originally sought to decide this question on the basis of the interaction and approach to equilibrium of a single resonator with a surrounding electromagnetic field. But he immediately recognized that, as with all systems subject to irreversible processes, the equilibrium state of electromagnetic radiation would itself be characterizable by a set of macroscopic properties. Among the equilibrium properties cited by Planck in 1897 (in addition to a definite value for the average energy of an oscillator) were: a spatially homogeneous distribution of electromagnetic energy; and a special sort of alteration of the "color" of the incident radiation.

The last named property indicates that Planck quickly anticipated that the new theory he was developing might provide a method of deriving the distribution law for "normal" radiation. "It is to be expected," wrote Planck, "that the resonator also acts in the direction of a certain equalization of the intensities of the different colors, from which result important conclusions for the energy distribution of a stationary radiation state inside the cavity being considered."[23]

Kirchhoff's law implied that the energy distribution within a closed cavity would be the same regardless of the specific characteristics of any emitting and absorbing materials present (as long as there were some such material). The simplicity of Planck's resonators did not exclude them from Kirchhoff's law; to the contrary, their simplicity appeared to promise a clearer, more direct view of fundamental principles than would be possible using a more complex model. Planck continued to believe that irreversibility was a universal phenomenon whose source had to lie in a general and fundamental aspect of the most basic laws of nature. Success in establishing the correct equilibrium distribution law would be persuasive evidence in favor of those who believed that all phenomena would ultimately be reducible to electromagnetism.

But success was not to come. The laws of electromagnetic theory were subject to objections similar to those Planck had earlier raised to discredit the kinetic theory.[24] Boltzmann pointed this and more out to Planck. In a note published in the *Proceedings of the Prussian Academy of Sciences* in Berlin later in 1897, Boltzmann addressed himself directly to Planck's earlier general approach to thermodynamics and to his more recent work on radiation processes:

> To accept the dissipation of energy [i.e., irreversibility] as a fact of experience and to be satisfied with those formulas which the general

mechanical theory of heat sets out as the best expression of this empirical fact, is an attitude which permits of no objections.... If one, however, wishes to derive the dissipation of energy by means of a picture of electromagnetic waves, this can be done only by means of...an hypothesis...that the initial states are arranged in a special manner.[25]

Alternatively, in a subsequent note, Boltzmann suggested the applicability of his own probabilistic definition of entropy which he had formulated in 1877 and included in his 1896 text:

Just as [has been done] in the gas theory, one could also determine the most probable state of the radiation; or more validly, a general rule embracing all of the many states in which the waves are not ordered in any way...[26]

Planck was gradually convinced that his own oscillator-based derivation of irreversibility was in error and that his attempt to reduce irreversibility to electrodynamics would have to be set aside. Abandoning this program was made easier, however, because the comments made by Boltzmann pointed the way to what Planck hoped would be a general means of deriving the entropy of an oscillator and, as a result, the law of blackbody radiation. The derivation also provided Planck with some entirely new insights into the nature of irreversibility.

Guided by the assumption of "molecular disorder" used by Boltzmann in his Lectures on Gas Theory (1896), Planck formulated his own special assumption of "natural radiation." Both assumptions presented specific limitations on the allowable initial states of thermodynamic processes and eliminated the possibility of behavior in violation of the second law. In the gas theory, the assumption of molecular disorder ruled out correlations between the motions of neighboring particles, even immediately after one or more of those molecules had undergone a collision; while in Planck's theory of cavity radiation the assumption of natural radiation ruled out correlations between the phases of the various waves (Fourier components) that made up the electromagnetic state of the system.[27] Planck viewed these assumptions as examples of a general principle of elementary disorder, a physical principle he now believed essential for understanding the second law and one that was not a direct consequence of either the laws of mechanics or the laws of electrodynamics. Significantly, the principle of elementary disorder—a statement about the actual states achievable by physical systems—is given great emphasis in the *Wärmestrahlung* (1906) [§133–§137, §146; pp. 135–140, 149–150] and in Planck's other writings on irreversibility up to 1914.

Planck still considered the approach of a resonator to a definite average energy to be an irreversible process, and to Planck that continued to imply that a proper entropy function could be defined. The fact that the irreversibility of the processes depended essentially on the assumption of disorder among the waves making up the surrounding radiation did not alter the role of entropy as the quantitative indicator of the approach of the system to an equilibrium state. It did imply that the concepts of disorder and entropy were closely related.

In 1899 detailed electromagnetic calculations combined with the hypothesis of natural radiation led Planck to a simple linear equation relating the equilibrium radiation energy density (at a given frequency) to the time-average of the energy of an interacting ideal oscillator (of the same frequency). To determine the equilibrium value of the oscillator's average energy, Planck sought to provide an appropriate definition of oscillator entropy. As Martin Klein has described, Planck apparently worked backward from the so-called Wien distribution law to obtain a consistent oscillator entropy function. After he defined a similar expression for the entropy of the surrounding radiation Planck showed that the total entropy of the system could only increase with time. Then, using a variational principle involving virtual exchanges of energy between oscillators of differing frequencies, Planck derived the Wien distribution law.[28]

The story of how, once experimental evidence of discrepancies between new, long-wavelength measurements and the Wien distribution law were made known to him, Planck revised his oscillator entropy expression to obtain a new distribution law has been told in detail.[29] Also told, but with far more controversy, is the story of how Planck adopted at least some aspects of Boltzmann's combinatorial method of defining entropy to obtain the result he required.[30]

The crucial differences in the accounts concern the following questions: (1) Did Planck actually postulate discrete energy values for the oscillator in the course of defining entropy; and (2) how closely and faithfully was Planck following Boltzmann's statistical methods.

Referring the reader to the relevant literature,[31] I would limit myself to the following points, which will be elaborated in my discussion of Planck's subsequent research: (1) Planck was not concerned with describing details of the motion or behavior of an oscillator. The oscillator was meant to be an abstraction, and Planck admitted at the outset that it was a fundamentally incomplete model for an actual center of emission and absorption. (2) Combinatorial methods were used to obtain a quantitative measure of the disorder that presumably arose from sources outside of electrodynamic theory, a measure that could be related to entropy. (3) Planck did not accept a probabilistic view of irreversibility. He believed irreversibility was a universal characteristic of natural processes. And therefore, (4) Planck's entropy calculations were not based on rigorous probability considerations. Any well-defined combinatorial expression would have sufficed, so long as that expression demonstrated the monotonic character of all proper entropy functions and proved consistent with existing energy distribution measurements. Once established, the form of the entropy function, it could be hoped, would provide insight into the physical source of irreversibility.

In my account, the question of whether Planck introduced discontinuities in 1900 is not a major issue; the answer depends on the attitude toward physical concepts and the level of commitment one requires to credit someone with such an introduction. Planck, I suspect, would have been the first to admit that he had no idea what went on at an actual center of the emission and absorption of electromagnetic radiation. He believed that physicists were well-advised to refrain from such speculation.

The question of how closely Planck understood and followed Boltzmann's methods is more difficult and probably should be expanded to embrace the larger question of the status of statistical physics at the turn-of-the century. Whatever can be said about this larger question (and much research needs to be done), it would be a mistake to assume that generally, behind the use of statistics in the kinetic theory of gases, there lay a complete, logically consistent theoretical structure.[32] The more specific question of what Boltzmann himself believed or understood has been addressed by historians, but it is beyond the scope of this essay.[33]

Most important for understanding Planck's work on the quantum theory, the recognition that the second law of thermodynamics is of a different sort than the first (probabilistic rather than absolute) was not shared widely until well into the second decade of the century.[34] Planck, for one, continued to believe in the absolute validity of the second law (without even overwhelmingly improbable exceptions) at least until 1912, the year he finished the second edition of the *Wärmestrahlung*.[35]

To document this claim, one must look carefully at Planck's subsequent writings and other physicists' comments on them. In addition to the above mentioned sections of the *Wärmestrahlung* and the detailed physics papers of the period, one can cite Planck's published reactions to the publication of J. W. Gibbs's *Elementary Principles in Statistical Mechanics*, which appeared in 1902.[36] Also supportive are the semi-popular lectures on physics Planck delivered in Leiden in 1908 and at Columbia University in New York in 1909.[37] The continued insistence on the absolute validity of the second law is central to all of these writings. Similarly, that this was his view was clear to his contemporary readers. The most important reader for our purpose was Einstein, who in 1906 published a review of the *Wärmestrahlung* (1906) that commented, if indirectly, on this aspect of Planck's procedure.[38]

In 1909, Einstein had not changed his view of Planck's procedure, criticizing it more directly and in more detail. In a paper entitled, "On the Current State of the Radiation Problem," Einstein included a detailed discussion of Planck's use of probabilities in the calculation of entropy. Immediately upon taking up Planck's derivation of the radiation law, Einstein isolated the basic difference between his and Planck's understanding of irreversibility:

> If one proceeds from the standpoint that the irreversibility of natural processes is only apparent and that an irreversible process consists of a transformation to a more probable state, one must first give a definition of the probability W of a state. The single definition that in my opinion can be considered would be the following:
> Let $A_1, A_2, \ldots, A_f$ be all the states accessible to a closed system of definite energy, or more precisely, all the states that we can differentiate by certain auxiliary means. According to the classical theory, after a certain time the system assumes one specific state from among these (e.g., A_f) and remains in this state (thermodynamic equilibrium). According to the statistical theory, however, the system assumes in a random sequence all of the states $A_1, A_2, \ldots, A_f$ over and over.[39]

Einstein urged the latter view in no uncertain terms. In a footnote he cited as proof, "...the properties of Brownian motion."

After he had presented his own definition of probability for thermodynamic systems, Einstein added that "neither Herr Boltzmann nor Herr Planck has given a definition of W. They purely formally set W [equal to] the number of complexions of the state considered." Einstein admitted that if one required each of the microstates to be equally probable, one could be led indirectly but equivalently to the correct expression for the probability of a given macrostate. He insisted, however, that in that case one was not free in the choice of how a complexion was to be defined. Einstein claimed that while this had been "clearly recognized by Boltzmann," it had been ignored by Planck.[39]

The distinction being made by Einstein was between his insistence that entropy had to be properly "defined" as proportional to an actual probability, as opposed to Planck's procedure of "setting" entropy proportional to an *ad hoc* "thermodynamic probability." For Einstein, entropy was a statistical quantity that reflected the probabilistic nature of irreversibility and of equilibrium. For Planck, entropy was a special quantitative measure of disorder. The whole procedure of counting complexions was justified solely because it produced a well-behaved function that could be shown to act like thermodynamic entropy.

As early as his discussion of Gibbs's *Statistical Mechanics,*[40] Planck had rejected the notion that, for natural systems, deviations from the average values predicted by statistical calculations could ever be observable. Whereas fluctuations to Einstein were quantitative, measurable consequences of the statistical nature of the second law of thermodynamics, in 1909—and for some years thereafter—fluctuations continued to have no place in Planck's view of the world. Planck remained committed to the absolute validity of the second law. For him combinatorials or "probabilities" were only to be applied as a means of calculating entropy and then only to those *hidden* variables (such as the positions and velocities of individual molecules or the instantaneous values for the energy of an electromagnetic oscillator) that were inaccessible to observation and measurement. A necessary prerequisite for such a use of combinatorials was the physical hypothesis of elementary disorder. The source of that disorder was still the great unsolved mystery.

By the time he delivered his famous Leiden address, "On the Unity of the Physical World Picture" in 1908, Planck had finally come to recognize that his view of the physical significance of the principle of elementary disorder was not the same as Boltzmann's. He had not been clear on this point earlier.[41] The difference between his view and Boltzmann's stemmed from the fact that in Boltzmann's presentation of the kinetic theory of gases certain conceivable arrangements of atoms led directly to violations of the second law. Because of that, according to Planck, Boltzmann had left a place open for such behavior in his physical world picture. Planck disagreed because, as he insisted, "a nature in which the reverse flow of heat into a warmer body...occurs, would no longer be our nature."[42]

There remained, for Planck, a fundamental difference between reversible and irreversible behavior, just as there was a fundamental difference between the kinds of mechanical or electrodynamic models generally used to represent matter and the actual systems that existed in nature. Irreversibility provided one of the few direct clues for physicists to follow in constructing better microscopic theories.

It was his special understanding of the second law that guided Planck's original formulation of a quantum hypothesis in 1900, and it was that basically unchanged view that continued to guide his work on the quantum theory up to and including the second edition of the *Wärmestrahlung*. As has been indicated, the failure of mechanics or electrodynamics to account for irreversibility was, for Planck, an indication of their incompleteness. And that incompleteness is what left room for, and indeed required, the quantum constant or some other assumption (assumptions he would eventually refer to as "unclassical").

The Status of "Microphysics": An Aside on the Term "Classical"

Historians have recently begun to emphasize an important development that took place beginning in the 19th century and continuing into our own.[43] The development was from experimental physics being a science limited to macroscopic observations of the "properties of bulk matter" (e.g., temperatures, pressures, light intensity, etc.) to a science in which— with the discovery, for example, of cathode rays, x-rays, and the Zeeman effect— investigators began to argue with more and more confidence from their macroscopic measurements to conclusions about the hitherto unobservable microworld. The continuation of this development saw, with the discovery of radioactivity and later of cosmic-ray experiments, the advent of experimental techniques thought to reveal characteristics of individual "atomic" processes more and more directly. The microworld gradually became populated with more specially conceived entities like electrons, ions, atoms, and assorted kinds of radiation.

The transformation of the experimental art and its increasing focus on smaller and smaller scale processes bears a direct, if complex, relation to contemporary ideas and theoretical assumptions. Maxwell recognized early on that the information about microscopic physical processes available from experiment, especially those assumed to involve atoms or molecules, was of a fundamentally different sort (i.e., statistical) than that made use of in the so-called dynamical (Maxwell used the term historical) branches of physics epitomized by Newtonian mechanics. In response, the Maxwellians of the latter part of the century chose to reject all special atomistic assumptions in favor of an attempt "to describe the world in terms of continuous values of field variables,"[44] while others, most notably Maxwell himself and Ludwig Boltzmann, went on to develop the kinetic theory of gases. In the kinetic theory, certain macroscopic behavior was explained as the aggregate effect of atoms or molecules supposed to obey simple dynamical laws (they were imagined to obey Newton's laws of motion in free space and to interact by means of central attractive or repulsive forces whose means of action or propagation was not clearly spelled out).[45] But both the electromagnetic and kinetic programs were vulnerable to the enormous philosophical gap between the macroscopic evidence gathered and analyzed and the microscopic assumptions that were made. That gap was clearly recognized and deeply concerned many late 19th-century physicists.[46]

The scientific literature (articles, lectures, and texts) prior to 1900 is filled with lively discussion of the proper status to be given to scientists' microscopic assumptions (whether they were thought of as based on atomistic or continuum mechanics, mechanical analogies, electricity, or on combinations of those or others).[47] Indeed, it has been noted that around the turn of the century, in Europe and in England, many physicists began shrinking away from claims that their theories were sufficient to allow them to actually know and understand the microscopic processes of nature.[48]

In Germany, among academics, such discussions often focussed on the physicist as potential provider of a world-view and on various alternatives (e.g., mechanical, energetic, electromagnetic, etc.). Taking part, in addition to Planck, were virtually all leaders in the profession.[49] It is easy for the modern reader to overlook the extent to which the many research programs of the late 19th-century physics were viewed, even by their most prolific proponents, as *ad hoc* and essentially incomplete. But to miss such difficulties is to miss an important part of the background necessary for understanding Planck's work and his decision to focus on irreversibility and the second law of thermodynamics.

At the turn of the century, neither electrodynamics nor even atomistic mechanics held "classical" status, as we now apply the term. Given the successes and the limitations of Maxwellian electrodynamics, on the one hand, and the kinetic theory of gases, on the other, the problem of how the ponderable atoms of matter were to be imagined to interact with the electromagnetic field remained especially vexing. There was no widely accepted approach. The electron theory of H. A. Lorentz was seen as an extremely promising development, but it was still difficult to see how all of the complexities displayed by matter (chemical, spectroscopic, etc.) might be explained on the basis of the theory of electrons.[50] It was only later in this century, as new experimental evidence combined with increased confidence in their ability to propose precise, mathematically described models of atomic entities and processes, that physicists—like Planck in the second edition of the *Wärmestrahlung* [Introduction, p. ix]—could speak at all of a "classical" approach and mean theories characterized by an insistence on the general applicability of the continuous partial differential equations of Maxwellian electrodynamics. And even in those cases the classical theories were meant more as a foil for the new quantum assumptions than as historically meaningful descriptions of earlier approaches.

In fact, during the first decade of the twentieth century when physicists spoke of classical physics, they often meant one of several different things. Some meant physics based on the mechanics of individual corpuscles; some, the physics of the aether before acceptance of the theory of relativity; and some, simply the standard treatment of thermodynamic or macroscopic phenomena like temperature, elasticity, sound, etc. In the latter case, physicists continued to believe in the utility of such theories, but they were not nearly so confident that they could see how they could be referred back to microscopic processes, or that they knew what sort of microscopic processes were required.[51]

Planck was impressed with the accomplishments of mechanical and electromagnetic theory. But his commitment to those theories was always on a lesser level than his commitment to requiring physical theory to conform with the absolute, unifying regularities he believed existed in nature. Later in life, Planck remarked that he was first attracted to theoretical physics by the unexpected happenstance that the processes of human thought and the idealizations of the theorist could reflect the lawfulness of nature.[52] For him, the job of the theorist was to explore as fully as possible the implications of this lawfulness. Accordingly, the two laws of thermodynamics, in Planck's hands, were used as tools for selecting the proper models, analogies, or assumptions to use to better understand physical processes. And they provided the guidance necessary to prevent the errors that unavoidably result from pursuing theories beyond their establishable realms of validity.

Planck, Irreversibility, and Quanta to 1911

Between 1902 and the appearance of the original *Wärmestrahlung*, Planck did not publish specifically on quanta. First, he turned to several studies to test the ability of idealized electrodynamic systems to account for subtle macroscopic phenomena (like dispersion and the reflection of light from conducting surfaces). Afterwards, Planck only gradually returned to the subject of quanta. Instead he spent a great deal of his creative energy exploring the significance of one of Einstein's famous papers of 1905, the paper on the electrodynamics of moving bodies.

Planck was open to drastic reformulations of physical theory, especially reformulations that would unify previously separate concepts. That is clearly demonstrated by Planck's being one of the first (and certainly the most influential) physicists to embrace Einstein's 1905 theory of special relativity—especially the new relations among the concepts of time, space, mass, and energy.[53]

Planck's work on the theory of relativity focussed on determining which of the basic principles of physics were consistent with the new theory. Entropy and the action integral of Hamiltonian mechanics were both shown to be invariant, i.e., independent of inertial coordinate system if the proper relativistic coordinate transformations were carefully applied.

But in addition, the mass–energy relationship, which he had given a prominent place among the implications of Einstein's work, provided

Planck, in 1908, with further evidence of the unknown processes taking place within the atom. He thought that these extraordinarily energetic processes were responsible for at least part of the atom's mass. Planck viewed the results of recent experiments with natural radioactivity as supporting evidence for such a vast reservoir of "latent energy within atoms." He even speculated on the possible form that energy might take:

> The latent energy is independent of the temperature and of the motions of the chemical atoms. Its form can be potential just as well as kinetic, for nothing stands in the way of assuming what anyway appears reasonable from an electromagnetic standpoint; that is, that within the chemical atom there occur certain stationary motions, like standing vibrations, that are associated with no, or very little external radiation. The energy of these vibrations, which can be appreciable, cannot be detected as long as the atom remains undisturbed, except by the inertia with which the oscillating system resists a translational acceleration and by the gravitational action directly associated with it.[54]

Planck believed that the quantum constant was somehow characteristic of these subatomic processes. This was suggested by dimensional considerations he had first described in the *Wärmestrahlung* (1906) [§149, pp. 153–54] and that now assumed added significance. The constant h (6.55×10^{-27} erg sec) has the same dimensions as the action integral that Planck had just shown to be Lorentz invariant. In his 1908 paper on relativity, Planck remarked that if the existence of this quantum of action were accepted, it would be possible to maintain that "each and every change in nature corresponds to a definite number of action elements completely independent of the choice of coordinate system."[55]

Although hardly a detailed proposal for the physical significance of quanta, this statement clearly indicates that Planck did not believe that there was any contradiction between electrodynamics in free space, generalized and extended by the principle of relativity, and the existence of a realm of microscopic processes governed by an as yet unknown set of laws. Planck's earlier work on dispersion and on the optical properties of conductors can in fact be seen as, in part, a conscious effort to provide information on the extent to which physicists could safely ignore such unknown microscopic processes when trying to account for macroscopic behavior.

Prior to the spring of 1908, Planck discussed quanta and their possible physical significance in only a few of the letters that have been preserved. The most important of his exchanges on the subject came prior to his completing the first edition of the *Wärmestrahlung* and was with Paul

Ehrenfest. The subject of this exchange has been described by Martin J. Klein and by Thomas S. Kuhn. As they point out, the net result was to raise serious questions about the ability of the undamped oscillators of Planck's radiation theory to account for the actual redistribution of energy among the various frequencies of radiation in a blackbody cavity.[56] Planck did not immediately address the implications of that result in print. He simply included a remark describing the difficulty in the conclusion of the *Wärmestrahlung* (1906) [§190, pp. 220–21]. Apparently Planck was inclined to set considerations of the entire matter aside until such time as more general insight into the still mysterious aspects of the small-scale interactions of radiation and matter was at hand.

Another mention of the possible physical significance of quanta in Planck's correspondence is contained in a letter to Wilhelm Wien dated 2 March 1907. The discussion indicates that by early 1907 Planck had begun to consider possible ways to associate the quantum hypothesis with the behavior of actual microscopic particles of matter.

The discussion of quanta was placed at the end of the letter, which was otherwise devoted to the evaluation of a paper that had been submitted to the *Annalen der Physik* for publication. The relevant section reads:

> Your measurement of the radiation emitted from the individual molecules naturally interests me a great deal. However, shouldn't the circumstances that this radiation amounts to considerably less than the elementary quantum of energy be explained by that fact that not all molecules emit uniformly but, to the contrary, in general quite a few molecules do not emit at all while others emit either one full energy quantum or several energy quanta? The entire calculation of the radiation entropy and the elementary quantum is based on the unequal and random distribution of energy among the individual molecules.[57]

This suggestion is especially interesting because it foreshadows the major reformulation of the radiation theory made by Planck in 1911 and included in the revised edition of the *Wärmestrahlung* (1913). As has been indicated, in the original quantum papers and in the *Wärmestrahlung* (1906) [§146, pp. 149–50] Planck had associated the disorder that gave rise to entropy with the irregular changes in the state of oscillation of a single oscillator as it interacted with the "natural radiation" that surrounded it. Except when meant to serve as an artificial aid to calculation, Planck's collection of oscillators was used to represent the rapidly changing "hidden" states of a single oscillator.[58] The suggestion made above, although it scarcely addresses the question of the laws governing the internal motions of molecules, was that a collection of oscillators, beyond being a convenient calculational device, might serve as a model for the molecules of a material body and that elementary disorder was to be associated with the entire system rather than with each molecule separately. Apparently this idea was not pursued further by Planck until, in 1911, it was made a part of a second version of his so-called second quantum theory.

As Kuhn has discussed in detail, Planck's return in 1908 to the issue of quanta was spurred by an address in Rome delivered by H. A. Lorentz.[59] Lorentz was hardly new to the theory of heat radiation. In 1903 he had published a spectral distribution law (later referred to as the Rayleigh–Jeans distribution law). He had derived this law by considering the emission and absorption of long-wavelength radiation by free electrons within a conducting material. In the years between 1903 and 1908 Lorentz sought in vain for a means of generalizing his own approach. He wanted to find a way to lift the restriction to long waves and thereby derive a formula that, like Planck's, would exhibit a maximum at a wavelength whose variation with temperature obeyed the well-known Wien displacement law. By April of 1908, Lorentz had concluded that his search was doomed to failure. He was able to show that the Rayleigh–Jeans law resulted directly from the application of the Hamiltonian formalism of mechanics and the theorems of statistical mechanics (especially, the so-called equipartition theorem) to a general system composed of aether, electrons, and molecules.[60]

Lorentz's approach to the blackbody problem had been directly through the theory of electrons. The fundamental assumption of that theory was that the interaction between radiation and matter could be completely described by the forces exerted on electrons by the local electromagnetic fields. The electrons were thought to be embedded in or otherwise associated with ponderable atoms. How Planck's theory could be applied to electrons, and especially "free" electrons in conductors, was not at all obvious.

There was, however, strong theoretical and experimental evidence for the existence of "free" electrons in metals. Drude's successful theory of the electrical and thermal conductivity was based on that assumption.[61] And of even greater relevance to the radiation problem, the recent measurements of Hagen and Rubens indicated that the absorption of long-wavelength radiation by metals agreed very well with the predictions of Drude's theory.[62] The question that was formulated by Lorentz was: Why didn't the electrons in metals emit radiation in accordance with the predictions of the electron theory in the region of the spectrum where the Rayleigh–Jeans formula called for a high emissivity and the Planck formula called for virtually none? As he was quickly forced to admit, the information given by the experiment was unequivocal. They did not emit.

As far as Lorentz could determine the situation had nothing to do with resonators of fixed frequency or, therefore, with energy elements of a definite size.

In a letter to Wien, Lorentz revealed that he had reservations about the consistency of Planck's use of the equations of electrodynamics and the introduction of energy elements. In the high frequency end of the spectrum, the average energy of one of Planck's resonators is considerably less than one of the corresponding energy elements. It was difficult for Lorentz to understand how this could happen unless, as he remarked, "several of these resonators (under appropriate circumstances the majority) possess no energy whatsoever." Lorentz stated that this fact contradicted the assumptions that had been made by Planck in the original derivation of the relationship between the average energy of an oscillator and the energy density of the surrounding radiation where, in his words, "...the resonators absorb or give up energy to the aether in an entirely continuous manner (without a mention of a finite energy quantum)." Because the resonators were thought of as continuously exposed to external radiation, why they should remain dormant for any significant length of time was a mystery to Lorentz. He concluded his letter to Wien by announcing that he hoped soon to determine Planck's own opinion on the matter.[63]

Lorentz pondered the problems he had outlined for Wien for almost two months before writing to Planck. He used this time to develop and present a sketch of what he believed was implicit in Planck's radiation theory. Unfortunately, Lorentz's letter has not survived. However, its contents can be largely reconstructed from Planck's reply.[64]

Thomas Kuhn has pointed out that Planck's letter, written in October of 1908, contains one of his earliest explicit attempts to provide a specific physical interpretation of energy quanta. But even so, this interpretation was quite consistent with Planck's long-term recognition that the detailed behavior at centers of emission and absorption of radiation was not fully described by existing theories. It was also consistent with his earlier insistence on retaining the experimentally well-confirmed Maxwell's theory (in free space), and Planck's habitual avoidance of detailed models of atoms and molecules.[65]

In his letter, Planck postulated the existence of two separate natural realms governed by two sets of physical laws. One of those realms was at least accessible, in principle, to direct observation and measurement. It was characterized by the continuity of the processes that took place within it. The other realm was microscopic and in general was inaccessible to such measurement. This second realm was the presumed source of irreversibility and the home of the quantum of action.

Planck was convinced that ordinary mechanical models would always be inadequate to represent the processes that went on in the interior of atoms and molecules during "mathematically" infinitesimal time periods. No matter how ingenious the model, it would presumably have to obey Hamilton's equations; and therefore, as Lorentz had shown, it would necessarily lead to the discredited law of Rayleigh and Jeans. For that reason, Planck remained willing to concede that entirely new sorts of laws were required to describe the detailed behavior of microscopic systems, including the requirement that their energy was limited to integral multiples of the elementary quantum of energy $E = h\nu$. What he was not willing to concede was that those laws, whatever their eventual form, would contradict theories that led to predictions confirmed by macroscopic measurements. He insisted that proper consideration need always be given to the distinction between "mathematical" and "physical" infinitesimals. Planck's basic strategy for coping with the problem of interpreting the quantum hypothesis was to preserve tested theories by carefully demarcating the region of the validity of their laws. When eventually his attempt to demarcate those realms on the basis of time scales failed, he was free to explore alternatives, as we shall see.

By the end of 1909 Planck decided to publish once again on the quantum theory. This time it was in response to Einstein's continuing attempt to demonstrate that even the propagation of light in free space could not escape the quantum hypothesis. Of Einstein's planned publication, Planck wrote to Wien:

> ...Soon I will again begin to speak on the radiation theory, particularly because Einstein now intends to publish several considerations (in the *Physikalische Zeitschrift*). He has arrived at the supposition that the elementary quantum h also has significance for processes in the pure vacuum. That, for the present, I do not believe; nor do you; and Lorentz, quite certainly not. Why should one complicate the theory unnecessarily? There are already enough difficulties with it and one can be entirely content to lump them all together in a single place—inside of the molecule.[66]

How one was to imagine limiting the quantum phenomena to "inside of the molecule" was the question that led Planck to his so-called second quantum theory, or the theory of quantum emission. The major difficulty to be overcome was the one that had been emphasized by Lorentz. At high frequencies it was difficult to imagine, even given large field fluctuations in "mathematically infinitesimal time elements," how electromagnetic waves consistent with Maxwell's equations in free space could provide a full-energy quantum to an oscillator in a reasonable period of time.

Planck's Second Quantum Theory

The consideration of the problem pointed to by Lorentz clearly demonstrated to Planck that his earlier formulation of blackbody theory was not internally consistent. The electrodynamic and the statistical portions of the derivation of that law, the ones provided in the original edition of the *Wärmestrahlung*, were not fully reconcilable, even if different time scales were posited for the two components of the derivation.

The solution that Planck arrived at (in several steps) was to assume that his oscillators absorbed electromagnetic energy continuously, thereby relegating the quantum hypothesis to the process of emission. The virtue of that approach was that it allowed him to continue to reject Einstein's light-quantum hypothesis.

As claimed earlier, Planck's view of the quantum hypothesis is so closely associated with his understanding of irreversibility and the second law that it is not surprising to find that this shift in interpreting the physical significance of quanta was accompanied by a shift in the method he used to calculate entropy. It should also come as no surprise that Planck came to view the new way of calculating entropy as far more significant than the detailed assumptions he had made about the emission process.

There were two versions of the theory of quantum emission. The first version of the "second theory" was presented to the German Physical Society on 3 February 1911.[67] In it, Planck assumed that an oscillator absorbed energy continuously, precisely according to the laws of Maxwellian electrodynamics. At least with respect to absorption, he made no distinction between mathematically and physically infinitesimal time elements. Emission, on the other hand, was imagined to take place one quantum at a time according to a newly formulated statistical law that attributed a fixed probability of emission for any specified time element (as long as that element was small compared to the average time between emissions). It should perhaps be emphasized that in both this and the later version of the second theory, Planck did not assume that the process of emission was inherently probabilistic, just that the laws governing emission were as yet unknown.[68] In the aggregate, the emission process was thought to obey his statistical law.

One result of the new assumptions was that the average value of oscillator energy could be divided into two parts. One part involved the randomly varying number of energy elements contained by an oscillator; the other part the continuously increasing value of oscillatory energy that proceeded independent of the quantum emission process. Planck argued that only the first part was characterized by disorder and hence only it could contribute to the entropy of the oscillator. He then showed that the second part contributed a fixed average energy (equal to $h\nu/2$) to the energy that resulted from the random portion. The fixed component was completely independent of temperature and was quickly dubbed the zero-point energy.[69]

The statistical portion of Planck's new derivation at first followed closely the one that is given in the *Wärmestrahlung* (1906). That is, Planck assumed that for the purpose of calculating the average value and for obtaining a quantitative measure of the disorder to be associated with that value, a collection of N oscillators could be thought to represent a *single* oscillator over an extended period of time. Although there were certain physical difficulties with that assumption, Planck was able to carry them forward and derive the now standard radiation formula, albeit using an altered formula for oscillator entropy.[70]

Five months later, Planck abandoned his earlier statistical procedures and published a new version of the theory of quantum emission.[71] In this version Planck assumed that a given oscillator within a collection of identical oscillators absorbed energy continuously, but could emit energy only at those instants when it possessed an integral multiple of the energy element $h\nu$. At those instants the probability that it emitted was related in an unspecified way to the state of equilibrium achieved by the entire collection of oscillators and, therefore, to the properties of the electromagnetic field through which those oscillators were thought to interact. When an individual oscillator emitted, it emitted all of its energy at once. In making these assumptions Planck discarded the notion that assumptions about the behavior of a single oscillator were the appropriate basis for a statistical derivation of the state of thermal equilibrium between matter and radiation. He turned instead to a derivation based not on random changes in the energy of a single oscillator, but rather on the random distribution of energy among an actual collection.

As can easily be imagined, this change also had a number of important implications for the attempt to discover the dynamical significance of the quantum constant. First of all, the shift from time averages to averages over a collection of oscillators effectively ruled out the continued reliance on the distinction between physically and mathematically infinitesimal time periods for distinguishing between the realms of electrodynamics and quantum physics. That distinction had in fact already become quite fuzzy with the assumption of continuous absorption for even the smallest of time intervals. With the second version of the theory of quantum emission, Planck was forced to seriously consider alternative methods of demarcating those realms.

But the most important implication of the shift to ensemble averages is that, for the first time, it furnished Planck with a physical picture of the distribution of oscillators over the phase space defined by the equations that governed the "motion" of a single oscillator. The distribution function itself had previously been irrelevant to Planck's theory, as he had emphasized in the first edition of the *Wärmestrahlung* [§147, pp. 150–51]. Now, with the exception of his use of the quantum constant to fix the size of the cells of "equal probability" in phase space (something he had done already in 1906 [§150, pp. 155–56]), Planck was able to formulate the statistical portion of his radiation theory precisely as Boltzmann had formulated the derivation of the entropy of a gas in the *Lectures* (1896).

What he derived in the second paper on quantum emission was the distribution of N oscillators over a quantized phase space that would remain stationary over time (given the new form of the hypothesis of quantum emission). He then calculated the entropy of that distribution, using the formula:

$$S = -Nk \sum_{i=1}^{\infty} p_i \ln(p_i)$$

(where the sum is over cells in phase space and the p_i are the proper fractions of the N oscillators located in each cell). The calculation was exactly equivalent to the way Boltzmann had calculated the entropy of an ideal gas in his 1896 textbook.

In his original presentation[72] of his new theory, the derivation of the blackbody distribution law from the stationary distribution was not even attempted. Planck was satisfied to note that his equation for the average energy of an oscillator was the same as that given in his earlier version of the theory of quantum emission. He admitted that, because of the proposed alteration in the law of emission, an additional assumption was required to complete the derivation. He only hinted at what it might be. That alone indicates that Planck was much more concerned with his new method of calculating entropy. For him, as always, his specific dynamical assumptions were tentative, if suggestive, and not meant to be centrally important.

Entropy in the New *Wärmestrahlung*

Planck continued to believe that providing a general method of calculating entropy was the important goal and that speculations about dynamical laws could best be done when guided by a rigorous thermodynamic treatment of the system being considered. The emphasis given the new method in the revised edition of the *Wärmestrahlung* (1913) confirms that he believed he had found a new and promising way of calculating entropy. Although the new edition and its translation [§147, pp. 153–54] contain a full account of the latest version of Planck's theory of quantum emission, the details of the theory are presented following the general discussion of how the entropy of any system composed of a large number of identical entities (including simple harmonic oscillators) is to be calculated. When the new theory is presented, it is with many qualifications. The conclusion to the new *Wärmestrahlung* [§190, pp. 214–15] is likewise filled with qualifications and warnings, most notably about the failure of the theory to account for the redistribution of energy between oscillators and radiation of different frequencies. These cautions indicate that Planck believed the significance of the revised version lay elsewhere than his special dynamical assumptions.

The revised *Wärmestrahlung* (1913) is organized to emphasize the general procedure of determining what Planck calls states of "thermodynamic equilibrium" for physical systems (mechanical or otherwise) composed of large numbers of identical systems, each describable in terms of generalized coordinates and momenta of the sort used in the Hamiltonian formalism of mechanics. This is a marked change of emphasis from the original text, which gives far more emphasis to the detailed electrodynamic calculations and assumptions characteristic of the earliest, stillborn attempt by Planck to explain irreversibility on the basis of electrodynamics.[73]

An important influence leading up to this new emphasis, and one given a prominent place in the new edition, was the "new heat theorem" (now referred to as the third law of thermodynamics) that had been proposed by Walther Nernst in 1906.[74] In fact, Planck was the first physicist to emphasize that the new heat theorem could be considered an entropy theorem; that is, that it supplied a means of fixing the otherwise undetermined additive constant in expression for the entropy of a condensed system. Planck published that interpretation of the new heat theorem in a new and revised edition of his textbook on thermodynamics. In the preface to that edition (dated 1910) Planck indicated that he had begun to consider the physical significance of that theorem in areas outside the scope of thermodynamics.[75]

Early in December 1910, Otto Sackur, a physical chemist at the University of Breslau, submitted a paper to the *Annalen* in which he argued for a connection between the new law of thermodynamics and the role played by the quantum constant in statistical calculations of entropy.[76] Planck seized upon that idea, as he seized upon the method of associating entropy and disorder with the distribution of a collection of identical systems over phase space.

In fact, in the introduction to the new *Wärmestrahlung*, Planck asserted that both the quantum hypothesis and Nernst's heat theorem could be viewed as a direct consequence of a proper understanding of entropy [Preface, p. vii]. In the text, Planck promised to show that the elimination of the arbitrary additive constant generally associated with the statistical definition of entropy "leads necessarily to the 'hypothesis of quanta' and moreover it also leads, as regards radiant heat, to a definite law of distribution of energy for black radiation, and, as regards heat energy of bodies, to Nernst's heat theorem" [§120, pp. 119–20].

The basic thrust of his method was to separate the requirements and implications of the approach of natural systems to states of thermodynamic equilibrium from the assumptions necessary to describe the dynamics that gave rise to that irreversible behavior. Fixed, physically determined regions in phase space were necessary to correctly derive states of thermodynamic equilibrium. The dynamic explanation of the form taken by those regions need not be determined at the outset. As he had written in an earlier paper on Nernst's theorem, "my theory requires only that certain discontinuities are contained in the elementary laws that govern atomic forces out of which the discrete regions of probability result."[77] It is a familiar approach.

After Sections I and II of the new *Wärmestrahlung* introduce the subject of heat radiation and the results obtainable from the basic equations of electrodynamics and thermodynamics, Section III presents the general procedure for calculating entropy. The quantum hypothesis is introduced as a consequence of the general requirement that entropy has to be a *unique*, single-valued state function. That is, the size and shape of the cells in phase space used to define the p_i required to calculate entropy are not arbitrary but fixed by the physical characteristics of the systems being considered.[78]

What is changed in this approach is the way of calculating entropy, not its function as a guide for seeking to understand natural systems and to avoid errors traceable to the unwarranted extrapolation of otherwise successful dynamical theories. There is only one potentially important concession, and it is the same concession he had made in his Leiden address. Planck now recognized that there was some disagreement among physicists on the question of whether irreversibility was a universal characteristic of all natural systems [§117, p. 117, especially footnote 1].

But regardless of whether others suspected that violations of the second law could sometimes occur, Planck viewed his new entropy function as a mathematical expression of the overwhelming predominance of irreversible behavior in nature and therefore still a powerful tool for use in investigating the dynamics of small-scale physical processes.

The *Wärmestrahlung* (1913) was meant as a model for Planck's thermodynamic approach. In it, the treatment of ideal oscillators obeying the assumptions of the latest theory of quantum emission is first given as an application of the general procedure for defining entropy [§135–§140, pp. 135–43]. Later the Planck distribution law for blackbody radiation is shown to follow as a characteristic of the state of thermal equilibrium of a collection of such oscillators, given the general electrodynamic results obtained in the introductory chapters. Finally, in Part V, Planck investigates the dynamical implications of his earlier results; that is, the question of the nature of the processes that are actually responsible for the irreversible approach of radiation systems to their characteristic states of thermal equilibrium [§170–§190, pp. 189–215]. His language clearly shows that his strategy and levels of commitment have not radically changed.

> To this much more difficult question only a partial answer can, at present, be given. In the first place, it is evident from the extensive discussion... [earlier in the text] that, since irreversible processes are to be dealt with, the principles of pure electrodynamics alone will not suffice. For the second principle of thermodynamics or the principle of increase of entropy is foreign to the contents of pure electrodynamics as well as of pure mechanics.[79]

Shortly after the appearance of the *Wärmestrahlung* (1913) Planck would finally come to accept the probabilistic nature of irreversibility and that fluctuation phenomena did not contradict the second law, but rather they were a potentially measurable consequence of it. In part that realization came from consideration of the physical implications of his new method for defining entropy. In any case, the formalism contained in the *Wärmestrahlung* (1913), with little modification, would prove applicable to a new view of irreversibility.

The Theory of Quantum Collisions

After completing the second edition of the *Wärmestrahlung*, Planck continued his two-pronged (statistical-thermodynamic and dynamic) investigation of irreversible radiation processes. A detailed account of that work is beyond the scope of this introduction. A brief description of how those investigations proceeded and of Planck's final acceptance of a probabilistic interpretation of irreversibility, however, will reinforce the point of the extraordinarily close coupling of the quantum hypothesis and the special understanding of irreversibility presented by Planck's scientific work.

After 1913, the major question that remained open for Planck was whether the new definition of entropy offered important new insights into either the nature of irreversibility or the dynamics of the interaction of radiation and matter. One hope for further progress had arisen earlier in correspondence and other exchanges with Arnold Sommerfeld[80] and with Wilhelm Wien.

At the first Solvay conference, held in Brussels late in 1911, Planck responded to a question posed by Wien about how Planck thought his oscillators accounted for the redistribution of energy among the various radiation frequencies. Planck admitted that electrons undoubtedly played a significant role. He expressed optimism that his most recent formulation of the way to calculate the statistical equilibrium of a collection of oscillators (that is, the final version of the second theory) could be extended and shown to be consistent with a system that included oscillators, electrons, and molecules—all free to exchange energy with one another during collisions.[81] If so, it might be hoped that the laws of electromagnetic theory could be presumed valid for small-scale oscillatory behavior, except perhaps during the process of electron emission or absorption, or during molecular collisions. That way, the realm of the quantum could be associated with small-scale atomic processes and be thought of as completely independent of electrodynamics. Because, in Planck's words, "up to now we know so enormously little about the processes that take place within a molecule, there is, in spite of everything, plenty of room for the play of fantasy."[82]

The other hope was that a modified electrodynamic theory could be developed, one that retained Maxwell's equations for the behavior of radiation in free space but that would account for the quantum aspects of the behavior at the centers of emission and absorption. This second approach was not new. The use in the *Wärmestrahlung* (1913) of the simple equations of an undamped harmonic oscillator [§135, p. 137] itself contradicted Maxwell's theory near the oscillator. (In Maxwell's theory an oscillating dipole must emit radiant energy continuously.) Planck recognized this and expressed the hope that the inconsistency could be localized "...to those regions of space in the interior or on the surface of the oscillator...." For, still according to Planck, "it is precisely in those regions that new hypotheses are to be admitted in the first place."[83]

Planck was incorrect, as was quickly pointed out to him. The asymmetry between emission and absorption in his theory was inconsistent with Maxwell's equations and this inconsistency could be limited to a small sphere surrounding one of his non-Maxwellian oscillators.[84] Accepting that conclusion, Planck apparently chose to concentrate first on the theory of quantum collisions, a theory that is sometimes referred to as Planck's "third" quantum theory.[85] In fact, in his conclusion to the revised *Wärmestrahlung* [§190, p. 215] Planck suggested that such a theory would be required to explain the redistribution of energy between oscillators and radiation of different frequencies.

Planck published three papers on the theory of quantum collisions. The first was presented to the Prussian Academy in April 1913,[86] the second later that month at the Wolfskehl Symposium on the Kinetic Theory in Göttingen,[87] and the third to the Academy in July 1914.[88] All three papers shared the assumption that for the interaction of electrons and atoms and for the collisions of one atom with another, there existed a well-defined "action sphere" associated with each atom and that the magnitude of the sphere was somehow related to the quantum of action *h*. Although the equations of Maxwell's electrodynamics were used to describe both the process of absorption and emission of radiation, Planck made extraordinarily complex assumptions about what went on inside the action spheres during collisions. The assumptions were required to ensure that his approach would remain consistent with the statistical formulation of the quantum hypothesis described in the *Wärmestrahlung* (1913).

What caused Planck to give up on that approach was his and others' continuing work on the interaction of radiation with a system composed of rotating electric dipoles.[89] Naturally, for his part, Planck chose to apply the statistical portion of his quantum theory and then examine the possible dynamical implications. The net result was, on the one hand, that Planck revised his understanding of irreversibility in such a way to enable him to make use of quantitatively described fluctuations in the energy of such dipoles; and on the other hand, that he used the results of the statistical calculations to demonstrate, once and for all, that the equations of electrodynamics could not be assumed valid for small-scale processes regardless of what went on during collisions with electrons or between molecules. By July 1915 Planck could write:

> Whether the quantum hypothesis plays any role in pure electrodynamics, that is in the interactions between the waves of an electromagnetic field and an electric pole or dipole..., or whether its significance is limited to the laws governing the collisions of ponderable particles could, over the past year, have still been considered an open question. Today, it appears to have received a definitive answer, and to be sure it is the first alternative.... This is because a calculation by A. D. Fokker of the stationary state of a system of rotating dipoles in a given radiation field, rigorously based on the fundamental principles of classical electrodynamics, has led in a completely unambiguous way to a result that directly contradicts experience.[90]

Though the details cannot be provided here, Planck's work from 1913 to 1915 reveals once again just how closely associated were his understandings of irreversibility and of quanta.

As it turned out, in the course of that work Planck was reexposed to the ideas of Einstein and quantitative methods of treating the fluctuations one would expect a rotating electric dipole to undergo in a state of thermal equilibrium. Unlike the earlier encounters, this time Planck recognized that fluctuation approach developed by Einstein and his young assistant, A. D. Fokker,[91] could easily be accommodated within his own statistical formalism. Simultaneously, probably aided by discussions with Paul Ehrenfest, Planck's views about irreversibility changed considerably.[92]

Although a single source of Planck's "conversion" cannot be precisely determined, that his understanding changed dramatically cannot be doubted. By 1914 Planck had a new tool for examining the dynamical significance of the natural behavior described by the now probabilistically understood second law of thermodynamics.

Planck's Acceptance of the Statistical View of Irreversibility in 1914

None of Planck's published papers and none of the correspondence I know of prior to 1914 reveals any dramatic new insights into the nature of irreversibility.[93] However, in 1914 his view was remarkably different from the one he had repeatedly expressed over the previous three decades. New insights into the statistical foundations of thermodynamics provided the theme for the Founder's Day address that Planck, as Rector of the University of Berlin, delivered on 3 August. That was the day that Germany declared war on France. It is difficult to imagine more impressive testimony to the seriousness with which Planck approached his work or to the importance he attributed to the philosophical and ideological implications of the physicists' world view and especially to the whole question of irreversibility.

The title of Planck's address was "Dynamical and Statistical Lawfulness." In it he spoke to the epistemological significance of his then explicit belief that an entire class of physical laws reflected regularities that were not absolute, occurring in nature without exception, but were in fact essentially statistical. "That liquid flows from a higher to a lower level is necessary" declared Planck, "but that heat flows from a higher temperature to a lower temperature is only probable." And later he asserted that "there remains no doubt for the physicist who by nature is bound to inductive reasoning that matter is atomistically constituted, that heat is the motion of molecules, and that heat conduction, as well as all other irreversible processes, obeys not dynamical, but statistical, that is to say probabilistic laws."

The Founder's Day address contained the first public statement by Planck that deviations from the average behavior of thermal systems were in themselves to be considered as regularities rather than as anomalies to be excluded from the outset by special hypothesis. Even more striking, Brownian motion for the first time assumed the same importance for Planck that had already been attributed to it by Einstein:

> I believe that no one, once having viewed such a preparation [small particles suspended in water] in good light under a microscope will ever forget the first impression of the spectacle that thus presents itself. It is a view into a new world....The final proof for the correctness of this view [that the motion represents fluctuations attributable directly to the random motion of water molecules interacting with the particles] has only been achieved in recent times, in that the statistical laws for the distribution density, the velocities, the paths, indeed even the rotations of these microscopic particles derived by Einstein and Smoluchowski have in all details, received brilliant experimental confirmation through the work of Jean Perrin.[94]

As has already been indicated, after 1914, fluctuation calculations—combined with his only slightly reinterpreted statistical formalism for calculating entropy—formed the basis of Planck's continuing efforts to determine the kinds of alterations that would be required to reconcile the physicists' unified understanding of nature with the quantum of action.

References

1. John Heilbron, *The Dilemmas of an Upright Man: Max Planck as Spokesman for German Science* (Berkeley: University of California Press, 1986), pp. 1–5, which contains an excellent bibliography of biographical sources. Political references are based on C. Brinton, J. B. Christopher, and R. L. Wolff, *A History of Civilization,* 2 Vols. (Englewood Cliffs, N.J.: Prentice-Hall, 1967), Vol. 2, pp. 253–259. For interesting comments on the possible significance for Planck's political beliefs of his being born in a city then under foreign occupation, see Lewis Pyenson's review of Heilbron's *The Dilemmas...*, *ISIS 79* (1988), pp. 122–126.
2. On the emergence of theoretical physics as a specialty and the programs current in period 1870–1925, see Christa Jungnickel and Russell McCormmach, *Intellectual Mastery of Nature: Theoretical Physics from Ohm to Einstein,* 2 Vols. (Chicago: Univ. of Chicago Press, 1986). See also, Heilbron (note 1), p. 10.

3. Hans Kangro, *Vorgeschichte des Planckschen Strahlungsgesetzes* (Wiesbaden: Franz Steiner Verlag, 1970); translated as *The Early History of Planck's Radiation Law* (London: Taylor & Francis Ltd., 1976), pp. 111–116. On Planck's reaction to the lectures of Helmholtz and Kirchhoff, see Thomas S. Kuhn, *Black-body Theory and the Quantum Discontinuity: 1894–1912* (New York: Oxford Univ. Press, 1978), p. 14. Kuhn's study of Planck and the Quantum theory has been reprinted in paperback (Chicago: Univ. of Chicago Press, 1987) with a new afterword (which was originally published as an article in *Historical Studies in the Physical Sciences* ***14***, pp. 231–52). The pagination of the original edition has been retained.
4. Unless noted otherwise, this and subsequent information on Planck's positions at Kiel and Berlin are from Jungnickel and McCormmach (note 2), Vol. 2, pp. 43–52.
5. Fifteen years later, in the Winter of 1905, Planck revised the regular three-year cycle he had developed to include the subject of heat radiation. It was those lectures that were published in 1906 and that are reprinted as part 2 of this volume. See below.
6. Kangro (note 3), p. 121.
7. Planck was especially suspicious of complex molecular theories and also of the entire "energetics" research program. See Heilbron (note 1), pp. 9–17; and Jungnickel and McCormmach (note 2), Vol. 2, pp. 217–227.
8. The importance of thermodynamics in Planck's work was first emphasized by Martin J. Klein. See M. J. Klein, "Max Planck and the Beginnings of the Quantum Theory," Archive for History of Exact Sciences **1** (1962), pp. 459–479; "Planck, Entropy and Quanta, 1900–1906," The Natural Philosopher **1** (1963), pp. 83–108; "Thermodynamics and quanta in Planck's work," Physics Today **19**:11 (1966), 23–32; and "The Beginnings of the Quantum Theory," in C. Weiner (ed.) *History of Twentieth Century Physics: Proceedings of the International School of Physics "Enrico Fermi", Course LVII* (New York: Academic Press, 1977), pp. 1–39. A clear and concise summary of much of this work is contained in chapter 10 of Klein's biographical study, *Paul Ehrenfest: Volume 1, The Making of a Theoretical Physicist* (New York: American Elsevier, 1970), 217–230. See also Kuhn (note 3), pp. 13–18.
9. Klein [1970] (note 8), pp. 228–232; Kuhn (note 3), pp. 134–140. The original papers are: (a) Max Planck, "Zur Theorie des Gesetzes der Energieverteilung im Normalspectrum," Verh. d. D. Phys. Ges. **2** (1900), pp. 237–245; (b) "Uber das Gesetz der Energieverteilung im Normalspectrum," Ann. Phys. **4** (1901), pp. 553–563; and (c) "Uber irreversible Strahlungsvorgänge (Nachtrag)," Ann. Phys. **6** (1901), pp. 818–831. Planck's first quantum paper (a) has been translated into English and reprinted in D. ter Haar, *The Old Quantum Theory* (Oxford: Pergamon Press, 1967) pp. 79–90. All of Planck's major scientific papers have been reprinted in *Physikalische Abhandlungen und Vortrage*, 3 Vols. (Braunschweig: F. Vieweg, 1988). [Henceforth references to Planck's papers will be followed by a roman numeral and arabic number(s) indicating the volume and page(s), respectively, that the cited passage can be found in Planck's collected works.]
10. Jungnickel and McCormmach (note 2), Vol. 2, p. 123. Planck had already (1897) published an extremely important and influential textbook summarizing his understanding of thermodynamics. The most accessible version is: Max Planck, *Treatise on Thermodynamics*, A. Ogg (transl.), (New York: Dover, 1945), which is a reprinting of the English translation (1926) of the textbook's seventh (1922) German edition, and which includes the extremely instructive author's preface from 1897. See also, Heilbron (note 1), p. 13.
11. Unsigned obituary of Morton Masius, in Physics Today (April 1980), p. 71. I thank Katherine Sopka for this reference.

12. Gerald Holton, "On the Hesitant Rise of Quantum Physics Research in the United States," in *Thematic Origins of Scientific Thought: Kepler to Einstein,* revised edition (Cambridge: Harvard University Press, 1988), pp. 147–187. The citation is from p. 148.
13. See, for example, Max Jammer, *The Conceptual Development of Quantum Mechanics* (New York: McGraw-Hill, 1966), pp. 46–50; and Klein [1966] (note 8), pp. 28–29.
14. Kuhn (note 3), p. 235. There is also considerable disagreement in the historical literature on how to interpret the 1906 edition of the *Wärmestrahlung.* Differences between the treatment of energy elements in the text with that found in Planck's original quantum papers (note 9) are viewed either as representing the first stage in the retreat from quantization, or as simply a more reliable indicator of what Planck actually had in mind in 1900/1901. On this issue, see Peter Galison, "Kuhn and the Quantum Controversy," British Journal for the Philosophy of Science **32** (1981), pp. 71–85.
15. Max Planck, "Über den zweiten Hauptsatz der mechanische Wärmetheorie," Inauguraldissertation, München (1879): I, 1–61.
16. Max Planck, "Über das Princip der Vermehrung der Entropie, I," Ann. Phys. **30** (1887), pp. 562–582: I, 198.
17. Klein [1970] (note 8), pp. 218–19.
18. Kangro (note 3), pp. 110–124.
19. See Klein [1962] (note 8), pp. 2–4; Kuhn (note 3), pp. 29–37, 72–91; and Jungnickel and McCormmach (note 2), Vol. 2, pp. 228-31.
20. On Planck's library holdings on the new electrodynamics, see Kangro (note 3), pp. 119–120. On the possible mechanical interpretation of Planck's work, see Kuhn (note 3), p. 31.
21. Several authors have commented that although initially suspicious, Planck became quite familiar with the methods and assumptions of the kinetic theory, probably as a result of his serving as editor of Kirchhoff's lectures on that subject. See Stephen G. Brush, *The Kind of Motion we Call Heat: A History of the Kinetic Theory of Gases in the 19th Century* (New York: Elsevier, 1976), p. 643. For an account and for specific references on the debate between Boltzmann, Zermelo, and Planck over the adequacy of the kinetic theory as an explanation of irreversibility see Jungnickel and McCormmach (note 2), Vol. 2, pp. 214–215. For more detailed discussions, see M. J. Klein, "The Development of Boltzmann's Statistical Ideas," *The Boltzmann Equation: Theory and Applications,* E. G. D. Cohen and W. Thirring (eds), Acta Physica Austraica, Suppl. X (New York: Springer, 1973), pp. 53–106; and Kuhn (note 3), pp. 60–67.
22. Kangro (note 3), pp. 222, asserts that Planck "considered resonators as real physical entities" referring to an earlier cited (p. 181) statement of Planck to Wien that the elementary oscillators "one may conceive as being in some relationship with the ponderable atoms of radiating bodies." My conclusion is that the relationship was that of an abstract model for the emission/absorption process, which was in turn thought to be somehow associated with ponderable matter. The oscillators themselves were never thought to be real.
23. Max Planck, "Über irreversible Strahlungsvorgänge, Erste Mittheilung," Berl. Ber. (1897): I, 496. See also Kangro (note 3), pp. 128–133.
24. Max Planck (note 23): I, 494. See Klein [1962] (note 8), p. 462, and Kuhn (note 3), p. 77.
25. L. Boltzmann, "Über irreversible Strahlungsvorgänge," Berl. Ber. (1897), pp. 660–662. Reprinted in L. Boltzmann, *Wissenschaftliche Abhandlungen,* 3 Vols., F. Hasenohrl (ed.) (New York: Chelsea, 1968): III, 616. This and subsequent translations are my own unless specified otherwise.

26. L. Boltzmann, "Über irreversible Strahlungsvorgänge, II," Berl. Ber. (1897), pp. 1016–1018: III, 621.

27. For an extended discussion of the origins of Boltzmann's concept of molecular disorder, see Kuhn (note 3), pp. 60–67 and on the precise definition of "natural radiation," pp. 76–82. On the interaction of Planck and Boltzmann, see Kangro (note 3), pp. 130–136.

28. Klein [1962] (note 8), pp. 462–463.

29. Klein [1962] (note 8), pp. 464–465; Kangro (note 3), pp. 187–212; Kuhn (note 3), pp. 92–97.

30. Klein [1970] (note 8), pp. 222–223; Kuhn (note 3), pp. 97–110. The method used by Planck involved the counting as one "complexion" each possible way of distributing an appropriate number of energy units among a hypothetical collection of oscillators. The oscillators were meant to represent an individual oscillator over a period of time. The entropy of a given time-average value for oscillator energy is then set to be proportional to the logarithm of the number of complexions consistent with that average value.

31. For a detailed recapitulation and analysis (with bibliography) of the differences and agreements between the accounts of Kuhn and Klein, see Galison (note 14), pp. 71–85. See also: "Paradigm Lost? A Review Symposium" (with reviews by Martin J. Klein, Abner Shimony, and Trevor J. Pinch) ISIS **70** (1979), pp. 429–440; and Thomas S. Kuhn, "Revisiting Planck," Historical Studies in the Physical Sciences **14**, pp. 231–52 (see note 3).

32. See Brush (note 21); Kuhn (note 3), pp. 20–21 (especially his note 45), pp. 42–43 (especially his note 16) and 67–71; and Martin J. Klein, "Mechanical Explanation at the End of the Nineteenth Century," Centaurus **17** (1972), pp. 58–82.

33. See Klein [1973] (note 21), pp. 53–106; and Kuhn (note 3), pp. 38–71.

34. One influential source, in Germany, of the increasing precision in the application of statistical reasoning to physical problems was the publication of P. and T. Ehrenfest, "Begriffliche Grundlagen der statistischen Auffasung in der Mechanik," *Encyklopädie der mathematischen Wissenschaften* (Leipzig: B. G. Teubner), Vol. IV, Part 32 (1911). On Paul Ehrenfest's work in statistical mechanics see Klein [1970] (note 8), pp. 94–140.

35. This claim is documented in Chapter 2 of my doctoral dissertation: A. A. Needell, *Irreversibility and the Failure of Classical Dynamics: Max Planck's work on the Quantum Theory, 1900–1915* (Ann Arbor: Univ. Microfilms, 1980). Some of the evidence is also presented below.

36. Max Planck's review of J. W. Gibbs, Elementary Principles of Statistical Mechanics..., Beiblätter z. d. Ann. Phys. **27** (1903), pp. 748–753. Planck's reaction to Gibbs is discussed briefly in Jagdish Mehra and Helmut Rechenberg, *The Historical Development of the Quantum Theory,* Vol. 1, Part 1 (New York: Springer-Verlag, 1982), pp. 56-59. They emphasize Planck's adoption in the 1906 *Wärmestrahlung* of some aspects of Gibbs' "phase- space" formalism for representing the state of a physical system. See also Kangro (note 3), pp. 238–39.

37. M. Planck, "Die Einheit des Physikalischen Weltbildes," Phys. Zeitschr. **10** (1909), pp. 62–75: III, 6–29; *Eight Lectures on Theoretical Physics* (New York: Columbia Univ. Press, 1915).

38. A. Einstein, review of M. Planck, Vorlesungen ... Wärmestrahlung, Ann. Phys. Beilbl. **30** (1906), pp. 764–766. On the significance of this review, see T. S. Kuhn (note 3), pp. 139; and M. J. Klein and A. A. Needell, "Some Unnoticed Publications by Einstein," ISIS **68** (1977), pp. 601–04.

39. A. Einstein, "Zum gegenwärtigen Stand des Strahlung- problems," Phys. ZS. **10** (1909), p. 187.
40. See notes 36 and 38.
41. See especially, Max Planck, "Über die Verteilung der Energie zwischen Aether und Materie," Jubelband fur J. Bosscha, Archive Néelandaise (1901); republished Ann. Phys. **9** (1902), pp. 629–641: I, 731–743.
42. M. Planck [1909] (note 37): III, 20.
43. The citations and the thrust of the argument that follows are taken largely from: Peter Galison, *How Experiments End* (Chicago: Univ. of Chicago Press, 1987), pp. 21–33.
44. Galison (note 43), p. 32, who closely follows Jed Z. Buchwald, *From Maxwell to Microphysics: Aspects of the Electromagnetic Theory in the Last Quarter of the 19th Century* (Chicago: Univ. of Chicago Press, 1985).
45. See Brush (note 21) and Kuhn (note 3), 18–19.
46. Jungnickel and McCormmach (note 2), Vol. 2, especially pp. 12–13; on Kirchhoff, 126–129; on Helmholtz, 134–43; on Boltzmann, 154–57; and on Drude, 167–71. See also Kuhn (note 3), pp. 29–30 for a suggestive discussion of the status of theories of mechanical continua.
47. Jungnickel and McCormmach (note 2), Vol. 2, p. 212.
48. J. L. Heilbron, "Fin-de-Siecle Physics," in Bernhard *et al.* (eds.), *Science, Technology and Society in the Time of Alfred Nobel* (New York: Pergamon Press, 1982), pp. 51–73. Heilbron traces that tendency, and especially its manifestation in the more popular lectures and writings, both to the desire of physicists to achieve an internal consensus and provide a consistent philosophical foundation for their work, and to their perceived need to "secure their place in the wider society"—a place threatened to some degree by antimodernist cultural tendencies and by their association, for good or evil, with technological developments and the economic and social disruptions attributed to them.
49. See Jungnickel and McCormmach (note 2), Vol. 2, pp. 211–253.
50. On the problem of associating electrodynamics and matter and on Lorentz's theory, see Kuhn (note 3), pp. 112–13 and Jungnickel and McCormmach (note 2), Vol. 2, pp. 231–245.
51. Jungnickel and McCormmach (note 2), Vol. 2, pp. 189 and especially 313. My own dissertation (note 35), Kuhn's account of Planck's work on blackbody theory (note 3), and most earlier commentators have used the word "classical" to describe, in general, turn-of-the-century physical theory. I now believe that that is anachronistic and quite misleading.
52. Max Planck, *Wissenschaftliche Selbstbiographie* (Leipzig: J. A. Barth, 1948), p. 7: III, 374.
53. See Stanley Goldberg, "Max Planck's Philosophy of Nature and his Elaboration of the Special Theory of Relativity," Historical Studies in the Physical Sciences **7** (1976), pp. 125–60; Jungnickel and McCormmach (note 2), Vol. 2, pp. 245–253; and Heilbron (note 1), pp. 28–32.
54. Max Planck, "Zur Dynamik bewegter Systeme," Ann. Phys. **26** (1908), pp. 1–34: II, 205–06.
55. M. Planck (note 54): II, 198.
56. Klein [1970] (note 8), pp. 232–240; Kuhn (note 3), pp. 130–140, 152–163. The electrodynamic equations governing undamped oscillators have independent solutions for each frequency component of the spectrum. They do not disperse radiation at one frequency arbitrarily across that spectrum.

57. Planck to Wien, 2 March 1907. In Wien microfilm collection, Yale University Library Manuscript Collection and other repositories. This paragraph is also transcribed in Kuhn (note 3), pp. 305–06 in his footnote 44 and is mentioned briefly in Kangro (note 3), pp. 220–222.

58. For example, see Planck's first quantum paper (note 9A): I, 700. The basic irreversible process he concerned himself with was the approach of a single oscillator to equilibrium. His several separate collections of oscillators of differing frequencies were used to illustrate a calculational method that he didn't even carry all the way through. See T. S. Kuhn [1987] (note 3), pp. 351–357.

59. T. S. Kuhn (note 3), pp. 188–96.

60. For a concise description of Lorentz's work, see Klein [1970] (note 8), pp. 230–237. For a discussion of the status of the equipartition theorem and for additional references see Kuhn (note 3), p. 151. In Lorentz's treatment, the electrons interacted with radiation and with the ponderable molecules, thereby providing the means for establishing a state of equilibrium between radiation and matter.

61. Paul Drude, "Zur Elektronentheorie der Metalle," Ann. Phys. **1** (1900), pp. 566–613.

62. E. Hagen and H. Rubens, "Das Reflexionsvermögen von Metallen und belegten Glasspiegeln, Mitteilung aus der PTR," Ann. Phys. **306** (1900), pp. 352–375.

63. Lorentz to Wien, 6 June 1908. Different portions of this letter have been published by Armin Hermann, *The Genesis of the Quantum Theory,* transl. C. W. Nash (Cambridge: MIT Press, 1971), p. 41; and Kuhn (note 3), p. 194.

64. Lorentz's letter to Planck was dated 2 August 1908 as is revealed in Planck's reply dated 7 October: Lorentz Correspondence, Archive for History of Quantum Physics (AHQP) Microfilm, Center for History of Physics, American Institute of Physics, New York and other repositories. Portions of Planck's letter are translated in Kuhn (note 3), pp. 197–198 with the original German provided in his footnotes 32 and 33.

65. Kuhn (note 3), pp. 196–199.

66. Planck to Wien, 27 February 1909: Wien Microfilm, Yale University Library Manuscript Collection and other repositories. The original German is transcribed in Needell (note 35), footnote 112, p. 157.

67. M. Planck, "Eine neue Strahlungshypothese," Verh. d. Deutsch. Phys. Ges. **13** (1911), pp. 138–148: II, 249–259.

68. For example, see §147, p. 153 of revised *Wärmestrahlung.*

69. At first Planck assumed that the zero-point energy would have no observable consequences. That turned out not to be the case for diatomic hydrogen, the molecules of which could be thought of as capable of rotating like pinwheels whose rotation frequencies depended on their energies. See Needell (note 35), pp. 252–268.

70. Another important aspect of this approach is that to eliminate two oscillator constants Planck made use of the earlier electrodynamic equation for emission that he had rejected in favor of his quantum hypothesis. In so doing he developed an argument whose formulation in the next version of his theory of quantum emission became a model for Bohr's "correspondence principle," a cornerstone of the so-called "old quantum theory." See Max Jammer (note 13), p. 50; and T. S. Kuhn (note 3), p. 240 for references.

71. M. Planck, "Zur Hypothese der Quantenemission," Berl. Ber. (1911), pp. 723–731: II, 260–268.

72. M. Planck (note 71).

73. For a more detailed description of the differing structure of the two editions of the *Wärmestrahlung* see Kuhn (note 3), pp. 240–244.

74. See Klein [1966] (note 8), pp. 29–30.

75. M. Planck, *Treatise on Thermodynamics,* transl. A. Ogg, reprinted (New York: Dover, 1945), preface to the third edition dated November 1910.

76. O. Sackur, "Zur kinetischen Begründung des Nernstschen Wärmetheorems," Ann. Phys. **34** (1910), pp. 455-468.

77. M. Planck, "Über neuere thermodynamische Theorie (Nernstsches Wärmetheorie und Quantenhypothese)," Phys. ZS. **13** (1912), pp. 165–175: III, 63. See also §133, pp. 132–34 of revised *Wärmestrahlung.*

78. An assumption he would later refer to as "the physical structure of phase space."

79. Revised *Wärmestrahlung,* pp. 189.

80. Planck to Sommerfeld, 6 April 1911; 29 July 1911: AHQP Microfilm Collection, Center for History of Physics, AIP, New York and other repositories. See also S. Nisio, "The Formation of the Sommerfeld Quantum Theory of 1916," Japanese Studies in the History of Science **12** (1973), pp. 39–78.

81. The Proceedings of the first Solvay Conference were originally published as *La Théorie du rayonnment et les quanta: Rapports et discussions de la réunion tenue à Bruxells, du 30 octobre au 3 novembre 1911,* eds. P. Langevin and M. de Broglie (Paris: Gauthier-Villars, 1912). A German translation was published in 1914 with an extremely valuable summary of work done on the quantum theory up to the summer of 1913. That edition *Die Theorie der Strahlung und der Quanten,* ed. A. Eucken (Halle: W. Knapp, 1914) was the edition consulted for this study.

82. M. Planck, "Energie und Temperatur," Phys. ZS. **12** (1911), pp. 681–687: III, 52.

83. M. Planck, "Über die Begründung des Gesetzes der schwarzen Strahlung," Ann. Phys. **37** (1912), pp. 642–656: II, 289.

84. Carl W. Oseen, "Über die Möglichkeit ungedämpfter Schwingungen nach der Maxwell–Lorentzschen Theorie und uber die Plancksche Strahlungstheorie," Ann. Phys. **43** (1913), pp. 639–651. See Needell (note 35), pp. 235–36 for a description of the exchange between Planck and Oseen and for other references.

85. For example, in Jammer (note 13), p. 49.

86. M. Planck, "Über das Gleichgewicht zwischen Oszillatoren, freien Electronen und Strahlender Wärme," Berl Ber. (1913), pp. 350–363: II, 302–315.

87. M. Planck, "Die Gegenwärtige Bedeutung der Quantenhypothese für die kinetische Gastheorie," Math. Vorl. a. d. Univ. Göttingen **6** (1914), pp. 3–16: II, 323.

88. M. Planck, "Eine veränderte Formulierung der Quantenhypothese," Berl. Ber. (1914), pp. 918–923: II, 330–335.

89. This work is described in detail in Needell (note 35), pp. 263–274.

90. M. Planck, "Über Quantenwirkungen in der Elektrodynamik," Berl Ber. (1915), pp. 512–519: II, 341–348.

91. A. D. Fokker, "Über Brownsche Bewegungen im Strahlungsfeld," Phys. ZS. **15** (1914), pp. 96–98. And see Needell (note 35), pp. 263–268.

92. For the influence of Ehrenfest on Planck's understanding of irreversibility, see Needell (note 35), pp. 63–68.

93. One possible exception is the footnote added to the description of "natural radiation" in the *Wärmestrahlung* (1913) referred to earlier [§177, p. 117].

94. Max Planck, "Dynamische und statistische Gesetzmässigkeit" (Leipzig: J. A. Barth, 1914): III, 81.

I have received extremely valuable comments and criticisms of earlier drafts of this introduction from Stephen Brush, Ron Doel, Paul Forman, Peter Galison, John Heilbron, Gerald Holton, Helge Kragh, Martin Klein, Thomas Kuhn, and Katherine Sopka. Their contributions are gratefully acknowledged.

THE THEORY
OF
HEAT RADIATION

BY

DR. MAX PLANCK
PROFESSOR OF THEORETICAL PHYSICS IN THE UNIVERSITY OF BERLIN

AUTHORISED TRANSLATION

BY

MORTON MASIUS, M. A., Ph. D. (Leipzig)
INSTRUCTOR IN PHYSICS IN THE WORCESTER POLYTECHNIC INSTITUTE

WITH 7 ILLUSTRATIONS

PHILADELPHIA
P. BLAKISTON'S SON & CO.
1012 WALNUT STREET

TRANSLATOR'S PREFACE

The present volume is a translation of the second edition of Professor *Planck's* WAERMESTRAHLUNG (1913). The profoundly original ideas introduced by *Planck* in the endeavor to reconcile the electromagnetic theory of radiation with experimental facts have proven to be of the greatest importance in many parts of physics. Probably no single book since the appearance of *Clerk Maxwell's* ELECTRICITY AND MAGNETISM has had a deeper influence on the development of physical theories. The great majority of English-speaking physicists are, of course, able to read the work in the language in which it was written, but I believe that many will welcome the opportunity offered by a translation to study the ideas set forth by *Planck* without the difficulties that frequently arise in attempting to follow a new and somewhat difficult line of reasoning in a foreign language.

Recent developments of physical theories have placed the quantum of action in the foreground of interest. Questions regarding the bearing of the quantum theory on the law of equipartition of energy, its application to the theory of specific heats and to photoelectric effects, attempts to form some concrete idea of the physical significance of the quantum, that is, to devise a "model" for it, have created within the last few years a large and ever increasing literature. Professor *Planck* has, however, in this book confined himself exclusively to radiation phenomena and it has seemed to me probable that a brief résumé of this literature might prove useful to the reader who wishes to pursue the subject further. I have, therefore, with Professor *Planck's* permission, given in an appendix a list of the most important papers on the subjects treated of in this book and others closely related to them. I have also added a short note on one or two derivations of formulæ where the treatment in the book seemed too brief or to present some difficulties.

In preparing the translation I have been under obligation for advice and helpful suggestions to several friends and colleagues and especially to Professor A. W. Duff who has read the manuscript and the galley proof.

MORTON MASIUS.

WORCESTER, MASS.,
February, 1914.

PREFACE TO SECOND EDITION

Recent advances in physical research have, on the whole, been favorable to the special theory outlined in this book, in particular to the hypothesis of an elementary quantity of action. My radiation formula especially has so far stood all tests satisfactorily, including even the refined systematic measurements which have been carried out in the Physikalisch-technische Reichsanstalt at Charlottenburg during the last year. Probably the most direct support for the fundamental idea of the hypothesis of quanta is supplied by the values of the elementary quanta of matter and electricity derived from it. When, twelve years ago, I made my first calculation of the value of the elementary electric charge and found it to be $4.69 \cdot 10^{-10}$ electrostatic units, the value of this quantity deduced by *J. J. Thomson* from his ingenious experiments on the condensation of water vapor on gas ions, namely $6.5 \cdot 10^{-10}$ was quite generally regarded as the most reliable value. This value exceeds the one given by me by 38 per cent. Meanwhile the experimental methods, improved in an admirable way by the labors of *E. Rutherford*, *E. Regener*, *J. Perrin*, *R. A. Millikan*, *The Svedberg* and others, have without exception decided in favor of the value deduced from the theory of radiation which lies between the values of *Perrin* and *Millikan*.

To the two mutually independent confirmations mentioned, there has been added, as a further strong support of the hypothesis of quanta, the heat theorem which has been in the meantime announced by *W. Nernst*, and which seems to point unmistakably to the fact that, not only the processes of radiation, but also the molecular processes take place in accordance with certain elementary quanta of a definite finite magnitude. For the hypothesis of quanta as well as the heat theorem of *Nernst* may be reduced to the simple proposition that the thermodynamic probability (Sec. 120) of a physical state is a definite integral number, or, what amounts to the same thing, that the entropy of a state has a quite definite, positive value, which, as a minimum, becomes zero, while in contrast therewith the entropy may, according to the classical thermodynamics, decrease without limit to minus infinity. For the present, I would consider this proposition as the very quintessence of the hypothesis of quanta.

In spite of the satisfactory agreement of the results mentioned with one another as well as with experiment, the ideas from which they originated have met with wide interest but, so far as I am able to judge, with little general acceptance, the reason probably being that the hypothesis of quanta has not as yet been satisfactorily completed. While many physicists, through conservatism, reject the ideas developed by me, or, at any rate, maintain an expectant attitude, a few authors have attacked them for the opposite reason, namely, as being inadequate, and have felt compelled to supplement them by assumptions of a still more radical nature, for example, by the assumption that any radiant energy whatever, even though it travel freely in a vacuum, consists of indivisible quanta or cells. Since nothing probably is a greater drawback to the successful development of a new hypothesis than overstepping its boundaries, I have always stood for making as close a connection between the hypothesis of quanta and the classical dynamics as possible, and for not stepping outside of the boundaries of the latter until the experimental facts leave no other course open. I have attempted to keep to this standpoint in the revision of this treatise necessary for a new edition.

The main fault of the original treatment was that it began with the classical electrodynamical laws of emission and absorption, whereas later on it became evident that, in order to meet the demand of experimental measurements, the assumption of finite energy elements must be introduced, an assumption which is in direct contradiction to the fundamental ideas of classical electrodynamics. It is true that this inconsistency is greatly reduced by the fact that, in reality, only mean values of energy are taken from classical electrodynamics, while, for the statistical calculation, the real values are used; nevertheless the treatment must, on the whole, have left the reader with the unsatisfactory feeling that it was not clearly to be seen, which of the assumptions made in the beginning could, and which could not, be finally retained.

In contrast thereto I have now attempted to treat the subject from the very outset in such a way that none of the laws stated need, later on, be restricted or modified. This presents the advantage that the theory, so far as it is treated here, shows no contradiction in itself, though certainly I do not mean that it does not seem to call for improvements in many respects, as regards both its internal structure and its external form. To treat of the numerous applications, many of them very important, which the hypothesis of quanta has already found in other parts of physics, I have not regarded as part of my task, still less to discuss all differing opinions.

Thus, while the new edition of this book may not claim to bring the theory of heat radiation to a conclusion that is satisfactory in all respects, this deficiency will not be of decisive importance in judging the theory. For any one who would make his attitude concerning the hypothesis of quanta depend on whether the significance of the quantum of action for the elementary physical processes is made clear in every respect or may be demonstrated by some simple dynamical model, misunderstands, I believe, the character and the meaning of the hypothesis of quanta. It is impossible to express a really new principle in terms of a model following old laws. And, as regards the final formulation of the hypothesis, we should not forget that, from the classical point of view, the physics of the atom really has always remained a very obscure, inaccessible region, into which the introduction of the elementary quantum of action promises to throw some light.

Hence it follows from the nature of the case that it will require painstaking experimental and theoretical work for many years to come to make gradual advances in the new field. Any one who, at present, devotes his efforts to the hypothesis of quanta, must, for the time being, be content with the knowledge that the fruits of the labor spent will probably be gathered by a future generation.

THE AUTHOR.

BERLIN,
November, 1912.

PREFACE TO FIRST EDITION

In this book the main contents of the lectures which I gave at the University of Berlin during the winter semester 1906–07 are presented. My original intention was merely to put together in a connected account the results of my own investigations, begun ten years ago, on the theory of heat radiation; it soon became evident, however, that it was desirable to include also the foundation of this theory in the treatment, starting with Kirchhoff's Law on emitting and absorbing power; and so I attempted to write a treatise which should also be capable of serving as an introduction to the study of the entire theory of radiant heat on a consistent thermodynamic basis. Accordingly the treatment starts from the simple known experimental laws of optics and advances, by gradual extension and by the addition of the results of electrodynamics and thermodynamics, to the problems of the spectral distribution of energy and of irreversibility. In doing this I have deviated frequently from the customary methods of treatment, wherever the matter presented or considerations regarding the form of presentation seemed to call for it, especially in deriving Kirchhoff's laws, in calculating Maxwell's radiation pressure, in deriving Wien's displacement law, and in generalizing it for radiations of any spectral distribution of energy whatever.

I have at the proper place introduced the results of my own investigations into the treatment. A list of these has been added at the end of the book to facilitate comparison and examination as regards special details.

I wish, however, to emphasize here what has been stated more fully in the last paragraph of this book, namely, that the theory thus developed does not by any means claim to be perfect or complete, although I believe that it points out a possible way of accounting for the processes of radiant energy from the same point of view as for the processes of molecular motion.

TABLE OF CONTENTS

PART I

FUNDAMENTAL FACTS AND DEFINITIONS

PART II

DEDUCTIONS FROM ELECTRODYNAMICS AND THERMODYNAMICS

PART III

ENTROPY AND PROBABILITY

PART IV

A SYSTEM OF OSCILLATORS IN A STATIONARY FIELD OF RADIATION

PART V

IRREVERSIBLE RADIATION PROCESSES

PART I

FUNDAMENTAL FACTS AND DEFINITIONS

RADIATION OF HEAT

CHAPTER I

GENERAL INTRODUCTION

1. Heat may be propagated in a stationary medium in two entirely different ways, namely, by conduction and by radiation. Conduction of heat depends on the temperature of the medium in which it takes place, or more strictly speaking, on the non-uniform distribution of the temperature in space, as measured by the temperature gradient. In a region where the temperature of the medium is the same at all points there is no trace of heat conduction.

Radiation of heat, however, is in itself entirely independent of the temperature of the medium through which it passes. It is possible, for example, to concentrate the solar rays at a focus by passing them through a converging lens of ice, the latter remaining at a constant temperature of 0°, and so to ignite an inflammable body. Generally speaking, radiation is a far more complicated phenomenon than conduction of heat. The reason for this is that the state of the radiation at a given instant and at a given point of the medium cannot be represented, as can the flow of heat by conduction, by a single vector (that is, a single directed quantity). All heat rays which at a given instant pass through the same point of the medium are perfectly independent of one another, and in order to specify completely the state of the radiation the intensity of radiation must be known in all the directions, infinite in number, which pass through the point in question; for this purpose two opposite directions must be considered as distinct, because the radiation in one of them is quite independent of the radiation in the other.

1

2. Putting aside for the present any special theory of heat radiation, we shall state for our further use a law supported by a large number of experimental facts. This law is that, so far as their physical properties are concerned, heat rays are identical with light rays of the same wave length. The term "heat radiation," then, will be applied to all physical phenomena of the same nature as light rays. Every light ray is simultaneously a heat ray. We shall also, for the sake of brevity, occasionally speak of the "color" of a heat ray in order to denote its wave length or period. As a further consequence of this law we shall apply to the radiation of heat all the well-known laws of experimental optics, especially those of reflection and refraction, as well as those relating to the propagation of light. Only the phenomena of diffraction, so far at least as they take place in space of considerable dimensions, we shall exclude on account of their rather complicated nature. We are therefore obliged to introduce right at the start a certain restriction with respect to the size of the parts of space to be considered. Throughout the following discussion it will be assumed that the linear dimensions of all parts of space considered, as well as the radii of curvature of all surfaces under consideration, are large compared with the wave lengths of the rays considered. With this assumption we may, without appreciable error, entirely neglect the influence of diffraction caused by the bounding surfaces, and everywhere apply the ordinary laws of reflection and refraction of light. To sum up: We distinguish once for all between two kinds of lengths of entirely different orders of magnitude—dimensions of bodies and wave lengths. Moreover, even the differentials of the former, *i.e.*, elements of length, area and volume, will be regarded as large compared with the corresponding powers of wave lengths. The greater, therefore, the wave length of the rays we wish to consider, the larger must be the parts of space considered. But, inasmuch as there is no other restriction on our choice of size of the parts of space to be considered, this assumption will not give rise to any particular difficulty.

3. Even more essential for the whole theory of heat radiation than the distinction between large and small lengths, is the distinction between long and short intervals of time. For the definition of intensity of a heat ray, as being the energy transmitted by the ray per unit time, implies the assumption that the unit of time chosen is large compared with the period of vibration corresponding to the color of the ray. If this were not so, obviously the value of the intensity of the radiation would, in general, depend upon the particular phase of vibration at which the measurement of the energy of the ray was begun, and the intensity of a ray of constant period and amplitude would not be independent of the initial phase, unless by chance the unit of time were an integral multiple of the period. To avoid this difficulty, we are obliged to postulate quite generally that the unit of time, or rather that element of time used in defining the intensity, even if it appear in the form of a differential, must be large compared with the period of all colors contained in the ray in question.

The last statement leads to an important conclusion as to radiation of variable intensity. If, using an acoustic analogy, we speak of "beats" in the case of intensities undergoing periodic changes, the "unit" of time required for a definition of the instantaneous intensity of radiation must necessarily be small compared with the period of the beats. Now, since from the previous statement our unit must be large compared with a period of vibration, it follows that the period of the beats must be large compared with that of a vibration. Without this restriction it would be impossible to distinguish properly between "beats" and simple "vibrations." Similarly, in the general case of an arbitrarily variable intensity of radiation, the vibrations must take place very rapidly as compared with the relatively slower changes in intensity. These statements imply, of course, a certain far-reaching restriction as to the generality of the radiation phenomena to be considered.

It might be added that a very similar and equally essential restriction is made in the kinetic theory of gases by dividing the motions of a chemically simple gas into two classes: visible, coarse, or molar, and invisible, fine, or molecular. For, since the velocity of a single molecule is a perfectly unambiguous quantity, this distinction cannot be drawn unless the assumption be made that the velocity-components of the molecules contained in sufficiently small volumes have certain mean values, independent of the size of the volumes. This in general need not by any means be the case. If such a mean value, including the value zero, does not exist, the distinction between motion of the gas as a whole and random undirected heat motion cannot be made.

Turning now to the investigation of the laws in accordance with which the phenomena of radiation take place in a medium supposed to be at rest, the problem may be approached in two ways: We must either select a certain point in space and investigate the different rays passing through this one point as time goes on, or we must select one distinct ray and inquire into its history, that is, into the way in which it was created, propagated, and finally destroyed. For the following discussion, it will be advisable to start with the second method of treatment and to consider first the three processes just mentioned.

4. Emission.—The creation of a heat ray is generally denoted by the word emission. According to the principle of the conservation of energy, emission always takes place at the expense of other forms of energy (heat,[1] chemical or electric energy, etc.) and hence it follows that only material particles, not geometrical volumes or surfaces, can emit heat rays. It is true that for the sake of brevity we frequently speak of the surface of a body as radiating heat to the surroundings, but this form of expression does not imply that the surface actually emits heat rays. Strictly speaking, the surface of a body never emits rays, but rather it allows part of the rays coming from the interior to pass through. The other part is reflected inward and according as the fraction transmitted is larger or smaller the surface seems to emit more or less intense radiations.

We shall now consider the interior of an emitting substance assumed to be physically homogeneous, and in it we shall select any volume-element $d\tau$ of not too small size. Then the energy which is emitted by radiation in unit time by all particles in this volume-element will be proportional to $d\tau$. Should we attempt a closer analysis of the process of emission and resolve it into its elements, we should undoubtedly meet very complicated conditions, for then it would be necessary to consider elements of space of such small size that it would no longer be admissible to think of the substance as homogeneous, and we would have to allow for the atomic constitution. Hence the finite quantity

[1] Here as in the following the German "Körperwärme" will be rendered simply as "heat." (Tr.)

obtained by dividing the radiation emitted by a volume-element $d\tau$ by this element $d\tau$ is to be considered only as a certain mean value. Nevertheless, we shall as a rule be able to treat the phenomenon of emission as if all points of the volume-element $d\tau$ took part in the emission in a uniform manner, thereby greatly simplifying our calculation. Every point of $d\tau$ will then be the vertex of a pencil of rays diverging in all directions. Such a pencil coming from one single point of course does not represent a finite amount of energy, because a finite amount is emitted only by a finite though possibly small volume, not by a single point.

We shall next assume our substance to be isotropic. Hence the radiation of the volume-element $d\tau$ is emitted uniformly in all directions of space. Draw a cone in an arbitrary direction, having any point of the radiating element as vertex, and describe around the vertex as center a sphere of unit radius. This sphere intersects the cone in what is known as the solid angle of the cone, and from the isotropy of the medium it follows that the radiation in any such conical element will be proportional to its solid angle. This holds for cones of any size. If we take the solid angle as infinitely small and of size $d\Omega$ we may speak of the radiation emitted in a certain direction, but always in the sense that for the emission of a finite amount of energy an infinite number of directions are necessary and these form a finite solid angle.

5. The distribution of energy in the radiation is in general quite arbitrary; that is, the different colors of a certain radiation may have quite different intensities. The color of a ray in experimental physics is usually denoted by its wave length, because this quantity is measured directly. For the theoretical treatment, however, it is usually preferable to use the frequency ν instead, since the characteristic of color is not so much the wave length, which changes from one medium to another, as the frequency, which remains unchanged in a light or heat ray passing through stationary media. We shall, therefore, hereafter denote a certain color by the corresponding value of ν, and a certain interval of color by the limits of the interval ν and ν', where $\nu' > \nu$. The radiation lying in a certain interval of color divided by the magnitude $\nu'-\nu$ of the interval, we shall call the mean radiation in the interval ν to ν'. We shall then assume that if, keeping ν constant, we take the interval $\nu'-\nu$ sufficiently small and denote it by $d\nu$ the value of the mean radiation approaches a definite limiting value, independent of the size of $d\nu$, and this we shall briefly call the "radiation of frequency ν." To produce a finite intensity of radiation, the frequency interval, though perhaps small, must also be finite.

We have finally to allow for the polarization of the emitted radiation. Since the medium was assumed to be isotropic the emitted rays are unpolarized. Hence every ray has just twice the intensity of one of its plane polarized components, which could, *e.g.*, be obtained by passing the ray through a *Nicol's* prism.

6. Summing up everything said so far, we may equate the total energy in a range of frequency from ν to $\nu+d\nu$ emitted in the time dt in the direction of the conical element $d\Omega$ by a volume element $d\tau$ to

$$dt \cdot d\tau \cdot d\Omega \cdot d\nu \cdot 2\epsilon_\nu. \tag{1}$$

The finite quantity ϵ_ν is called the coefficient of emission of the medium for the frequency ν. It is a positive function of ν and refers to a plane polarized ray of definite color and direction. The total emission of the volume-element $d\tau$ may be obtained from this by integrating over all directions and all frequencies. Since ϵ_ν is independent of the direction, and since the integral over all conical elements $d\Omega$ is 4π, we get:

$$dt \cdot d\tau . 8\pi \int_0^\infty \epsilon_\nu d\nu. \tag{2}$$

7. The coefficient of emission ϵ depends, not only on the frequency ν, but also on the condition of the emitting substance contained in the volume-element $d\tau$, and, generally speaking, in a very complicated way, according to the physical and chemical processes which take place in the elements of time and volume in question. But the empirical law that the emission of any volume-element depends entirely on what takes place inside of this element holds true in all cases (*Prevost's* principle). A body A at 100° C. emits toward a body B at 0° C. exactly the same amount of radiation as toward an equally large and similarly situated body B' at 1000° C. The fact that the body A is cooled by B and heated by B' is due entirely to the fact that B is a weaker, B' a stronger emitter than A.

We shall now introduce the further simplifying assumption that the physical and chemical condition of the emitting substance depends on but a single variable, namely, on its absolute temperature T. A necessary consequence of this is that the coefficient of emission ϵ depends, apart from the frequency ν and the nature of the medium, only on the temperature T. The last statement excludes from our consideration a number of radiation phenomena, such as fluorescence, phosphorescence, electrical and chemical luminosity, to which *E. Wiedemann* has given the common name "phenomena of luminescence." We shall deal with pure "temperature radiation" exclusively.

A special case of temperature radiation is the case of the chemical nature of the emitting substance being invariable. In this case the emission takes place entirely at the expense of the heat of the body. Nevertheless, it is possible, according to what has been said, to have temperature radiation while chemical changes are taking place, provided the chemical condition is completely determined by the temperature.

8. Propagation.—The propagation of the radiation in a medium assumed to be homogeneous, isotropic, and at rest takes place in straight lines and with the same velocity in all directions, diffraction phenomena being entirely excluded. Yet, in general, each ray suffers during its propagation a certain weakening, because a certain fraction of its energy is continuously deviated from its original direction and scattered in all directions. This phenomenon of "scattering," which means neither a creation nor a destruction of radiant energy but simply a change in distribution, takes place, generally speaking, in all media differing from an absolute vacuum, even in substances which are perfectly pure chemically.[1] The cause of this is that no substance is homogeneous in the absolute sense of the word. The smallest elements of space always exhibit some discontinuities on account of their atomic structure. Small impurities, as, for instance, particles of dust, increase the influence of scattering without, however, appreciably affecting its general character. Hence, so-called "turbid"

[1] See, *e.g.*, *Lobry de Bruyn* and *L. K. Wolff*, Rec. des Trav. Chim. des Pays-Bas **23**, p. 155, 1904.

media, *i.e.*, such as contain foreign particles, may be quite properly regarded as optically homogeneous,[1] provided only that the linear dimensions of the foreign particles as well as the distances of neighboring particles are sufficiently small compared with the wave lengths of the rays considered. As regards optical phenomena, then, there is no fundamental distinction between chemically pure substances and the turbid media just described. No space is optically void in the absolute sense except a vacuum. Hence a chemically pure substance may be spoken of as a vacuum made turbid by the presence of molecules.

A typical example of scattering is offered by the behavior of sunlight in the atmosphere. When, with a clear sky, the sun stands in the zenith, only about two-thirds of the direct radiation of the sun reaches the surface of the earth. The remainder is intercepted by the atmosphere, being partly absorbed and changed into heat of the air, partly, however, scattered and changed into diffuse skylight. This phenomenon is produced probably not so much by the particles suspended in the atmosphere as by the air molecules themselves.

Whether the scattering depends on reflection, on diffraction, or on a resonance effect on the molecules or particles is a point that we may leave entirely aside. We only take account of the fact that every ray on its path through any medium loses a certain fraction of its intensity. For a very small distance, s, this fraction is proportional to s, say

$$\beta_\nu s \tag{3}$$

where the positive quantity β_ν is independent of the intensity of radiation and is called the "coefficient of scattering" of the medium. Inasmuch as the medium is assumed to be isotropic, β_ν is also independent of the direction of propagation and polarization of the ray. It depends, however, as indicated by the subscript ν, not only on the physical and chemical constitution of the body but also to a very marked degree on the frequency. For certain values of ν, β_ν may be so large that the straight-line propagation of the rays is virtually destroyed. For other values of ν, however, β_ν may become so small that the scattering can

[1] To restrict the word homogeneous to its absolute sense would mean that it could not be applied to any material substance.

be entirely neglected. For generality we shall assume a mean value of β_ν. In the cases of most importance β_ν increases quite appreciably as ν increases, *i.e.*, the scattering is noticeably larger for rays of shorter wave length;[1] hence the blue color of diffuse skylight.

The scattered radiation energy is propagated from the place where the scattering occurs in a way similar to that in which the emitted energy is propagated from the place of emission, since it travels in all directions in space. It does not, however, have the same intensity in all directions, and moreover is polarized in some special directions, depending to a large extent on the direction of the original ray. We need not, however, enter into any further discussion of these questions.

9. While the phenomenon of scattering means a continuous modification in the interior of the medium, a discontinuous change in both the direction and the intensity of a ray occurs when it reaches the boundary of a medium and meets the surface of a second medium. The latter, like the former, will be assumed to be homogeneous and isotropic. In this case, the ray is in general partly reflected and partly transmitted. The reflection and refraction may be "regular," there being a single reflected ray according to the simple law of reflection and a single transmitted ray, according to *Snell's* law of refraction, or, they may be "diffuse," which means that from the point of incidence on the surface the radiation spreads out into the two media with intensities that are different in different directions. We accordingly describe the surface of the second medium as "smooth" or "rough" respectively. Diffuse reflection occurring at a rough surface should be carefully distinguished from reflection at a smooth surface of a turbid medium. In both cases part of the incident ray goes back to the first medium as diffuse radiation. But in the first case the scattering occurs on the surface, in the second in more or less thick layers entirely inside of the second medium.

10. When a smooth surface completely reflects all incident rays, as is approximately the case with many metallic surfaces, it is termed "reflecting." When a rough surface reflects all incident rays completely and uniformly in all directions, it is

[1] *Lord Rayleigh*, Phil. Mag., **47**, p. 379, 1899.

called "white." The other extreme, namely, complete transmission of all incident rays through the surface never occurs with smooth surfaces, at least if the two contiguous media are at all optically different. A rough surface having the property of completely transmitting the incident radiation is described as "black."

In addition to "black surfaces" the term "black body" is also used. According to *G. Kirchhoff*[1] it denotes a body which has the property of allowing all incident rays to enter without surface reflection and not allowing them to leave again. Hence it is seen that a black body must satisfy three independent conditions. First, the body must have a black surface in order to allow the incident rays to enter without reflection. Since, in general, the properties of a surface depend on both of the bodies which are in contact, this condition shows that the property of blackness as applied to a body depends not only on the nature of the body but also on that of the contiguous medium. A body which is black relatively to air need not be so relatively to glass, and *vice versa.* Second, the black body must have a certain minimum thickness depending on its absorbing power, in order to insure that the rays after passing into the body shall not be able to leave it again at a different point of the surface. The more absorbing a body is, the smaller the value of this minimum thickness, while in the case of bodies with vanishingly small absorbing power only a layer of infinite thickness may be regarded as black. Third, the black body must have a vanishingly small coefficient of scattering (Sec. 8). Otherwise the rays received by it would be partly scattered in the interior and might leave again through the surface.[2]

11. All the distinctions and definitions mentioned in the two preceding paragraphs refer to rays of one definite color only. It might very well happen that, *e.g.*, a surface which is rough for a certain kind of rays must be regarded as smooth for a different kind of rays. It is readily seen that, in general, a surface shows

[1] *G. Kirchhoff*, Pogg. Ann., **109,** p. 275, 1860. Gesammelte Abhandlungen, J. A. Barth, Leipzig, 1882, p. 573. In defining a black body *Kirchhoff* also assumes that the absorption of incident rays takes place in a layer "infinitely thin." We do not include this in our definition.

[2] For this point see especially *A. Schuster*, Astrophysical Journal, **21,** p. 1, 1905, who has pointed out that an infinite layer of gas with a black surface need by no means be a black body.

decreasing degrees of roughness for increasing wave lengths Now, since smooth non-reflecting surfaces do not exist (Sec. 10), it follows that all approximately black surfaces which may be realized in practice (lamp black, platinum black) show appreciable reflection for rays of sufficiently long wave lengths.

12. Absorption.—Heat rays are destroyed by "absorption." According to the principle of the conservation of energy the energy of heat radiation is thereby changed into other forms of energy (heat, chemical energy). Thus only material particles can absorb heat rays, not elements of surfaces, although sometimes for the sake of brevity the expression absorbing surfaces is used.

Whenever absorption takes place, the heat ray passing through the medium under consideration is weakened by a certain fraction of its intensity for every element of path traversed. For a sufficiently small distance s this fraction is proportional to s, and may be written

$$\alpha_\nu s \tag{4}$$

Here α_ν is known as the "coefficient of absorption" of the medium for a ray of frequency ν. We assume this coefficient to be independent of the intensity; it will, however, depend in general in non-homogeneous and anisotropic media on the position of s and on the direction of propagation and polarization of the ray (example: tourmaline). We shall, however, consider only homogeneous isotropic substances, and shall therefore suppose that α_ν has the same value at all points and in all directions in the medium, and depends on nothing but the frequency ν, the temperature T, and the nature of the medium.

Whenever α_ν does not differ from zero except for a limited range of the spectrum, the medium shows "selective" absorption. For those colors for which $\alpha_\nu=0$ and also the coefficient of scattering $\beta_\nu=0$ the medium is described as perfectly "transparent" or "diathermanous." But the properties of selective absorption and of diathermancy may for a given medium vary widely with the temperature. In general we shall assume a mean value for α_ν. This implies that the absorption in a distance equal to a single wave length is very small, because the distance s, while small, contains many wave lengths (Sec. 2).

13. The foregoing considerations regarding the emission, the propagation, and the absorption of heat rays suffice for a mathematical treatment of the radiation phenomena. The calculation requires a knowledge of the value of the constants and the initial and boundary conditions, and yields a full account of the changes the radiation undergoes in a given time in one or more contiguous media of the kind stated, including the temperature changes caused by it. The actual calculation is usually very complicated. We shall, however, before entering upon the treatment of special cases discuss the general radiation phenomena from a different point of view, namely by fixing our attention not on a definite ray, but on a definite position in space.

14. Let $d\sigma$ be an arbitrarily chosen, infinitely small element of area in the interior of a medium through which radiation passes. At a given instant rays are passing through this element in many different directions. The energy radiated through it in an element of time dt in a definite direction is proportional to the area $d\sigma$, the length of time dt and to the cosine of the angle θ made by the normal of $d\sigma$ with the direction of the radiation. If we make $d\sigma$ sufficiently small, then, although this is only an approximation to the actual state of affairs, we can think of all points in $d\sigma$ as being affected by the radiation in the same way. Then the energy radiated through $d\sigma$ in a definite direction must be proportional to the solid angle in which $d\sigma$ intercepts that radiation and this solid angle is measured by $d\sigma\ \cos\ \theta$. It is readily seen that, when the direction of the element is varied relatively to the direction of the radiation, the energy radiated through it vanishes when

$$\theta = \frac{\pi}{2}.$$

Now in general a pencil of rays is propagated from every point of the element $d\sigma$ in all directions, but with different intensities in different directions, and any two pencils emanating from two points of the element are identical save for differences of higher order. A single one of these pencils coming from a single point does not represent a finite quantity of energy, because a finite amount of energy is radiated only through a finite area. This holds also for the passage of rays through a so-called focus. For example, when sunlight passes through a converging lens and is concentrated in the focal plane of the lens, the solar rays do not converge to a single point, but each pencil of parallel rays forms a separate focus and all these foci together constitute a surface representing a small but finite image of the sun. A finite amount of energy does not pass through less than a finite portion of this surface.

15. Let us now consider quite generally the pencil, which is propagated from a point of the element $d\sigma$ as vertex in all directions of space and on both sides of $d\sigma$. A certain direction may be specified by the angle θ (between 0 and π), as already used, and by an azimuth ϕ (between 0 and 2π). The intensity in this direction is the energy propagated in an infinitely thin cone limited by θ and $\theta+d\theta$ and ϕ and $\phi+d\phi$. The solid angle of this cone is

$$d\Omega = \sin\theta \cdot d\theta \cdot d\phi. \tag{5}$$

Thus the energy radiated in time dt through the element of area $d\sigma$ in the direction of the cone $d\Omega$ is:

$$dt\, d\sigma \cos\theta\, d\Omega\, K = K \sin\theta \cos\theta\, d\theta\, d\phi\, d\sigma\, dt. \tag{6}$$

The finite quantity K we shall term the "specific intensity" or the "brightness," $d\Omega$ the "solid angle" of the pencil emanating from a point of the element $d\sigma$ in the direction (θ, ϕ). K is a positive function of position, time, and the angles θ and ϕ. In general the specific intensities of radiation in different directions are entirely independent of one another. For example, on substituting $\pi-\theta$ for θ and $\pi+\phi$ for ϕ in the function K, we obtain the specific intensity of radiation in the diametrically opposite direction, a quantity which in general is quite different from the preceding one.

For the total radiation through the element of area $d\sigma$ toward one side, say the one on which θ is an acute angle, we get, by integrating with respect to ϕ from 0 to 2π and with respect to θ from 0 to $\frac{\pi}{2}$

$$\int_0^{2\pi} d\phi \int_0^{\frac{\pi}{2}} d\theta K \sin\theta \cos\theta\, d\sigma\, dt.$$

Should the radiation be uniform in all directions and hence K be a constant, the total radiation on one side will be

$$\pi K \, d\sigma \, dt. \tag{7}$$

16. In speaking of the radiation in a definite direction (θ, ϕ) one should always keep in mind that the energy radiated in a cone is not finite unless the angle of the cone is finite. No finite radiation of light or heat takes place in one definite direction only, or expressing it differently, in nature there is no such thing as absolutely parallel light or an absolutely plane wave front. From a pencil of rays called "parallel" a finite amount of energy of radiation can only be obtained if the rays or wave normals of the pencil diverge so as to form a finite though perhaps exceedingly narrow cone.

17. The specific intensity K of the whole energy radiated in a certain direction may be further divided into the intensities of the separate rays belonging to the different regions of the spectrum which travel independently of one another. Hence we consider the intensity of radiation within a certain range of frequencies, say from ν to ν'. If the interval $\nu' - \nu$ be taken sufficiently small and be denoted by $d\nu$, the intensity of radiation within the interval is proportional to $d\nu$. Such radiation is called homogeneous or monochromatic.

A last characteristic property of a ray of definite direction, intensity, and color is its state of polarization. If we break up a ray, which is in any state of polarization whatsoever and which travels in a definite direction and has a definite frequency ν, into two plane polarized components, the sum of the intensities of the components will be just equal to the intensity of the ray as a whole, independently of the direction of the two planes, provided the two planes of polarization, which otherwise may be taken at random, are at right angles to each other. If their position be denoted by the azimuth ψ of one of the planes of vibration (plane of the electric vector), then the two components of the intensity may be written in the form

$$\mathsf{K}_\nu \cos^2\psi + \mathsf{K}_\nu' \sin^2\psi$$

and

$$\mathsf{K}_\nu \sin^2\psi + \mathsf{K}_\nu' \cos^2\psi \tag{8}$$

Herein K is independent of ψ. These expressions we shall call the "components of the specific intensity of radiation of frequency ν." The sum is independent of ψ and is always equal to the intensity of the whole ray $\mathsf{K}_\nu + \mathsf{K}_\nu'$. At the same time K_ν and K_ν' represent respectively the largest and smallest values which either of the components may have, namely, when $\psi = 0$ and $\psi = \frac{\pi}{2}$. Hence we call these values the "principal values of the intensities," or the "principal intensities," and the corresponding planes of vibration we call the "principal planes of vibration" of the ray. Of course both, in general, vary with the time. Thus we may write generally

$$K = \int_c^\infty d\nu \; (\mathsf{K}_\nu + \mathsf{K}_\nu') \tag{9}$$

where the positive quantities K_ν and K_ν', the two principal values of the specific intensity of the radiation (brightness) of frequency ν, depend not only on ν but also on their position, the time, and on the angles θ and ϕ. By substitution in (6) the energy radiated in the time dt through the element of area $d\sigma$ in the direction of the conical element $d\Omega$ assumes the value

$$dt\, d\sigma \cos\theta\, d\Omega \int_0^\infty d\nu \; (\mathsf{K}_\nu + \mathsf{K}_\nu') \tag{10}$$

and for monochromatic plane polarized radiation of brightness K_ν:

$$dt\, d\sigma \cos\theta\, d\Omega\, \mathsf{K}_\nu\, d\nu = dt\, d\sigma \sin\theta \cos\theta\, d\theta\, d\phi\, \mathsf{K}_\nu\, d\nu. \tag{11}$$

For unpolarized rays $\mathsf{K}_\nu = \mathsf{K}_\nu'$, and hence

$$K = 2\int_0^\infty d\nu\; \mathsf{K}_\nu, \tag{12}$$

and the energy of a monochromatic ray of frequency ν will be:

$$2dt\, d\sigma \cos\theta\, d\Omega\; \mathsf{K}_\nu\, d\nu = 2dt\, d\sigma \sin\theta \cos\theta\, d\theta\, d\phi\; \mathsf{K}_\nu\, d\nu. \tag{13}$$

When, moreover, the radiation is uniformly distributed in all directions, the total radiation through $d\sigma$ toward one side may be found from (7) and (12); it is

$$2\pi\, d\sigma\, dt \int_0^\infty \mathsf{K}_\nu d\nu. \tag{14}$$

18. Since in nature K_ν can never be infinitely large, K will not have a finite value unless K_ν differs from zero over a finite range of frequencies. Hence there exists in nature no absolutely homogeneous or monochromatic radiation of light or heat. A finite amount of radiation contains always a finite although possibly very narrow range of the spectrum. This implies a fundamental difference from the corresponding phenomena of acoustics, where a finite intensity of sound may correspond to a single definite frequency. This difference is, among other things, the cause of the fact that the second law of thermodynamics has an important bearing on light and heat rays, but not on sound waves. This will be further discussed later on.

19. From equation (9) it is seen that the quantity K_ν, the intensity of radiation of frequency ν, and the quantity K, the intensity of radiation of the whole spectrum, are of different dimensions. Further it is to be noticed that, on subdividing the spectrum according to wave lengths λ, instead of frequencies ν, the intensity of radiation E_λ of the wave lengths λ corresponding to the frequency ν is not obtained simply by replacing ν in the expression for K_ν by the corresponding value of λ deduced from

$$\nu = \frac{q}{\lambda} \tag{15}$$

where q is the velocity of propagation. For if $d\lambda$ and $d\nu$ refer to the same interval of the spectrum, we have, not $E_\lambda = \mathsf{K}_\nu$, but $E_\lambda \, d\lambda = \mathsf{K}_\nu \, d\nu$. By differentiating (15) and paying attention to the signs of corresponding values of $d\lambda$ and $d\nu$ the equation

$$d\nu = \frac{q d\lambda}{\lambda^2}$$

is obtained. Hence we get by substitution:

$$E_\lambda = \frac{q\mathsf{K}_\nu}{\lambda^2}. \tag{16}$$

This relation shows among other things that in a certain spectrum the maxima of E_λ and K_ν lie at different points of the spectrum.

20. When the principal intensities K_ν and K_ν' of all monochromatic rays are given at all points of the medium and for all directions, the state of radiation is known in all respects and all questions regarding it may be answered. We shall show this by one or two applications to special cases. Let us first find the amount of energy which is radiated through any element of area $d\sigma$ toward any other element $d\sigma'$. The distance r between the two elements may be thought of as large compared with the linear dimensions of the elements $d\sigma$ and $d\sigma'$ but still so small that no appreciable amount of radiation is absorbed or scattered along it. This condition is, of course, superfluous for diathermanous media.

From any definite point of $d\sigma$ rays pass to all points of $d\sigma'$. These rays form a cone whose vertex lies in $d\sigma$ and whose solid angle is

$$d\Omega = \frac{d\sigma' \cos(\nu', r)}{r^2}$$

where ν' denotes the normal of $d\sigma'$ and the angle (ν', r) is to be taken as an acute angle. This value of $d\Omega$ is, neglecting small quantities of higher order, independent of the particular position of the vertex of the cone on $d\sigma$.

If we further denote the normal to $d\sigma$ by ν the angle θ of (14) will be the angle (ν, r) and hence from expression (6) the energy of radiation required is found to be:

$$K \cdot \frac{d\sigma\, d\sigma' \cos(\nu, r) \cdot \cos(\nu', r)}{r^2}\, dt. \tag{17}$$

For monochromatic plane polarized radiation of frequency ν the energy will be, according to equation (11),

$$\mathsf{K}_\nu\, d\nu \cdot \frac{d\sigma\, d\sigma' \cos(\nu, r) \cos(\nu', r)}{r^2} \cdot dt. \tag{18}$$

The relative size of the two elements $d\sigma$ and $d\sigma'$ may have any value whatever. They may be assumed to be of the same or of a different order of magnitude, provided the condition remains satisfied that r is large compared with the linear dimensions of each of them. If we choose $d\sigma$ small compared with $d\sigma'$, the rays diverge from $d\sigma$ to $d\sigma'$, whereas they converge from $d\sigma$ to $d\sigma'$, if we choose $d\sigma$ large compared with $d\sigma'$.

21. Since every point of $d\sigma$ is the vertex of a cone spreading out toward $d\sigma'$, the whole pencil of rays here considered, which is

2

defined by $d\sigma$ and $d\sigma'$, consists of a double infinity of point pencils or of a fourfold infinity of rays which must all be considered equally for the energy radiation. Similarly the pencil of rays may be thought of as consisting of the cones which, emanating from all points of $d\sigma$, converge in one point of $d\sigma'$ respectively as a vertex. If we now imagine the whole pencil of rays to be cut by a plane at any arbitrary distance from the elements $d\sigma$ and $d\sigma'$ and lying either between them or outside, then the cross-sections of any two point pencils on this plane will not be identical, not even approximately. In general they will partly overlap and partly lie outside of each other, the amount of overlapping being different for different intersecting planes. Hence it follows that there is no definite cross-section of the pencil of rays so far as the uniformity of radiation is concerned. If, however, the intersecting plane coincides with either $d\sigma$ or $d\sigma'$, then the pencil has a definite cross-section. Thus these two planes show an exceptional property. We shall call them the two "focal planes" of the pencil.

In the special case already mentioned above, namely, when one of the two focal planes is infinitely small compared with the other, the whole pencil of rays shows the character of a point pencil inasmuch as its form is approximately that of a cone having its vertex in that focal plane which is small compared with the other. In that case the "cross-section" of the whole pencil at a definite point has a definite meaning. Such a pencil of rays, which is similar to a cone, we shall call an elementary pencil, and the small focal plane we shall call the first focal plane of the elementary pencil. The radiation may be either converging toward the first focal plane or diverging from the first focal plane. All the pencils of rays passing through a medium may be considered as consisting of such elementary pencils, and hence we may base our future considerations on elementary pencils only, which is a great convenience, owing to their simple nature.

As quantities necessary to define an elementary pencil with a given first focal plane $d\sigma$, we may choose not the second focal plane $d\sigma'$ but the magnitude of that solid angle $d\Omega$ under which $d\sigma'$ is seen from $d\sigma$. On the other hand, in the case of an arbitrary pencil, that is, when the two focal planes are of the same order of magnitude, the second focal plane in general cannot be replaced by the solid angle $d\Omega$ without the pencil changing markedly in character. For if, instead of $d\sigma'$ being given, the magnitude and direction of $d\Omega$, to be taken as constant for all points of $d\sigma$, is given, then the rays emanating from $d\sigma$ do not any longer form the original pencil, but rather an elementary pencil whose first focal plane is $d\sigma$ and whose second focal plane lies at an infinite distance.

22. Since the energy radiation is propagated in the medium with a finite velocity q, there must be in a finite space a finite amount of energy. We shall therefore speak of the "space density of radiation," meaning thereby the ratio of the total quantity of energy of radiation contained in a volume-element to the magnitude of the latter. Let us now calculate the space density of radiation u at any arbitrary point of the medium. When we consider an infinitely small element of volume v at the point in question, having any shape whatsoever, we must allow for all rays passing through the volume-element v. For this purpose we shall construct about any point O of v as center a sphere of radius r, r being large compared with the linear dimensions of v but still so small that no appreciable absorption or scattering of the radiation takes place in the distance r (Fig. 1). Every ray which reaches v must then come from some point on the surface of the sphere. If, then, we at first consider only all the rays that come from the points of an infinitely small element of area $d\sigma$ on the surface of the sphere, and reach v, and then sum up for all elements of the spherical surface, we shall have accounted for all rays and not taken any one more than once.

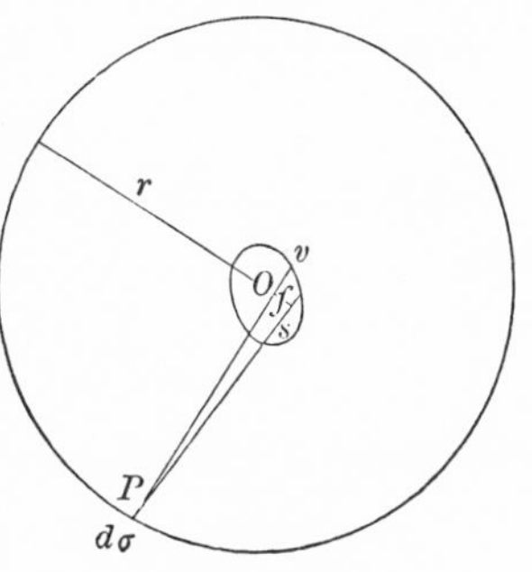

FIG. 1.

Let us then calculate first the amount of energy which is contributed to the energy contained in v by the radiation sent from such an element $d\sigma$ to v. We choose $d\sigma$ so that its linear dimensions are small compared with those of v and consider the cone of rays which, starting at a point of $d\sigma$, meets the volume v. This cone consists of an infinite number of conical elements with the common vertex at P, a point of $d\sigma$, each cutting out of the volume v a certain element of length, say s. The solid angle of such a conical element is $\frac{f}{r^2}$ where f denotes the area of cross-section normal to the axis of the cone at a distance r from the vertex. The time required for the radiation to pass through the distance s is:

$$\tau = \frac{s}{q}$$

From expression (6) we may find the energy radiated through a certain element of area. In the present case $d\Omega = \frac{f}{r^2}$ and $\theta = 0$; hence the energy is:

$$\tau d\sigma \frac{f}{r^2} K = \frac{fs}{r^2 q} \cdot K\, d\sigma. \tag{19}$$

This energy enters the conical element in v and spreads out into the volume fs. Summing up over all conical elements that start from $d\sigma$ and enter v we have

$$\frac{K d\sigma}{r^2 q} \Sigma\, fs = \frac{K d\sigma}{r^2 q}\, v.$$

This represents the entire energy of radiation contained in the volume v, so far as it is caused by radiation through the element $d\sigma$. In order to obtain the total energy of radiation contained in v we must integrate over all elements $d\sigma$ contained in the surface of the sphere. Denoting by $d\Omega$ the solid angle $\frac{d\sigma}{r^2}$ of a cone which has its center in O and intersects in $d\sigma$ the surface of the sphere, we get for the whole energy:

$$\frac{v}{q} \int K\, d\Omega.$$

The volume density of radiation required is found from this by dividing by v. It is

$$u = \frac{1}{q} \int K\, d\Omega. \tag{20}$$

Since in this expression r has disappeared, we can think of K as the intensity of radiation at the point O itself. In integrating, it is to be noted that K in general depends on the direction (θ, ϕ). For radiation that is uniform in all directions K is a constant and on integration we get:

$$u=\frac{4\pi K}{q} \tag{21}$$

23. A meaning similar to that of the volume density of the total radiation u is attached to the volume density of radiation of a definite frequency u_ν. Summing up for all parts of the spectrum we get:

$$u=\int_0^\infty \mathsf{u}_\nu d\nu. \tag{22}$$

Further by combining equations (9) and (20) we have:

$$\mathsf{u}_\nu=\frac{1}{q}\int (\mathsf{K}_\nu+\mathsf{K}_\nu')\, d\Omega, \tag{23}$$

and finally for unpolarized radiation uniformly distributed in all directions:

$$\mathsf{u}_\nu=\frac{8\pi\ \mathsf{K}_\nu}{q} \tag{24}$$

CHAPTER II

RADIATION AT THERMODYNAMIC EQUILIBRIUM. KIRCHHOFF'S LAW. BLACK RADIATION

24. We shall now apply the laws enunciated in the last chapter to the special case of thermodynamic equilibrium, and hence we begin our consideration by stating a certain consequence of the second principle of thermodynamics: A system of bodies of arbitrary nature, shape, and position which is at rest and is surrounded by a rigid cover impermeable to heat will, no matter what its initial state may be, pass in the course of time into a permanent state, in which the temperature of all bodies of the system is the same. This is the state of thermodynamic equilibrium, in which the entropy of the system has the maximum value compatible with the total energy of the system as fixed by the initial conditions. This state being reached, no further increase in entropy is possible.

In certain cases it may happen that, under the given conditions, the entropy can assume not only one but several maxima, of which one is the absolute one, the others having only a relative significance.[1] In these cases every state corresponding to a maximum value of the entropy represents a state of thermodynamic equilibrium of the system. But only one of them, the one corresponding to the absolute maximum of entropy, represents the absolutely stable equilibrium. All the others are in a certain sense unstable, inasmuch as a suitable, however small, disturbance may produce in the system a permanent change in the equilibrium in the direction of the absolutely stable equilibrium. An example of this is offered by supersaturated steam enclosed in a rigid vessel or by any explosive substance. We shall also meet such unstable equilibria in the case of radiation phenomena (Sec. 52).

[1] See, *e.g.*, *M. Planck*, Vorlesungen über Thermodynamik, Leipzig, Veit and Comp., 1911 (or English Translation, Longmans Green & Co.), Secs. 165 and 189, *et seq.*

25. We shall now, as in the previous chapter, assume that we are dealing with homogeneous isotropic media whose condition depends only on the temperature, and we shall inquire what laws the radiation phenomena in them must obey in order to be consistent with the deduction from the second principle mentioned in the preceding section. The means of answering this inquiry is supplied by the investigation of the state of thermodynamic equilibrium of one or more of such media, this investigation to be conducted by applying the conceptions and laws established in the last chapter.

We shall begin with the simplest case, that of a single medium extending very far in all directions of space, and, like all systems we shall here consider, being surrounded by a rigid cover impermeable to heat. For the present we shall assume that the medium has finite coefficients of absorption, emission, and scattering.

Let us consider, first, points of the medium that are far away from the surface. At such points the influence of the surface is, of course, vanishingly small and from the homogeneity and the isotropy of the medium it will follow that in a state of thermodynamic equilibrium the radiation of heat has everywhere and in all directions the same properties. Then K_ν, the specific intensity of radiation of a plane polarized ray of frequency ν (Sec. 17), must be independent of the azimuth of the plane of polarization as well as of position and direction of the ray. Hence to each pencil of rays starting at an element of area $d\sigma$ and diverging within a conical element $d\Omega$ corresponds an exactly equal pencil of opposite direction converging within the same conical element toward the element of area.

Now the condition of thermodynamic equilibrium requires that the temperature shall be everywhere the same and shall not vary in time. Therefore in any given arbitrary time just as much radiant heat must be absorbed as is emitted in each volume-element of the medium. For the heat of the body depends only on the heat radiation, since, on account of the uniformity in temperature, no conduction of heat takes place. This condition is not influenced by the phenomenon of scattering, because scattering refers only to a change in direction of the energy radiated, not to a creation or destruction of it. We shall, therefore, calculate the energy emitted and absorbed in the time dt by a volume-element v.

According to equation (2) the energy emitted has the value

$$dt\, v \cdot 8\pi \int_0^\infty \epsilon_\nu\, d\nu$$

where ϵ_ν, the coefficient of emission of the medium, depends only on the frequency ν and on the temperature in addition to the chemical nature of the medium.

26. For the calculation of the energy absorbed we shall employ the same reasoning as was illustrated by Fig. 1 (Sec. 22) and shall retain the notation there used. The radiant energy absorbed by the volume-element v in the time dt is found by considering the intensities of all the rays passing through the element v and taking that fraction of each of these rays which is absorbed in v. Now, according to (19), the conical element that starts from $d\sigma$ and cuts out of the volume v a part equal to fs has the intensity (energy radiated per unit time)

$$d\sigma \cdot \frac{f}{r^2} \cdot K$$

or, according to (12), by considering the different parts of the spectrum separately:

$$2\, d\sigma \frac{f}{r^2} \int_0^\infty \mathsf{K}_\nu\, d\nu.$$

Hence the intensity of a monochromatic ray is:

$$2\, d\sigma \frac{f}{r^2}\, \mathsf{K}_\nu\, d\nu.$$

The amount of energy of this ray absorbed in the distance s in the time dt is, according to (4),

$$dt\, \alpha_\nu s\, 2\, d\sigma \frac{f}{r^2}\, \mathsf{K}_\nu\, d\nu.$$

Hence the absorbed part of the energy of this small cone of rays, as found by integrating over all frequencies, is:

$$dt\, 2\, d\sigma \frac{fs}{r^2} \int_0^\infty \alpha_\nu\, \mathsf{K}_\nu\, d\nu.$$

When this expression is summed up over all the different cross-sections f of the conical elements starting at $d\sigma$ and passing through v, it is evident that $\Sigma fs = v$, and when we sum up over all elements $d\sigma$ of the spherical surface of radius r we have

$$\int \frac{d\sigma}{r^2} = 4\pi.$$

Thus for the total radiant energy absorbed in the time dt by the volume-element v the following expression is found:

$$dt\, v\, 8\pi \int_0^\infty \alpha_\nu\, \mathsf{K}_\nu\, d\nu. \tag{25}$$

By equating the emitted and absorbed energy we obtain:

$$\int_0^\infty \epsilon_\nu\, d\nu = \int_0^\infty \alpha_\nu\, \mathsf{K}_\nu\, d\nu.$$

A similar relation may be obtained for the separate parts of the spectrum. For the energy emitted and the energy absorbed in the state of thermodynamic equilibrium are equal, not only for the entire radiation of the whole spectrum, but also for each monochromatic radiation. This is readily seen from the following. The magnitudes of ϵ_ν, α_ν, and K_ν are independent of position. Hence, if for any single color the absorbed were not equal to the emitted energy, there would be everywhere in the whole medium a continuous increase or decrease of the energy radiation of that particular color at the expense of the other colors. This would be contradictory to the condition that K_ν for each separate frequency does not change with the time. We have therefore for each frequency the relation:

$$\epsilon_\nu = \alpha_\nu\, \mathsf{K}_\nu, \text{ or} \tag{26}$$

$$\mathsf{K}_\nu = \frac{\epsilon_\nu}{\alpha_\nu}, \tag{27}$$

i.e.: in the interior of a medium in a state of thermodynamic equilibrium the specific intensity of radiation of a certain frequency is equal to the coefficient of emission divided by the coefficient of absorption of the medium for this frequency.

27. Since ϵ_ν and α_ν depend only on the nature of the medium, the temperature, and the frequency ν, the intensity of radiation of a definite color in the state of thermodynamic equilibrium is completely defined by the nature of the medium and the temperature. An exceptional case is when $\alpha_\nu = 0$, that is, when the medium does not at all absorb the color in question. Since K_ν cannot become infinitely large, a first consequence of this is that in that case $\epsilon_\nu = 0$ also, that is, a medium does not emit any color which it does not absorb. A second consequence is that if ϵ_ν and α_ν both vanish, equation (26) is satisfied by every value of K_ν. *In a medium which is diathermanous for a certain color thermodynamic equilibrium can exist for any intensity of radiation whatever of that color.*

This supplies an immediate illustration of the cases spoken of before (Sec. 24), where, for a given value of the total energy of a system enclosed by a rigid cover impermeable to heat, several states of equilibrium can exist, corresponding to several relative maxima of the entropy. That is to say, since the intensity of radiation of the particular color in the state of thermodynamic equilibrium is quite independent of the temperature of a medium which is diathermanous for this color, the given total energy may be arbitrarily distributed between radiation of that color and the heat of the body, without making thermodynamic equilibrium impossible. Among all these distributions there is one particular one, corresponding to the absolute maximum of entropy, which represents absolutely stable equilibrium. This one, unlike all the others, which are in a certain sense unstable, has the property of not being appreciably affected by a small disturbance. Indeed we shall see later (Sec. 48) that among the infinite number of values, which the quotient $\frac{\epsilon_\nu}{\alpha_\nu}$ can have, if numerator and denominator both vanish, there exists one particular one which depends in a definite way on the nature of the medium, the frequency ν, and the temperature. This distinct value of the fraction is accordingly called the stable intensity of radiation K_ν in the medium, which at the temperature in question is diathermanous for rays of the frequency ν.

Everything that has just been said of a medium which is diathermanous for a certain kind of rays holds true for an absolute vacuum, which is a medium diathermanous for rays of all kinds, the only difference being that one cannot speak of the heat and the temperature of an absolute vacuum in any definite sense.

For the present we again shall put the special case of diathermancy aside and assume that all the media considered have a finite coefficient of absorption.

28. Let us now consider briefly the phenomenon of scattering at thermodynamic equilibrium. Every ray meeting the volume-element v suffers there, apart from absorption, a certain weakening of its intensity because a certain fraction of its energy is diverted in different directions. The value of the total energy of scattered radiation received and diverted, in the time dt by the volume-element v in all directions, may be calculated from expression (3) in exactly the same way as the value of the absorbed energy was calculated in Sec. 26. Hence we get an expression similar to (25), namely,

$$dt\ v\ 8\pi \int_0^\infty \beta_\nu\ \mathsf{K}_\nu\ d\nu. \tag{28}$$

The question as to what becomes of this energy is readily answered. On account of the isotropy of the medium, the energy scattered in v and given by (28) is radiated uniformly in all directions just as in the case of the energy entering v. Hence that part of the scattered energy received in v which is radiated out in a cone of solid angle $d\Omega$ is obtained by multiplying the last expression by $\frac{d\Omega}{4\pi}$. This gives

$$2\ dt\ v\ d\Omega \int_0^\infty \beta_\nu\ \mathsf{K}_\nu\ d\nu,$$

and, for monochromatic plane polarized radiation,

$$dt\ v\ d\Omega\ \beta_\nu\ \mathsf{K}_\nu\ d\nu. \tag{29}$$

Here it must be carefully kept in mind that this uniformity of radiation in all directions holds only for all rays striking the element v taken together; a single ray, even in an isotropic medium, is scattered in different directions with different intensities and different directions of polarization. (See end of Sec. 8.)

It is thus found that, when thermodynamic equilibrium of radiation exists inside of the medium, the process of scattering produces, on the whole, no effect. The radiation falling on a volume-element from all sides and scattered from it in all directions behaves exactly as if it had passed directly through the volume-element without the least modification. Every ray loses by scattering just as much energy as it regains by the scattering of other rays.

29. We shall now consider from a different point of view the radiation phenomena in the interior of a very extended homogeneous isotropic medium which is in thermodynamic equilibrium. That is to say, we shall confine our attention, not to a definite volume-element, but to a definite pencil, and in fact to an elementary pencil (Sec. 21). Let this pencil be specified by the infinitely small focal plane $d\sigma$ at the point O (Fig. 2), perpendicular to the axis of the pencil, and by the solid angle $d\Omega$, and let the radiation take place toward the focal plane in the direction of the arrow. We shall consider exclusively rays which belong to this pencil.

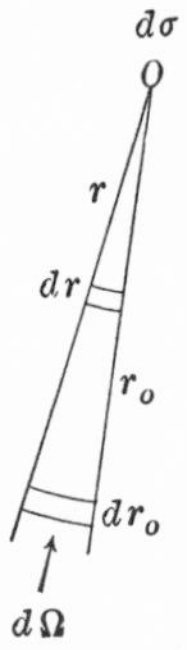

FIG. 2.

The energy of monochromatic plane polarized radiation of the pencil considered passing in unit time through $d\sigma$ is represented, according to (11), since in this case $dt=1$, $\theta=0$, by

$$d\sigma\, d\Omega\, \mathsf{K}_\nu\, d\nu. \tag{30}$$

The same value holds for any other cross-section of the pencil. For first, $\mathsf{K}_\nu\, d\nu$ has everywhere the same magnitude (Sec. 25), and second, the product of any right section of the pencil and the solid angle at which the focal plane $d\sigma$ is seen from this section has the constant value $d\sigma\, d\Omega$, since the magnitude of the cross-section increases with the distance from the vertex O of the pencil in the proportion in which the solid angle decreases. Hence the radiation inside of the pencil takes place just as if the medium were perfectly diathermanous.

On the other hand, the radiation is continuously modified along its path by the effect of emission, absorption, and scattering. We shall consider the magnitude of these effects separately.

30. Let a certain volume-element of the pencil be bounded by two cross-sections at distances equal to r_o (of arbitrary length) and $r_o + dr_o$ respectively from the vertex O. The volume will be represented by $dr_o \cdot r_o^2 \, d\Omega$. It emits in unit time toward the focal plane $d\sigma$ at O a certain quantity E of energy of monochromatic plane polarized radiation. E may be obtained from (1) by putting

$$dt = 1, \; d\tau = dr_o \, r_o^2 \, d\Omega, \; d\Omega = \frac{d\sigma}{r_o^2}$$

and omitting the numerical factor 2. We thus get

$$E = dr_o \cdot d\Omega \, d\sigma \, \epsilon_\nu \, d\nu. \tag{31}$$

Of the energy E, however, only a fraction E_o reaches O, since in every infinitesimal element of distance s which it traverses before reaching O the fraction $(\alpha_\nu + \beta_\nu)s$ is lost by absorption and scattering. Let E_r represent that part of E which reaches a cross-section at a distance $r(< r_o)$ from O. Then for a small distance $s = dr$ we have

$$E_{r+dr} - E_r = E_r(\alpha_\nu + \beta_\nu)dr,$$

or,

$$\frac{dE_r}{dr} = E_r(\alpha_\nu + \beta_\nu),$$

and, by integration,

$$E_r = Ee^{(\alpha_\nu + \beta_\nu)(r - r_o)}$$

since, for $r = r_o$, $E_r = E$ is given by equation (31). From this, by putting $r = 0$, the energy emitted by the volume-element at r_o which reaches O is found to be

$$E_o = Ee^{-(\alpha_\nu + \beta_\nu)r_o} = dr_o \, d\Omega \, d\sigma \, \epsilon_\nu \, e^{-(\alpha_\nu + \beta_\nu)r_o} \, d\nu. \tag{32}$$

All volume-elements of the pencils combined produce by their emission an amount of energy reaching $d\sigma$ equal to

$$d\Omega \, d\sigma \, d\nu \, \epsilon_\nu \int_0^\infty dr_o \, e^{-(\alpha_\nu + \beta_\nu)r_o} = d\Omega \, d\sigma \frac{\epsilon_\nu}{\alpha_\nu + \beta_\nu} \, d\nu. \tag{33}$$

31. If the scattering did not affect the radiation, the total energy reaching $d\sigma$ would necessarily consist of the quantities of energy emitted by the different volume-elements of the pencil, allowance being made, however, for the losses due to absorption on the way. For $\beta_\nu = 0$ expressions (33) and (30) are identical, as may be seen by comparison with (27). Generally, however, (30) is larger than (33) because the energy reaching $d\sigma$ contains also some rays which were not at all emitted from elements inside of the pencil, but somewhere else, and have entered later on by scattering. In fact, the volume-elements of the pencil do not merely scatter outward the radiation which is being transmitted inside the pencil, but they also collect into the pencil rays coming from without. The radiation E' thus collected by the volume-element at r_o is found, by putting in (29),

$$dt = 1,\ v = dr_o\, d\Omega\, r_o^2,\ d\Omega = \frac{d\sigma}{r_o^2},$$

to be

$$E' = dr_o\, d\Omega\, d\sigma\, \beta_\nu\, \mathsf{K}_\nu\, d\nu.$$

This energy is to be added to the energy E emitted by the volume-element, which we have calculated in (31). Thus for the total energy contributed to the pencil in the volume-element at r_o we find:

$$E + E' = dr_o\, d\Omega\, d\sigma\, (\epsilon_\nu + \beta_\nu\, \mathsf{K}_\nu)\, d\nu.$$

The part of this reaching O is, similar to (32):

$$dr_o\, d\Omega\, d\sigma\, (\epsilon_\nu + \beta_\nu\, \mathsf{K}_\nu)\, d\nu\, e^{-r_o(\alpha_\nu + \beta_\nu)}$$

Making due allowance for emission and collection of scattered rays entering on the way, as well as for losses by absorption and scattering, all volume-elements of the pencil combined give for the energy ultimately reaching $d\sigma$

$$d\Omega\, d\sigma\, (\epsilon_\nu + \beta_\nu\, \mathsf{K}_\nu)\, d\nu \int_o^\infty dr_o\, e^{-r_o(\alpha_\nu + \beta_\nu)} = d\Omega\, d\sigma \frac{\epsilon_\nu + \beta_\nu\, \mathsf{K}_\nu}{\alpha_\nu + \beta_\nu} d\nu,$$

and this expression is really exactly equal to that given by (30), as may be seen by comparison with (26).

32. The laws just derived for the state of radiation of a homogeneous isotropic medium when it is in thermodynamic equilibrium hold, so far as we have seen, only for parts of the medium which lie very far away from the surface, because for such parts only may the radiation be considered, by symmetry, as independent of position and direction. A simple consideration, however, shows that the value of K_ν, which was already calculated and given by (27), and which depends only on the temperature and the nature of the medium, gives the correct value of the intensity of radiation of the frequency considered for all directions up to points directly below the surface of the medium. For in the state of thermodynamic equilibrium every ray must have just the same intensity as the one travelling in an exactly opposite direction, since otherwise the radiation would cause a unidirectional transport of energy. Consider then any ray coming from the surface of the medium and directed inward; it must have the same intensity as the opposite ray, coming from the interior. A further immediate consequence of this is *that the total state of radiation of the medium is the same on the surface as in the interior.*

33. While the radiation that starts from a surface element and is directed toward the interior of the medium is in every respect equal to that emanating from an equally large parallel element of area in the interior, it nevertheless has a different history. That is to say, since the surface of the medium was assumed to be impermeable to heat, it is produced only by reflection at the surface of radiation coming from the interior. So far as special details are concerned, this can happen in very different ways, depending on whether the surface is assumed to be smooth, *i.e.*, in this case reflecting, or rough, *e.g.*, white (Sec. 10). In the first case there corresponds to each pencil which strikes the surface another perfectly definite pencil, symmetrically situated and having the same intensity, while in the second case every incident pencil is broken up into an infinite number of reflected pencils, each having a different direction, intensity, and polarization. While this is the case, nevertheless the rays that strike a surface-element from all different directions with the same intensity K_ν also produce, all taken together, a uniform radiation of the same intensity K_ν, directed toward the interior of the medium.

34. Hereafter there will not be the slightest difficulty in dispensing with the assumption made in Sec. 25 that the medium in question extends very far in all directions. For after thermodynamic equilibrium has been everywhere established in our medium, the equilibrium is, according to the results of the last paragraph, in no way disturbed, if we assume any number of rigid surfaces impermeable to heat and rough or smooth to be inserted in the medium. By means of these the whole system is divided into an arbitrary number of perfectly closed separate systems, each of which may be chosen as small as the general restrictions stated in Sec. 2 permit. It follows from this that the value of the specific intensity of radiation K_ν given in (27) remains valid for the thermodynamic equilibrium of a substance enclosed in a space as small as we please and of any shape whatever.

35. From the consideration of a system consisting of a single homogeneous isotropic substance we now pass on to the treatment of a system consisting of two different homogeneous isotropic substances contiguous to each other, the system being, as before, enclosed by a rigid cover impermeable to heat. We consider the state of radiation when thermodynamic equilibrium exists, at first, as before, with the assumption that the media are of considerable extent. Since the equilibrium is nowise disturbed, if we think of the surface separating the two media as being replaced for an instant by an area entirely impermeable to heat radiation, the laws of the last paragraphs must hold for each of the two substances separately. Let the specific intensity of radiation of frequency ν polarized in an arbitrary plane be K_ν in the first substance (the upper one in Fig. 3), and K_ν' in the second, and, in general, let all quantities referring to the second substance be indicated by the addition of an accent. Both of the quantities K_ν and K_ν' depend, according to equation (27), only on the temperature, the frequency ν, and the nature of the two substances, and these values of the intensities of radiation hold up to very small distances from the bounding surface of the substances, and hence are entirely independent of the properties of this surface.

36. We shall now suppose, to begin with, that the bounding surface of the media is smooth (Sec. 9). Then every ray coming from the first medium and falling on the bounding surface is divided into two rays, the reflected and the transmitted ray. The directions of these two rays vary with the angle of incidence and the color of the incident ray; the intensity also varies with its polarization. Let us denote by ρ (coefficient of reflection) the fraction of the energy reflected, then the fraction transmitted is $(1-\rho)$, ρ depending on the angle of incidence, the frequency, and the polarization of the incident ray. Similar remarks apply to

ρ' the coefficient of reflection of a ray coming from the second medium and falling on the bounding surface.

Now according to (11) we have for the monochromatic plane polarized radiation of frequency ν, emitted in time dt toward the first medium (in the direction of the feathered arrow upper left

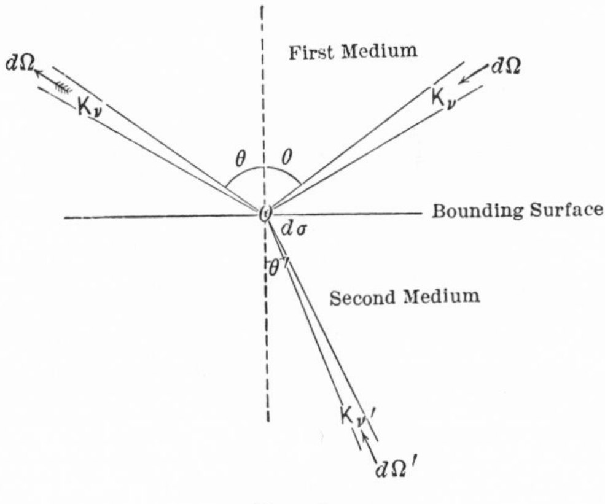

FIG. 3.

hand in Fig. 3), from an element $d\sigma$ of the bounding surface and contained in the conical element $d\Omega$,

$$dt\ d\sigma\ \cos\theta\ d\Omega\ \mathsf{K}_\nu\ d\nu, \tag{34}$$

where

$$d\Omega = \sin\theta\ d\theta\ d\phi. \tag{35}$$

This energy is supplied by the two rays which come from the first and the second medium and are respectively reflected from or transmitted by the element $d\sigma$ in the corresponding direction (the unfeathered arrows). (Of the element $d\sigma$ only the one point O is indicated.) The first ray, according to the law of reflection, continues in the symmetrically situated conical element $d\Omega$, the second in the conical element

$$d\Omega' = \sin\theta'\ d\theta'\ d\phi' \tag{36}$$

where, according to the law of refraction,

$$\phi' = \phi \text{ and } \frac{\sin\theta}{\sin\theta'} = \frac{q}{q'} \tag{37}$$

3

If we now assume the radiation (34) to be polarized either in the plane of incidence or at right angles thereto, the same will be true for the two radiations of which it consists, and the radiation coming from the first medium and reflected from $d\sigma$ contributes the part

$$\rho \, dt \, d\sigma \cos \theta \, d\Omega \, \mathsf{K}_\nu \, d\nu \tag{38}$$

while the radiation coming from the second medium and transmitted through $d\sigma$ contributes the part

$$(1-\rho') \, dt \, d\sigma \cos \theta' \, d\Omega' \, \mathsf{K}_\nu' \, d\nu. \tag{39}$$

The quantities dt, $d\sigma$, ν and $d\nu$ are written without the accent, because they have the same values in both media.

By adding (38) and (39) and equating their sum to the expression (34) we find

$$\rho \cos \theta \, d\Omega \, \mathsf{K}_\nu + (1-\rho') \cos \theta' \, d\Omega' \, \mathsf{K}_\nu' = \cos \theta \, d\Omega \, \mathsf{K}_\nu.$$

Now from (37) we have

$$\frac{\cos \theta \, d\theta}{q} = \frac{\cos \theta' \, d\theta'}{q'}$$

and further by (35) and (36)

$$d\Omega' \cos \theta' = \frac{d\Omega \cos \theta \, q'^2}{q^2}.$$

Therefore we find

$$\rho \, \mathsf{K}_\nu + (1-\rho') \frac{q'^2}{q^2} \mathsf{K}_\nu' = \mathsf{K}$$

or

$$\frac{\mathsf{K}_\nu}{\mathsf{K}_\nu'} \cdot \frac{q^2}{q'^2} = \frac{1-\rho'}{1-\rho}.$$

37. In the last equation the quantity on the left side is independent of the angle of incidence θ and of the particular kind of polarization; hence the same must be true for the right side. Hence, whenever the value of this quantity is known for a single angle of incidence and any definite kind of polarization, this value will remain valid for all angles of incidence and all kinds of polarization. Now in the special case when the rays are polarized at right angles to the plane of incidence and strike the bounding surface at the angle of polarization, $\rho=0$, and $\rho'=0$. The expression on the right side of the last equation then becomes 1; hence it must always be 1 and we have the general relations:

$$\rho=\rho' \tag{40}$$

and

$$q^2\, \mathsf{K}_\nu = q'^2\, \mathsf{K}_\nu' \tag{41}$$

38. The first of these two relations, which states that the coefficient of reflection of the bounding surface is the same on both sides, is a special case of a general law of reciprocity first stated by *Helmholtz*.[1] According to this law the loss of intensity which a ray of definite color and polarization suffers on its way through any media by reflection, refraction, absorption, and scattering is exactly equal to the loss suffered by a ray of the same intensity, color, and polarization pursuing an exactly opposite path. An immediate consequence of this law is that the radiation striking the bounding surface of any two media is always transmitted as well as reflected equally on both sides, for every color, direction, and polarization.

39. The second formula (41) establishes a relation between the intensities of radiation in the two media, for it states that, when thermodynamic equilibrium exists, *the specific intensities of radiation of a certain frequency in the two media are in the inverse ratio of the squares of the velocities of propagation or in the direct ratio of the squares of the indices of refraction.*[2]

By substituting for K_ν its value from (27) we obtain the following theorem: *The quantity*

$$q^2\, \mathsf{K}_\nu = q^2\, \frac{\epsilon_\nu}{\alpha_\nu} \tag{42}$$

does not depend on the nature of the substance, and is, therefore, a universal function of the temperature T and the frequency ν alone.

The great importance of this law lies evidently in the fact that it states a property of radiation which is the same for all bodies

[1] *H. v. Helmholtz*, Handbuch d. physiologischen Optik 1. Lieferung, Leipzig, Leop. Voss, 1856, p. 169. See also *Helmholtz*, Vorlesungen über die Theorie der Wärme herausgegeben von *F. Richarz*, Leipzig, J. A. Barth, 1903, p. 161. The restrictions of the law of reciprocity made there do not bear on our problems, since we are concerned with temperature radiation only (Sec. 7).

[2] *G. Kirchhoff*, Gesammelte Abhandlungen, Leipzig, J. A. Barth, 1882, p. 594. *R. Clausius*, Pogg. Ann. **121**, p. 1, 1864.

in nature, and which need be known only for a single arbitrarily chosen body, in order to be stated quite generally for all bodies. We shall later on take advantage of the opportunity offered by this statement in order actually to calculate this universal function (Sec. 165).

40. We now consider the other case, that in which the bounding surface of the two media is rough. This case is much more general than the one previously treated, inasmuch as the energy of a pencil directed from an element of the bounding surface into the first medium is no longer supplied by two definite pencils, but by an arbitrary number, which come from both media and strike the surface. Here the actual conditions may be very complicated according to the peculiarities of the bounding surface, which moreover may vary in any way from one element to another. However, according to Sec. 35, the values of the specific intensities of radiation K_ν and K_ν' remain always the same in all directions in both media, just as in the case of a smooth bounding surface. That this condition, necessary for thermodynamic equilibrium, is satisfied is readily seen from *Helmholtz's* law of reciprocity, according to which, in the case of stationary radiation, for each ray striking the bounding surface and diffusely reflected from it on both sides, there is a corresponding ray at the same point, of the same intensity and opposite direction, produced by the inverse process at the same point on the bounding surface, namely by the gathering of diffusely incident rays into a definite direction, just as is the case in the interior of each of the two media.

41. We shall now further generalize the laws obtained. First, just as in Sec. 34, the assumption made above, namely, that the two media extend to a great distance, may be abandoned since we may introduce an arbitrary number of bounding surfaces without disturbing the thermodynamic equilibrium. Thereby we are placed in a position enabling us to pass at once to the case of any number of substances of any size and shape. For when a system consisting of an arbitrary number of contiguous substances is in the state of thermodynamic equilibrium, the equilibrium is in no way disturbed, if we assume one or more of the surfaces of contact to be wholly or partly impermeable to heat. Thereby we can always reduce the case of any number of substances to that of two substances in an enclosure impermeable to heat, and, therefore, the law may be stated quite generally, that, when any arbitrary system is in the state of thermodynamic equilibrium, the specific intensity of radiation K_ν is determined in each separate substance by the universal function (42).

42. We shall now consider a system in a state of thermodynamic equilibrium, contained within an enclosure impermeable to heat and consisting of n emitting and absorbing adjacent bodies of any size and shape whatever. As in Sec. 36, we again confine our attention to a monochromatic plane polarized pencil which proceeds from an element $d\sigma$ of the bounding surface of the two media in the direction toward the first medium (Fig. 3, feathered arrow) within the conical element $d\Omega$. Then, as in (34), the energy supplied by the pencil in unit time is

$$d\sigma \cos \theta \, d\Omega \, \mathsf{K}_\nu \, d\nu = I. \tag{43}$$

This energy of radiation I consists of a part coming from the first medium by regular or diffuse reflection at the bounding surface and of a second part coming through the bounding surface from the second medium. We shall, however, not stop at this mode of division, but shall further subdivide I according to that one of the n media from which the separate parts of the radiation I have been emitted. This point of view is distinctly different from the preceding, since, *e.g.*, the rays transmitted from the second medium through the bounding surface into the pencil considered have not necessarily been emitted in the second medium, but may, according to circumstances, have traversed a long and very complicated path through different media and may have undergone therein the effect of refraction, reflection, scattering, and partial absorption any number of times. Similarly the rays of the pencil, which coming from the first medium are reflected at $d\sigma$, were not necessarily all emitted in the first medium. It may even happen that a ray emitted from a certain medium, after passing on its way through other media, returns to the original one and is there either absorbed or emerges from this medium a second time.

We shall now, considering all these possibilities, denote that part of I which has been emitted by volume-elements of the first medium by I_1 no matter what paths the different constituents have pursued, that which has been emitted by volume-elements of the second medium by I_2, etc. Then since every part of I must have been emitted by an element of some body, the following equation must hold,

$$I = I_1 + I_2 + I_3 + \ldots\ldots I_n. \tag{44}$$

43. The most adequate method of acquiring more detailed information as to the origin and the paths of the different rays of which the radiations I_1, I_2, I_3, ... I_n consist, is to pursue the opposite course and to inquire into the future fate of that pencil, which travels exactly in the opposite direction to the pencil I and which therefore comes from the first medium in the cone $d\Omega$ and falls on the surface element $d\sigma$ of the second medium. For since every optical path may also be traversed in the opposite direction, we may obtain by this consideration all paths along which rays can pass into the pencil I, however complicated they may otherwise be. Let J represent the intensity of this inverse pencil, which is directed toward the bounding surface and is in the same state of polarization. Then, according to Sec. 40,

$$J = I. \tag{45}$$

At the bounding surface $d\sigma$ the rays of the pencil J are partly reflected and partly transmitted regularly or diffusely, and thereafter, travelling in both media, are partly absorbed, partly scattered, partly again reflected or transmitted to different media, etc., according to the configuration of the system. But finally the whole pencil J after splitting into many separate rays will be completely absorbed in the n media. Let us denote that part of J which is finally absorbed in the first medium by J_1, that which is finally absorbed in the second medium by J_2, etc., then we shall have

$$J = J_1 + J_2 + J_3 + \ldots\ldots + J_n.$$

Now the volume-elements of the n media, in which the absorption of the rays of the pencil J takes place, are precisely the same as those in which takes place the emission of the rays constituting the pencil I, the first one considered above. For, according to *Helmholtz's* law of reciprocity, no appreciable radiation of the pencil J can enter a volume-element which contributes no appreciable radiation to the pencil I and *vice versa.*

Let us further keep in mind that the absorption of each volume-element is, according to (42), proportional to its emission and that, according to *Helmholtz's* law of reciprocity, the decrease which the energy of a ray suffers on any path is always equal to the decrease suffered by the energy of a ray pursuing the opposite path. It will then be clear that the volume-elements considered absorb the rays of the pencil J in just the same ratio as they contribute by their emission to the energy of the opposite pencil I. Since, moreover, the sum I of the energies given off by emission by all volume-elements is equal to the sum J of the energies absorbed by all elements, the quantity of energy absorbed by each separate volume-element from the pencil J must be equal to the quantity of energy emitted by the same element into the pencil I. In other words: *the part of a pencil I which has been emitted from a certain volume of any medium is equal to the part of the pencil* $J(=I)$ *oppositely directed, which is absorbed in the same volume.*

Hence not only are the sums I and J equal, but their constituents are also separately equal or

$$J_1=I_1,\ J_2=I_2,\ \ldots\ldots\ J_n=I_n. \tag{46}$$

44. Following *G. Kirchhoff*[1] we call the quantity I_2, *i.e.*, the intensity of the pencil emitted from the second medium into the first, the *emissive power* E of the second medium, while we call the ratio of J_2 to J, *i.e.*, that fraction of a pencil incident on the second medium which is absorbed in this medium, the *absorbing power* A of the second medium. Therefore

$$E=I_2(\leqq I),\quad A=\frac{J_2}{J}(\leqq 1). \tag{47}$$

The quantities E and A depend (a) on the nature of the two media, (b) on the temperature, the frequency ν, and the direction and the polarization of the radiation considered, (c) on the nature of the bounding surface and on the magnitude of the surface element $d\sigma$ and that of the solid angle $d\Omega$, (d) on the geometrical extent and the shape of the total surface of the two media, (e) on the nature and form of all other bodies of the system. For a ray may pass from the first into the second medium, be partly transmitted by the latter, and then, after reflection somewhere else,

[1] *G. Kirchhoff*, Gesammelte Abhandlungen, 1882, p. 574.

may return to the second medium and may be there entirely absorbed.

With these assumptions, according to equations (46), (45), and (43), *Kirchhoff's* law holds,

$$\frac{E}{A}=I=d\sigma\ \cos\theta\, d\Omega\, \mathsf{K}_\nu\, d\nu, \tag{48}$$

i.e., the ratio of the emissive power to the absorbing power of any body is independent of the nature of the body. For this ratio is equal to the intensity of the pencil passing through the *first* medium, which, according to equation (27), does not depend on the second medium at all. The value of this ratio does, however, depend on the nature of the first medium, inasmuch as, according to (42), it is not the quantity K_ν but the quantity $q^2\mathsf{K}_\nu$, which is a universal function of the temperature and frequency. The proof of this law given by *G. Kirchhoff l.c.* was later greatly simplified by *E. Pringsheim.*[1]

45. When in particular the second medium is a black body (Sec. 10) it absorbs all the incident radiation. Hence in that case $J_2=J$, $A=1$, and $E=A$, *i.e., the emissive power of a black body is independent of its nature. Its emissive power is larger than that of any other body at the same temperature and, in fact, is just equal to the intensity of radiation in the contiguous medium.*

46. We shall now add, without further proof, another general law of reciprocity, which is closely connected with that stated at the end of Sec. 43 and which may be stated thus: *When any emitting and absorbing bodies are in the state of thermodynamic equilibrium, the part of the energy of definite color emitted by a body A, which is absorbed by another body B, is equal to the part of the energy of the same color emitted by B which is absorbed by A.* Since a quantity of energy emitted causes a decrease of the heat of the body, and a quantity of energy absorbed an increase of the heat of the body, it is evident that, when thermodynamic equilibrium exists, any two bodies or elements of bodies selected at random exchange by radiation equal amounts of heat with each other. Here, of course, care must be taken to distinguish between the radiation emitted and the total radiation which reaches one body from the other.

[1] *E. Pringsheim,* Verhandlungen der Deutschen Physikalischen Gesellschaft, **3**, p. 81, 1901.

47. The law holding for the quantity (42) can be expressed in a different form, by introducing, by means of (24), the volume density u_ν of monochromatic radiation instead of the intensity of radiation K_ν. We then obtain the law that, for radiation in a state of thermodynamic equilibrium, the quantity

$$\mathsf{u}_\nu\, q^3 \tag{49}$$

is a function of the temperature T and the frequency ν, and is the same for all substances.[1] This law becomes clearer if we consider that the quantity

$$\mathsf{u}_\nu\, d\nu \frac{q^3}{\nu^3} \tag{50}$$

also is a universal function of T, ν, and $\nu+d\nu$, and that the product $\mathsf{u}_\nu\, d\nu$ is, according to (22), the volume density of the radiation whose frequency lies between ν and $\nu+d\nu$, while the quotient $\frac{q}{\nu}$ represents the wave length of a ray of frequency ν in the medium in question. The law then takes the following simple form: *When any bodies whatever are in thermodynamic equilibrium, the energy of monochromatic radiation of a definite frequency, contained in a cubical element of side equal to the wave length, is the same for all bodies.*

48. We shall finally take up the case of diathermanous (Sec. 12) media, which has so far not been considered. In Sec. 27 we saw that, in a medium which is diathermanous for a given color and is surrounded by an enclosure impermeable to heat, there can be thermodynamic equilibrium for any intensity of radiation of this color. There must, however, among all possible intensities of radiation be a definite one, corresponding to the absolute maximum of the total entropy of the system, which designates the absolutely stable equilibrium of radiation. In fact, in equation (27) the intensity of radiation K_ν for $\alpha_\nu=0$ and $\epsilon_\nu=0$ assumes the value $\frac{0}{0}$, and hence cannot be calculated from this equation. But we see also that this indeterminateness is removed by equation (41), which states that in the case of thermodynamic

[1] In this law it is assumed that the quantity q in (24) is the same as in (37). This does not hold for strongly dispersing or absorbing substances. For the generalization applying to such cases see *M. Laue*, Annalen d. Physik, **32**, p. 1085, 1910.

equilibrium the product $q^2 \mathsf{K}_\nu$ has the same value for all substances. From this we find immediately a definite value of K_ν which is thereby distinguished from all other values. Furthermore the physical significance of this value is immediately seen by considering the way in which that equation was obtained. It is that intensity of radiation which exists in a diathermanous medium, if it is in thermodynamic equilibrium when in contact with an arbitrary absorbing and emitting medium. The volume and the form of the second medium do not matter in the least, in particular the volume may be taken as small as we please. Hence we can formulate the following law: *Although generally speaking thermodynamic equilibrium can exist in a diathermanous medium for any intensity of radiation whatever, nevertheless there exists in every diathermanous medium for a definite frequency at a definite temperature an intensity of radiation defined by the universal function* (42). *This may be called the stable intensity, inasmuch as it will always be established, when the medium is exchanging stationary radiation with an arbitrary emitting and absorbing substance.*

49. According to the law stated in Sec. 45, the intensity of a pencil, when a state of stable heat radiation exists in a diathermanous medium, is equal to the emissive power E of a black body in contact with the medium. On this fact is based the possibility of measuring the emissive power of a black body, although absolutely black bodies do not exist in nature.[1] A diathermanous cavity is enclosed by strongly emitting walls[2] and the walls kept at a certain constant temperature T. Then the radiation in the cavity, when thermodynamic equilibrium is established for every frequency ν, assumes the intensity corresponding to the velocity of propagation q in the diathermanous medium, according to the universal function (42). Then any element of area of the walls radiates into the cavity just as if the wall were a black body of temperature T. The amount lacking in the intensity of the rays actually emitted by the walls as compared with the emission of a black body is supplied by rays

[1] *W. Wien* and *O. Lummer*, Wied. Annalen, **56**, p. 451, 1895.

[2] The strength of the emission influences only the time required to establish stationary radiation, but not its character. It is essential, however, that the walls transmit no radiation to the exterior.

which fall on the wall and are reflected there. Similarly every element of area of a wall receives the same radiation.

In fact, the radiation I starting from an element of area of a wall consists of the radiation E emitted by the element of area and of the radiation reflected from the element of area from the incident radiation I, *i.e.*, the radiation which is not absorbed $(1-A)I$. We have, therefore, in agreement with *Kirchhoff's* law (48),

$$I=E+(1-A)I.$$

If we now make a hole in one of the walls of a size $d\sigma$, so small that the intensity of the radiation directed toward the hole is not changed thereby, then radiation passes through the hole to the exterior where we shall suppose there is the same diathermanous medium as within. This radiation has exactly the same properties as if $d\sigma$ were the surface of a black body, and this radiation may be measured for every color together with the temperature T.

50. Thus far all the laws derived in the preceding sections for diathermanous media hold for a definite frequency, and it is to be kept in mind that a substance may be diathermanous for one color and adiathermanous for another. Hence the radiation of a medium completely enclosed by absolutely reflecting walls is, when thermodynamic equilibrium has been established for all colors for which the medium has a finite coefficient of absorption, always the stable radiation corresponding to the temperature of the medium such as is represented by the emission of a black body. Hence this is briefly called "black" radiation.[1] On the other hand, the intensity of colors for which the medium is diathermanous is not necessarily the stable black radiation, unless the medium is in a state of stationary exchange of radiation with an absorbing substance.

There is but one medium that is diathermanous for all kinds of rays, namely, the absolute vacuum, which to be sure cannot be produced in nature except approximately. However, most gases, *e.g.*, the air of the atmosphere, have, at least if they are not too dense, to a sufficient approximation the optical properties of a vacuum with respect to waves of not too short length. So far as

[1] *M. Thiesen*, Verhandlungen d. Deutschen Physikal. Gesellschaft, **2**, p. 65, 1900.

this is the case the velocity of propagation q may be taken as the same for all frequencies, namely,

$$c=3\times10^{10}\frac{\text{cm}}{\text{sec}} \tag{51}$$

51. Hence in a vacuum bounded by totally reflecting walls any state of radiation may persist. But as soon as an arbitrarily small quantity of matter is introduced into the vacuum, a stationary state of radiation is gradually established. In this the radiation of every color which is appreciably absorbed by the substance has the intensity K_ν corresponding to the temperature of the substance and determined by the universal function (42) for $q=c$, the intensity of radiation of the other colors remaining indeterminate. If the substance introduced is not diathermanous for any color, *e.g.*, a piece of carbon however small, there exists at the stationary state of radiation in the whole vacuum for all colors the intensity K_ν of black radiation corresponding to the temperature of the substance. The magnitude of K_ν regarded as a function of ν gives the spectral distribution of black radiation in a vacuum, or the so-called *normal energy spectrum*, which depends on nothing but the temperature. In the normal spectrum, since it is the spectrum of emission of a black body, the intensity of radiation of every color is the largest which a body can emit at that temperature at all.

52. It is therefore possible to change a perfectly arbitrary radiation, which exists at the start in the evacuated cavity with perfectly reflecting walls under consideration, into black radiation by the introduction of a minute particle of carbon. The characteristic feature of this process is that the heat of the carbon particle may be just as small as we please, compared with the energy of radiation contained in the cavity of arbitrary magnitude. Hence, according to the principle of the conservation of energy, the total energy of radiation remains essentially constant during the change that takes place, because the changes in the heat of the carbon particle may be entirely neglected, even if its changes in temperature should be finite. Herein the carbon particle exerts only a releasing (auslösend) action. Thereafter the intensities of the pencils of different frequencies originally present and having different frequencies, directions, and different states of polarization change at the expense of one another, corresponding to the passage of the system from a less to a more stable state of radiation or from a state of smaller to a state of larger entropy. From a thermodynamic point of view this process is perfectly analogous, since the time necessary for the process is not essential, to the change produced by a minute spark in a quantity of oxy-hydrogen gas or by a small drop of liquid in a quantity of super-saturated vapor. In all these cases the magnitude of the disturbance is exceedingly small and cannot be compared with the magnitude of the energies undergoing the resultant changes, so that in applying the two principles of thermodynamics the cause of the disturbance of equilibrium, *viz.*, the carbon particle, the spark, or the drop, need not be considered. It is always a case of a system passing from a more or less unstable into a more stable state, wherein, according to the first principle of thermodynamics, the energy of the system remains constant, and, according to the second principle, the entropy of the system increases.

PART II

DEDUCTIONS FROM ELECTRODYNAMICS AND THERMODYNAMICS

CHAPTER I

MAXWELL'S RADIATION PRESSURE

53. While in the preceding part the phenomena of radiation have been presented with the assumption of only well known elementary laws of optics summarized in Sec. 2, which are common to all optical theories, we shall hereafter make use of the electromagnetic theory of light and shall begin by deducing a consequence characteristic of that theory. We shall, namely, calculate the magnitude of the mechanical force, which is exerted by a light or heat ray passing through a vacuum on striking a reflecting (Sec. 10) surface assumed to be at rest.

For this purpose we begin by stating Maxwell's general equations for an electromagnetic process in a vacuum. Let the vector E denote the electric field-strength (intensity of the electric field) in electric units and the vector H the magnetic field-strength in magnetic units. Then the equations are, in the abbreviated notation of the vector calculus,

$$\begin{aligned} \dot{\mathsf{E}} &= c \operatorname{curl} \mathsf{H} & \dot{\mathsf{H}} &= -c \operatorname{curl} \mathsf{E} \\ \text{div. } \mathsf{E} &= 0 & \text{div. } \mathsf{H} &= 0 \end{aligned} \qquad (52)$$

Should the reader be unfamiliar with the symbols of this notation, he may readily deduce their meaning by working backward from the subsequent equations (53).

54. In order to pass to the case of a plane wave in any direction we assume that all the quantities that fix the state depend only on the time t and on one of the coordinates x', y', z', of an orthogonal right-handed system of coordinates, say on x'. Then the equations (52) reduce to

$$\frac{\partial \mathsf{E}_{x'}}{\partial t} = 0 \qquad \frac{\partial \mathsf{H}_{x'}}{\partial t} = 0$$

$$\frac{\partial \mathsf{E}_{y'}}{\partial t} = -c\frac{\partial \mathsf{H}_{z'}}{\partial x'} \qquad \frac{\partial \mathsf{H}_{y'}}{\partial t} = c\frac{\partial \mathsf{E}_{z'}}{\partial x'}$$

$$\frac{\partial \mathsf{E}_{z'}}{\partial t} = c\frac{\partial \mathsf{H}_{y'}}{\partial x'} \qquad \frac{\partial \mathsf{H}_{z'}}{\partial t} = -c\frac{\partial \mathsf{E}_{y'}}{\partial x'} \tag{53}$$

$$\frac{\partial \mathsf{E}_{x'}}{\partial x'} = 0 \qquad \frac{\partial \mathsf{H}_{x'}}{\partial x'} = 0$$

Hence the most general expression for a plane wave passing through a vacuum in the direction of the positive x'-axis is

$$\begin{aligned} \mathsf{E}_{x'} &= 0 & \mathsf{H}_{x'} &= 0 \\ \mathsf{E}_{y'} &= f\left(t-\frac{x'}{c}\right) & \mathsf{H}_{y'} &= -g\left(t-\frac{x'}{c}\right) \\ \mathsf{E}_{z'} &= g\left(t-\frac{x'}{c}\right) & \mathsf{H}_{z'} &= f\left(t-\frac{x'}{c}\right) \end{aligned} \tag{54}$$

where f and g represent two arbitrary functions of the same argument.

55. Suppose now that this wave strikes a reflecting surface, *e.g.*, the surface of an absolute conductor (metal) of infinitely large conductivity. In such a conductor even an infinitely small electric field-strength produces a finite conduction current; hence the electric field-strength E in it must be always and everywhere infinitely small. For simplicity we also suppose the conductor to be non-magnetizable, *i.e.*, we assume the magnetic induction B in it to be equal to the magnetic field-strength H, just as is the case in a vacuum.

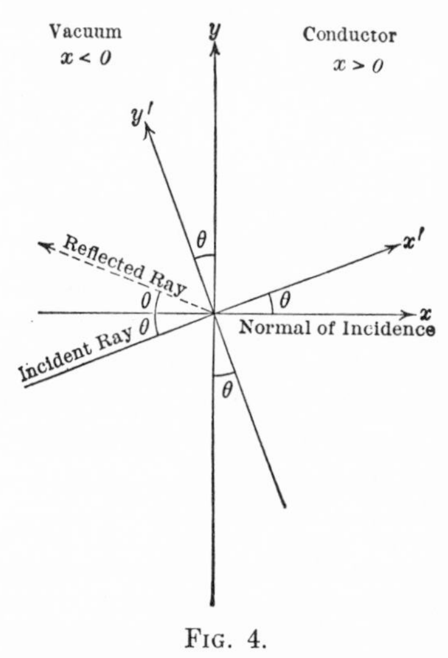

Fig. 4.

If we place the x-axis of a right-handed coordinate system (xyz) along the normal of the surface directed toward the interior of the conductor, the x-axis is the normal of incidence. We place the $(x'y')$ plane in the plane of incidence and take this as the plane of the figure (Fig. 4). Moreover, we can also, without any restriction of generality, place the y-axis in the plane of the figure, so that the z-axis coincides with the z'-axis (directed from the figure toward the observer). Let the common origin O of the two coordinate systems lie in the surface. If finally θ represents the angle of incidence, the coordinates with and without accent are related to each other by the following equations:

$$x = x' \cos\theta - y' \sin\theta \qquad x' = x \cos\theta + y \sin\theta$$
$$y = x' \sin\theta + y' \cos\theta \qquad y' = -x \sin\theta + y \cos\theta$$
$$z = z' \qquad z' = z$$

By the same transformation we may pass from the components of the electric or magnetic field-strength in the first coordinate system to their components in the second system. Performing this transformation the following values are obtained from (54) for the components of the electric and magnetic field-strengths of the incident wave in the coordinate system without accent,

$$\begin{aligned} \mathsf{E}_x &= -\sin\theta \cdot f & \qquad \mathsf{H}_x &= \sin\theta \cdot g \\ \mathsf{E}_y &= \cos\theta \cdot f & \qquad \mathsf{H}_y &= -\cos\theta \cdot g \\ \mathsf{E}_z &= g & \qquad \mathsf{H}_z &= f \end{aligned} \tag{55}$$

Herein the argument of the functions f and g is

$$t - \frac{x'}{c} = t - \frac{x \cos\theta + y \sin\theta}{c} \tag{56}$$

56. In the surface of separation of the two media $x = 0$. According to the general electromagnetic boundary conditions the components of the field-strengths in the surface of separation, *i.e.*, the four quantities E_y, E_z, H_y, H_z must be equal to each other on the two sides of the surface of separation for this value of x. In the conductor the electric field-strength E is infinitely small in accordance with the assumption made above. Hence E_y and E_z must vanish also in the vacuum for $x = 0$. This condition cannot be satisfied unless we assume in the vacuum, besides the incident, also a reflected wave superposed on the former in such a way that the components of the electric field of the two waves in the y and z direction just cancel at every instant and at every point in the surface of separation. By this assumption and the condition that the reflected wave is a plane wave returning into the interior of the vacuum, the other four components of the reflected wave are also completely determined. They are all functions of the single argument

$$t-\frac{-x \cos \theta+y \sin \theta}{c}. \tag{57}$$

The actual calculation yields as components of the total electromagnetic field produced in the vacuum by the superposition of the two waves, the following expressions valid for points of the surface of separation $x=0$,

$$\begin{aligned}
\mathsf{E}_x &= -\sin\theta\cdot f - \sin\theta\cdot f = -\ 2 \sin\theta\cdot f \\
\mathsf{E}_y &= \cos\theta\cdot f\ -\ \cos\theta\cdot f\ =\ 0 \\
\mathsf{E}_z &= g\ - g\ =\ 0 \\
\mathsf{H}_x &= \sin\theta\cdot g\ -\ \sin\theta\cdot g\ =\ 0 \\
\mathsf{H}_y &= -\ \cos\theta\cdot g\ -\ \cos\theta\cdot g = \ -2\ \cos\theta\cdot g \\
\mathsf{H}_z &= f\ + f\ =\ 2f.
\end{aligned} \tag{58}$$

In these equations the argument of the functions f and g is, according to (56) and (57),

$$t-\frac{y \sin \theta}{c}$$

From these values the electric and magnetic field-strength within the conductor in the immediate neighborhood of the separating surface $x=0$ is obtained:

$$\begin{aligned}
\mathsf{E}_x &= 0 \qquad & \mathsf{H}_x &= 0 \\
\mathsf{E}_y &= 0 \qquad & \mathsf{H}_y &= -2\ \cos\theta\cdot g \\
\mathsf{E}_z &= 0 \qquad & \mathsf{H}_z &= 2f
\end{aligned} \tag{59}$$

where again the argument $t-\frac{y \sin \theta}{c}$ is to be substituted in the functions f and g. For the components of E all vanish in an absolute conductor and the components H_x, H_y, H_z are all continuous at the separating surface, the two latter since they are tangential components of the field-strength, the former since it is the normal component of the magnetic induction B (Sec. 55), which likewise remains continuous on passing through any surface of separation.

On the other hand, the normal component of the electric field-strength E_x is seen to be discontinuous; the discontinuity shows the existence of an electric charge on the surface, the surface density of which is given in magnitude and sign as follows:

$$\frac{1}{4\pi}\ 2\ \sin\theta\cdot f=\frac{1}{2\pi}\ \sin\theta\cdot f. \qquad (60)$$

In the interior of the conductor at a finite distance from the bounding surface, *i.e.*, for $x>0$, all six field components are infinitely small. Hence, on increasing x, the values of H_y and H_z, which are finite for $x=0$, approach the value 0 at an infinitely rapid rate.

57. A certain mechanical force is exerted on the substance of the conductor by the electromagnetic field considered. We shall calculate the component of this force normal to the surface. It is partly of electric, partly of magnetic, origin. Let us first consider the former, F_e. Since the electric charge existing on the surface of the conductor is in an electric field, a mechanical force equal to the product of the charge and the field-strength is exerted on it. Since, however, the field-strength is discontinuous, having the value $-2\sin\theta f$ on the side of the vacuum and 0 on the side of the conductor, from a well-known law of electrostatics the magnitude of the mechanical force F_e acting on an element of surface $d\sigma$ of the conductor is obtained by multiplying the electric charge of the element of area calculated in (60) by the arithmetic mean of the electric field-strength on the two sides. Hence

$$\mathsf{F}_e=\frac{\sin\theta}{2\pi}\,f\,d\sigma(-\sin\theta f)=-\frac{\sin^2\theta}{2\pi}\,f^2\,d\sigma$$

This force acts in the direction toward the vacuum and therefore exerts a tension.

58. We shall now calculate the mechanical force of magnetic origin F_m. In the interior of the conducting substance there are certain conduction currents, whose intensity and direction are determined by the vector I of the current density

$$\mathsf{I}=\frac{c}{4\pi}\ \text{curl}\ \mathsf{H}. \qquad (61)$$

A mechanical force acts on every element of space $d\tau$ of the conductor through which a conduction current flows, and is given by the vector product

$$\frac{d\tau}{c}\,[\mathsf{IH}] \qquad (62)$$

Hence the component of this force normal to the surface of the conductor $x=0$ is equal to

$$\frac{d\tau}{c}(\mathsf{I}_y\mathsf{H}_z-\mathsf{I}_z\mathsf{H}_y).$$

On substituting the values of I_y and I_z from (61) we obtain

$$\frac{d\tau}{4\pi}\left[\mathsf{H}_z\left(\frac{\partial\mathsf{H}_x}{\partial z}-\frac{\partial\mathsf{H}_z}{\partial x}\right)-\mathsf{H}_y\left(\frac{\partial\mathsf{H}_y}{\partial x}-\frac{\partial\mathsf{H}_x}{\partial y}\right)\right].$$

In this expression the differential coefficients with respect to y and z are negligibly small in comparison to those with respect to x, according to the remark at the end of Sec. 56; hence the expression reduces to

$$-\frac{d\tau}{4\pi}\left(\mathsf{H}_y\frac{\partial\mathsf{H}_y}{\partial x}+\mathsf{H}_z\frac{\partial\mathsf{H}_z}{\partial x}\right).$$

Let us now consider a cylinder cut out of the conductor perpendicular to the surface with the cross-section $d\sigma$, and extending from $x=0$ to $x=\infty$. The entire mechanical force of magnetic origin acting on this cylinder in the direction of the x-axis, since $d\tau=d\sigma\, x$, is given by

$$\mathsf{F}_m=-\frac{d\sigma}{4\pi}\int_0^\infty dx\left(H_y\frac{\partial\mathsf{H}_y}{\partial x}+\mathsf{H}_z\frac{\partial\mathsf{H}_z}{\partial x}\right).$$

On integration, since H vanishes for $x=\infty$, we obtain

$$\mathsf{F}_m=\frac{d\sigma}{8\pi}(\mathsf{H}^2_y+\mathsf{H}^2_z)_{x=0}$$

or by equation (59)

$$\mathsf{F}_m=\frac{d\sigma}{2\pi}(\cos^2\theta\cdot g^2+f^2).$$

By adding F_e and F_m the total mechanical force acting on the cylinder in question in the direction of the x-axis is found to be

$$\mathsf{F}=\frac{d\sigma}{2\pi}\cos^2\theta\,(f^2+g^2). \tag{63}$$

This force exerts on the surface of the conductor a pressure, which acts in a direction normal to the surface toward the interior and is called "*Maxwell's* radiation pressure." The existence and the magnitude of the radiation pressure as predicted by the theory was first found by delicate measurements with the radiometer by *P. Lebedew.*[1]

59. We shall now establish a relation between the radiation pressure and the energy of radiation $I dt$ falling on the surface element $d\sigma$ of the conductor in a time element dt. The latter from *Poynting's* law of energy flow is

$$I dt = \frac{c}{4\pi}(\mathsf{E}_y\mathsf{H}_z - \mathsf{E}_z\mathsf{H}_y)\; d\sigma\; dt,$$

hence from (55)

$$I dt = \frac{c}{4\pi} \cos\theta\, (f^2 + g^2)\; d\sigma\; dt.$$

By comparison with (63) we obtain

$$\mathsf{F} = \frac{2\cos\theta}{c} I. \tag{64}$$

From this we finally calculate the total pressure p, *i.e.*, that mechanical force, which an arbitrary radiation proceeding from the vacuum and totally reflected upon incidence on the conductor exerts in a normal direction on a unit surface of the conductor. The energy radiated in the conical element

$$d\Omega = \sin\theta\; d\theta\; d\phi$$

in the time dt on the element of area $d\sigma$ is, according to (6),

$$I dt = K \cos\theta\; d\Omega\; d\sigma\; dt,$$

where K represents the specific intensity of the radiation in the direction $d\Omega$ toward the reflector. On substituting this in (64) and integrating over $d\Omega$ we obtain for the total pressure of all pencils which fall on the surface and are reflected by it

$$p = \frac{2}{c}\int K \cos^2\theta\; d\Omega, \tag{65}$$

the integration with respect to ϕ extending from 0 to 2π and with respect to θ from 0 to $\frac{\pi}{2}$.

[1] *P. Lebedew,* Annalen d. Phys., **6,** p. 433, 1901. See also *E. F. Nichols* and *G. F. Hull,* Annalen d. Phys., **12,** p. 225, 1903.

In case K is independent of direction as in the case of black radiation, we obtain for the radiation pressure

$$p = \frac{2K}{c} \int_0^{2\pi} d\phi \int_0^{\frac{\pi}{2}} d\theta \cos^2 \theta \sin \theta = \frac{4\pi K}{3c}$$

or, if we introduce instead of K the volume density of radiation u from (21)

$$p = \frac{u}{3}. \qquad (66)$$

This value of the radiation pressure holds only when the reflection of the radiation occurs at the surface of an absolute non-magnetizable conductor. Therefore we shall in the thermodynamic deductions of the next chapter make use of it only in such cases. Nevertheless it will be shown later on (Sec. 66) that equation (66) gives the pressure of uniform radiation against any totally reflecting surface, no matter whether it reflects uniformly or diffusely.

60. In view of the extraordinarily simple and close relation between the radiation pressure and the energy of radiation, the question might be raised whether this relation is really a special consequence of the electromagnetic theory, or whether it might not, perhaps, be founded on more general energetic or thermodynamic considerations. To decide this question we shall calculate the radiation pressure that would follow by Newtonian mechanics from *Newton's* (emission) theory of light, a theory which, in itself, is quite consistent with the energy principle. According to it the energy radiated onto a surface by a light ray passing through a vacuum is equal to the kinetic energy of the light particles striking the surface, all moving with the constant velocity c. The decrease in intensity of the energy radiation with the distance is then explained simply by the decrease of the volume density of the light particles.

Let us denote by n the number of the light particles contained in a unit volume and by m the mass of a particle. Then for a beam of parallel light the number of particles impinging in unit time on the element $d\sigma$ of a reflecting surface at the angle of incidence θ is

$$nc \cos \theta \, d\sigma. \qquad (67)$$

Their kinetic energy is given according to Newtonian mechanics by

$$I = nc \cos \theta \, d\sigma \frac{mc^2}{2} = nm \cos \theta \frac{c^3}{2} d\sigma. \tag{68}$$

Now, in order to determine the normal pressure of these particles on the surface, we may note that the normal component of the velocity $c \cos \theta$ of every particle is changed on reflection into a component of opposite direction. Hence the normal component of the momentum of every particle (impulse-coordinate) is changed through reflection by $-2mc \cos \theta$. Then the change in momentum for all particles considered will be, according to (67),

$$-2nm \cos^2 \theta \, c^2 \, d\sigma. \tag{69}$$

Should the reflecting body be free to move in the direction of the normal of the reflecting surface and should there be no force acting on it except the impact of the light particles, it would be set into motion by the impacts. According to the law of action and reaction the ensuing motion would be such that the momentum acquired in a certain interval of time would be equal and opposite to the change in momentum of all the light particles reflected from it in the same time interval. But if we allow a separate constant force to act from outside on the reflector, there is to be added to the change in momenta of the light particles the impulse of the external force, *i.e.*, the product of the force and the time interval in question.

Therefore the reflector will remain continuously at rest, whenever the constant external force exerted on it is so chosen that its impulse for any time is just equal to the change in momentum of all the particles reflected from the reflector in the same time. Thus it follows that the force F itself which the particles exert by their impact on the surface element $d\sigma$ is equal and opposite to the change of their momentum in unit time as expressed in (69)

$$\mathsf{F} = 2 \, nm \cos^2 \theta \, c^2 \, d\sigma$$

and by making use of (68),

$$\mathsf{F} = \frac{4 \cos \theta}{c} I.$$

On comparing this relation with equation (64) in which all symbols have the same physical significance, it is seen that

Newton's radiation pressure is twice as large as *Maxwell's* for the same energy radiation. A necessary consequence of this is that the magnitude of *Maxwell's* radiation pressure cannot be deduced from general energetic considerations, but is a special feature of the electromagnetic theory and hence all deductions from *Maxwell's* radiation pressure are to be regarded as consequences of the electromagnetic theory of light and all confirmations of them are confirmations of this special theory.

CHAPTER II

STEFAN-BOLTZMANN LAW OF RADIATION

61. For the following we imagine a perfectly evacuated hollow cylinder with an absolutely tight-fitting piston free to move in a vertical direction with no friction. A part of the walls of the cylinder, say the rigid bottom, should consist of a black body, whose temperature T may be regulated arbitrarily from the outside. The rest of the walls including the inner surface of the piston may be assumed as totally reflecting. Then, if the piston remains stationary and the temperature, T, constant, the radiation in the vacuum will, after a certain time, assume the character of black radiation (Sec. 50) uniform in all directions. The specific intensity, K, and the volume density, u, depend only on the temperature, T, and are independent of the volume, V, of the vacuum and hence of the position of the piston.

If now the piston is moved downward, the radiation is compressed into a smaller space; if it is moved upward the radiation expands into a larger space. At the same time the temperature of the black body forming the bottom may be arbitrarily changed by adding or removing heat from the outside. This always causes certain disturbances of the stationary state. If, however, the arbitrary changes in V and T are made sufficiently slowly, the departure from the conditions of a stationary state may always be kept just as small as we please. Hence the state of radiation in the vacuum may, without appreciable error, be regarded as a state of thermodynamic equilibrium, just as is done in the thermodynamics of ordinary matter in the case of so-called infinitely slow processes, where, at any instant, the divergence from the state of equilibrium may be neglected, compared with the changes which the total system considered undergoes as a result of the entire process.

If, *e.g.*, we keep the temperature T of the black body forming the bottom constant, as can be done by a suitable connection between it and a heat reservoir of large capacity, then, on raising the piston, the black body will emit more than it absorbs, until the newly made space is filled with the same density of radiation as was the original one. *Vice versa,* on lowering the piston the black body will absorb the superfluous radiation until the original radiation corresponding to the temperature T is again established. Similarly, on raising the temperature T of the black body, as can be done by heat conduction from a heat reservoir which is slightly warmer, the density of radiation in the vacuum will be correspondingly increased by a larger emission, etc. To accelerate the establishment of radiation equilibrium the reflecting mantle of the hollow cylinder may be assumed white (Sec. 10), since by diffuse reflection the predominant directions of radiation that may, perhaps, be produced by the direction of the motion of the piston, are more quickly neutralized. The reflecting surface of the piston, however, should be chosen for the present as a perfect metallic reflector, to make sure that the radiation pressure (66) on the piston is *Maxwell's.* Then, in order to produce mechanical equilibrium, the piston must be loaded by a weight equal to the product of the radiation pressure p and the cross-section of the piston. An exceedingly small difference of the loading weight will then produce a correspondingly slow motion of the piston in one or the other direction.

Since the effects produced from the outside on the system in question, the cavity through which the radiation travels, during the processes we are considering, are partly of a mechanical nature (displacement of the loaded piston), partly of a thermal nature (heat conduction away from and toward the reservoir), they show a certain similarity to the processes usually considered in thermodynamics, with the difference that the system here considered is not a material system, *e.g.*, a gas, but a purely energetic one. If, however, the principles of thermodynamics hold quite generally in nature, as indeed we shall assume, then they must also hold for the system under consideration. That is to say, in the case of any change occurring in nature the energy of all systems taking part in the change must remain constant (first principle), and, moreover, the entropy of all systems taking part in the change must increase, or in the limiting case of reversible processes must remain constant (second principle).

62. Let us first establish the equation of the first principle for an infinitesimal change of the system in question. That the cavity enclosing the radiation has a certain energy we have already (Sec. 22) deduced from the fact that the energy radiation is propagated with a finite velocity. We shall denote the energy by U. Then we have

$$U = Vu, \tag{70}$$

where u the volume density of radiation depends only on the temperature of T the black body at the bottom.

The work done by the system, when the volume V of the cavity increases by dV against the external forces of pressure (weight of the loaded piston), is pdV, where p represents *Maxwell's* radiation pressure (66). This amount of mechanical energy is therefore gained by the surroundings of the system, since the weight is raised. The error made by using the radiation pressure on a stationary surface, whereas the reflecting surface moves during the volume change, is evidently negligible, since the motion may be thought of as taking place with an arbitrarily small velocity.

If, moreover, Q denotes the infinitesimal quantity of heat in mechanical units, which, owing to increased emission, passes from the black body at the bottom to the cavity containing the radiation, the bottom or the heat reservoir connected to it loses this heat Q, and its internal energy is decreased by that amount. Hence, according to the first principle of thermodynamics, since the sum of the energy of radiation and the energy of the material bodies remains constant, we have

$$dU + pdV - Q = 0. \tag{71}$$

According to the second principle of thermodynamics the cavity containing the radiation also has a definite entropy. For when the heat Q passes from the heat reservoir into the cavity, the entropy of the reservoir decreases, the change being

$$-\frac{Q}{T}$$

Therefore, since no changes occur in the other bodies—inasmuch as the rigid absolutely reflecting piston with the weight on it does not change its internal condition with the motion—there must somewhere in nature occur a compensation of entropy having at least the value $\frac{Q}{T}$, by which the above diminution is compensated, and this can be nowhere except in the entropy of the cavity containing the radiation. Let the entropy of the latter be denoted by S.

Now, since the processes described consist entirely of states of equilibrium, they are perfectly reversible and hence there is no increase in entropy. Then we have

$$dS-\frac{Q}{T}=0, \tag{72}$$

or from (71)

$$dS=\frac{dU+pdV}{T} \tag{73}$$

In this equation the quantities U, p, V, S represent certain properties of the heat radiation, which are completely defined by the instantaneous state of the radiation. Therefore the quantity T is also a certain property of the state of the radiation, *i.e.*, the black radiation in the cavity has a certain temperature T and this temperature is that of a body which is in heat equilibrium with the radiation.

63. We shall now deduce from the last equation a consequence which is based on the fact that the state of the system considered, and therefore also its entropy, is determined by the values of two independent variables. As the first variable we shall take V, as the second either T, u, or p may be chosen. Of these three quantities any two are determined by the third together with V. We shall take the volume V and the temperature T as independent variables. Then by substituting from (66) and (70) in (73) we have

$$dS=\frac{V}{T}\frac{du}{dT}dT+\frac{4u}{3T}dV. \tag{74}$$

From this we obtain

$$\left(\frac{\partial S}{\partial T}\right)_V=\frac{V}{T}\frac{du}{dT} \qquad \left(\frac{\partial S}{\partial V}\right)_T=\frac{4u}{3T}$$

On partial differentiation of these equations, the first with respect to V, the second with respect to T, we find

$$\frac{\partial^2 S}{\partial T \partial V} = \frac{1}{T}\frac{du}{dT} = \frac{4}{3T}\frac{du}{dT} - \frac{4u}{3T^2}$$

or

$$\frac{du}{dT} = \frac{4u}{T}$$

and on integration

$$u = aT^4 \tag{75}$$

and from (21) for the specific intensity of black radiation

$$K = \frac{c}{4\pi} \cdot u = \frac{ac}{4\pi} T^4. \tag{76}$$

Moreover for the pressure of black radiation

$$p = \frac{a}{3} T^4, \tag{77}$$

and for the total radiant energy

$$U = aT^4 \cdot V. \tag{78}$$

This law, which states that the volume density and the specific intensity of black radiation are proportional to the fourth power of the absolute temperature, was first established by *J. Stefan*[1] on a basis of rather rough measurements. It was later deduced by *L. Boltzmann*[2] on a thermodynamic basis from *Maxwell's* radiation pressure and has been more recently confirmed by *O. Lummer* and *E. Pringsheim*[3] by exact measurements between 100° and 1300° C., the temperature being defined by the gas thermometer. In ranges of temperature and for requirements of precision for which the readings of the different gas thermometers no longer agree sufficiently or cannot be obtained at all, the *Stefan-Boltzmann* law of radiation can be used for an absolute definition of temperature independent of all substances.

64. The numerical value of the constant a is obtained from measurements made by *F. Kurlbaum*.[4] According to them, if

[1] *J. Stefan*, Wien. Berichte, **79**, p. 391, 1879.
[2] *L. Boltzmann*, Wied. Annalen, **22**, p. 291, 1884.
[3] *O. Lummer* und *E. Pringsheim*, Wied. Annalen, **63**, p. 395, 1897. Annalen d. Physik, **3**, p. 159, 1900.
[4] *F. Kurlbaum*, Wied. Annalen, **65**, p. 759, 1898.

we denote by S_t the total energy radiated in one second into air by a square centimeter of a black body at a temperature of $t°$ C., the following equation holds

$$S_{100}-S_o=0.0731\frac{\text{watt}}{\text{cm}^2}=7.31\times10^5\frac{\text{erg}}{\text{cm}^2\text{ sec}}$$

Now, since the radiation in air is approximately identical with the radiation into a vacuum, we may according to (7) and (76) put

$$S_t=\pi K=\frac{ac}{4}(273+t)^4$$

and from this

$$S_{100}-S_o=\frac{ac}{4}(373^4-273^4),$$

therefore

$$a=\frac{4\times7.31\times10^5}{3\times10^{10}\times(373^4-273^4)}=7.061\times10^{-15}\frac{\text{erg}}{\text{cm}^3\text{ degree}^4}$$

Recently *Kurlbaum* has increased the value measured by him by 2.5 per cent.,[1] on account of the bolometer used being not perfectly black, whence it follows that $a=7.24\cdot10^{-15}$.

Meanwhile the radiation constant has been made the object of as accurate measurements as possible in various places. Thus it was measured by *Féry, Bauer* and *Moulin, Valentiner, Féry* and *Drecq, Shakespear, Gerlach,* with in some cases very divergent results, so that a mean value may hardly be formed.

For later computations we shall use the most recent determination made in the physical laboratory of the University of Berlin[2]

$$\frac{ac}{4}=\sigma=5.46\cdot10^{-12}\frac{\text{watt}}{\text{cm}^2\text{ degree}^4}$$

From this a is found to be

$$a=\frac{4\cdot5.46\cdot10^{-12}\cdot10^7}{3\cdot10^{10}}=7.28\cdot10^{-15}\frac{\text{erg}}{\text{cm}^3\text{ degree}^4}$$

which agrees rather closely with *Kurlbaum's* corrected value.

[1] *F. Kurlbaum,* Verhandlungen d. Deutsch. physikal. Gesellschaft, **14,** p. 580, 1912.

[2] According to private information kindly furnished by my colleague *H. Rubens* (July, 1912). (These results have since been published. See *W. H. Westphal,* Verhandlungen d. Deutsch. physikal. Gesellschaft, **14,** p. 987, 1912, Tr.)

65. The magnitude of the entropy S of black radiation found by integration of the differential equation (73) is

$$S=\frac{4}{3}aT^3V. \tag{80}$$

In this equation the additive constant is determined by a choice that readily suggests itself, so that at the zero of the absolute scale of temperature, that is to say, when u vanishes, S shall become zero. From this the entropy of unit volume or the volume density of the entropy of black radiation is obtained,

$$\frac{S}{V}=s=\frac{4}{3}aT^3. \tag{81}$$

66. We shall now remove a restricting assumption made in order to enable us to apply the value of *Maxwell's* radiation pressure, calculated in the preceding chapter. Up to now we have assumed the cylinder to be fixed and only the piston to be free to move. We shall now think of the whole of the vessel, consisting of the cylinder, the black bottom, and the piston, the latter attached to the walls in a definite height above the bottom, as being free to move in space. Then, according to the principle of action and reaction, the vessel as a whole must remain constantly at rest, since no external force acts on it. This is the conclusion to which we must necessarily come, even without, in this case, admitting *a priori* the validity of the principle of action and reaction. For if the vessel should begin to move, the kinetic energy of this motion could originate only at the expense of the heat of the body forming the bottom or the energy of radiation, as there exists in the system enclosed in a rigid cover no other available energy; and together with the decrease of energy the entropy of the body or the radiation would also decrease, an event which would contradict the second principle, since no other changes of entropy occur in nature. Hence the vessel as a whole is in a state of mechanical equilibrium. An immediate consequence of this is that the pressure of the radiation on the black bottom is just as large as the oppositely directed pressure of the radiation on the reflecting piston. Hence the pressure of black radiation is the same on a black as on a reflecting body of the same temperature and the same may be readily proven

5

for any completely reflecting surface whatsoever, which we may assume to be at the bottom of the cylinder without in the least disturbing the stationary state of radiation. Hence we may also in all the foregoing considerations replace the reflecting metal by any completely reflecting or black body whatsoever, at the same temperature as the body forming the bottom, and it may be stated as a quite general law that the radiation pressure depends only on the properties of the radiation passing to and fro, not on the properties of the enclosing substance.

67. If, on raising the piston, the temperature of the black body forming the bottom is kept constant by a corresponding addition of heat from the heat reservoir, the process takes place isothermally. Then, along with the temperature T of the black body, the energy density u, the radiation pressure p, and the density of the entropy s also remain constant; hence the total energy of radiation increases from $U=uV$ to $U'=uV'$, the entropy from $S=sV$ to $S'=sV'$ and the heat supplied from the heat reservoir is obtained by integrating (72) at constant T,

$$Q=T(S'-S)=Ts(V'-V)$$

or, according to (81) and (75),

$$Q=\frac{4}{3}aT^4(V'-V)=\frac{4}{3}(U'-U).$$

Thus it is seen that the heat furnished from the outside exceeds the increase in energy of radiation $(U'-U)$ by $\frac{1}{3}(U'-U)$. This excess in the added heat is necessary to do the external work accompanying the increase in the volume of radiation.

68. Let us also consider a reversible adiabatic process. For this it is necessary not merely that the piston and the mantle but also that the bottom of the cylinder be assumed as completely reflecting, *e.g.*, as white. Then the heat furnished on compression or expansion of the volume of radiation is $Q=0$ and the energy of radiation changes only by the value pdV of the external work. To insure, however, that in a finite adiabatic process the radiation shall be perfectly stable at every instant, *i.e.*, shall have the character of black radiation, we may assume that inside the evacuated cavity there is a carbon particle of minute size. This particle, which may be assumed to possess an absorbing power differing from zero for all kinds of rays, serves merely to produce stable equilibrium of the radiation in the cavity (Sec. 51 *et seq.*) and thereby to insure the reversibility of the process, while its heat contents may be taken as so small compared with the energy of radiation, U, that the addition of heat required for an appreciable temperature change of the particle is perfectly negligible. Then at every instant in the process there exists absolutely stable equilibrium of radiation and the radiation has the temperature of the particle in the cavity. The volume, energy, and entropy of the particle may be entirely neglected.

On a reversible adiabatic change, according to (72), the entropy S of the system remains constant. Hence from (80) we have as a condition for such a process

$$T^3 V = \text{const.},$$

or, according to (77),

$$pV^{\frac{4}{3}} = \text{const.},$$

i.e., on an adiabatic compression the temperature and the pressure of the radiation increase in a manner that may be definitely stated. The energy of the radiation, U, in such a case varies according to the law

$$\frac{U}{T} = \frac{3}{4} S = \text{const.},$$

i.e., it increases in proportion to the absolute temperature, although the volume becomes smaller.

69. Let us finally, as a further example, consider a simple case of an irreversible process. Let the cavity of volume V, which is everywhere enclosed by absolutely reflecting walls, be uniformly filled with black radiation. Now let us make a small hole through any part of the walls, *e.g.*, by opening a stopcock, so that the radiation may escape into another completely evacuated space, which may also be surrounded by rigid, absolutely reflecting walls. The radiation will at first be of a very irregular character; after some time, however, it will assume a stationary condition and will fill both communicating spaces uniformly, its total volume being, say, V'. The presence of a carbon particle will cause all conditions of black radiation to be satisfied in the new state. Then, since there is neither external work nor addition of heat from the outside, the energy of the new state is, according to the first principle, equal to that of the original one, or $U' = U$ and hence from (78)

$$T'^4 V' = T^4 V$$

$$\frac{T'}{T} = \sqrt[4]{\frac{V}{V'}}$$

which defines completely the new state of equilibrium. Since $V' > V$ the temperature of the radiation has been lowered by the process.

According to the second principle of thermodynamics the entropy of the system must have increased, since no external changes have occurred; in fact we have from (80)

$$\frac{S'}{S} = \frac{T'^3 V'}{T^3 V} = \sqrt[4]{\frac{V'}{V}} > 1. \qquad (82)$$

70. If the process of irreversible adiabatic expansion of the radiation from the volume V to the volume V' takes place as just described with the single difference that there is no carbon particle present in the vacuum, after the stationary state of radiation is established, as will be the case after a certain time on account of the diffuse reflection from the walls of the cavity, the radiation in the new volume V' will not any longer have the character of black radiation, and hence no definite temperature. Nevertheless the radiation, like every system in a definite physical state, has a definite entropy, which, according to the second principle, is larger than the original S, but not as large as the S' given in (82). The calculation cannot be performed without the use of laws to be taken up later (see Sec. 103). If a carbon particle is afterward introduced into the vacuum, absolutely stable equilibrium is established by a second irreversible process, and, the total energy as well as the total volume remaining constant, the radiation assumes the normal energy distribution of black radiation and the entropy increases to the maximum value S' given by (82).

CHAPTER III

WIEN'S DISPLACEMENT LAW

71. Though the manner in which the volume density u and the specific intensity K of black radiation depend on the temperature is determined by the *Stefan-Boltzmann* law, this law is of comparatively little use in finding the volume density $\mathfrak{u}_\nu$ corresponding to a definite frequency ν, and the specific intensity of radiation $\mathfrak{K}_\nu$ of monochromatic radiation, which are related to each other by equation (24) and to u and K by equations (22) and (12). There remains as one of the principal problems of the theory of heat radiation the problem of determining the quantities $\mathfrak{u}_\nu$ and $\mathfrak{K}_\nu$ for black radiation in a vacuum and hence, according to (42), in any medium whatever, as functions of ν and T, or, in other words, to find the distribution of energy in the normal spectrum for any arbitrary temperature. An essential step in the solution of this problem is contained in the so-called "displacement law" stated by *W. Wien*,[1] the importance of which lies in the fact that it reduces the functions $\mathfrak{u}_\nu$ and $\mathfrak{K}_\nu$ of the two arguments ν and T to a function of a single argument.

The starting point of *Wien's* displacement law is the following theorem. If the black radiation contained in a perfectly evacuated cavity with absolutely reflecting walls is compressed or expanded adiabatically and infinitely slowly, as described above in Sec. 68, *the radiation always retains the character of black radiation, even without the presence of a carbon particle.* Hence the process takes place in an absolute vacuum just as was calculated in Sec. 68 and the introduction, as a precaution, of a carbon particle is shown to be superfluous. But this is true only in this special case, not at all in the case described in Sec. 70.

The truth of the proposition stated may be shown as follows:

[1] *W. Wien*, Sitzungsberichte d. Akad. d. Wissensch. Berlin, Febr. 9, 1893, p. 55. Wiedemann's Annal., **52**, p. 132, 1894. See also among others *M. Thiesen*, Verhandl. d. Deutsch. phys. Gesellsch, **2**, p. 65, 1900. *H. A. Lorentz*, Akad. d. Wissensch. Amsterdam, May 18, 1901, p. 607. *M. Abraham*, Annal. d. Physik. **14**, p. 236, 1904.

Let the completely evacuated hollow cylinder, which is at the start filled with black radiation, be compressed adiabatically and infinitely slowly to a finite fraction of the original volume. If, now, the compression being completed, the radiation were no longer black, there would be no stable thermodynamic equilibrium of the radiation (Sec. 51). It would then be possible to produce a finite change at constant volume and constant total energy of radiation, namely, the change to the absolutely stable state of radiation, which would cause a finite increase of entropy. This change could be brought about by the introduction of a carbon particle, containing a negligible amount of heat as compared with the energy of radiation. This change, of course, refers only to the spectral density of radiation u_ν, whereas the total density of energy u remains constant. After this has been accomplished, we could, leaving the carbon particle in the space, allow the hollow cylinder to return adiabatically and infinitely slowly to its original volume and then remove the carbon particle. The system will then have passed through a cycle without any external changes remaining. For heat has been neither added nor removed, and the mechanical work done on compression has been regained on expansion, because the latter, like the radiation pressure, depends only on the total density u of the energy of radiation, not on its spectral distribution. Therefore, according to the first principle of thermodynamics, the total energy of radiation is at the end just the same as at the beginning, and hence also the temperature of the black radiation is again the same. The carbon particle and its changes do not enter into the calculation, for its energy and entropy are vanishingly small compared with the corresponding quantities of the system. The process has therefore been reversed in all details; it may be repeated any number of times without any permanent change occurring in nature. This contradicts the assumption, made above, that a finite increase in entropy occurs; for such a finite increase, once having taken place, cannot in any way be completely reversed. Therefore no finite increase in entropy can have been produced by the introduction of the carbon particle in the space of radiation, but the radiation was, before the introduction and always, in the state of stable equilibrium.

72. In order to bring out more clearly the essential part of this important proof, let us point out an analogous and more or less obvious consideration. Let a cavity containing originally a vapor in a state of saturation be compressed adiabatically and infinitely slowly.

"Then on an arbitrary adiabatic compression the vapor remains always just in the state of saturation. For let us suppose that it becomes supersaturated on compression. After the compression to an appreciable fraction of the original volume has taken place, condensation of a finite amount of vapor and thereby a change into a more stable state, and hence a finite increase of entropy of the system, would be produced at constant volume and constant total energy by the introduction of a minute drop of liquid, which has no appreciable mass or heat capacity. After this has been done, the volume could again be increased adiabatically and infinitely slowly until again all liquid is evaporated and thereby the process completely reversed, which contradicts the assumed increase of entropy."

Such a method of proof would be erroneous, because, by the process described, the change that originally took place is not at all completely reversed. For since the mechanical work expended on the compression of the supersaturated steam is not equal to the amount gained on expanding the saturated steam, there corresponds to a definite volume of the system when it is being compressed an amount of energy different from the one during expansion and therefore the volume at which all liquid is just vaporized cannot be equal to the original volume. The supposed analogy therefore breaks down and the statement made above in quotation marks is incorrect.

73. We shall now again suppose the reversible adiabatic process described in Sec. 68 to be carried out with the black radiation contained in the evacuated cavity with white walls and white bottom, by allowing the piston, which consists of absolutely reflecting metal, to move downward infinitely slowly, with the single difference that now there shall be no carbon particle in the cylinder. The process will, as we now know, take place exactly as there described, and, since no absorption or emission of radiation takes place, we can now give an account of the changes of color and intensity which the separate pencils of the system undergo. Such changes will of course occur only on reflection from the moving metallic reflector, not on reflection from the stationary walls and the stationary bottom of the cylinder.

If the reflecting piston moves down with the constant, infinitely small, velocity v, the monochromatic pencils striking it during the motion will suffer on reflection a change of color, intensity, and direction. Let us consider these different influences in order.[1]

74. To begin with, we consider the *change of color* which a monochromatic ray suffers by reflection from the reflector, which is moving with an infinitely small velocity. For this purpose we consider first the case of a ray which falls normally from below on the reflector and hence is reflected normally downward. Let the plane A (Fig. 5) represent the position of the reflector at the time t, the plane A' the position at the time $t+\delta t$, where the distance AA' equals $v\delta t$, v denoting the velocity of the reflector.

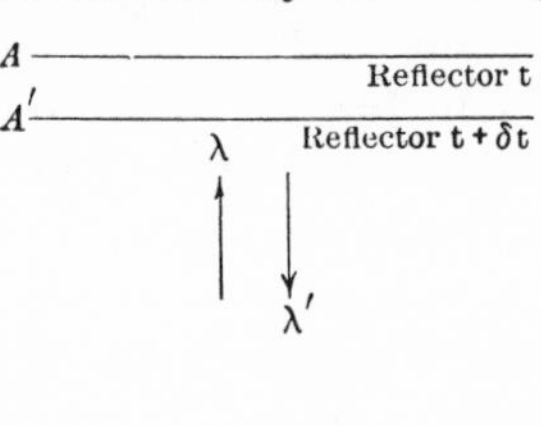

Fig. 5.

Let us now suppose a stationary plane B to be placed parallel to A at a suitable distance and let us denote by λ the wave length of the ray incident on the reflector and by λ' the wave length of the ray reflected from it. Then at a time t there are in the interval AB in the vacuum containing the radiation $\frac{AB}{\lambda}$ waves of the incident and $\frac{AB}{\lambda'}$ waves of the reflected ray, as can be seen, *e.g.*, by thinking of the electric field-strength as being drawn at the different points of each of the two rays at the time t in the form of a sine curve. Reckoning both incident and reflected ray there are at the time t

$$AB\left(\frac{1}{\lambda}+\frac{1}{\lambda'}\right)$$

waves in the interval between A and B. Since this is a large number, it is immaterial whether the number is an integer or not.

[1] The complete solution of the problem of reflection of a pencil from a moving absolutely reflecting surface including the case of an arbitrarily large velocity of the surface may be found in the paper by *M. Abraham* quoted in Sec. 71. See also the text-book by the same author. Electromagnetische Theorie der Strahlung, 1908 (Leipzig, B. G. Teubner).

Similarly at the time $t+\delta t$, when the reflector is at A', there are

$$A'B\left(\frac{1}{\lambda}+\frac{1}{\lambda'}\right)$$

waves in the interval between A' and B all told.

The latter number will be smaller than the former, since in the shorter distance $A'B$ there is room for fewer waves of both kinds than in the longer distance AB. The remaining waves must have been expelled in the time δt from the space between the stationary plane B and the moving reflector, and this must have taken place through the plane B downward; for in no other way could a wave disappear from the space considered.

Now $\nu\delta t$ waves pass in the time δt through the stationary plane B in an upward direction and $\nu'\delta t$ waves in a downward direction; hence we have for the difference

$$(\nu'-\nu)\ \delta t=(AB-A'B)\left(\frac{1}{\lambda}+\frac{1}{\lambda'}\right)$$

or, since

$$AB-A'B=v\delta t,$$

and

$$\lambda=\frac{c}{\nu}\quad \lambda'=\frac{c}{\nu'}$$

$$\nu'=\frac{c+v}{c-v}\nu$$

or, since v is infinitely small compared with c,

$$\nu'=\nu\left(1+\frac{2v}{c}\right)$$

75. When the radiation does not fall on the reflector normally but at an acute angle of incidence θ, it is possible to pursue a very similar line of reasoning, with the difference that then A, the point of intersection of a definite ray BA with the reflector at the time t, has not the same position on the reflector as the point of intersection, A', of the same ray with the reflector at the time $t+\delta t$ (Fig. 6). The number of waves which lie in the interval BA at the time t is $\frac{BA}{\lambda}\cdot$ Similarly, at the time t the number of waves in the interval AC representing the distance of the point

A from a wave plane CC', belonging to the reflected ray and stationary in the vacuum, is $\frac{AC}{\lambda'}$.

Hence there are, all told, at the time t in the interval BAC

$$\frac{BA}{\lambda}+\frac{AC}{\lambda'}$$

waves of the ray under consideration. We may further note that the angle of reflection θ' is not exactly equal to the angle

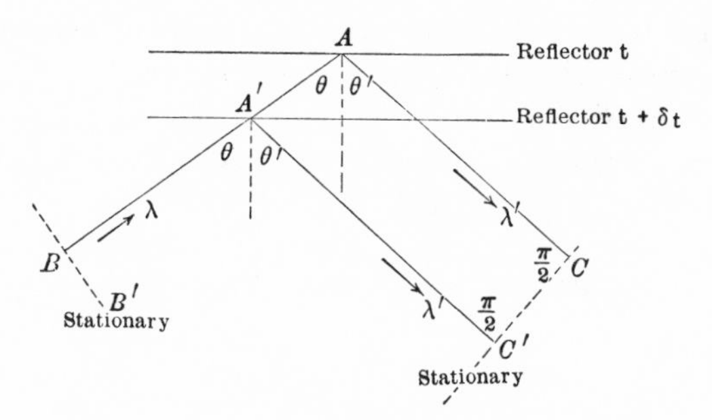

Fig. 6.

of incidence, but is a little smaller as can be shown by a simple geometric consideration based on *Huyghens'* principle. The difference of θ and θ', however, will be shown to be non-essential for our calculation. Moreover there are at the time $t+\delta t$, when the reflector passes through A',

$$\frac{BA'}{\lambda}+\frac{A'C'}{\lambda'}$$

waves in the distance $BA'C'$. The latter number is smaller than the former and the difference must equal the total number of waves which are expelled in the time δt from the space which is bounded by the stationary planes BB' and CC'.

Now $\nu\delta t$ waves enter into the space through the plane BB' in the time δt and $\nu'\delta t$ waves leave the space through the plane CC' Hence we have

$$(\nu'-\nu)\,\delta t=\left(\frac{BA}{\lambda}+\frac{AC}{\lambda'}\right)-\left(\frac{BA'}{\lambda}+\frac{A'C'}{\lambda'}\right)$$

but

$$BA-BA'=AA'=\frac{v\delta t}{\cos\theta}$$

$$AC-A'C'=AA'\cos(\theta+\theta')$$

$$\lambda=\frac{c}{\nu},\ \lambda'=\frac{c}{\nu'}.$$

Hence

$$\nu'=\frac{c\cos\theta+v}{c\cos\theta-v\cos(\theta+\theta')}\nu.$$

This relation holds for any velocity v of the moving reflector. Now, since in our case v is infinitely small compared with c, we have the simpler expression

$$\nu'=\nu(1+\frac{v}{c\cos\theta}[1+\cos(\theta+\theta')])$$

The difference between the two angles θ and θ' is in any case of the order of magnitude $\frac{v}{c}$; hence we may without appreciable error replace θ' by θ, thereby obtaining the following expression for the frequency of the reflected ray for oblique incidence

$$\nu'=\nu\left(1+\frac{2v\cos\theta}{c}\right) \tag{83}$$

76. From the foregoing it is seen that the frequency of all rays which strike the moving reflector are increased on reflection, when the reflector moves toward the radiation, and decreased, when the reflector moves in the direction of the incident rays ($v<0$). However, the total radiation of a definite frequency ν striking the moving reflector is by no means reflected as monochromatic radiation but the change in color on reflection depends also essentially on the angle of incidence θ. Hence we may not speak of a certain spectral "displacement" of color except in the case of a single pencil of rays of definite direction, whereas in the case of the entire monochromatic radiation we must refer to a spectral "dispersion.". The change in color is the largest for normal incidence and vanishes entirely for grazing incidence.

77. Secondly, let us calculate the *change in energy*, which the moving reflector produces in the incident radiation, and let us consider from the outset the general case of oblique incidence. Let a monochromatic, infinitely thin, unpolarized pencil of rays. which falls on a surface element of the reflector at the angle of incidence θ, transmit the energy $I\delta t$ to the reflector in the time δt. Then, ignoring vanishingly small quantities, the mechanical pressure of the pencil of rays normally to the reflector is, according to equation (64),

$$\mathsf{F}=\frac{2\cos\theta}{c}I,$$

and to the same degree of approximation the work done from the outside on the incident radiation in the time δt is

$$\mathsf{F}v\delta t=\frac{2v\cos\theta}{c}I\delta t. \tag{84}$$

According to the principle of the conservation of energy this amount of work must reappear in the energy of the reflected radiation. Hence the reflected pencil has a larger intensity than the incident one. It produces, namely, in the time δt the energy[1]

$$I\delta t+\mathsf{F}v\delta t=I\left(1+\frac{2v\cos\theta}{c}\right)\delta t=I'\delta t. \tag{85}$$

Hence we may summarize as follows: By the reflection of a monochromatic unpolarized pencil, incident at an angle θ on a reflector moving toward the radiation with the infinitely small velocity v, the radiant energy $I\delta t$, whose frequencies extend from ν to $\nu+d\nu$, is in the time δt changed into the radiant energy $I'\delta t$ with the interval of frequency $(\nu', \nu'+d\nu')$, where I' is given by (85), ν' by (83), and accordingly $d\nu'$, the spectral breadth of the reflected pencil, by

$$d\nu'=d\nu\left(1+\frac{2v\cos\theta}{c}\right) \tag{86}$$

A comparison of these values shows that

$$\frac{I'}{I}=\frac{\nu'}{\nu}=\frac{d\nu'}{d\nu} \tag{87}$$

[1] It is clear that the change in intensity of the reflected radiation caused by the motion of the reflector can also be derived from purely electrodynamical considerations, since electrodynamics are consistent with the energy principle. This method is somewhat lengthy, but it affords a deeper insight into the details of the phenomenon of reflection.

The absolute value of the radiant energy which has disappeared in this change is, from equation (13),

$$I\delta t = 2\mathsf{K}_\nu \, d\sigma \cos\theta \, d\Omega \, d\nu \, \delta t, \tag{88}$$

and hence the absolute value of the radiant energy which has been formed is, according to (85),

$$I'\delta t = 2\mathsf{K}_\nu d\sigma \cos\theta \, d\Omega \, d\nu \left(1 + \frac{2v \cos\theta}{c}\right) \delta t. \tag{89}$$

Strictly speaking these last two expressions would require an infinitely small correction, since the quantity I from equation (88) represents the energy radiation on a stationary element of area $d\sigma$, while, in reality, the incident radiation is slightly increased by the motion of $d\sigma$ toward the incident pencil. The additional terms resulting therefrom may, however, be omitted here without appreciable error.

78. As regards finally the *changes in direction*, which are imparted to the incident ray by reflection from the moving reflector, we need not calculate them at all at this stage. For if the motion of the reflector takes place sufficiently slowly, all irregularities in the direction of the radiation are at once equalized by further reflection from the walls of the vessel. We may, indeed, think of the whole process as being accomplished in a very large number of short intervals, in such a way that the piston, after it has moved a very small distance with very small velocity, is kept at rest for a while, namely, until all irregularities produced in the directions of the radiation have disappeared as the result of the reflection from the white walls of the hollow cylinder. If this procedure be carried on sufficiently long, the compression of the radiation may be continued to an arbitrarily small fraction of the original volume, and while this is being done, the radiation may be always regarded as uniform in all directions. This continuous process of equalization refers, of course, only to difference in the direction of the radiation; for changes in the color or intensity of the radiation of however small size, having once occurred, can evidently never be equalized by reflection from totally reflecting stationary walls but continue to exist forever.

79. With the aid of the theorems established we are now in a position to calculate the change of the density of radiation for every frequency for the case of infinitely slow adiabatic compression of the perfectly evacuated hollow cylinder, which is filled with uniform radiation. For this purpose we consider the radiation at the time t in a definite infinitely small interval of frequencies, from ν to $\nu+d\nu$, and inquire into the change which the total energy of radiation contained in this definite constant interval suffers in the time δt.

At the time t this radiant energy is, according to Sec. 23, $V\,\mathsf{u}d\nu$; at the time $t+\delta t$ it is $(V\mathsf{u}+\delta\,(V\mathsf{u}))d\nu$, hence the change to be calculated is

$$\delta(V\mathsf{u})d\nu. \tag{90}$$

In this the density of monochromatic radiation u is to be regarded as a function of the mutually independent variables ν and t, the differentials of which are distinguished by the symbols d and δ.

The change of the energy of monochromatic radiation is produced only by the reflection from the moving reflector, that is to say, firstly by certain rays, which at the time t belong to the interval $(\nu,d\nu)$, leaving this interval on account of the change in color suffered by reflection, and secondly by certain rays, which at the time t do not belong to the interval $(\nu,d\nu)$, coming into this interval on account of the change in color suffered on reflection. Let us calculate these influences in order. The calculation is greatly simplified by taking the width of this interval $d\nu$ so small that

$$d\nu \text{ is small compared with } \frac{v}{c}\nu, \tag{91}$$

a condition which can always be satisfied, since $d\nu$ and v are mutually independent.

80. The rays which at the time t belong to the interval $(\nu,d\nu)$ and leave this interval in the time δt on account of reflection from the moving reflector, are simply those rays which strike the moving reflector in the time δt. For the change in color which such a ray undergoes is, from (83) and (91), large compared with $d\nu$, the width of the whole interval. Hence we need only calculate the energy, which in the time δt is transmitted to the reflector by the rays in the interval $(\nu,d\nu)$.

For an elementary pencil, which falls on the element $d\sigma$ of the reflecting surface at the angle of incidence θ, this energy is, according to (88) and (5),

$$I\delta t = 2\mathsf{K}_\nu d\sigma \cos\theta\, d\Omega\, d\nu\, \delta t = 2\mathsf{K}_\nu\, d\sigma \sin\theta \cos\theta\, d\theta\, d\phi\, d\nu\, \delta t.$$

Hence we obtain for the total monochromatic radiation, which falls on the whole surface F of the reflector, by integration with respect to ϕ from 0 to 2π, with respect to θ from 0 to $\frac{\pi}{2}$, and with respect to $d\sigma$ from 0 to F,

$$2\pi\, F\, \mathsf{K}_\nu\, d\nu\, \delta t. \qquad (92)$$

Thus this radiant energy leaves, in the time δt, the interval of frequencies $(\nu, d\nu)$ considered.

81. In calculating the radiant energy which enters the interval $(\nu, d\nu)$ in the time δt on account of reflection from the moving reflector, the rays falling on the reflector at different angles of incidence must be considered separately. Since in the case of a positive v, the frequency is increased by the reflection, the rays which must be considered have, at the time t, the frequency $\nu_1 < \nu$. If we now consider at the time t a monochromatic pencil of frequency $(\nu_1, d\nu_1)$, falling on the reflector at an angle of incidence θ, a necessary and sufficient condition for its entrance, by reflection, into the interval $(\nu, d\nu)$ is

$$\nu = \nu_1\left(1 + \frac{2v\cos\theta}{c}\right) \text{ and } d\nu = d\nu_1\left(1 + \frac{2v\cos\theta}{c}\right)$$

These relations are obtained by substituting ν_1 and ν respectively in the equations (83) and (86) in place of the frequencies before and after reflection ν and ν'.

The energy which this pencil carries into the interval $(\nu_1, d\nu)$ in the time δt is obtained from (89), likewise by substituting ν_1 for ν. It is

$$2\mathsf{K}_{\nu 1}\, d\sigma \cos\theta\, d\Omega\, d\nu_1\left(1 + \frac{2v\cos\theta}{c}\right)\delta t = 2\mathsf{K}_{\nu 1}\, d\sigma \cos\theta\, d\Omega\, d\nu\, \delta t.$$

Now we have

$$\mathsf{K}_{\nu 1} = \mathsf{K}_\nu + (\nu_1 - \nu)\frac{\partial \mathsf{K}}{\partial \nu} + \quad . \quad . \quad . \quad . \quad .$$

where we shall assume $\frac{\partial \mathsf{K}}{\partial \nu}$ to be finite.

Hence, neglecting small quantities of higher order,

$$\mathsf{K}_{\nu_1} = \mathsf{K}_\nu - \frac{2\nu v \cos\theta}{c} \frac{\partial \mathsf{K}}{\partial \nu}$$

Thus the energy required becomes

$$2d\sigma\left(\mathsf{K}_\nu - \frac{2\nu v \cos\theta}{c} \frac{\partial \mathsf{K}}{\partial \nu}\right) \sin\theta \cos\theta \, d\theta \, d\phi \, d\nu \, \delta t,$$

and, integrating this expression as above, with respect to $d\sigma$, ϕ, and θ, the total radiant energy which enters into the interval $(\nu, d\nu)$ in the time δt becomes

$$2\pi F\left(\mathsf{K}_\nu - \frac{4}{3} \frac{\nu v}{c} \frac{\partial \mathsf{K}}{\partial \nu}\right) d\nu \, \delta t. \tag{93}$$

82. The difference of the two expressions (93) and (92) is equal to the whole change (90), hence

$$-\frac{8\pi}{3} F \frac{\nu v}{c} \frac{\partial \mathsf{K}}{\partial \nu} \delta t = \delta(V\mathsf{u}),$$

or, according to (24),

$$-\frac{1}{3} F \nu \, v \frac{\partial \mathsf{u}}{\partial \nu} \delta t = \delta(V\mathsf{u}),$$

or, finally, since $Fv\delta t$ is equal to the decrease of the volume V,

$$\frac{1}{3} \nu \frac{\partial \mathsf{u}}{\partial \nu} \delta V = \delta(V\mathsf{u}) = \mathsf{u}\delta V + V\delta \mathsf{u}, \tag{94}$$

whence it follows that

$$\delta \mathsf{u} = \left(\frac{\nu}{3} \frac{\partial \mathsf{u}}{\partial \nu} - \mathsf{u}\right) \frac{\delta V}{V}. \tag{95}$$

This equation gives the change of the energy density of any definite frequency ν, which occurs on an infinitely slow adiabatic compression of the radiation. It holds, moreover, not only for black radiation, but also for *radiation originally of a perfectly arbitrary distribution of energy*, as is shown by the method of derivation.

Since the changes taking place in the state of the radiation in the time δt are proportional to the infinitely small velocity v and are reversed on changing the sign of the latter, this equation holds for any sign of δV; *hence the process is reversible.*

83. Before passing on to the general integration of equation (95) let us examine it in the manner which most easily suggests itself. According to the energy principle, the change in the radiant energy

$$U = Vu = V \int_0^\infty \mathsf{u} \, d\nu,$$

occurring on adiabatic compression, must be equal to the external work done against the radiation pressure

$$-p\delta V = -\frac{u}{3}\,\delta V = -\frac{\delta V}{3}\int_0^\infty \mathsf{u}\,d\nu. \tag{96}$$

Now from (94) the change in the total energy is found to be

$$\delta U = \int_0^\infty d\nu \; \delta\,(V\mathsf{u}) = \frac{\delta V}{3}\int_0^\infty \nu \frac{\partial \mathsf{u}}{\partial \nu}\,d\nu,$$

or, by partial integration,

$$\delta U = \frac{\delta V}{3}\left([\nu \mathsf{u}]_0^\infty - \int_0^\infty \mathsf{u}\,d\nu\right),$$

and this expression is, in fact, identical with (96), since the product $\nu\mathsf{u}$ vanishes for $\nu = 0$ as well as for $\nu = \infty$. The latter might at first seem doubtful; but it is easily seen that, if $\nu\mathsf{u}$ for $\nu = \infty$ had a value different from zero, the integral of u with respect to ν taken from 0 to ∞ could not have a finite value, which, however, certainly is the case.

84. We have already emphasized (Sec. 79) that u must be regarded as a function of two independent variables, of which we have taken as the first the frequency ν and as the second the time t. Since, now, in equation (95) the time t does not explicitly appear, it is more appropriate to introduce the volume V, which depends only on t, as the second variable instead of t itself. Then equation (95) may be written as a partial differential equation as follows:

$$V\frac{\partial \mathsf{u}}{\partial V} = \frac{\nu}{3}\frac{\partial \mathsf{u}}{\partial \nu} - \mathsf{u}. \tag{97}$$

From this equation, if, for a definite value of V, u is known as a function of ν, it may be calculated for all other values of V as a

6

function of ν. The general integral of this differential equation, as may be readily seen by substitution, is

$$\mathsf{u}=\frac{1}{V}\,\phi(\nu^3\,V), \tag{98}$$

where ϕ denotes an arbitrary function of the single argument $\nu^3 V$. Instead of this we may, on substituting $\nu^3 V\phi(\nu^3 V)$ for $\phi(\nu^3 V)$, write

$$\mathsf{u}=\nu^3\phi(\nu^3 V). \tag{99}$$

Either of the last two equations is the general expression of *Wien's* displacement law.

If for a definitely given volume V the spectral distribution of energy is known (*i.e.*, u as a function of ν), it is possible to deduce therefrom the dependence of the function ϕ on its argument, and thence the distribution of energy for any other volume V', into which the radiation filling the hollow cylinder may be brought by a reversible adiabatic process.

84a. The characteristic feature of this new distribution of energy may be stated as follows: If we denote all quantities referring to the new state by the addition of an accent, we have the following equation in addition to (99)

$$\mathsf{u}'=\nu'^3\phi\,(\nu'^3 V').$$

Therefore, if we put

$$\nu'^3 V'=\nu^3 V, \tag{99a}$$

we shall also have

$$\frac{\mathsf{u}'}{\nu'^3}=\frac{\mathsf{u}}{\nu^3} \text{ and } \mathsf{u}'V'=\mathsf{u}V, \tag{99b}$$

i.e., if we coordinate with every frequency ν in the original state that frequency ν' which is to ν in the inverse ratio of the cube roots of the respective volumes, the corresponding energy densities u' and u will be in the inverse ratio of the volumes.

The meaning of these relations will be more clearly seen, if we write

$$\frac{V'}{\lambda'^3}=\frac{V}{\lambda^3}$$

This is the number of the cubes of the wave lengths, which correspond to the frequency ν and are contained in the volume of the radiation. Moreover $\mathsf{u}d\nu V = \mathsf{U}d\nu$ denotes the radiant energy lying between the frequencies ν and $\nu + d\nu$, which is contained in the volume V. Now since, according to (99a),

$$\sqrt[3]{V'}\ d\nu' = \sqrt[3]{V}\ \ d\nu \text{ or } \frac{d\nu'}{\nu'} = \frac{d\nu}{\nu} \tag{99c}$$

we have, taking account of (99b),

$$\mathsf{U}'\ \frac{d\nu'}{\nu'} = \mathsf{U}\ \frac{d\nu}{\nu}$$

These results may be summarized thus: On an infinitely slow reversible adiabatic change in volume of radiation contained in a cavity and uniform in all directions, the frequencies change in such a way that the number of cubes of wave lengths of every frequency contained in the total volume remains unchanged, and the radiant energy of every infinitely small spectral interval changes in proportion to the frequency.

85. Returning now to the discussion of Sec. 73 we introduce the assumption that at first the spectral distribution of energy is the normal one, corresponding to black radiation. Then, according to the law there proven, the radiation retains this property without change during a reversible adiabatic change of volume and the laws derived in Sec. 68 hold for the process. The radiation then possesses in every state a definite temperature T, which depends on the volume V according to the equation derived in that paragraph,

$$T^3 V = \text{const.} = T'^3 V'. \tag{100}$$

Hence we may now write equation (99) as follows:

$$\mathsf{u} = \nu^3 \phi\left(\frac{\nu^3}{T^3}\right)$$

or

$$\mathsf{u} = \nu^3 \phi\left(\frac{T}{\nu}\right)$$

Therefore, if for a single temperature the spectral distribution of black radiation, *i.e.*, u as a function of ν, is known, the dependence of the function ϕ on its argument, and hence the spectral distribution for any other temperature, may be deduced therefrom.

If we also take into account the law proved in Sec. 47, that, for the black radiation of a definite temperature, the product $\mathsf{u}q^3$ has for all media the same value, we may also write

$$\mathsf{u}=\frac{\nu^3}{c^3}F\left(\frac{T}{\nu}\right) \tag{101}$$

where now the function F no longer contains the velocity of propagation.

86. For the total radiation density in space of the black radiation in the vacuum we find

$$u=\int_0^\infty \mathsf{u}\,d\nu=\frac{1}{c^3}\int_0^\infty \nu^3 F\left(\frac{T}{\nu}\right)d\nu, \tag{102}$$

or, on introducing $\frac{T}{\nu}=x$ as the variable of integration instead of ν,

$$u=\frac{T^4}{c^3}\int_0^\infty \frac{F(x)}{x^5}dx. \tag{103}$$

If we let the absolute constant

$$\frac{1}{c^3}\int_0^\infty \frac{F(x)}{x^5}dx=a \tag{104}$$

the equation reduces to the form of the *Stefan-Boltzmann* law of radiation expressed in equation (75).

87. If we combine equation (100) with equation (99a) we obtain

$$\frac{\nu'}{T'}=\frac{\nu}{T} \tag{105}$$

Hence the laws derived at the end of Sec. 84a assume the following form: On infinitely slow reversible adiabatic change in volume of black radiation contained in a cavity, the temperature T varies in the inverse ratio of the cube root of the volume V, the frequencies ν vary in proportion to the temperature, and the radiant energy $\mathsf{U}d\nu$ of an infinitely small spectral interval varies in the same ratio. Hence the total radiant energy U as the sum of the energies of all spectral intervals varies also in proportion to the temperature, a statement which agrees with the conclusion arrived at already at the end of Sec. 68, while the space density of radiation, $u=\frac{U}{V}$, varies in proportion to the fourth power of the temperature, in agreement with the *Stefan-Boltzmann* law.

88. *Wien's* displacement law may also in the case of black radiation be stated for the specific intensity of radiation K_ν of a plane polarized monochromatic ray. In this form it reads according to (24)

$$\mathsf{K}_\nu = \frac{\nu^3}{c^3} F\left(\frac{T}{\nu}\right) \tag{106}$$

If, as is usually done in experimental physics, the radiation intensity is referred to wave lengths λ instead of frequencies ν, according to (16), namely

$$\mathsf{E}_\lambda = \frac{c\mathsf{K}_\nu}{\lambda^2}$$

equation (106) takes the following form:

$$E_\lambda = \frac{c^2}{\lambda^5} F\left(\frac{\lambda T}{c}\right). \tag{107}$$

This form of *Wien's* displacement law has usually been the starting-point for an experimental test, the result of which has in all cases been a fairly accurate verification of the law.[1]

89. Since E_λ vanishes for $\lambda=0$ as well as for $\lambda=\infty$, E_λ must have a maximum with respect to λ, which is found from the equation

$$\frac{dE_\lambda}{d\lambda} = 0 = -\frac{5}{\lambda^6} F\left(\frac{\lambda T}{c}\right) + \frac{1}{\lambda^5} \frac{T}{c} \dot{F}\left(\frac{\lambda T}{c}\right)$$

where $\dot{F}$ denotes the differential coefficient of F with respect to its argument. Or

$$\frac{\lambda T}{c} \dot{F}\left(\frac{\lambda T}{c}\right) - 5F\left(\frac{\lambda T}{c}\right) = 0. \tag{108}$$

This equation furnishes a definite value for the argument $\frac{\lambda T}{c}$, so

[1] *E.g.*, *F. Paschen*, Sitzungsber. d. Akad. d. Wissensch. Berlin, pp. 405 and 959, 1899. *O. Lummer* und *E. Pringsheim*, Verhandlungen d. Deutschen physikalischen Gesellschaft **1**, pp. 23 and 215, 1899. Annal. d. Physik **6**, p. 192, 1901.

that for the wave length λ_m corresponding to the maximum of the radiation intensity E_λ the relation holds

$$\lambda_m T = b. \tag{109}$$

With increasing temperature the maximum of radiation is therefore displaced in the direction of the shorter wave lengths.

The numerical value of the constant b as determined by *Lummer* and *Pringsheim*[1] is

$$b = 0.294 \text{ cm. degree.} \tag{110}$$

Paschen[2] has found a slightly smaller value, about 0.292.

We may emphasize again at this point that, according to Sec. 19, the maximum of E_λ does not by any means occur at the same point in the spectrum as the maximum of K_ν and that hence the significance of the constant b is essentially dependent on the fact that the intensity of monochromatic radiation is referred to wave lengths, not to frequencies.

90. The value also of the maximum of E_λ is found from (107) by putting $\lambda = \lambda_m$. Allowing for (109) we obtain

$$E_{max} = \text{const. } T^5, \tag{111}$$

i.e., the value of the maximum of radiation in the spectrum of the black radiation is proportional to the fifth power of the absolute temperature.

Should we measure the intensity of monochromatic radiation not by E_λ but by K_ν, we would obtain for the value of the radiation maximum a quite different law, namely,

$$\mathsf{K}_{max} = \text{const. } T^3. \tag{112}$$

[1] *O. Lummer* und *E. Pringsheim*, l. c.

[2] *F. Paschen*, Annal. d. Physik, **6**, p. 657, 1901.

CHAPTER IV

RADIATION OF ANY ARBITRARY SPECTRAL DISTRIBUTION OF ENERGY. ENTROPY AND TEMPERATURE OF MONOCHROMATIC RADIATION

91. We have so far applied *Wien's* displacement law only to the case of black radiation; it has, however, a much more general importance. For equation (95), as has already been stated, gives, for any original spectral distribution of the energy radiation contained in the evacuated cavity and radiated uniformly in all directions, the change of this energy distribution accompanying a reversible adiabatic change of the total volume. Every state of radiation brought about by such a process is perfectly stationary and can continue infinitely long, subject, however, to the condition that no trace of an emitting or absorbing substance exists in the radiation space. For otherwise, according to Sec. 51, the distribution of energy would, in the course of time, change through the releasing action of the substance irreversibly, *i.e.*, with an increase of the total entropy, into the stable distribution correponding to black radiation.

The difference of this general case from the special one dealt with in the preceding chapter is that we can no longer, as in the case of black radiation, speak of a definite temperature of the radiation. Nevertheless, since the second principle of thermodynamics is supposed to hold quite generally, the radiation, like every physical system which is in a definite state, has a definite entropy, $S=Vs$. This entropy consists of the entropies of the monochromatic radiations, and, since the separate kinds of rays are independent of one another, may be obtained by addition. Hence

$$s=\int_0^\infty \mathfrak{s}\,d\nu, \qquad S=V\int_0^\infty \mathfrak{s}\,d\nu, \tag{113}$$

where $\mathfrak{s}\,d\nu$ denotes the entropy of the radiation of frequencies between ν and $\nu+d\nu$ contained in unit volume. $\mathfrak{s}$ is a definite function of the two independent variables ν and u and in the following will always be treated as such.

92. If the analytical expression of the function s were known, the law of energy distribution in the normal spectrum could immediately be deduced from it; for the normal spectral distribution of energy or that of black radiation is distinguished from all others by the fact that it has the maximum of the entropy of radiation S.

Suppose then we take s to be a known function of ν and u. Then as a condition for black radiation we have

$$\delta S = 0, \tag{114}$$

for any variations of energy distribution, which are possible with a constant total volume V and constant total energy of radiation U. Let the variation of energy distribution be characterized by making an infinitely small change $\delta\mathsf{u}$ in the energy u of every separate definite frequency ν. Then we have as fixed conditions

$$\delta V = 0 \text{ and } \int_0^\infty \delta\mathsf{u}\, d\nu = 0. \tag{115}$$

The changes d and δ are of course quite independent of each other.

Now since $\delta V = 0$, we have from (114) and (113)

$$\int_0^\infty \delta\mathsf{s}\, d\nu = 0,$$

or, since ν remains unvaried

$$\int_0^\infty \frac{\partial\mathsf{s}}{\partial\mathsf{u}} \delta\mathsf{u}\, d\nu = 0,$$

and, by allowing for (115), the validity of this equation for all values of $\delta\mathsf{u}$ whatever requires that

$$\frac{\partial\mathsf{s}}{\partial\mathsf{u}} = \text{const.} \tag{116}$$

for all different frequencies. This equation states the law of energy distribution in the case of black radiation.

93. The constant of equation (116) bears a simple relation to the temperature of black radiation. For if the black radiation, by conduction into it of a certain amount of heat at constant volume V, undergoes an infinitely small change in energy δU, then, according to (73), its change in entropy is

$$\delta S = \frac{\delta U}{T} \cdot$$

However, from (113) and (116),

$$\delta S = V \int_0^\infty \frac{\partial \mathfrak{s}}{\partial \mathfrak{u}} \, \delta \mathfrak{u} \, d\nu = \frac{\partial \mathfrak{s}}{\partial \mathfrak{u}} V \int_0^\infty \delta \mathfrak{u} \, d\nu = \frac{\partial \mathfrak{s}}{\partial \mathfrak{u}} \delta U$$

hence

$$\frac{\partial \mathfrak{s}}{\partial \mathfrak{u}} = \frac{1}{T} \tag{117}$$

and the above quantity, which was found to be the same for all frequencies in the case of black radiation, is shown to be the reciprocal of the temperature of black radiation.

Through this law the concept of temperature gains significance also for radiation of a quite arbitrary distribution of energy. For since $\mathfrak{s}$ depends only on $\mathfrak{u}$ and ν, *monochromatic radiation, which is uniform in all directions and has a definite energy density* $\mathfrak{u}$, *has also a definite temperature given by* (117), *and, among all conceivable distributions of energy, the normal one is characterized by the fact that the radiations of all frequencies have the same temperature.*

Any change in the energy distribution consists of a passage of energy from one monochromatic radiation into another, and, if the temperature of the first radiation is higher, the energy transformation causes an increase of the total entropy and is hence possible in nature without compensation; on the other hand, if the temperature of the second radiation is higher, the total entropy decreases and therefore the change is impossible in nature, unless compensation occurs simultaneously, just as is the case with the transfer of heat between two bodies of different temperatures.

94. Let us now investigate *Wien's* displacement law with regard to the dependence of the quantity $\mathfrak{s}$ on the variables $\mathfrak{u}$ and ν.

From equation (101) it follows, on solving for T and substituting the value given in (117), that

$$\frac{1}{T}=\frac{1}{\nu}F\left(\frac{c^3\mathsf{u}}{\nu^3}\right)=\frac{\partial \mathsf{s}}{\partial \mathsf{u}} \tag{118}$$

where again F represents a function of a single argument and the constants do not contain the velocity of propagation c. On integration with respect to the argument we obtain

$$\mathsf{s}=\frac{\nu^2}{c^3}F_1\left(\frac{c^3\mathsf{u}}{\nu^3}\right) \tag{119}$$

the notation remaining the same. In this form *Wien*'s displacement law has a significance for every separate monochromatic radiation and hence also for radiations of any arbitrary energy distribution.

95. According to the second principle of thermodynamics, the total entropy of radiation of quite arbitrary distribution of energy must remain constant on adiabatic reversible compression. We are now able to give a direct proof of this proposition on the basis of equation (119). For such a process, according to equation (113), the relation holds:

$$\delta S=\int_0^\infty d\nu(V\delta\mathsf{s}+\mathsf{s}\delta V)$$

$$=\int_0^\infty d\nu\left(V\frac{\partial \mathsf{s}}{\partial \mathsf{u}}\delta\mathsf{u}+\mathsf{s}\delta V\right)\cdot \tag{120}$$

Here, as everywhere, s should be regarded as a function of u and ν, and $\delta\nu=0$.

Now for a reversible adiabatic change of state the relation (95) holds. Let us take from the latter the value of $\delta\mathsf{u}$ and substitute. Then we have

$$\delta S=\delta V\int_0^\infty d\nu\left\{\frac{\partial \mathsf{s}}{\partial \mathsf{u}}\left(\frac{\nu}{3}\frac{d\mathsf{u}}{d\nu}-\mathsf{u}\right)+\mathsf{s}\right\}.$$

In this equation the differential coefficient of u with respect to ν refers to the spectral distribution of energy originally assigned arbitrarily and is therefore, in contrast to the partial differential coefficients, denoted by the letter d.

Now the complete differential is:

$$\frac{ds}{d\nu}=\frac{\partial s}{\partial u}\frac{du}{d\nu}+\frac{\partial s}{\partial \nu}$$

Hence by substitution:

$$\delta S=\delta V\int_0^\infty d\nu\left\{\frac{\nu}{3}\left(\frac{ds}{d\nu}-\frac{\partial s}{\partial \nu}\right)-u\frac{\partial s}{\partial u}+s\right\}. \qquad (121)$$

But from equation (119) we obtain by differentiation

$$\frac{\partial s}{\partial u}=\frac{1}{\nu}\dot{F}\left(\frac{c^3u}{\nu^3}\right) \text{ and } \frac{\partial s}{\partial \nu}=\frac{2\nu}{c^3}F\left(\frac{c^3u}{\nu^3}\right)-\frac{3u}{\nu^2}\dot{F}\left(\frac{c^3u}{\nu^3}\right) \qquad (122)$$

Hence

$$\frac{\nu\partial s}{\partial \nu}=2s-3u\frac{\partial s}{\partial u} \qquad (123)$$

On substituting this in (121), we obtain

$$\delta S=\delta V\int_0^\infty d\nu\left(\frac{\nu}{3}\frac{ds}{d\nu}+\frac{1}{3}s\right) \qquad (124)$$

or,

$$\delta S=\frac{\delta V}{3}[\nu s]_o^\infty=0,$$

as it should be. That the product νs vanishes also for $\nu=\infty$ may be shown just as was done in Sec. 83 for the product νu.

96. By means of equations (118) and (119) it is possible to give to the laws of reversible adiabatic compression a form in which their meaning is more clearly seen and which is the generalization of the laws stated in Sec. 87 for black radiation and a supplement to them. It is, namely, possible to derive (105) again from (118) and (99b). Hence the laws deduced in Sec. 87 for the change of frequency and temperature of the monochromatic radiation energy remain valid for a radiation of an originally quite arbitrary distribution of energy. The only difference as compared with the black radiation consists in the fact that now every frequency has its own distinct temperature.

Moreover it follows from (119) and (99b) that

$$\frac{s'}{\nu'^2}=\frac{s}{\nu^2} \qquad (125)$$

Now $\mathsf{s}d\nu V=\mathsf{S}d\nu$ denotes the radiation entropy between the frequencies ν and $\nu+d\nu$ contained in the volume V. Hence on account of (125), (99a), and (99c)

$$\mathsf{S}'d\nu'=\mathsf{S}d\nu, \tag{126}$$

i.e., the radiation entropy of an infinitely small spectral interval remains constant. This is another statement of the fact that the total entropy of radiation, taken as the sum of the entropies of all monochromatic radiations contained therein, remains constant.

97. We may go one step further, and, from the entropy s and the temperature T of an unpolarized monochromatic radiation which is uniform in all directions, draw a certain conclusion regarding the entropy and temperature of a single, plane polarized, monochromatic pencil. That every separate pencil also has a certain entropy follows by the second principle of thermodynamics from the phenomenon of emission. For since, by the act of emission, heat is changed into radiant heat, the entropy of the emitting body decreases during emission, and, along with this decrease, there must be, according to the principle of increase of the total entropy, an increase in a different form of entropy as a compensation. This can only be due to the energy of the emitted radiation. Hence every separate, plane polarized, monochromatic pencil has its definite entropy, which can depend only on its energy and frequency and which is propagated and spreads into space with it. We thus gain the idea of entropy radiation, which is measured, as in the analogous case of energy radiation, by the amount of entropy which passes in unit time through unit area in a definite direction. Hence statements, exactly similar to those made in Sec. 14 regarding energy radiation, will hold for the radiation of entropy, inasmuch as every pencil possesses and conveys, not only its energy, but also its entropy. Referring the reader to the discussions of Sec. 14, we shall, for the present, merely enumerate the most important laws for future use.

98. In a space filled with any radiation whatever the entropy radiated in the time dt through an element of area $d\sigma$ in the direction of the conical element $d\Omega$ is given by an expression of the form

$$dt\, d\sigma \cos\theta\, d\Omega\, L=L \sin\theta \cos\theta\, d\theta\, d\phi\, d\sigma\, dt. \tag{127}$$

The positive quantity L we shall call the "specific intensity of entropy radiation" at the position of the element of area $d\sigma$ in the direction of the solid angle $d\Omega$. L is, in general, a function of position, time, and direction.

The total radiation of entropy through the element of area $d\sigma$ toward one side, say the one where θ is an acute angle, is obtained by integration with respect to ϕ from 0 to 2π and with respect to θ from 0 to $\frac{\pi}{2}$. It is

$$d\sigma\, dt \int_0^{2\pi} d\phi \int_0^{\frac{\pi}{2}} d\theta\, L \sin\theta \cos\theta.$$

When the radiation is uniform in all directions, and hence L constant, the entropy radiation through $d\sigma$ toward one side is

$$\pi L\, d\sigma\, dt. \tag{128}$$

The specific intensity L of the entropy radiation in every direction consists further of the intensities of the separate rays belonging to the different regions of the spectrum, which are propagated independently of one another. Finally for a ray of definite color and intensity the nature of its polarization is characteristic. When a monochromatic ray of frequency ν consists of two mutually independent[1] components, polarized at right angles to each other, with the principal intensities of energy radiation (Sec. 17) K_ν and K_ν', the specific intensity of entropy radiation is of the form

$$L = \int_0^\infty d\nu (\mathsf{L}_\nu + \mathsf{L}'_\nu). \tag{129}$$

The positive quantities L_ν and L'_ν in this expression, the principal intensities of entropy radiation of frequency ν, are determined by the values of K_ν and K_ν'. By substitution in (127), this gives for the entropy which is radiated in the time

[1] "Independent" in the sense of "noncoherent." If, *e.g.*, a ray with the principal intensities K and K' is elliptically polarized, its entropy is not equal to $\mathsf{L} + \mathsf{L}'$, but equal to the entropy of a plane polarized ray of intensity $\mathsf{K} + \mathsf{K}'$. For an elliptically polarized ray may be transformed at once into a plane polarized one, *e.g.*, by total reflection. For the entropy of a ray with coherent components see below Sec. 104, *et seq.*

dt through the element of area $d\sigma$ in the direction of the conical element $d\Omega$ the expression

$$dt\, d\sigma \cos\theta\, d\Omega \int_0^\infty d\nu (\mathsf{L}_\nu + \mathsf{L}'_\nu),$$

and, for monochromatic plane polarized radiation,

$$dt\, d\sigma \cos\theta\, d\Omega\, \mathsf{L}_\nu\, d\nu = \mathsf{L}_\nu\, d\nu \sin\theta \cos\theta\, d\theta\, d\phi\, d\sigma\, dt. \quad (130)$$

For unpolarized rays $\mathsf{L}_\nu = \mathsf{L}'_\nu$ and (129) becomes

$$L = 2\int_0^\infty \mathsf{L}_\nu\, d\nu.$$

For radiation which is uniform in all directions the total entropy radiation toward one side is, according to (128),

$$2\pi\, d\sigma\, dt \int_0^\infty \mathsf{L}_\nu\, d\nu.$$

99. From the intensity of the propagated entropy radiation the expression for the *space density* of the radiant entropy may also be obtained, just as the space density of the radiant energy follows from the intensity of the propagated radiant energy. (Compare Sec. 22.) In fact, in analogy with equation (20), the space density, s, of the entropy of radiation at any point in a vacuum is

$$s = \frac{1}{c}\int L\, d\Omega, \quad (131)$$

where the integration is to be extended over the conical elements which spread out from the point in question in all directions. L is constant for uniform radiation and we obtain

$$s = \frac{4\pi L}{c} \quad (132)$$

By spectral resolution of the quantity L, according to equation (129), we obtain from (131) also the space density of the monochromatic radiation entropy:

$$\mathsf{s} = \frac{1}{c}\int (\mathsf{L} + \mathsf{L}')\, d\Omega,$$

and for unpolarized radiation, which is uniform in all directions

$$\mathsf{s} = \frac{8\pi \mathsf{L}}{c} \quad (133)$$

100. As to how the entropy radiation L depends on the energy radiation K *Wien's* displacement law in the form of (119) affords immediate information. It follows, namely, from it, considering (133) and (24), that

$$\mathsf{L}=\frac{\nu^2}{c^2}F\left(\frac{c^2\mathsf{K}}{\nu^3}\right) \tag{134}$$

and, moreover, on taking into account (118),

$$\frac{\partial \mathsf{L}}{\partial \mathsf{K}}=\frac{\partial \mathsf{s}}{\partial \mathsf{u}}=\frac{1}{T} \tag{135}$$

Hence also

$$T=\nu F_1\left(\frac{c^2\mathsf{K}}{\nu^3}\right) \tag{136}$$

or

$$\mathsf{K}=\frac{\nu^3}{c^2}F_2\left(\frac{T}{\nu}\right) \tag{137}$$

It is true that these relations, like the equations (118) and (119), were originally derived for radiation which is unpolarized and uniform in all directions. They hold, however, generally in the case of any radiation whatever for each separate monochromatic plane polarized ray. For, since the separate rays behave and are propagated quite independently of one another, the intensity, L, of the entropy radiation of a ray can depend only on the intensity of the energy radiation, K, of the same ray. Hence every separate monochromatic ray has not only its energy but also its entropy defined by (134) and its temperature defined by (136).

101. The extension of the conception of temperature to a single monochromatic ray, just discussed, implies that at the same point in a medium, through which any rays whatever pass, there exist in general an infinite number of temperatures, since every ray passing through the point has its separate temperature, and, moreover, even the rays of different color traveling in the same direction show temperatures that differ according to the spectral distribution of energy. In addition to all these temperatures there is finally the temperature of the medium itself, which at the outset is entirely independent of the temperature of the radiation. This complicated method of consideration lies in the nature of the case and corresponds to the complexity of the physical processes in a medium through which radiation travels in such a way. It is only in the case of stable thermodynamic equilibrium that there is but one temperature, which then is common to the medium itself and to all rays of whatever color crossing it in different directions.

In practical physics also the necessity of separating the conception of radiation temperature from that of body temperature has made itself felt to a continually increasing degree. Thus it has for some time past been found advantageous to speak, not only of the real temperature of the sun, but also of an "apparent" or "effective" temperature of the sun, *i.e.*, that temperature which the sun would need to have in order to send to the earth the heat radiation actually observed, if it radiated like a black body. Now the apparent temperature of the sun is obviously nothing but the actual temperature of the solar rays,[1] depending entirely on the nature of the rays, and hence a property of the rays and not a property of the sun itself. Therefore it would be, not only more convenient, but also more correct, to apply this notation directly, instead of speaking of a fictitious temperature of the sun, which can be made to have a meaning only by the introduction of an assumption that does not hold in reality.

Measurements of the brightness of monochromatic light have recently led *L. Holborn* and *F. Kurlbaum*[2] to the introduction of the concept of "black" temperature of a radiating surface. The black temperature of a radiating surface is measured by the brightness of the rays which it emits. It is in general a separate one for each ray of definite color, direction, and polarization, which the surface emits, and, in fact, merely represents the temperature of such a ray. It is, according to equation (136), determined by its brightness (specific intensity), K, and its frequency, ν, without any reference to its origin and previous states. The definite numerical form of this equation will be given below in Sec. 166. Since a black body has the maximum emissive power, the temperature of an emitted ray can never be higher than that of the emitting body.

[1] On the average, since the solar rays of different color do not have exactly the same temperature.

[2] *L. Holborn* und *F. Kurlbaum*, Annal. d. Physik., **10**, p. 229, 1903.

102. Let us make one more simple application of the laws just found to the special case of black radiation. For this, according to (81), the total space density of entropy is

$$s = \frac{4}{3} a^3 T. \tag{138}$$

Hence, according to (132), the specific intensity of the total entropy radiation in any direction is

$$L = \frac{c}{3\pi} a T^3, \tag{139}$$

and the total entropy radiation through an element of area $d\sigma$ toward one side is, according to (128),

$$\frac{c}{3} a T^3 d\sigma\, dt. \tag{140}$$

As a special example we shall now apply the two principles of thermodynamics to the case in which the surface of a black body of temperature T and of infinitely large heat capacity is struck by black radiation of temperature T' coming from all directions. Then, according to (7) and (76), the black body emits per unit area and unit time the energy

$$\pi K = \frac{ac}{4} T^4,$$

and, according to (140), the entropy

$$\frac{ac}{3} T^3.$$

On the other hand, it absorbs the energy

$$\frac{ac}{4} T'^4$$

and the entropy

$$\frac{ac}{3} T'^3.$$

Hence, according to the first principle, the total heat added to the body, positive or negative according as T' is larger or smaller than T, is

$$Q = \frac{ac}{4} T'^4 - \frac{ac}{4} T^4 = \frac{ac}{4} (T'^4 - T^4),$$

7

and, according to the second principle, the change of the entire entropy is positive or zero. Now the entropy of the body changes by $\frac{Q}{T}$, the entropy of the radiation in the vacuum by

$$\frac{ac}{3}(T^3-T'^3).$$

Hence the change per unit time and unit area of the entire entropy of the system considered is

$$\frac{ac}{4}\,\frac{T'^4-T^4}{T}+\frac{ac}{3}(T^3-T'^3)\geq 0.$$

In fact this relation is satisfied for all values of T and T'. The minimum value of the expression on the left side is zero; this value is reached when $T=T'$. In that case the process is reversible. If, however, T differs from T', we have an appreciable increase of entropy; hence the process is irreversible. In particular we find that if $T=0$ the increase in entropy is ∞, *i.e.*, the absorption of heat radiation by a black body of vanishingly small temperature is accompanied by an infinite increase in entropy and cannot therefore be reversed by any finite compensation. On the other hand for $T'=0$, the increase in entropy is only equal to $\frac{ac}{12}T^3$, *i.e.*, the emission of a black body of temperature T without simultaneous absorption of heat radiation is irreversible without compensation, but can be reversed by a compensation of at least the stated finite amount. For example, if we let the rays emitted by the body fall back on it, say by suitable reflection, the body, while again absorbing these rays, will necessarily be at the same time emitting new rays, and this is the compensation required by the second principle.

Generally we may say: Emission without simultaneous absorption is irreversible, while the opposite process, absorption without emission, is impossible in nature.

103. A further example of the application of the two principles of thermodynamics is afforded by the irreversible expansion of originally black radiation of volume V and temperature T to the larger volume V' as considered above in Sec. 70, but in the absence of any absorbing or emitting substance whatever. Then not only the total energy but also the energy of every separate frequency ν remains constant; hence, when on account of diffuse reflection from the walls the radiation has again become uniform in all directions, $\mathsf{u}_\nu V = \mathsf{u}'_\nu V'$; moreover by this relation, according to (118), the temperature T'_ν of the monochromatic radiation of frequency ν in the final state is determined. The actual calculation, however, can be performed only with the help of equation (275) (see below). The total entropy of radiation, *i.e.*, the sum of the entropies of the radiations of all frequencies,

$$V' \int_0^\infty \mathsf{s}'_\nu \, d\nu,$$

must, according to the second principle, be larger in the final state than in the original state. Since T'_ν has different values for the different frequencies ν, the final radiation is no longer black. Hence, on subsequent introduction of a carbon particle into the cavity, a finite change of the distribution of energy is obtained, and simultaneously the entropy increases further to the value S' calculated in (82).

104. In Sec. 98 we have found the intensity of entropy radiation of a definite frequency in a definite direction by adding the entropy radiations of the two independent components K and K', polarized at right angles to each other, or

$$\mathsf{L}(\mathsf{K}) + \mathsf{L}(\mathsf{K}'), \tag{141}$$

where L denotes the function of K given in equation (134). This method of procedure is based on the general law that the entropy of two mutually independent physical systems is equal to the sum of the entropies of the separate systems.

If, however, the two components of a ray, polarized at right angles to each other, are not independent of each other, this method of procedure no longer remains correct. This may be seen, *e.g.*, on resolving the radiation intensity, not with reference to the two principal planes of polarization with the principal intensities K and K', but with reference to any other two planes at right angles to each other, where, according to equation (8), the intensities of the two components assume the following values

$$\begin{aligned} \mathsf{K}\cos^2\psi + \mathsf{K}'\sin^2\psi &= \mathsf{K}'' \\ \mathsf{K}\sin^2\psi + \mathsf{K}'\cos^2\psi &= \mathsf{K}'''. \end{aligned} \tag{142}$$

In that case, of course, the entropy radiation is not equal to $\mathsf{L}(\mathsf{K}'')+\mathsf{L}(\mathsf{K}''')$.

Thus, while the energy radiation is always obtained by the summation of any two components which are polarized at right angles to each other, no matter according to which azimuth the resolution is performed, since always

$$\mathsf{K}''+\mathsf{K}'''=\mathsf{K}+\mathsf{K}', \tag{143}$$

a corresponding equation does not hold in general for the entropy radiation. The cause of this is that the two components, the intensities of which we have denoted by K'' and K''', are, unlike K and K', not independent or noncoherent in the optic sense. In such a case

$$\mathsf{L}(\mathsf{K}'')+\mathsf{L}(\mathsf{K}''')>\mathsf{L}(\mathsf{K})+\mathsf{L}(\mathsf{K}'), \tag{144}$$

as is shown by the following consideration.

Since in the state of thermodynamic equilibrium all rays of the same frequency have the same intensity of radiation, the intensities of radiation of any two plane polarized rays will tend to become equal, *i.e.*, the passage of energy between them will be accompanied by an increase of entropy, when it takes place in the direction from the ray of greater intensity toward that of smaller intensity. Now the left side of the inequality (144) represents the entropy radiation of two noncoherent plane polarized rays with the intensities K'' and K''', and the right side the entropy radiation of two noncoherent plane polarized rays with the intensities K and K'. But, according to (142), the values of K'' and K''' lie between K and K'; therefore the inequality (144) holds.

At the same time it is apparent that the error committed, when the entropy of two coherent rays is calculated as if they were noncoherent, is always in such a sense that the entropy found is too large. The radiations K'' and K''' are called "partially coherent," since they have some terms in common. In the special case when one of the two principal intensities K and K' vanishes entirely, the radiations K'' and K''' are said to be "completely coherent," since in that case the expression for one radiation may be completely reduced to that for the other. The entropy of two completely coherent plane polarized rays is equal to the entropy of a single plane polarized ray, the energy of which is equal to the sum of the two separate energies.

105. Let us for future use solve also the more general problem of calculating the entropy radiation of a ray consisting of an arbitrary number of plane polarized noncoherent components K_1, K_2, K_3, ... , the planes of vibration (planes of the electric vector) of which are given by the azimuths ψ_1, ψ_2, ψ_3, ... This problem amounts to finding the principal intensities K_0 and K_0' of the whole ray; for the ray behaves in every physical respect as if it consisted of the noncoherent components K_0 and K_0'. For this purpose we begin by establishing the value K_ψ of the component of the ray for an azimuth ψ taken arbitrarily. Denoting by f the electric vector of the ray in the direction ψ, we obtain this value K_ψ from the equation

$$f=f_1 \cos(\psi_1-\psi)+f_2 \cos(\psi_2-\psi)+f_3 \cos(\psi_3-\psi)+ \ldots \ldots,$$

where the terms on the right side denote the projections of the vectors of the separate components in the direction ψ, by squaring and averaging and taking into account the fact that $f_1, f_2, f_3, \ldots$ are noncoherent

$$\mathsf{K}_\psi=\mathsf{K}_1 \cos^2(\psi_1-\psi)+\mathsf{K}_2 \cos^2(\psi_2-\psi)+ \ldots \ldots$$

or

$$\mathsf{K}_\psi=A \cos^2\psi+B \sin^2\psi+C \sin\psi \cos\psi$$

where

$$\begin{aligned} A&=\mathsf{K}_1 \cos^2\psi_1+\mathsf{K}_2 \cos^2\psi_2+ \ldots \ldots \\ B&=\mathsf{K}_1 \sin^2\psi_1+\mathsf{K}_2 \sin^2\psi_2+ \ldots \ldots \\ C&=2(\mathsf{K}_1 \sin\psi_1 \cos\psi_1+\mathsf{K}_2 \sin\psi_2 \cos\psi_2+ \ldots \ldots). \end{aligned} \tag{145}$$

The principal intensities K_0 and K_0' of the ray follow from this expression as the maximum and the minimum value of K_ψ according to the equation

$$\frac{d\mathsf{K}_\psi}{d\psi}=0 \text{ or, } \tan 2\psi=\frac{C}{A-B}$$

Hence it follows that the principal intensities are

$$\left.\begin{matrix}\mathsf{K}_0\\ K_0'\end{matrix}\right\}=\tfrac{1}{2}\left(A+B\pm\sqrt{(A-B)^2+C^2}\right), \tag{146}$$

or, by taking (145) into account,

$$\left.\begin{matrix}\mathsf{K}_0\\ \mathsf{K}_0'\end{matrix}\right\}=\frac{1}{2}\Big(\mathsf{K}_1+\mathsf{K}_2+ \ldots$$
$$\pm\sqrt{(\mathsf{K}_1 \cos 2\psi_1+\mathsf{K}_2 \cos 2\psi_2+\ldots)^2+(\mathsf{K}_1 \operatorname{ins} 2\psi_1+\mathsf{K}_2 \sin 2\psi_2+\ldots)^2}.\Big) \tag{147}$$

Then the entropy radiation required becomes:

$$\mathsf{L}(\mathsf{K}_0)+\mathsf{L}(\mathsf{K}_0'). \tag{148}$$

106. When two ray components K and K', polarized at right angles to each other, are noncoherent, K and K' are also the principal intensities, and the entropy radiation is given by (141). The converse proposition, however, does not hold in general, that is to say, the two components of a ray polarized at right angles to each other, which correspond to the principal intensities K and K', are not necessarily noncoherent, and hence the entropy radiation is not always given by (141).

This is true, *e.g.*, in the case of elliptically polarized light. There the radiations K and K' are completely coherent and their entropy is equal to $\mathsf{L}(\mathsf{K}+\mathsf{K}')$. This is caused by the fact that it is possible to give the two ray components an arbitrary displacement of phase in a reversible manner, say by total reflection. Thereby it is possible to change elliptically polarized light to plane polarized light and *vice versa.*

The entropy of completely or partially coherent rays has been investigated most thoroughly by *M. Laue.*[1] For the significance of optical coherence for thermodynamic probability see the next part, Sec. 119.

[1] *M. Laue,* Annalen d. Phys., **23,** p. 1, 1907.

CHAPTER V

ELECTRODYNAMICAL PROCESSES IN A STATIONARY FIELD OF RADIATION

107. We shall now consider from the standpoint of pure electrodynamics the processes that take place in a vacuum, which is bounded on all sides by reflecting walls and through which heat radiation passes uniformly in all directions, and shall then inquire into the relations between the electrodynamical and the thermodynamic quantities.

The electrodynamical state of the field of radiation is determined at every instant by the values of the electric field-strength E and the magnetic field-strength H at every point in the field, and the changes in time of these two vectors are completely determined by *Maxwell's* field equations (52), which we have already used in Sec. 53, together with the boundary conditions, which hold at the reflecting walls. In the present case, however, we have to deal with a solution of these equations of much greater complexity than that expressed by (54), which corresponds to a plane wave. For a plane wave, even though it be periodic with a wave length lying within the optical or thermal spectrum, can never be interpreted as heat radiation. For, according to Sec. 16, a finite intensity K of heat radiation requires a finite solid angle of the rays and, according to Sec. 18, a spectral interval of finite width. But an absolutely plane, absolutely periodic wave has a zero solid angle and a zero spectral width. Hence in the case of a plane periodic wave there can be no question of either entropy or temperature of the radiation.

108. Let us proceed in a perfectly general way to consider the components of the field-strengths E and H as functions of the time at a definite point, which we may think of as the origin of the coordinate system. Of these components, which are produced by all rays passing through the origin, there are six; we select one of them, say E_z, for closer consideration. However complicated it may be, it may under all circumstances be written as a *Fourier's* series for a limited time interval, say from $t=0$ to $t=\mathsf{T}$; thus

$$\mathsf{E}_z = \sum_{n=1}^{n=\infty} C_n \cos\left(\frac{2\pi n t}{\mathsf{T}} - \theta_n\right) \tag{149}$$

where the summation is to extend over all positive integers n, while the constants C_n (positive) and θ_n may vary arbitrarily from term to term. The time interval T, the fundamental period of the *Fourier's* series, we shall choose so large that all times t which we shall consider hereafter are included in this time interval, so that $0<t<\mathsf{T}$. Then we may regard E_z as identical in all respects with the *Fourier's* series, *i.e.*, we may regard E_z as consisting of "partial vibrations," which are strictly periodic and of frequencies given by

$$\nu = \frac{n}{\mathsf{T}}$$

Since, according to Sec. 3, the time differential dt required for the definition of the intensity of a heat ray is necessarily large compared with the periods of vibration of all colors contained in the ray, a single time differential dt contains a large number of vibrations, *i.e.*, the product νdt is a large number. Then it follows *a fortiori* that νt and, still more,

$$\nu\mathsf{T} = n \text{ is enormously large} \tag{150}$$

for all values of ν entering into consideration. From this we must conclude that all amplitudes C_n with a moderately large value for the ordinal number n do not appear at all in the *Fourier's* series, that is to say, they are negligibly small.

109. Though we have no detailed special information about the function E_z, nevertheless its relation to the radiation of heat affords some important information as to a few of its general properties. Firstly, for the space density of radiation in a vacuum we have, according to Maxwell's theory,

$$u = \frac{1}{8\pi}\left(\overline{\mathsf{E}_x^2} + \overline{\mathsf{E}_y^2} + \overline{\mathsf{E}_z^2} + \overline{\mathsf{H}_x^2} + \overline{\mathsf{H}_y^2} + \overline{\mathsf{H}_z^2}\right).$$

Now the radiation is uniform in all directions and in the stationary state, hence the six mean values named are all equal to one another, and it follows that

$$u=\frac{3}{4\pi}\overline{\mathsf{E}_z^2}. \tag{151}$$

Let us substitute in this equation the value of E_z as given by (149). Squaring the latter and integrating term by term through a time interval, from 0 to t, assumed large in comparison with all periods of vibration $\frac{1}{\nu}$ but otherwise arbitrary, and then dividing by t, we obtain, since the radiation is perfectly stationary,

$$u=\frac{3}{8\pi}\sum C_n^2. \tag{152}$$

From this relation we may at once draw an important conclusion as to the nature of E_z as a function of time. Namely, since the *Fourier's* series (149) consists, as we have seen, of a great many terms, the squares, C_n^2, of the separate amplitudes of vibration the sum of which gives the space density of radiation, must have exceedingly small values. Moreover in the integral of the square of the *Fourier's* series the terms which depend on the time t and contain the products of any two different amplitudes all cancel; hence the amplitudes C_n and the phase-constants θ_n must vary from one ordinal number to another in a quite irregular manner. We may express this fact by saying that the separate partial vibrations of the series are very small and in a "chaotic"[1] state.

For the specific intensity of the radiation travelling in any direction whatever we obtain from (21)

$$K=\frac{cu}{4\pi}=\frac{3c}{32\pi^2}\sum C_n^2. \tag{153}$$

110. Let us now perform the spectral resolution of the last two equations. To begin with we have from (22):

$$u=\int_0^\infty \mathsf{u}_\nu d\nu=\frac{3}{8\pi}\sum_1^\infty C_n^2. \tag{154}$$

On the right side of the equation the sum $\sum$ consists of separate

[1] Compare footnote to page 116 (Tr.).

terms, every one of which corresponds to a separate ordinal number n and to a simple periodic partial vibration. Strictly speaking this sum does not represent a continuous sequence of frequencies ν, since n is an integral number. But n is, according to (150), so enormously large for all frequencies which need be considered that the frequencies ν corresponding to the successive values of n lie very close together. Hence the interval $d\nu$, though infinitesimal compared with ν, still contains a large number of partial vibrations, say n', where

$$d\nu = \frac{n'}{\mathsf{T}} \tag{155}$$

If now in (154) we equate, instead of the total energy densities, the energy densities corresponding to the interval $d\nu$ only, which are independent of those of the other spectral regions, we obtain

$$\mathsf{u}_\nu \, d\nu = \frac{3}{8\pi} \sum_n^{n+n'} C_n^2,$$

or, according to (155),

$$\mathsf{u}_\nu = \frac{3\mathsf{T}}{8\pi} \cdot \frac{1}{n'} \sum_n^{n+n'} C_n^2 = \frac{3\mathsf{T}}{8\pi} \cdot \overline{C_n^2}, \tag{156}$$

where we denote by $\overline{C_n^2}$ the average value of C_n^2 in the interval from n to $n+n'$. The existence of such an average value, the magnitude of which is independent of n, provided n' be taken small compared with n, is, of course, not self-evident at the outset, but is due to a special property of the function E_z which is peculiar to stationary heat radiation. On the other hand, since many terms contribute to the mean value, nothing can be said either about the magnitude of a separate term C_n^2, or about the connection of two consecutive terms, but they are to be regarded as perfectly independent of each other.

In a very similar manner, by making use of (24), we find for the specific intensity of a monochromatic plane polarized ray, travelling in any direction whatever,

$$\mathsf{K}_\nu = \frac{3c\mathsf{T}}{64\pi^2} \overline{C_n^2}. \tag{157}$$

From this it is apparent, among other things, that, according to the electromagnetic theory of radiation, a monochromatic light or heat ray is represented, not by a simple periodic wave, but by a superposition of a large number of simple periodic waves, the mean value of which constitutes the intensity of the ray. In accord with this is the fact, known from optics, that two rays of the same color and intensity but of different origin never interfere with each other, as they would, of necessity, if every ray were a simple periodic one.

Finally we shall also perform the spectral resolution of the mean value of E_z^2, by writing

$$\overline{\mathsf{E}_z^2} = J = \int_0^\infty \mathsf{J}_\nu d\nu \tag{158}$$

Then by comparison with (151), (154), and (156) we find

$$\mathsf{J}_\nu = \frac{4\pi}{3}\mathsf{u}_\nu = \frac{\mathsf{T}}{2}\overline{C_n^2} \tag{159}$$

According to (157), J_ν is related to K_ν, the specific intensity of radiation of a plane polarized ray, as follows:

$$\mathsf{K}_\nu = \frac{3c}{32\pi^2}\mathsf{J}_\nu. \tag{160}$$

111. Black radiation is frequently said to consist of a large number of regular periodic vibrations. This method of expression is perfectly justified, inasmuch as it refers to the resolution of the total vibration in a *Fourier's* series, according to equation (149), and often is exceedingly well adapted for convenience and clearness of discussion. It should, however, not mislead us into believing that such a "regularity" is caused by a special physical property of the elementary processes of vibration. For the resolvability into a Fourier's series is mathematically self-evident and hence, in a physical sense, tells us nothing new. In fact, it is even always possible to regard a vibration which is damped to an arbitrary extent as consisting of a sum of regular periodic partial vibrations with constant amplitudes and constant phases. On the contrary, it may just as correctly be said that in all nature there is no process more complicated than the vibrations of black radiation. In particular, these vibrations do not depend in any characteristic manner on the special processes that take place in the centers of emission of the rays, say on the period or the damping of the emitting particles; for the normal spectrum is distinguished from all other spectra by the very fact that all individual differences caused by the special nature of the emitting substances are perfectly equalized and effaced. Therefore to attempt to draw conclusions concerning the special properties of the particles emitting the rays from the elementary vibrations in the rays of the normal spectrum would be a hopeless undertaking.

In fact, black radiation may just as well be regarded as consisting, not of regular periodic vibrations, but of absolutely irregular separate impulses. The special regularities, which we observe in monochromatic light resolved spectrally, are caused merely by the special properties of the spectral apparatus used, *e.g.*, the dispersing prism (natural periods of the molecules), or the diffraction grating (width of the slits). Hence it is also incorrect to find a characteristic difference between light rays and Roentgen rays (the latter assumed as an electromagnetic process in a vacuum) in the circumstance that in the former the vibrations take place with greater regularity. Roentgen rays may, under certain conditions, possess more selective properties than light rays. The resolvability into a *Fourier's* series of partial vibrations with constant amplitudes and constant phases exists for both kinds of rays in precisely the same manner. What especially distinguishes light vibrations from Roentgen vibrations is the much smaller frequency of the partial vibrations of the former. To this is due the possibility of their spectral resolution, and probably also the far greater regularity of the changes of the radiation intensity in every region of the spectrum in the course of time, which, however, is not caused by a special property of the elementary processes of vibration, but merely by the constancy of the mean values.

112. The elementary processes of radiation exhibit regularities only when the vibrations are restricted to a narrow spectral region, that is to say in the case of spectroscopically resolved light, and especially in the case of the natural spectral lines. If, *e.g.*, the amplitudes C_n of the *Fourier's* series (149) differ from zero only between the ordinal numbers $n=n_0$ and $n=n_1$, where $\frac{n_1-n_0}{n_0}$ is small, we may write

$$\mathsf{E}_z=C_0 \cos\left(\frac{2\pi n_0 t}{\mathsf{T}}-\theta_0\right), \tag{161}$$

where

$$C_0 \cos \theta_0=\sum_{n_0}^{n_1} C_n \cos\left(\frac{2\pi(n-n_0)t}{\mathsf{T}}-\theta_n\right)$$

$$C_0 \sin \theta_0=-\sum_{n_0}^{n_1} C_n \sin\left(\frac{2\pi(n-n_0)t}{\mathsf{T}}-\theta_n\right)$$

and E_z may be regarded as a single approximately periodic vibration of frequency $\nu_0=\frac{n_0}{\mathsf{T}}$ with an amplitude C_0 and a phase-constant θ_0 which vary slowly and irregularly.

The smaller the spectral region, and accordingly the smaller $\frac{n_1-n_0}{n_0}$, the slower are the fluctuations ("Schwankungen") of C_o and θ_o, and the more regular is the resulting vibration and also the larger is the difference of path for which radiation can interfere with itself. If a spectral line were absolutely sharp, the radiation would have the property of being capable of interfering with itself for differences of path of any size whatever. This case, however, according to Sec. 18, is an ideal abstraction, never occurring in reality.

PART III

ENTROPY AND PROBABILITY

CHAPTER I

FUNDAMENTAL DEFINITIONS AND LAWS. HYPOTHESIS OF QUANTA

113. Since a wholly new element, entirely unrelated to the fundamental principles of electrodynamics, enters into the range of investigation with the introduction of probability considerations into the electrodynamic theory of heat radiation, the question arises at the outset, whether such considerations are justifiable and necessary. At first sight we might, in fact, be inclined to think that in a purely electrodynamical theory there would be no room at all for probability calculations. For since, as is well known, the electrodynamic equations of the field together with the initial and boundary conditions determine uniquely the way in which an electrodynamical process takes place, in the course of time, considerations which lie outside of the equations of the field would seem, theoretically speaking, to be uncalled for and in any case dispensable. For either they lead to the same results as the fundamental equations of electrodynamics and then they are superfluous, or they lead to different results and in this case they are wrong.

In spite of this apparently unavoidable dilemma, there is a flaw in the reasoning. For on closer consideration it is seen that what is understood in electrodynamics by "initial and boundary" conditions, as well as by the "way in which a process takes place in the course of time," is entirely different from what is denoted by the same words in thermodynamics. In order to make this evident, let us consider the case of radiation *in vacuo*, uniform in all directions, which was treated in the last chapter.

From the standpoint of thermodynamics the state of radiation is completely determined, when the intensity of monochromatic radiation K_ν is given for all frequencies ν. The electrodynamical observer, however, has gained very little by this single statement; because for him a knowledge of the state requires that every one of the six components of the electric and magnetic field-strength be given at all points of the space; and, while from the thermodynamic point of view the question as to the way in which the process takes place in time is settled by the constancy of the intensity of radiation K_ν, from the electrodynamical point of view it would be necessary to know the six components of the field at every point as functions of the time, and hence the amplitudes C_n and the phase-constants θ_n of all the several partial vibrations contained in the radiation would have to be calculated. This, however, is a problem whose solution is quite impossible, for the data obtainable from the measurements are by no means sufficient. The thermodynamically measurable quantities, looked at from the electrodynamical standpoint, represent only certain mean values, as we saw in the special case of stationary radiation in the last chapter.

We might now think that, since in thermodynamic measurements we are always concerned with mean values only, we need consider nothing beyond these mean values, and, therefore, need not take any account of the particular values at all. This method is, however, impracticable, because frequently and that too just in the most important cases, namely, in the cases of the processes of emission and absorption, we have to deal with mean values which cannot be calculated unambiguously by electrodynamical methods from the measured mean values. For example, the mean value of C_n cannot be calculated from the mean value of C_n^2, if no special information as to the particular values of C_n is available.

Thus we see that the electrodynamical state is not by any means determined by the thermodynamic data and that in cases where, according to the laws of thermodynamics and according to all experience, an unambiguous result is to be expected, a purely electrodynamical theory fails entirely, since it admits not one definite result, but an infinite number of different results.

114. Before entering on a further discussion of this fact and of the difficulty to which it leads in the electrodynamical theory of heat radiation, it may be pointed out that exactly the same case and the same difficulty are met with in the mechanical theory of heat, especially in the kinetic theory of gases. For when, for example, in the case of a gas flowing out of an opening at the time

$t=0$, the velocity, the density, and the temperature are given at every point, and the boundary conditions are completely known, we should expect, according to all experience, that these data would suffice for a unique determination of the way in which the process takes place in time. This, however, from a purely mechanical point of view is not the case at all; for the positions and velocities of all the separate molecules are not at all given by the visible velocity, density, and temperature of the gas, and they would have to be known exactly, if the way in which the process takes place in time had to be completely calculated from the equations of motion. In fact, it is easy to show that, with given initial values of the visible velocity, density, and temperature, an infinite number of entirely different processes is mechanically possible, some of which are in direct contradiction to the principles of thermodynamics, especially the second principle.

115. From these considerations we see that, if we wish to calculate the way in which a thermodynamic process takes place in time, such a formulation of initial and boundary conditions as is perfectly sufficient for a unique determination of the process in thermodynamics, does not suffice for the mechanical theory of heat or for the electrodynamical theory of heat radiation. On the contrary, from the standpoint of pure mechanics or electrodynamics the solutions of the problem are infinite in number. Hence, unless we wish to renounce entirely the possibility of representing the thermodynamic processes mechanically or electrodynamically, there remains only one way out of the difficulty, namely, to supplement the initial and boundary conditions by special hypotheses of such a nature that the mechanical or electrodynamical equations will lead to an unambiguous result in agreement with experience. As to how such an hypothesis is to be formulated, no hint can naturally be obtained from the principles of mechanics or electrodynamics, for they leave the question entirely open. Just on that account any mechanical or electrodynamical hypothesis containing some further specialization of the given initial and boundary conditions, which cannot be tested by direct measurement, is admissible *a priori*. What hypothesis is to be preferred can be decided only by testing the results to which it leads in the light of the thermodynamic principles based on experience.

116. Although, according to the statement just made, a decisive test of the different admissible hypotheses can be made only *a posteriori*, it is nevertheless worth while noticing that it is possible to obtain *a priori*, without relying in any way on thermodynamics, a definite hint as to the nature of an admissible hypothesis. Let us again consider a flowing gas as an illustration (Sec. 114). The mechanical state of all the separate gas molecules is not at all completely defined by the thermodynamic state of the gas, as has previously been pointed out. If, however, we consider all conceivable positions and velocities of the separate gas molecules, consistent with the given values of the visible velocity, density, and temperature, and calculate for every combination of them the mechanical process, assuming some simple law for the impact of two molecules, we shall arrive at processes, the vast majority of which agree completely in the mean values, though perhaps not in all details. Those cases, on the other hand, which show appreciable deviations, are vanishingly few, and only occur when certain very special and far-reaching conditions between the coordinates and velocity-components of the molecules are satisfied. Hence, if the assumption be made that such special conditions do not exist, however different the mechanical details may be in other respects, a form of flow of gas will be found, which may be called quite definite with respect to all measurable mean values—and they are the only ones which can be tested experimentally—although it will not, of course, be quite definite in all details. And the remarkable feature of this is that it is just the motion obtained in this manner that satisfies the postulates of the second principle of thermodynamics.

117. From these considerations it is evident that the hypotheses whose introduction was proven above to be necessary completely answer their purpose, if they state nothing more than that exceptional cases, corresponding to special conditions which exist between the separate quantities determining the state and which cannot be tested directly, do not occur in nature. In mechanics this is done by the hypothesis[1] that the heat motion is a "molecular chaos";[2] in electrodynamics the same thing is accomplished

[1] *L. Boltzmann*, Vorlesungen über Gastheorie **1,** p. 21, 1896. Wiener Sitzungsberichte **78,** Juni, 1878, at the end. Compare also *S. H. Burbury*, Nature, **51,** p. 78, 1894.

[2] Hereafter *Boltzmann's* "Unordnung" will be rendered by chaos, "ungeordnet" by chaotic (Tr.).

by the hypothesis of "natural radiation," which states that there exist between the numerous different partial vibrations (149) of a ray no other relations than those caused by the measurable mean values (compare below, Sec. 148). If, for brevity, we denote any condition or process for which such an hypothesis holds as an "elemental chaos," the principle, *that in nature any state or any process containing numerous elements not in themselves measurable is an elemental chaos*, furnishes the necessary condition for a unique determination of the measurable processes in mechanics as well as in electrodynamics and also for the validity of the second principle of thermodynamics. This must also serve as a mechanical or electrodynamical explanation of the conception of entropy, which is characteristic of the second law and of the closely allied concept of temperature.[1] It also follows from this that the significance of entropy and temperature is, according to their nature, connected with the condition of an elemental chaos. The terms entropy and temperature do not apply to a purely periodic, perfectly plane wave, since all the quantities in such a wave are in themselves measurable, and hence cannot be an elemental chaos any more than a single rigid atom in motion can. The necessary condition for the hypothesis of an elemental chaos and with it for the existence of entropy and temperature can consist only in the irregular simultaneous effect of very many partial vibrations of different periods, which are propagated in the different directions in space independent of one another, or in the irregular flight of a multitude of atoms.

118. But what mechanical or electrodynamical quantity represents the entropy of a state? It is evident that this quantity depends in some way on the "probability" of the state. For since an elemental chaos and the absence of a record of any individual element forms an essential feature of entropy, the tendency to neutralize any existing temperature differences, which is connected with an increase of entropy, can mean nothing for the mechanical or electrodynamical observer but that uniform

[1] To avoid misunderstanding I must emphasize that the question, whether the hypothesis of elemental chaos is really everywhere satisfied in nature, is not touched upon by the preceding considerations. I intended only to show at this point that, wherever this hypothesis does not hold, the natural processes, if viewed from the thermodynamic (macroscopic) point of view, do not take place unambiguously.

distribution of elements in a chaotic state is more probable than any other distribution.

Now since the concept of entropy as well as the second principle of thermodynamics are of universal application, and since on the other hand the laws of probability have no less universal validity, it is to be expected that the connection between entropy and probability should be very close. Hence we make the following proposition the foundation of our further discussion: *The entropy of a physical system in a definite state depends solely on the probability of this state.* The fertility of this law will be seen later in several cases. We shall not, however, attempt to give a strict general proof of it at this point. In fact, such an attempt evidently would have no meaning at this point. For, so long as the "probability" of a state is not numerically defined, the correctness of the proposition cannot be quantitatively tested. One might, in fact, suspect at first sight that on this account the proposition has no definite physical meaning. It may, however, be shown by a simple deduction that it is possible by means of this fundamental proposition to determine quite generally the way in which entropy depends on probability, without any further discussion of the probability of a state.

119. For let S be the entropy, W the probability of a physical system in a definite state; then the propositon states that

$$S=f(W) \tag{162}$$

where $f(W)$ represents a universal function of the argument W. In whatever way W may be defined, it can be safely inferred from the mathematical concept of probability that the probability of a system which consists of two entirely independent[1] systems is equal to the product of the probabilities of these two systems separately. If we think, *e.g.*, of the first system as any body whatever on the earth and of the second system as a cavity containing radiation on Sirius, then the probability that the terrestrial body be in a certain state 1 and that simultaneously the radiation in the cavity in a definite state 2 is

$$W=W_1W_2, \tag{163}$$

[1] It is well known that the condition that the two systems be independent of each other is essential for the validity of the expression (163). That it is also a necessary condition for the additive combination of the entropy was proven first by *M. Laue* in the case of optically coherent rays. Annalen d. Physik, **20**, p. 365, 1906.

where W_1 and W_2 are the probabilities that the systems involved are in the states in question.

If now S_1 and S_2 are the entropies of the separate systems in the two states, then, according to (162), we have

$$S_1=f(W_1) \qquad S_2=f(W_2).$$

But, according to the second principle of thermodynamics, the total entropy of the two systems, which are independent (see footnote to preceding page) of each other, is $S=S_1+S_2$ and hence from (162) and (163)

$$f(W_1W_2)=f(W_1)+f(W_2).$$

From this functional equation f can be determined. For on differentiating both sides with respect to W_1, W_2 remaining constant, we obtain

$$W_2\dot{f}(W_1W_2)=\dot{f}(W_1).$$

On further differentiating with respect to W_2, W_1 now remaining constant, we get

$$\dot{f}(W_1W_2)+W_1W_2\ddot{f}(W_1W_2)=0$$

or

$$\dot{f}(W)+W\ddot{f}(W)=0.$$

The general integral of this differential equation of the second order is

$$f(W)=k \log W+\text{ const.}$$

Hence from (162) we get

$$S=k \log W+\text{ const.},$$

an equation which determines the general way in which the entropy depends on the probability. The universal constant of integration k is the same for a terrestrial as for a cosmic system, and its value, having been determined for the former, will remain valid for the latter. The second additive constant of integration may, without any restriction as regards generality, be included as a constant multiplier in the quantity W, which here has not yet been completely defined, so that the equation reduces to

$$S=k \log W.$$

120. The logarithmic connection between entropy and probability was first stated by *L. Boltzmann*[1] in his kinetic theory of

[1] *L. Boltzmann*, Vorlesungen über Gastheorie, **1**, Sec. 6.

gases. Nevertheless our equation (164) differs in its meaning from the corresponding one of Boltzmann in two essential points.

Firstly, *Boltzmann's* equation lacks the factor k, which is due to the fact that *Boltzmann* always used gram-molecules, not the molecules themselves, in his calculations. Secondly, and this is of greater consequence, *Boltzmann* leaves an additive constant undetermined in the entropy S as is done in the whole of classical thermodynamics, and accordingly there is a constant factor of proportionality, which remains undetermined in the value of the probability W.

In contrast with this we assign a definite absolute value to the entropy S. This is a step of fundamental importance, which can be justified only by its consequences. As we shall see later, this step leads necessarily to the "hypothesis of quanta" and moreover it also leads, as regards radiant heat, to a definite law of distribution of energy of black radiation, and, as regards heat energy of bodies, to *Nernst's* heat theorem.

From (164) it follows that with the entropy S the probability W is, of course, also determined in the absolute sense. We shall designate the quantity W thus defined as the "thermodynamic probability," in contrast to the "mathematical probability," to which it is proportional but not equal. For, while the mathematical probability is a proper fraction, the thermodynamic probability is, as we shall see, always an integer.

121. The relation (164) contains a general method for calculating the entropy S by probability considerations. This, however, is of no practical value, unless the thermodynamic probability W of a system in a given state can be expressed numerically. The problem of finding the most general and most precise definition of this quantity is among the most important problems in the mechanical or electrodynamical theory of heat. It makes it necessary to discuss more fully what we mean by the "state" of a physical system.

By the state of a physical system at a certain time we mean the aggregate of all those mutually independent quantities, which determine uniquely the way in which the processes in the system take place in the course of time for given boundary conditions. Hence a knowledge of the state is precisely equivalent to a knowledge of the "initial conditions." If we now take into account the considerations stated above in Sec. 113, it is evident that we must distinguish in the theoretical treatment two entirely different kinds of states, which we may denote as "microscopic" and "macroscopic" states. The microscopic state is the state as described by a mechanical or electrodynamical observer; it contains the separate values of all coordinates, velocities, and field-strengths. The microscopic processes, according to the laws of mechanics and electrodynamics, take place in a perfectly unambiguous way; for them entropy and the second principle of thermodynamics have no significance. The macroscopic state, however, is the state as observed by a thermodynamic observer; any macroscopic state contains a large number of microscopic ones, which it unites in a mean value. Macroscopic processes take place in an unambiguous way in the sense of the second principle, when, and only when, the hypothesis of the elemental chaos (Sec. 117) is satisfied.

122. If now the calculation of the probability W of a state is in question, it is evident that the state is to be thought of in the macroscopic sense. The first and most important question is now: How is a macroscopic state defined? An answer to it will dispose of the main features of the whole problem.

For the sake of simplicity, let us first consider a special case, that of a very large number, N, of simple similar molecules. Let the problem be solely the distribution of these molecules in space within a given volume, V, irrespective of their velocities, and further the definition of a certain macroscopic distribution in space. The latter cannot consist of a statement of the coordinates of all the separate molecules, for that would be a definite microscopic distribution. We must, on the contrary, leave the positions of the molecules undetermined to a certain extent, and that can be done only by thinking of the whole volume V as being divided into a number of small but finite *space elements*, G, each containing a specified number of molecules. By any such statement a definite macroscopic distribution in space is defined. The manner in which the molecules are distributed within every separate space element is immaterial, for here the hypothesis of elemental chaos (Sec. 117) provides a supplement, which insures the unambiguity of the macroscopic state, in spite of the microscopic indefiniteness. If we distinguish the space elements in order by the numbers 1, 2, 3, ... and, for any particular macroscopic distribution in space, denote the number of the molecules lying in the separate space elements by N_1, N_2, N_3 . . . . ., then to every definite system of values N_1, N_2, N_3 . . . . ., there corresponds a definite macroscopic distribution in space. We have of course always:

$$N_1+N_2+N_3+ \cdots = N \tag{165}$$

or if

$$\frac{N_1}{N}=w_1 \quad \frac{N_2}{N}=w_2, \ldots \tag{166}$$

$$w_1+w_2+w_3+ \cdots =1. \tag{167}$$

The quantity w_i may be called the density of distribution of the molecules, or the mathematical probability that any molecule selected at random lies in the ith space element.

If we now had, *e.g.*, only 10 molecules and 7 space elements, a definite space distribution would be represented by the values:

$$N_1=1,\ N_2=2,\ N_3=0,\ N_4=0,\ N_5=1,\ N_6=4,\ N_7=2, \tag{168}$$

which state that in the seven space elements there lie respectively 1, 2, 0, 0, 1, 4, 2 molecules.

123. The definition of a macroscopic distribution in space may now be followed immediately by that of its thermodynamic probability W. The latter is founded on the consideration that a certain distribution in space may be realized in many different ways, namely, by many different individual coordinations or "complexions," according as a certain molecule considered will happen to lie in one or the other space element. For, with a given distribution of space, it is of consequence only how many, not which, molecules lie in every space element.

The number of all complexions which are possible with a given distribution in space we equate to the thermodynamic probability W of the space distribution.

In order to form a definite conception of a certain complexion, we can give the molecules numbers, write these numbers in order from 1 to N, and place below the number of every molecule the number of that space element to which the molecule in question belongs in that particular complexion. Thus the following table represents one particular complexion, selected at random, for the distribution in the preceding illustration

$$\begin{matrix} 1 & 2 & 3 & 4 & 5 & 6 & 7 & 8 & 9 & 10 \\ 6 & 1 & 7 & 5 & 6 & 2 & 2 & 6 & 6 & 7 \end{matrix} \qquad (169)$$

By this the fact is exhibited that the

Molecule 2 lies in space element 1.
Molecules 6 and 7 lie in space element 2.
Molecule 4 lies in space element 5.
Molecules 1, 5, 8, and 9 lie in space element 6.
Molecules 3 and 10 lie in space element 7.

As becomes evident on comparison with (168), this complexion does, in fact, correspond in every respect to the space distribution given above, and in a similar manner it is easy to exhibit many other complexions, which also belong to the same space distribution. The number of all possible complexions required is now easily found by inspecting the lower of the two lines of figures in (169). For, since the number of the molecules is given, this line of figures contains a definite number of places. Since, moreover, the distribution in space is also given, the number of times that every figure (*i.e.*, every space element) appears in the line is equal to the number of molecules which lie in that particular space element. But every change in the table gives a new particular coordination between molecules and space elements and hence a new complexion. Hence the number of the possible complexions, or the thermodynamic probability, W, of the given space distribution, is equal to the number of "permutations with repetition" possible under the given conditions. In the simple numerical example chosen, we get for W, according to a well-known formula, the expression

$$\frac{10!}{1!\,2!\,0!\,0!\,1!\,4!\,2!} = 37{,}800.$$

The form of this expression is so chosen that it may be applied easily to the general case. The numerator is equal to factorial N, N being the total number of molecules considered, and the denominator is equal to the product of the factorials of the numbers, N_1, N_2, N_3, ... of the molecules, which lie in every separate space element and which, in the general case, must be thought of as large numbers. Hence we obtain for the required probability of the given space distribution

$$W = \frac{N!}{N_1!\, N_2!\, N_3! \, \ldots \ldots} \tag{170}$$

Since all the N's are large numbers, we may apply to their factorials Stirling's formula, which for a large number may be abridged[1] to[2]

$$n! = \left(\frac{n}{e}\right)^n \tag{171}$$

Hence, by taking account of (165), we obtain

$$W = \left(\frac{N}{N_1}\right)^{N_1} \left(\frac{N}{N_2}\right)^{N_2} \left(\frac{N}{N_3}\right)^{N_3} \cdot \cdot \cdot \cdot \cdot \tag{172}$$

124. Exactly the same method as in the case of the space distribution just considered may be used for the definition of a macroscopic state and of the thermodynamic probability in the general case, where not only the coordinates but also the velocities, the electric moments, etc., of the molecules are to be dealt with. Every thermodynamic state of a system of N molecules is, in the macroscopic sense, defined by the statement of the number of molecules, N_1, N_2, N_3, ... , which are contained in the region elements 1, 2, 3, ... of the "state space." This state space, however, is not the ordinary three-dimensional space, but an ideal space of as many dimensions as there are variables for every molecule. In other respects the definition and the calculation of the thermodynamic probability W are exactly the same as above and the entropy of the state is accordingly found from (164), taking (166) also into account, to be

$$S = -kN \Sigma w_1 \log w_1, \tag{173}$$

where the sum Σ is to be taken over all region elements. It is obvious from this expression that the entropy is in every case a *positive quantity*.

125. By the preceding developments the calculation of the

[1] Abridged in the sense that factors which in the logarithmic expression (173) would give rise to small additive terms have been omitted at the outset. A brief derivation of equation (173) may be found on p. 218 (Tr.).

[2] See for example *E. Czuber*, Wahrscheinlichkeitsrechnung (Leipzig, B. G. Teubner) p. 22, 1903; *H. Poincaré*, Calcul des Probabilités (Paris, Gauthier-Villars), p. 85, 1912.

entropy of a system of N molecules in a given thermodynamic state is, in general, reduced to the single problem of finding the magnitude G of the region elements in the state space. That such a definite finite quantity really exists is a characteristic feature of the theory we are developing, as contrasted with that due to *Boltzmann*, and forms the content of the so-called *hypothesis of quanta*. As is readily seen, this is an immediate consequence of the proposition of Sec. 120 that the entropy S has an absolute, not merely a relative, value; for this, according to (164), necessitates also an absolute value for the magnitude of the thermodynamic probability W, which, in turn, according to Sec. 123, is dependent on the number of complexions, and hence also on the number and size of the region elements which are used. Since all different complexions contribute uniformly to the value of the probability W, the region elements of the state space represent also *regions of equal probability*. If this were not so, the complexions would not be all equally probable.

However, not only the magnitude, but also the shape and position of the region elements must be perfectly definite. For since, in general, the distribution density w is apt to vary appreciably from one region element to another, a change in the shape of a region element, the magnitude remaining unchanged, would, in general, lead to a change in the value of w and hence to a change in S. We shall see that only in special cases, namely, when the distribution densities w are very small, may the absolute magnitude of the region elements become physically unimportant, inasmuch as it enters into the entropy only through an additive constant. This happens, *e.g.*, at high temperatures, large volumes, slow vibrations (state of an ideal gas, Sec. 132, *Rayleigh's* radiation law, Sec. 195). Hence it is permissible for such limiting cases to assume, without appreciable error, that G is infinitely small in the macroscopic sense, as has hitherto been the practice in statistical mechanics. As soon, however, as the distribution densities w assume appreciable values, the classical statistical mechanics fail.

126. If now the problem be to determine the magnitude G of the region elements of equal probability, the laws of the classical statistical mechanics afford a certain hint, since in certain limiting cases they lead to correct results.

Let $\phi_1, \phi_2, \phi_3, \ldots\ldots$ be the "generalized coordinates," $\psi_1, \psi_2, \psi_3, \ldots\ldots$ the corresponding "impulse coordinates" or "moments," which determine the microscopic state of a certain molecule; then the state space contains as many dimensions as there are coordinates ϕ and moments ψ for every molecule. Now the region element of probability, according to classical statistical mechanics, is identical with the infinitely small element of the state space (in the macroscopic sense)[1]

$$d\phi_1\, d\phi_2\, d\phi_3 \ldots\ldots \quad d\psi_1\, d\psi_2\, d\psi_3 \ldots\ldots \tag{174}$$

According to the hypothesis of quanta, on the other hand, every region element of probability has a definite finite magnitude

$$G = \int d\phi_1\, d\phi_2\, d\phi_3 \ldots\ldots \quad d\psi_1\, d\psi_2\, d\psi_3 \ldots\ldots \tag{175}$$

whose value is the same for all different region elements and, moreover, depends on the nature of the system of molecules considered. The shape and position of the separate region elements are determined by the limits of the integral and must be determined anew in every separate case.

[1] Compare, for example, *L. Boltzmann*, Gastheorie, **2**, p. 62 *et seq.*, 1898, or *J. W. Gibbs*, Elementary principles in statistical mechanics, Chapter I, 1902.

CHAPTER II

IDEAL MONATOMIC GASES

127. In the preceding chapter it was proven that the introduction of probability considerations into the mechanical and electrodynamical theory of heat is justifiable and necessary, and from the general connection between entropy S and probability W, as expressed in equation (164), a method was derived for calculating the entropy of a physical system in a given state. Before we apply this method to the determination of the entropy of radiant heat we shall in this chapter make use of it for calculating the entropy of an ideal monatomic gas in an arbitrarily given state. The essential parts of this calculation are already contained in the investigations of *L. Boltzmann*[1] on the mechanical theory of heat; it will, however, be advisable to discuss this simple case in full, firstly to enable us to compare more readily the method of calculation and physical significance of mechanical entropy with that of radiation entropy, and secondly, what is more important, to set forth clearly the differences as compared with *Boltzmann's* treatment, that is, to discuss the meaning of the universal constant k and of the finite region elements G. For this purpose the treatment of a special case is sufficient.

128. Let us then take N similar monatomic gas molecules in an arbitrarily given thermodynamic state and try to find the corresponding entropy. The state space is six-dimensional, with the three coordinates x, y, z, and the three corresponding moments $m\xi$, $m\eta$, $m\zeta$, of a molecule, where we denote the mass by m and velocity components by ξ, η, ζ. Hence these quantities are to be substituted for the ϕ and ψ in Sec. 126. We thus obtain for the size of a region element G the sextuple integral

$$G = m^3 \int d\sigma, \tag{176}$$

where, for brevity

$$dx\ dy\ dz\ d\xi\ d\eta\ d\zeta = d\sigma \tag{177}$$

[1] *L. Boltzmann*, Sitzungsber. d. Akad. d. Wissensch. zu Wien (II) **76**, p. 373, 1877. Compare also Gastheorie, **1**, p. 38, 1896.

127

If the region elements are known, then, since the macroscopic state of the system of molecules was assumed as known, the numbers N_1, N_2, N_3, ... of the molecules which lie in the separate region elements are also known, and hence the distribution densities w_1, w_2, w_3, ... (166) are given and the entropy of the state follows at once from (173).

129. The theoretical determination of G is a problem as difficult as it is important. Hence we shall at this point restrict ourselves from the very outset to the special case in which the distribution density varies but slightly from one region element to the next—the characteristic feature of the state of an ideal gas. Then the summation over all region elements may be replaced by the integral over the whole state space. Thus we have from (176) and (167)

$$\sum w_1 = \sum w_1 \frac{m^3}{G} \int d\sigma = \frac{m^3}{G} \int w d\sigma = 1, \qquad (178)$$

in which w is no longer thought of as a discontinuous function of the ordinal number, i, of the region element, where $i=1$, 2, 3, ... n, but as a continuous function of the variables, x, y, z, ξ, η, ζ, of the state space. Since the whole state region contains very many region elements, it follows, according to (167) and from the fact that the distribution density w changes slowly, that w has everywhere a small value.

Similarly we find for the entropy of the gas from (173):

$$S = -kN \sum w_1 \log w_1 = -kN \frac{m^3}{G} \int w \log w \, d\sigma. \qquad (179)$$

Of course the whole energy E of the gas is also determined by the distribution densities w. If w is sufficiently small in every region element, the molecules contained in any one region element are, on the average, so far apart that their energy depends only on the velocities. Hence:

$$E = \sum N_1 \frac{1}{2} m(\xi_1^2 + \eta_1^2 + \zeta_1^2) + E_0$$

$$= N \sum w_1 \frac{1}{2} m(\xi_1^2 + \eta_1^2 + \zeta_1^2) + E_0, \qquad (180)$$

where $\xi_1 \eta_1 \zeta_1$ denotes any velocity lying within the region element 1 and E_0 denotes the internal energy of the stationary molecules, which is assumed constant. In place of the latter expression we may write, again according to (176),

$$E=\frac{m^4N}{2G}\int(\xi^2+\eta^2+\zeta^2)w\,d\sigma+E_0. \tag{181}$$

130. Let us consider the state of thermodynamic equilibrium. According to the second principle of thermodynamics this state is distinguished from all others by the fact that, for a given volume V and a given energy E of the gas, the entropy S is a maximum. Let us then regard the volume

$$V=\int\int\int dx\,dy\,dz \tag{182}$$

and the energy E of the gas as given. The condition for equilibrium is $\delta S=0$, or, according to (179),

$$\Sigma(\log w_1+1)\delta w_1=0,$$

and this holds for any variations of the distribution densities whatever, provided that, according to (167) and (180), they satisfy the conditions

$$\Sigma\delta w_1=0 \text{ and } \Sigma(\xi_1{}^2+\eta_1{}^2+\zeta_1{}^2)\delta w_1=0.$$

This gives us as the necessary and sufficient condition for thermodynamic equilibrium for every separate distribution density w:

$$\log w+\beta(\xi^2+\eta^2+\zeta^2)+\text{ const. }=0$$

or

$$w=\alpha e^{-\beta(\xi^2+\eta^2+\zeta^2)}, \tag{183}$$

where α and β are constants. Hence in the state of equilibrium the distribution of the molecules in space is independent of x, y, z, that is, macroscopically uniform, and the distribution of velocities is the well-known one of *Maxwell*.

131. The values of the constants α and β may be found from those of V and E. For, on substituting the value of w just found in (178) and taking account of (177) and (182), we get

$$\frac{G}{m^3}=\alpha V\int\int\int_{-\infty}^{+\infty} e^{-\beta(\xi^2+\eta^2+\zeta^2)}d\xi\,d\eta\,d\zeta=\alpha V\left(\frac{\pi}{\beta}\right)^{\frac{3}{2}},$$

9

and on substituting w in (181) we get

$$E=E_o+\frac{\alpha m^4NV}{2G}\int\int\int_{-\infty}^{+\infty}(\xi^2+\eta^2+\zeta^2)e^{-\beta(\xi^2+\eta^2+\zeta^2)}d\xi\,d\eta\,d\zeta,$$

or

$$E=E_o+\frac{3\alpha m^4NV}{4G}\frac{1}{\beta}\left(\frac{\pi}{\beta}\right)^{\frac{3}{2}}.$$

Solving for α and β we have

$$\alpha=\frac{G}{V}\left(\frac{3N}{4\pi m(E-E_o)}\right)^{\frac{3}{2}} \qquad (184)$$

$$\beta=\frac{3}{4}\frac{Nm}{E-E_o}. \qquad (185)$$

From this finally we find, as an expression for the entropy S of the gas in the state of equilibrium with given values of N, V, and E,

$$S=kN\log\left\{\frac{V}{G}\left(\frac{4\pi\ em\ (E-E_o)}{3N}\right)^{\frac{3}{2}}\right\}. \qquad (186)$$

132. This determination of the entropy of an ideal monatomic gas is based solely on the general connection between entropy and probability as expressed in equation (164); in particular, we have at no stage of our calculation made use of any special law of the theory of gases. It is, therefore, of importance to see how the entire thermodynamic behavior of a monatomic gas, especially the equation of state and the values of the specific heats, may be deduced from the expression found for the entropy directly by means of the principles of thermodynamics. From the general thermodynamic equation defining the entropy, namely,

$$dS=\frac{dE+pdV}{T}, \qquad (187)$$

the partial differential coefficients of S with respect to E and V are found to be

$$\left(\frac{\partial S}{\partial E}\right)_V=\frac{1}{T}, \quad \left(\frac{\partial S}{\partial V}\right)_E=\frac{p}{T}$$

Hence, by using (186), we get for our gas

$$\left(\frac{\partial S}{\partial E}\right)_V = \frac{3}{2}\,\frac{kN}{E-E_o} = \frac{1}{T} \tag{188}$$

and

$$\left(\frac{\partial S}{\partial V}\right)_E = \frac{kN}{V} = \frac{p}{T}. \tag{189}$$

The second of these equations

$$p = \frac{kNT}{V} \tag{190}$$

contains the laws of *Boyle, Gay Lussac,* and *Avogadro,* the last named because the pressure depends only on the number N, not on the nature of the molecules. If we write it in the customary form:

$$p = \frac{RnT}{V}, \tag{191}$$

where n denotes the number of gram molecules or mols of the gas, referred to $O_2 = 32g$, and R represents the absolute gas constant

$$R = 831 \times 10^5 \frac{\text{erg}}{\text{degree}}, \tag{192}$$

we obtain by comparison

$$k = \frac{Rn}{N}. \tag{193}$$

If we now call the ratio of the number of mols to the number of molecules ω, or, what is the same thing, the ratio of the mass of a molecule to that of a mol, $\omega = \frac{n}{N}$, we shall have

$$k = \omega R. \tag{194}$$

From this the universal constant k may be calculated, when ω is given, and *vice versa.* According to (190) this constant k is nothing but the absolute gas constant, if it is referred to molecules instead of mols.

From equation (188)

$$E - E_0 = \tfrac{3}{2}\,kNT. \tag{195}$$

Now, since the energy of an ideal gas is also given by

$$E = Anc_vT + E_o \tag{196}$$

where c_v is the heat capacity of a mol at constant volume in calories and A is the mechanical equivalent of heat:

$$A = 419 \times 10^5 \frac{\text{erg}}{\text{cal}} \tag{197}$$

it follows that

$$c_v = \frac{3}{2} \frac{kN}{An}$$

and further, by taking account of (193)

$$c_v = \frac{3}{2} \frac{R}{A} = \frac{3}{2} \frac{831 \times 10^5}{419 \times 10^5} = 3.0 \tag{198}$$

as an expression for the heat capacity per mol of any monatomic gas at constant volume in calories.[1]

For the heat capacity per mol at constant pressure, c_p, we have as a consequence of the first principle of thermodynamics:

$$c_p - c_v = \frac{R}{A}$$

and hence by (198)

$$c_p = \frac{5}{2} \frac{R}{A}, \qquad \frac{c_p}{c_v} = \frac{5}{3}, \tag{199}$$

as is known to be the case for monatomic gases. It follows from (195) that the kinetic energy L of the gas molecules is equal to

$$L = E - E_o = \frac{3}{2} NkT \tag{200}$$

133. The preceding relations, obtained simply by identifying the mechanical expression of the entropy (186) with its thermodynamic expression (187), show the usefulness of the theory developed. In them an additive constant in the expression for the entropy is immaterial and hence the size G of the region element of probability does not matter. The hypothesis of quanta, however, goes further, since it fixes the absolute value of the entropy and thus leads to the same conclusion as the heat theorem

[1] Compare *F. Richarz*, Wiedemann's Annal., **67**, p. 705, 1899.

of *Nernst.* According to this theorem the "characteristic function" of an ideal gas[1] is in our notation

$$\Phi = S - \frac{E+pV}{T} = n\left(Ac_p \log T - R \log p + a - \frac{b}{T}\right),$$

where a denotes *Nernst's* chemical constant, and b the energy constant.

On the other hand, the preceding formulæ (186), (188), and (189) give for the same function Φ the following expression:

$$\Phi = N\left(\frac{5}{2}k \log T - k \log p + a'\right) - \frac{E_o}{T}$$

where for brevity a' is put for:

$$a' = k \log \left\{ \frac{kN}{eG} (2\pi mk)^{\frac{3}{2}} \right\}$$

From a comparison of the two expressions for Φ it is seen, by taking account of (199) and (193), that they agree completely, provided

$$a = \frac{N}{n} a' = R \log \left\{ \frac{Nk^{\frac{5}{2}}}{eG} (2\pi m)^{\frac{3}{2}} \right\}, \tag{201}$$

$$b = \frac{E_o}{n}$$

This expresses the relation between the chemical constant a of the gas and the region element G of the probability.[2]

It is seen that G is proportional to the total number, N, of the molecules. Hence, if we put $G = Ng$, we see that g, the molecular region element, depends only on the chemical nature of the gas.

Obviously the quantity g must be closely connected with the law, so far unknown, according to which the molecules act microscopically on one another. Whether the value of g varies with the nature of the molecules or whether it is the same for all kinds of molecules, may be left undecided for the present.

[1] *E.g., M. Planck,* Vorlesungen über Thermodynamik, Leipzig, Veit und Comp., 1911, Sec. 287, equation 267.

[2] Compare also *O. Sackur,* Annal. d. Physik, **36,** p. 958, 1911, Nernst-Festschrift, p. 405, 1912, and *H. Tetrode,* Annal. d. Physik, **38,** p. 434, 1912.

If g were known, *Nernst's* chemical constant, a, of the gas could be calculated from (201) and the theory could thus be tested. For the present the reverse only is feasible, namely, to calculate g from a. For it is known that a may be measured directly by the tension of the saturated vapor, which at sufficiently low temperatures satisfies the simple equation[1]

$$\log p = \frac{5}{2}\log T - \frac{A r_0}{RT} + \frac{a}{R} \tag{202}$$

(where r_0 is the heat of vaporization of a mol at 0° in calories). When a has been found by measurement, the size g of the molecular region element is found from (201) to be

$$g = (2\pi m)^{\frac{3}{2}} k^{\frac{5}{2}} e^{-\frac{a}{R}-1} \tag{203}$$

Let us consider the dimensions of g.

According to (176) g is of the dimensions [erg³sec³]. The same follows from the present equation, when we consider that the dimension of the chemical constant a is not, as might at first be thought, that of R, but, according to (202), that of $R \log \frac{p}{T^{\frac{5}{2}}}$.

134. To this we may at once add another quantitative relation. All the preceding calculations rest on the assumption that the distribution density w and hence also the constant α in (183) are small (Sec. 129). Hence, if we take the value of α from (184) and take account of (188), (189) and (201), it follows that

$$\frac{p}{T^{\frac{5}{2}}} e^{-\frac{a}{R}-1} \quad \text{must be small.}$$

When this relation is not satisfied, the gas cannot be in the ideal state. For the saturated vapor it follows then from (202) that $e^{-\frac{Ar_0}{RT}}$ is small. In order, then, that a saturated vapor may be assumed to be in the state of an ideal gas, the temperature T must certainly be less than $\frac{A}{R} r_0$ or $\frac{r_0}{2}$. Such a restriction is unknown to the classical thermodynamics.

[1] *M. Planck,* l. c., Sec. 288, equation 271.

CHAPTER III

IDEAL LINEAR OSCILLATORS

135. The main problem of the theory of heat radiation is to determine the energy distribution in the normal spectrum of black radiation, or, what amounts to the same thing, to find the function which has been left undetermined in the general expression of *Wien's* displacement law (119), the function which connects the entropy of a certain radiation with its energy. The purpose of this chapter is to develop some preliminary theorems leading to this solution. Now since, as we have seen in Sec. 48, the normal energy distribution in a diathermanous medium cannot be established unless the medium exchanges radiation with an emitting and absorbing substance, it will be necessary for the treatment of this problem to consider more closely the processes which cause the creation and the destruction of heat rays, that is, the processes of emission and absorption. In view of the complexity of these processes and the difficulty of acquiring knowledge of any definite details regarding them, it would indeed be quite hopeless to expect to gain any certain results in this way, if it were not possible to use as a reliable guide in this obscure region the law of *Kirchhoff* derived in Sec. 51. This law states that a vacuum completely enclosed by reflecting walls, in which any emitting and absorbing bodies are scattered in any arrangement whatever, assumes in the course of time the stationary state of black radiation, which is completely determined by one parameter only, namely, the temperature, and in particular does not depend on the number, the nature, and the arrangement of the material bodies present. Hence, for the investigation of the properties of the state of black radiation the nature of the bodies which are assumed to be in the vacuum is perfectly immaterial. In fact, it does not even matter whether such bodies really exist somewhere in nature, provided their existence and their properties are consistent with the laws of thermodynamics and electrodynamics. If, for any special arbitrary assumption regarding the nature and arrangement of emitting and absorbing systems, we can find a state of radiation in the surrounding vacuum which is distinguished by absolute stability, this state can be no other than that of black radiation.

Since, according to this law, we are free to choose any system whatever, we now select from all possible emitting and absorbing systems the simplest conceivable one, namely, one consisting of a large number N of similar stationary oscillators, each consisting of two poles, charged with equal quantities of electricity of opposite sign, which may move relatively to each other on a fixed straight line, the axis of the oscillator.

It is true that it would be more general and in closer accord with the conditions in nature to assume the vibrations to be those of an oscillator consisting of two poles, each of which has three degrees of freedom of motion instead of one, *i.e.*, to assume the vibrations as taking place in space instead of in a straight line only. Nevertheless we may, according to the fundamental principle stated above, restrict ourselves from the beginning to the treatment of one single component, without fear of any essential loss of generality of the conclusions we have in view.

It might, however, be questioned as a matter of principle, whether it is really permissib'e to think of the centers of mass of the oscillators as stationary, since, according to the kinetic theory of gases, all material particles which are contained in substances of finite temperature and free to move possess a certain finite mean kinetic energy of translatory motion. This objection, however, may also be removed by the consideration that the velocity is not fixed by the kinetic energy alone. We need only think of an oscillator as being loaded, say at its positive pole, with a comparatively large inert mass, which is perfectly neutral electrodynamically, in order to decrease its velocity for a given kinetic energy below any preassigned value whatever. Of course this consideration remains valid also, if, as is now frequently done, all inertia is reduced to electrodynamic action. For this action is at any rate of a kind quite different from the one to be considered in the following, and hence cannot influence it.

Let the state of such an oscillator be completely determined by its moment $f(t)$, that is, by the product of the electric charge of the pole situated on the positive side of the axis and the pole distance, and by the derivative of f with respect to the time or

$$\frac{df(t)}{dt}=\dot{f}(t). \tag{204}$$

Let the energy of the oscillator be of the following simple form:

$$U=\tfrac{1}{2}Kf^2+\tfrac{1}{2}L\dot{f}^2, \tag{205}$$

where K and L denote positive constants, which depend on the nature of the oscillator in some way that need not be discussed at this point.

If during its vibration an oscillator neither absorbed nor emitted any energy, its energy of vibration, U, would remain constant, and we would have:

$$dU=Kfdf+L\dot{f}d\dot{f}=0, \tag{205 a}$$

or, on account of (204),

$$Kf(t)+L\ddot{f}(t)=0. \tag{206}$$

The general solution of this differential equation is found to be a purely periodical vibration:

$$f=C\cos(2\pi\nu t-\theta) \tag{207}$$

where C and θ denote the integration constants and ν the number of vibrations per unit time:

$$\nu=\frac{1}{2\pi}\sqrt{\frac{K}{L}} \tag{208}$$

136. If now the assumed system of oscillators is in a space traversed by heat rays, the energy of vibration, U, of an oscillator will not in general remain constant, but will be always changing by absorption and emission of energy. Without, for the present, considering in detail the laws to which these processes are subject, let us consider any one arbitrarily given thermodynamic state of the oscillators and calculate its entropy, irrespective of the surrounding field of radiation. In doing this we proceed entirely according to the principle advanced in the two preceding chapters, allowing, however, at every stage for the conditions caused by the peculiarities of the case in question.

The first question is: What determines the thermodynamic state of the system considered? For this purpose, according to

Sec. 124, the numbers N_1, N_2, N_3, ... of the oscillators, which lie in the region elements 1, 2, 3, ... of the "state space" must be given. The state space of an oscillator contains those coordinates which determine the microscopic state of an oscillator. In the case in question these are only two in number, namely, the moment f and the rate at which it varies, $\dot{f}$, or instead of the latter the quantity

$$\psi = L\dot{f}, \tag{209}$$

which is of the dimensions of an impulse. The region element of the state plane is, according to the hypothesis of quanta (Sec. 126), the double integral

$$\int\int df\, d\psi = h. \tag{210}$$

The quantity h is the same for all region elements. *A priori*, it might, however, depend also on the nature of the system considered, for example, on the frequency of the oscillators. The following simple consideration, however, leads to the assumption that h is a universal constant. We know from the generalized displacement law of *Wien* (equation 119) that in the universal function, which gives the entropy radiation as dependent on the energy radiation, there must appear a universal constant of the dimension $\frac{c^3 \mathfrak{u}}{\nu^3}$ and this is of the dimension of a quantity of action[1] (erg sec.). Now, according to (210), the quantity h has precisely this dimension, on which account we may denote it as "element of action" or "quantity element of action." Hence, unless a second constant also enters, h cannot depend on any other physical quantities.

137. The principal difference, compared with the calculations for an ideal gas in the preceding chapter, lies in the fact that we do not now assume the distribution densities w_1, w_2, w_3 . . . . . of the oscillators among the separate region elements to vary but little from region to region as was assumed in Sec. 129. Accordingly the w's are not small, but finite proper fractions, and the summation over the region elements cannot be written as an integration.

[1] The quantity from which the principle of *least action* takes its name. (Tr.)

In the first place, as regards the shape of the region elements, the fact that in the case of undisturbed vibrations of an oscillator the phase is always changing, whereas the amplitude remains constant, leads to the conclusion that, for the macroscopic state of the oscillators, the amplitudes only, not the phases, must be considered, or in other words the region elements in the $f\psi$ plane are bounded by the curves $C=$const., that is, by ellipses, since from (207) and (209)

$$\left(\frac{f}{C}\right)^2+\left(\frac{\psi}{2\pi\nu LC}\right)^2=1. \tag{211}$$

The semi-axes of such an ellipse are:

$$a=C \text{ and } b=2\pi\nu LC. \tag{212}$$

Accordingly the region elements 1, 2, 3, ... n . . . . . . are the concentric, similar, and similarly situated elliptic rings, which are determined by the increasing values of C:

$$0,\ C_1,\ C_2,\ C_3,\ .\ .\ .\ .\ .\ C_{n-1},\ C_n\ .\ .\ .\ .\ . \tag{213}$$

The nth region element is that which is bounded by the ellipses $C=C_{n-1}$ and $C=C_n$. The first region element is the full ellipse C_1. All these rings have the same area h, which is found by subtracting the area of the full ellipse C_{n-1} from that of the full ellipse C_n; hence

$$h=(a_nb_n-a_{n-1}b_{n-1})\pi$$

or, according to (212),

$$h=(C_n^2-C_{n-1}^2)\ 2\pi^2\nu L,$$

where $n=1, 2, 3,$ . . . . .

From the additional fact that $C_0=0$, it follows that:

$$C_n^2=\frac{nh}{2\pi^2\nu L}. \tag{214}$$

Thus the semi-axes of the bounding ellipses are in the ratio of the square roots of the integral numbers.

138. The thermodynamic state of the system of oscillators is fixed by the fact that the values of the distribution densities w_1, w_2, w_3, ... of the oscillators among the separate region elements are given. *Within* a region element the distribution of the oscillators is according to the law of elemental chaos (Sec. 122), *i.e.*, it is approximately *uniform*.

These data suffice for calculating the entropy S as well as the energy E of the system in the given state, the former quantity directly from (173), the latter by the aid of (205). It must be kept in mind in the calculation that, since the energy varies appreciably within a region element, the energy E_n of all those oscillators which lie in the nth region element is to be found by an integration. Then the whole energy E of the system is:

$$E=E_1+E_2+ \quad . \quad . \quad . \quad . \quad . \quad E_n+ \quad . \quad . \quad . \quad . \quad . \tag{215}$$

E_n may be calculated with the help of the law that within every region element the oscillators are uniformly distributed. If the nth region element contains, all told, N_n oscillators, there are per unit area $\frac{N_n}{h}$ oscillators and hence $\frac{N_n}{h}\, df \cdot d\psi$ per element of area. Hence we have:

$$E_n=\frac{N_n}{h}\int\int U\ df\ d\psi.$$

In performing the integration, instead of f and ψ we take C and ϕ, as new variables, and since according to (211),

$$f=C\cos\phi \qquad \psi=2\pi\nu LC\sin\phi \tag{216}$$

we get:

$$E_n=2\pi\nu L\,\frac{N_n}{h}\int\int U\ C\ dC\ d\phi$$

to be integrated with respect to ϕ from 0 to 2π and with respect to C from C_{n-1} to C_n. If we substitute from (205), (209) and (216)

$$U=\tfrac{1}{2}KC^2, \tag{217}$$

we obtain by integration

$$E_n=\frac{\pi^2}{2}\nu LK\frac{N_n}{h}(C_n{}^4-C_{n-1}{}^4)$$

and from (214) and (208):

$$E_n=N_n(n-\tfrac{1}{2})h\nu=Nw_n(n-\tfrac{1}{2})h\nu,$$

that is, the mean energy of an oscillator in the nth region element is $(n-\frac{1}{2})h\nu$. This is exactly the arithmetic mean of the energies $(n-1)h\nu$ and $nh\nu$ which correspond to the two ellipses $C=C_{n-1}$ and $C=C_n$ bounding the region, as may be seen from (217), if the values of C_{n-1} and C_n are therein substituted from (214).

The total energy E is, according to (215),

$$E = Nh\nu \sum_{n=1}^{n=\infty} (n-\tfrac{1}{2}) w_n. \tag{219}$$

139. Let us now consider the state of thermodynamic equilibrium of the oscillators. According to the second principle of thermodynamics, the entropy S is in that case a maximum for a given energy E. Hence we assume E in (219) as given. Then from (179) we have for the state of equilibrium:

$$\delta S = 0 = \sum_1^\infty (\log w_n + 1)\delta w_n,$$

where according to (167) and (219)

$$\sum_1^\infty \delta w_n = 0 \text{ and } \sum_1^\infty (n-\tfrac{1}{2})\delta w_n = 0$$

From these relations we find:

$$\log w_n + \beta n + \text{const.} = 0$$

or

$$w_n = \alpha\gamma^n. \tag{220}$$

The values of the constants α and γ follow from equations (167) and (219):

$$\alpha = \frac{2Nh\nu}{2E - Nh\nu} \qquad \gamma = \frac{2E - Nh\nu}{2E + Nh\nu}. \tag{221}$$

Since w_n is essentially positive it follows that equilibrium is not possible in the system of oscillators considered unless the total energy E has a greater value than $\frac{Nh\nu}{2}$, that is unless the mean energy of the oscillators is at least $\frac{h\nu}{2}$. This, according to (218), is the mean energy of the oscillators lying in the first region element. In fact, in this extreme case all N oscillators lie in the first region element, the region of smallest energy; within this element they are arranged uniformly.

The entropy S of the system, which is in thermodynamic equilibrium, is found by combining (173) with (220) and (221)

$$S = kN\left\{\left(\frac{E}{Nh\nu}+\frac{1}{2}\right)\log\left(\frac{E}{Nh\nu}+\frac{1}{2}\right) - \left(\frac{E}{Nh\nu}-\frac{1}{2}\right)\log\left(\frac{E}{Nh\nu}-\frac{1}{2}\right)\right\} \tag{222}$$

140. The connection between energy and entropy just obtained allows furthermore a certain conclusion as regards the temperature. For from the equation of the second principle of thermodynamics, $dS=\frac{dE}{T}$ and from differentiation of (222) with respect to E it follows that

$$E=N\frac{h\nu}{2}\frac{1+e^{-\frac{h\nu}{kT}}}{1-e^{-\frac{h\nu}{kT}}}=Nh\nu\left(\frac{1}{2}+\frac{1}{e^{\frac{h\nu}{kT}}-1}\right)$$

Hence, for the zero point of the absolute temperature E becomes, not 0, but $N\frac{h\nu}{2}$. This is the extreme case discussed in the preceding paragraph, which just allows thermodynamic equilibrium to exist. That the oscillators are said to perform vibrations even at the temperature zero, the mean energy of which is as large as $\frac{h\nu}{2}$ and hence may become quite large for rapid vibrations, may at first sight seem strange. It seems to me, however, that certain facts point to the existence, inside the atoms, of vibrations independent of the temperature and supplied with appreciable energy, which need only a small suitable excitation to become evident externally. For example, the velocity, sometimes very large, of secondary cathode rays produced by Roentgen rays, and that of electrons liberated by photoelectric effect are independent of the temperature of the metal and of the intensity of the exciting radiation. Moreover the radioactive energies are also independent of the temperature. It is also well known that the close connection between the inertia of matter and its energy as postulated by the relativity principle leads to the assumption of very appreciable quantities of intra-atomic energy even at the zero of absolute temperature.

For the extreme case, $T=\infty$, we find from (223) that

$$E=NkT, \tag{224}$$

i.e., the energy is proportional to the temperature and independent of the size of the quantum of action, h, and of the nature of the oscillators. It is of interest to compare this value of the energy of vibration E of the system of oscillators, which holds at high temperatures, with the kinetic energy L of the molecular motion of an ideal monatomic gas at the same temperature as calculated in (200). From the comparison it follows that

$$E = \tfrac{2}{3} L \tag{225}$$

This simple relation is caused by the fact that for high temperatures the contents of the hypothesis of quanta coincide with those of the classical statistical mechanics. Then the absolute magnitude of the region element, G or h respectively, becomes physically unimportant (compare Sec. 125) and we have the simple law of equipartition of the energy among all variables in question (see below Sec. 169). The factor $\tfrac{2}{3}$ in equation (225) is due to the fact that the kinetic energy of a moving molecule depends on three variables (ξ, η, ζ,) and the energy of a vibrating oscillator on only two (f, ψ).

The heat capacity of the system of oscillators in question is, from (223),

$$\frac{dE}{dT} = Nk\left(\frac{h\nu}{kT}\right)^2 \frac{e^{\frac{h\nu}{kT}}}{\left(e^{\frac{h\nu}{kT}} - 1\right)^2} \tag{226}$$

It vanishes for $T=0$ and becomes equal to Nk for $T=\infty$. *A. Einstein*[1] has made an important application of this equation to the heat capacity of solid bodies, but a closer discussion of this would be beyond the scope of the investigations to be made in this book.

For the constants α and γ in the expression (220) for the distribution density w we find from (221):

$$\alpha = e^{\frac{h\nu}{kT}} - 1 \qquad \gamma = e^{-\frac{h\nu}{kT}} \tag{227}$$

and finally for the entropy S of our system as a function of temperature:

$$S = kN\left\{\frac{\frac{h\nu}{kT}}{e^{\frac{h\nu}{kT}} - 1} - \log\left(1 - e^{-\frac{h\nu}{kT}}\right)\right\} \tag{228}$$

[1] *A. Einstein*, Ann. d. Phys. **22**, p. 180, 1907. Compare also *M. Born* und *Th. von Kármán*, Phys. Zeitschr. **13**, p. 297, 1912.

CHAPTER IV

DIRECT CALCULATION OF THE ENTROPY IN THE CASE OF THERMODYNAMIC EQUILIBRIUM

141. In the calculation of the entropy of an ideal gas and of a system of resonators, as carried out in the preceding chapters, we proceeded in both cases, by first determining the entropy for an arbitrarily given state, then introducing the special condition of thermodynamic equilibrium, *i.e.*, of the maximum of entropy, and then deducing for this special case an expression for the entropy.

If the problem is only the determination of the entropy in the case of thermodynamic equilibrium, this method is a roundabout one, inasmuch as it requires a number of calculations, namely, the determination of the separate distribution densities w_1, w_2, w_3, ... which do not enter separately into the final result. It is therefore useful to have a method which leads directly to the expression for the *entropy* of a system in the state of thermodynamic equilibrium, without requiring any consideration of the *state* of thermodynamic equilibrium. This method is based on an important general property of the thermodynamic probability of a state of equilibrium.

We know that there exists between the entropy S and the thermodynamic probability W in any state whatever the general relation (164). In the state of thermodynamic equilibrium both quantities have maximum values; hence, if we denote the maximum values by a suitable index:

$$S_m = k \log W_m. \tag{229}$$

It follows from the two equations that:

$$\frac{W_m}{W} = e^{\frac{S_m - S}{k}}$$

Now, when the deviation from thermodynamic equilibrium is at all appreciable, $\frac{S_m - S}{k}$ is certainly a very large number. Accordingly W_m is not only large but of a very high order large, compared with W, that is to say: The thermodynamic probability of the state of equilibrium is enormously large compared with the thermodynamic probability of all states which, in the course of time, change into the state of equilibrium.

This proposition leads to the possibility of calculating W_m with an accuracy quite sufficient for the determination of S_m, without the necessity of introducing the special condition of equilibrium. According to Sec. 123, *et seq.*, W_m is equal to the number of all different complexions possible in the state of thermodynamic equilibrium. This number is so enormously large compared with the number of complexions of all states deviating from equilibrium that we commit no appreciable error if we think of the number of complexions of all states, which as time goes on change into the state of equilibrium, *i.e.*, all states which are at all possible under the given external conditions, as being included in this number. The total number of all possible complexions may be calculated much more readily and directly than the number of complexions referring to the state of equilibrium only.

142. We shall now use the method just formulated to calculate the entropy, in the state of equilibrium, of the system of ideal linear oscillators considered in the last chapter, when the total energy E is given. The notation remains the same as above.

We put then W_m equal to the number of complexions of all states which are at all possible with the given energy E of the system. Then according to (219) we have the condition:

$$E=h\nu\sum_{n=1}^{\infty}(n-\tfrac{1}{2})N_n. \tag{230}$$

Whereas we have so far been dealing with the number of complexions with given N_n, now the N_n are also to be varied in all ways consistent with the condition (230).

The total number of all complexions is obtained in a simple way by the following consideration. We write, according to (165), the condition (230) in the following form:

$$\frac{E}{h\nu}-\frac{N}{2}=\sum_{n=1}^{\infty}(n-1)N_n$$

10

or

$$0 \cdot N_1 + 1 \cdot N_2 + 2 \cdot N_3 + \quad . \quad . \quad . \quad . \quad . \quad + (n-1)N_n + \quad . \quad . \quad . \quad . \quad .$$
$$= \frac{E}{h\nu} - \frac{N}{2} = P. \qquad (231)$$

P is a given large positive number, which may, without restricting the generality, be taken as an integer.

According to Sec. 123 a complexion is a definite assignment of every individual oscillator to a definite region element 1, 2, 3, ... of the state plane (f, ψ). Hence we may characterize a certain complexion by thinking of the N oscillators as being numbered from 1 to N and, when an oscillator is assigned to the nth region element, writing down the number of the oscillator $(n-1)$ times. If in any complexion an oscillator is assigned to the first region element its number is not put down at all. Thus every complexion gives a certain row of figures, and *vice versa* to every row of figures there corresponds a certain complexion. The position of the figures in the row is immaterial.

What makes this form of representation useful is the fact that according to (231) the number of figures in such a row is always equal to P. Hence we have "combinations with repetitions of N elements taken P at a time," whose total number is

$$\frac{N(N+1)\ (N+2) \quad . \quad . \quad . \quad . \quad . \quad (N+P-1)}{1 \quad 2 \quad\quad 3 \quad\quad . \quad . \quad . \quad . \quad . \quad\quad P} = \frac{(N+P-1)!}{(N-1)!P!} \qquad (232)$$

If for example we had $N=3$ and $P=4$ all possible complexions would be represented by the rows of figures:

1111	1133	2222
1112	1222	2223
1113	1223	2233
1122	1233	2333
1123	1333	3333

The first row denotes that complexion in which the first oscillator lies in the 5th region element and the two others in the first. The number of complexions in this case is 15, in agreement with the formula.

143. For the entropy S of the system of oscillators which is in the state of thermodynamic equilibrium we thus obtain from equation (229) since N and P are large numbers:

$$S=k\log\frac{(N+P)!}{N!P!}$$

and by making use of Stirling's formula (171)[1]

$$S=kN\left\{\left(\frac{P}{N}+1\right)\log\left(\frac{P}{N}+1\right)-\frac{P}{N}\log\frac{P}{N}\right\}.$$

If we now replace P by E from (231) we find for the entropy exactly the same value as given by (222) and thus we have demonstrated in a special case both the admissibility and the practical usefulness of the method employed.[2]

[1] Compare footnote to page 124. See also page 218.

[2] A complete mathematical discussion of the subject of this chapter has been given by *H. A. Lorentz.* Compare, *e. g.*, Nature, **92**, p. 305, Nov. 6, 1913. (Tr.)

PART IV

A SYSTEM OF OSCILLATORS IN A STATIONARY FIELD OF RADIATION

CHAPTER I

THE ELEMENTARY DYNAMICAL LAW FOR THE VIBRATIONS OF AN IDEAL OSCILLATOR. HYPOTHESIS OF EMISSION OF QUANTA

144. All that precedes has been by way of preparation. Before taking the final step, which will lead to the law of distribution of energy in the spectrum of black radiation, let us briefly put together the essentials of the problem still to be solved. As we have already seen in Sec. 93, the whole problem amounts to the determination of the temperature corresponding to a monochromatic radiation of given intensity. For among all conceivable distributions of energy the normal one, that is, the one peculiar to black radiation, is characterized by the fact that in it the rays of all frequencies have the same temperature. But the temperature of a radiation cannot be determined unless it be brought into thermodynamic equilibrium with a system of molecules or oscillators, the temperature of which is known from other sources. For if we did not consider any emitting and absorbing matter there would be no possibility of defining the entropy and temperature of the radiation, and the simple propagation of free radiation would be a reversible process, in which the entropy and temperature of the separate pencils would not undergo any change. (Compare below Sec. 166.)

Now we have deduced in the preceding section all the characteristic properties of the thermodynamic equilibrium of a system of ideal oscillators. Hence, if we succeed in indicating a state of radiation which is in thermodynamic equilibrium with the system of oscillators, the temperature of the radiation can be no other than that of the oscillators, and therewith the problem is solved.

145. Accordingly we now return to the considerations of Sec. 135 and assume a system of ideal linear oscillators in a stationary field of radiation. In order to make progress along the line proposed, it is necessary to know the elementary dynamical law, according to which the mutual action between an oscillator and the incident radiation takes place, and it is moreover easy to see that this law cannot be the same as the one which the classical electrodynamical theory postulates for the vibrations of a linear Hertzian oscillator. For, according to this law, all the oscillators, when placed in a stationary field of radiation, would, since their properties are exactly similar, assume the same energy of vibration, if we disregard certain irregular variations, which, however, will be smaller, the smaller we assume the damping constant of the oscillators, that is, the more pronounced their natural vibration is. This, however, is in direct contradiction to the definite discrete values of the distribution densities w_1, w_2, w_3, ... which we have found in Sec. 139 for the stationary state of the system of oscillators. The latter allows us to conclude with certainty that in the dynamical law to be established the quantity element of action h must play a characteristic part. Of what nature this will be cannot be predicted *a priori;* this much, however, is certain, that the only type of dynamical law admissible is one that will give for the stationary state of the oscillators exactly the distribution densities w calculated previously. It is in this problem that the question of the dynamical significance of the quantum of action h stands for the first time in the foreground, a question the answer to which was unnecessary for the calculations of the preceding sections, and this is the principal reason why in our treatment the preceding section was taken up first.

146. In establishing the dynamical law, it will be rational to proceed in such a way as to make the deviation from the laws of classical electrodynamics, which was recognized as necessary, as slight as possible. Hence, as regards the influence of the field of radiation on an oscillator, we follow that theory closely. If the oscillator vibrates under the influence of any external electromagnetic field whatever, its energy U will not in general remain constant, but the energy equation (205 a) must be extended to include the work which the external electromagnetic field does on the oscillator, and, if the axis of the electric doublet coincides with the z-axis, this work is expressed by the term $\mathsf{E}_z\, df = \mathsf{E}_z \dot{f}\, dt$. Here E_z denotes the z component of the external electric field-strength at the position of the oscillator, that is, that electric field-strength which would exist at the position of the oscillator, if the latter were not there at all. The other components of the external field have no influence on the vibrations of the oscillator.

Hence the complete energy equation reads:

$$Kf\,df + L\dot{f}\,d\dot{f} = \mathsf{E}_z df$$

or:

$$Kf + L\ddot{f} = \mathsf{E}_z, \tag{233}$$

and the energy absorbed by the oscillator during the time element dt is:

$$\mathsf{E}_z \dot{f}\,dt \tag{234}$$

147. While the oscillator is absorbing it must also be emitting, for otherwise a stationary state would be impossible. Now, since in the law of absorption just assumed the hypothesis of quanta has as yet found no room, it follows that it must come into play in some way or other in the emission of the oscillator, and this is provided for by the introduction of the hypothesis of emission of quanta. That is to say, we shall assume that the emission does not take place continuously, as does the absorption, but that it occurs only at certain definite times, suddenly, in pulses, and in particular we assume that an oscillator can emit energy only at the moment when its energy of vibration, U, is an integral multiple n of the quantum of energy, $\epsilon = h\nu$. Whether it then really emits or whether its energy of vibration increases further by absorption will be regarded as a matter of chance. This will not be regarded as implying that there is no causality for emission; but the processes which cause the emission will be assumed to be of such a concealed nature that for the present their laws cannot be obtained by any but statistical methods. Such an assumption is not at all foreign to physics; it is, *e.g.*, made in the atomistic theory of chemical reactions and the disintegration theory of radioactive substances.

It will be assumed, however, that if emission does take place, the entire energy of vibration, U, is emitted, so that the vibration of the oscillator decreases to zero and then increases again by further absorption of radiant energy.

It now remains to fix the law which gives the probability that an oscillator will or will not emit at an instant when its energy has reached an integral multiple of ϵ. For it is evident that the statistical state of equilibrium, established in the system of oscillators by the assumed alternations of absorption and emission will depend on this law; and evidently the mean energy U of the oscillators will be larger, the larger the probability that in such a critical state no emission takes place. On the other hand, since the mean energy U will be larger, the larger the intensity of the field of radiation surrounding the oscillators, we shall state the law of emission as follows: *The ratio of the probability that no emission takes place to the probability that emission does take place is proportional to the intensity* I of the vibration which excites the oscillator and which was defined in equation (158). The value of the constant of proportionality we shall determine later on by the application of the theory to the special case in which the energy of vibration is very large. For in this case, as we know, the familiar formulæ of the classical dynamics hold for any period of the oscillator whatever, since the quantity element of action h may then, without any appreciable error, be regarded as infinitely small.

These statements define completely the way in which the radiation processes considered take place, as time goes on, and the properties of the stationary state. We shall now, in the first place, consider in the second chapter the absorption, and, then, in the third chapter the emission and the stationary distribution of energy, and, lastly, in the fourth chapter we shall compare the stationary state of the system of oscillators thus found with the thermodynamic state of equilibrium which was derived directly from the hypothesis of quanta in the preceding part. If we find them to agree, the hypothesis of emission of quanta may be regarded as admissible.

It is true that we shall not thereby prove that this hypothesis represents the only possible or even the most adequate expression of the elementary dynamical law of the vibrations of the oscillators. On the contrary I think it very probable that it may be greatly improved as regards form and contents. There is, however, no method of testing its admissibility except by the investigation of its consequences, and as long as no contradiction in itself or with experiment is discovered in it, and as long as no more adequate hypothesis can be advanced to replace it, it may justly claim a certain importance.

CHAPTER II

ABSORBED ENERGY

148. Let us consider an oscillator which has just completed an emission and which has, accordingly, lost all its energy of vibration. If we reckon the time t from this instant then for $t=0$ we have $f=0$ and $df/dt=0$, and the vibration takes place according to equation (233). Let us write E_z as in (149) in the form of a Fourier's series:

$$\mathsf{E}_z=\sum_{n=1}^{n=\infty}\left[A_n \cos \frac{2\pi nt}{\mathsf{T}}+B_n \sin \frac{2\pi nt}{\mathsf{T}}\right], \tag{235}$$

where T may be chosen very large, so that for all times t considered $t<\mathsf{T}$. Since we assume the radiation to be stationary, the constant coefficients A_n and B_n depend on the ordinal numbers n in a wholly irregular way, according to the hypothesis of natural radiation (Sec. 117). The partial vibration with the ordinal number n has the frequency ν, where

$$\omega=2\pi\nu=\frac{2\pi n}{\mathsf{T}}, \tag{236}$$

while for the frequency ν_o of the natural period of the oscillator

$$\omega_o=2\pi\nu_o=\sqrt{\frac{K}{L}}.$$

Taking the initial condition into account, we now obtain as the solution of the differential equation (233) the expression

$$f=\sum_{1}^{\infty}\left[a_n(\cos \omega t-\cos \omega_o t)+b_n\left(\sin \omega t-\frac{\omega}{\omega_o}\sin \omega_o t\right)\right], \tag{237}$$

where

$$a_n=\frac{A_n}{L(\omega_o{}^2-\omega^2)}, \quad b_n=\frac{B_n}{L(\omega_o{}^2-\omega^2)}. \tag{238}$$

This represents the vibration of the oscillator up to the instant when the next emission occurs.

The coefficients a_n and b_n attain their largest values when ω is nearly equal to ω_o. (The case $\omega=\omega_o$ may be excluded by assuming at the outset that $\nu_o\mathsf{T}$ is not an integer.)

149. Let us now calculate the total energy which is absorbed by the oscillator in the time from $t=0$ to $t=\tau$, where

$$\omega_o\,\tau \text{ is large.} \tag{239}$$

According to equation (234), it is given by the integral

$$\int_0^\tau \mathsf{E}_z \frac{df}{dt}\,dt, \tag{240}$$

the value of which may be obtained from the known expression for E_z (235) and from

$$\frac{df}{dt}=\sum_1^\infty [a_n(-\omega\sin\omega t+\omega_o\sin\omega_o t)+b_n(\omega\cos\omega t-\omega\cos\omega_o t)]. \tag{241}$$

By multiplying out, substituting for a_n and b_n their values from (238), and leaving off all terms resulting from the multiplication of two constants A_n and B_n, this gives for the absorbed energy the following value:

$$\frac{1}{L}\int_0^\tau dt\sum_1^\infty\left[\frac{A_n^2}{\omega_o^2-\omega^2}\cos\omega t(-\omega\sin\omega t+\omega_o\sin\omega_o t)+\right.$$
$$\left.\frac{B_n^2}{\omega_o^2-\omega^2}\sin\omega t(\omega\cos\omega t-\omega\cos\omega_o t)\right]. \tag{241a}$$

In this expression the integration with respect to t may be performed term by term. Substituting the limits τ and 0 it gives

$$\frac{1}{L}\sum_1^\infty\frac{A_n^2}{\omega_o^2-\omega^2}\left[-\frac{\sin^2\omega\tau}{2}+\omega_o\left(\frac{\sin^2\frac{\omega_o+\omega}{2}\tau}{\omega_o+\omega}+\frac{\sin^2\frac{\omega_o-\omega}{2}\tau}{\omega_o-\omega}\right)\right]$$
$$+\frac{1}{L}\sum_1^\infty\frac{B_n^2}{\omega_o^2-\omega^2}\left[\frac{\sin^2\omega\tau}{2}-\omega\left(\frac{\sin^2\frac{\omega_o+\omega}{2}\tau}{\omega_o+\omega}-\frac{\sin^2\frac{\omega_o-\omega}{2}\tau}{\omega_o-\omega}\right)\right]$$

In order to separate the terms of different order of magnitude, this expression is to be transformed in such a way that the difference $\omega_o - \omega$ will appear in all terms of the sum. This gives

$$\frac{1}{L}\sum_1^\infty \frac{A_n^2}{\omega_o^2-\omega^2}\left[\frac{\omega_o-\omega}{2(\omega_o+\omega)}\sin^2\omega\tau+\frac{\omega_o}{\omega_o+\omega}\sin\frac{\omega_o-\omega}{2}\tau\cdot\sin\frac{\omega_o+3\omega}{2}\tau\right.$$
$$\left.+\frac{\omega_o}{\omega_o-\omega}\sin^2\frac{\omega_o-\omega}{2}\tau\right]$$
$$+\frac{1}{L}\sum_1^\infty \frac{B_n{}^2}{\omega_o^2-\omega^2}\left[\frac{\omega_o-\omega}{2(\omega_o+\omega)}\sin^2\omega\tau\right.$$
$$\left.-\frac{\omega}{\omega_o+\omega}\sin\frac{\omega_o-\omega}{2}\tau\cdot\sin\frac{\omega_o+3\omega}{2}\tau+\frac{\omega}{\omega_o-\omega}\sin^2\frac{\omega_o-\omega}{2}\tau\right].$$

The summation with respect to the ordinal numbers n of the Fourier's series may now be performed. Since the fundamental period T of the series is extremely large, there corresponds to the difference of two consecutive ordinal numbers, $\Delta n=1$ only a very small difference of the corresponding values of $\omega, d\omega$, namely, according to (236),

$$\Delta n=1=\mathsf{T}d\nu=\frac{\mathsf{T}d\omega}{2\pi}, \tag{242}$$

and the summation with respect to n becomes an integration with respect to ω.

The last summation with respect to A_n may be rearranged as the sum of three series, whose orders of magnitude we shall first compare. So long as only the order is under discussion we may disregard the variability of the A_n^2 and need only compare the three integrals

$$\int_o^\infty d\omega\frac{\sin^2\omega\tau}{2(\omega_o+\omega)^2}=J_1,$$

$$\int_o^\infty d\omega\frac{\omega_o}{(\omega_o+\omega)^2(\omega_o-\omega)}\sin\frac{\omega_o-\omega}{2}\tau\cdot\sin\frac{\omega_o+3\omega}{2}\tau=J_2,$$

and

$$\int_o^\infty d\omega\frac{\omega_o}{(\omega_o+\omega)(\omega_o-\omega)^2}\sin^2\frac{\omega_o-\omega}{2}\tau=J_3.$$

The evaluation of these integrals is greatly simplified by the fact that, according to (239), $\omega_o\tau$ and therefore also $\omega\tau$ are large numbers, at least for all values of ω which have to be considered. Hence it is possible to replace the expression $\sin^2\omega\tau$ in the integral J_1 by its mean value $\frac{1}{2}$ and thus we obtain:

$$J_1 = \frac{1}{4\omega_o}$$

It is readily seen that, on account of the last factor, we obtain

$$J_2 = 0$$

for the second integral.

In order finally to calculate the third integral J_3 we shall lay off in the series of values of ω on both sides of ω_o an interval extending from $\omega_1(<\omega_o)$ to $\omega_2(>\omega_o)$ such that

$$\frac{\omega_o - \omega_1}{\omega_o} \text{ and } \frac{\omega_2 - \omega_o}{\omega_o} \text{ are small,} \tag{243}$$

and simultaneously

$$(\omega_o - \omega_1)\tau \text{ and } (\omega_2 - \omega_o)\tau \text{ are large.} \tag{244}$$

This can always be done, since $\omega_o\tau$ is large. If we now break up the integral J_3 into three parts, as follows:

$$J_3 = \int_0^\infty = \int_0^{\omega_1} + \int_{\omega_1}^{\omega_2} + \int_{\omega_2}^\infty,$$

it is seen that in the first and third partial integral the expression $\sin^2 \frac{\omega_o - \omega}{2}\tau$ may, because of the condition (244), be replaced by its mean value $\frac{1}{2}$. Then the two partial integrals become:

$$\int_0^{\omega_1} \frac{\omega_o d\omega}{2(\omega_o + \omega)(\omega_o - \omega)^2} \text{ and } \int_{\omega_2}^\infty \frac{\omega_o d\omega}{2(\omega_o + \omega)(\omega_o - \omega)^2}. \tag{245}$$

These are certainly smaller than the integrals:

$$\int_0^{\omega_1} \frac{d\omega}{2(\omega_o - \omega)^2} \text{ and } \int_{\omega_2}^\infty \frac{d\omega}{2(\omega_o - \omega)^2}$$

which have the values

$$\frac{1}{2}\,\frac{\omega_1}{\omega_o(\omega_o-\omega_1)} \quad \text{and} \quad \frac{1}{2(\omega_2-\omega_o)} \tag{246}$$

respectively. We must now consider the middle one of the three partial integrals:

$$\int_{\omega_1}^{\omega_2} d\omega \frac{\omega_o}{(\omega_o+\omega)(\omega_o-\omega)^2}\cdot \sin^2\frac{\omega_o-\omega}{2}\tau.$$

Because of condition (243) we may write instead of this:

$$\int_{\omega_1}^{\omega_2} d\omega \cdot \frac{\sin^2\dfrac{\omega_o-\omega}{2}\tau}{2(\omega_o-\omega)^2}$$

and by introducing the variable of integration x, where

$$x=\frac{\omega-\omega_o}{2}\tau$$

and taking account of condition (244) for the limits of the integral, we get:

$$\frac{\tau}{4}\int_{-\infty}^{+\infty}\frac{\sin^2 x\, dx}{x^2}=\frac{\tau}{4}\pi.$$

This expression is of a higher order of magnitude than the expressions (246) and hence of still higher order than the partial integrals (245) and the integrals J_1 and J_2 given above. Thus for our calculation only those values of ω will contribute an appreciable part which lie in the interval between ω_1 and ω_2, and hence we may, because of (243), replace the separate coefficients $A_n{}^2$ and $B_n{}^2$ in the expression for the total absorbed energy by their mean values $A_o{}^2$ and $B_o{}^2$ in the neighborhood of ω_o and thus, by taking account of (242), we shall finally obtain for the total value of the energy absorbed by the oscillator in the time τ:

$$\frac{1}{L}\frac{\tau}{8}(A_o{}^2+B_o{}^2)\ \mathsf{T} \tag{247}$$

If we now, as in (158), define I, the "intensity of the vibration exciting the oscillator," by spectral resolution of the mean value of the square of the exciting field-strength E_z:

$$\overline{\mathsf{E}_z^2} = \int_0^\infty \mathsf{I}_\nu \, d\nu \tag{248}$$

we obtain from (235) and (242):

$$\overline{\mathsf{E}_z^2} = \tfrac{1}{2} \sum_1^\infty (A_n^2 + B_n^2) = \tfrac{1}{2} \int_0^\infty (A_n^2 + B_n^2) \, \mathsf{T} \, d\nu,$$

and by comparison with (248):

$$\mathsf{I} = \tfrac{1}{2} (A_o^2 + B_o^2) \, \mathsf{T}.$$

Accordingly from (247) the energy absorbed in the time τ becomes:

$$\frac{\mathsf{I}}{4L}\tau,$$

that is, *in the time between two successive emissions, the energy U of the oscillator increases uniformly with the time*, according to the law

$$\frac{dU}{dt} = \frac{\mathsf{I}}{4L} = a. \tag{249}$$

Hence the energy absorbed by all N oscillators in the time dt is:

$$\frac{N\mathsf{I}}{4L} \, dt = Na \, dt. \tag{250}$$

CHAPTER III

EMITTED ENERGY. STATIONARY STATE

150. Whereas the absorption of radiation by an oscillator takes place in a perfectly continuous way, so that the energy of the oscillator increases continuously and at a constant rate, for its emission we have, in accordance with Sec. 147, the following law: The oscillator emits in irregular intervals, subject to the laws of chance; it emits, however, only at a moment when its energy of vibration is just equal to an integral multiple n of the elementary quantum $\epsilon = h\nu$, and then it always emits its whole energy of vibration $n\epsilon$.

We may represent the whole process by the following figure in which the abscissæ represent the time t and the ordinates the energy

$$U = n\epsilon + \rho, \ (\rho < \epsilon) \tag{251}$$

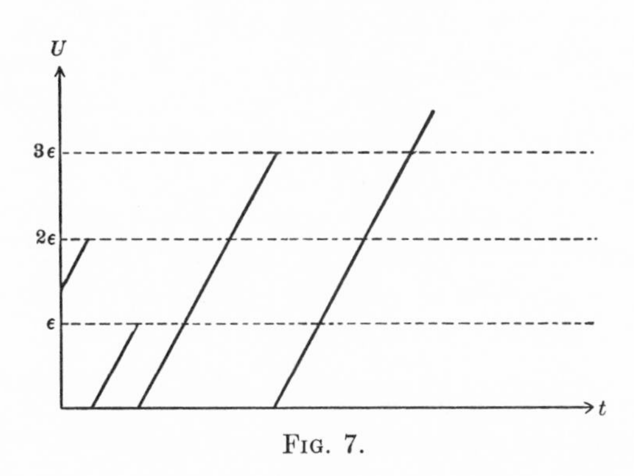

Fig. 7.

of a definite oscillator under consideration. The oblique parallel lines indicate the continuous increase of energy at a constant rate.

$$\frac{dU}{dt} = \frac{d\rho}{dt} = a, \tag{252}$$

11 161

which is, according to (249), caused by absorption at a constant rate. Whenever this straight line intersects one of the parallels to the axis of abscissæ $U=\epsilon$, $U=2\epsilon$, ... emission may possibly take place, in which case the curve drops down to zero at that point and immediately begins to rise again.

151. Let us now calculate the most important properties of the state of statistical equilibrium thus produced. Of the N oscillators situated in the field of radiation the number of those whose energy at the time t lies in the interval between $U=n\epsilon+\rho$ and $U+dU=n\epsilon+\rho+d\rho$ may be represented by

$$NR_{n,\rho}d\rho, \tag{253}$$

where R depends in a definite way on the integer n and the quantity ρ which varies continuously between 0 and ϵ.

After a time $dt=\frac{d\rho}{a}$ all the oscillators will have their energy increased by $d\rho$ and hence they will all now lie outside of the energy interval considered. On the other hand, during the same time dt, all oscillators whose energy at the time t was between $n\epsilon+\rho-d\rho$ and $n\epsilon+\rho$ will have entered that interval. The number of all these oscillators is, according to the notation used above,

$$NR_{n,\ \rho-d\rho}d\rho. \tag{254}$$

Hence this expression gives the number of oscillators which are at the time $t+dt$ in the interval considered.

Now, since we assume our system to be in a state of statistical equilibrium, the distribution of energy is independent of the time and hence the expressions (253) and (254) are equal, *i.e.*,

$$R_{n,\ \rho-d\rho}=R_{n,\ \rho}=R_n. \tag{255}$$

Thus R_n does not depend on ρ.

This consideration must, however, be modified for the special case in which $\rho=0$. For, in that case, of the oscillators, $N=R_{n-1}d\rho$ in number, whose energy at the time t was between $n\epsilon$ and $n\epsilon-d\rho$, during the time $dt=\frac{d\rho}{a}$ some enter into the energy interval (from $U=n\epsilon$ to $U+dU=n\epsilon+d\rho$) considered; but all of them do not necessarily enter, for an oscillator may possibly emit all its energy on passing through the value $U=n\epsilon$. If the probability that emission takes place be denoted by $\eta(<1)$ the number of oscillators which pass through the critical value without emitting will be

$$NR_{n-1}(1-\eta)d\rho, \tag{256}$$

and by equating (256) and (253) it follows that

$$R_n = R_{n-1}(1-\eta),$$

and hence, by successive reduction,

$$R_n = R_o(1-\eta)^n. \tag{257}$$

To calculate R_o we repeat the above process for the special case when $n=0$ and $\rho=0$. In this case the energy interval in question extends from $U=0$ to $dU=d\rho$. Into this interval enter in the time $dt=\frac{d\rho}{a}$ all the oscillators which perform an emission during this time, namely, those whose energy at the time t was between $\epsilon - d\rho$ and ϵ, $2\epsilon - d\rho$ and 2ϵ, $3\epsilon - d\rho$ and 3ϵ . . . . .
The numbers of these oscillators are respectively

$$NR_0 d\rho,\ NR_1 d\rho,\ NR_2 d\rho,$$

hence their sum multiplied by η gives the desired number of emitting oscillators, namely,

$$N\eta(R_o+R_1+R_2+ \ldots\ldots)\ d\rho, \tag{258}$$

and this number is equal to that of the oscillators in the energy interval between 0 and $d\rho$ at the time $t+dt$, which is $NR_o d\rho$. Hence it follows that

$$R_o = \eta(R_o+R_1+R_2+ \ldots\ldots). \tag{259}$$

Now, according to (253), the whole number of all the oscillators is obtained by integrating with respect to ρ from 0 to ϵ, and summing up with respect to n from 0 to ∞. Thus

$$N = N\sum_{n=0}^{n=\infty}\int_o^{\epsilon} R_{n,\rho}\ d\rho = N\ \Sigma\ R_n\epsilon \tag{260}$$

and

$$\Sigma\ R_n = \frac{1}{\epsilon}. \tag{261}$$

Hence we get from (257) and (259)

$$R_o = \frac{\eta}{\epsilon}\ ,\ R_n = \frac{\eta}{\epsilon}\ (1-\eta)^n. \tag{262}$$

152. The total energy emitted in the time element $dt=\frac{d\rho}{a}$ is found from (258) by considering that every emitting oscillator expends all its energy of vibration and is

$$\begin{aligned}&N\eta\, d\rho(R_o+2R_1+3R_2+\ \ldots\ldots)\epsilon\\ &=N\eta\, d\rho\, \eta(1+2(1-\eta)+3(1-\eta)^2+\ \ldots\ldots)\\ &=N\, d\rho=Na\, dt.\end{aligned}$$

It is therefore equal to the energy absorbed in the same time by all oscillators (250), as is necessary, since the state is one of statistical equilibrium.

Let us now consider the *mean energy* $\overline{U}$ of an oscillator. It is evidently given by the following relation, which is derived in the same way as (260):

$$N\overline{U}=N\sum_0^\infty \int_o^\epsilon (n\epsilon+\rho)R_n\, d\rho. \qquad (263)$$

From this it follows by means of (262), that

$$\overline{U}=\left(\frac{1}{\eta}-\frac{1}{2}\right)\epsilon=\left(\frac{1}{\eta}-\frac{1}{2}\right)h\nu \qquad (264)$$

Since $\eta<1$, $\overline{U}$ lies between $\frac{h\nu}{2}$ and ∞. Indeed, it is immediately evident that $\overline{U}$ can never become less than $\frac{h\nu}{2}$ since the energy of *every* oscillator, however small it may be, will assume the value $\epsilon=h\nu$ within a time limit, which can be definitely stated.

153. The probability constant η contained in the formulæ for the stationary state is determined by the law of emission enunciated in Sec. 147. According to this, the ratio of the probability that no emission takes place to the probability that emission does take place is proportional to the intensity I of the vibration exciting the oscillator, and hence

$$\frac{1-\eta}{\eta}=p\mathsf{I} \qquad (265)$$

where the constant of proportionality is to be determined in such a way that for very large energies of vibration the familiar formulæ of classical dynamics shall hold.

Now, according to (264), η becomes small for large values of $\overline{U}$ and for this special case the equations (264) and (265) give

$$\overline{U}=ph\nu\mathsf{I},$$

and the energy emitted or absorbed respectively in the time dt by all N oscillators becomes, according to (250),

$$\frac{N\mathsf{I}}{4L}dt=\frac{N\overline{U}}{4Lph\nu}dt. \tag{266}$$

On the other hand, *H. Hertz* has already calculated from *Maxwell's* theory the energy emitted by a linear oscillator vibrating periodically. For the energy emitted in the time of one-half of one vibration he gives the expression[1]

$$\frac{\pi^4E^2l^2}{3\lambda^3}$$

where λ denotes half the wave length, and the product El (the C of our notation) denotes the amplitude of the moment f (Sec. 135) of the vibrations. This gives for the energy emitted in the time of a whole vibration

$$\frac{16\pi^4C^2}{3\lambda^3}$$

where λ denotes the whole wave length, and for the energy emitted by N similar oscillators in the time dt

$$N\frac{16\pi^4C^2\nu^4}{3c^3}dt$$

since $\lambda=\frac{c}{\nu}$. On introducing into this expression the energy U of an oscillator from (205), (207), and (208), namely

$$U=2\pi^2\nu^2LC^2,$$

we have for the energy emitted by the system of oscillators

$$N\frac{8\pi^2\nu^2U}{3c^3L}dt \tag{267}$$

[1] *H. Hertz,* Wied. Ann. **36**, p. 12, 1889.

and by equating the expressions (266) and (267) we find for the factor of proportionality p

$$p=\frac{3c^3}{32\pi^2 h\nu^3}. \tag{268}$$

154. By the determination of p the question regarding the properties of the state of statistical equilibrium between the system of the oscillators and the vibration exciting them receives a general answer. For from (265) we get

$$\eta=\frac{1}{1+p\mathfrak{I}}$$

and further from (262)

$$R_n=\frac{1}{\epsilon}\,\frac{(p\mathfrak{I})^n}{(1+p\mathfrak{I})^{n+1}}. \tag{269}$$

Hence in the state of stationary equilibrium the number of oscillators whose energy lies between $nh\nu$ and $(n+1)h\nu$ is, from equation (253),

$$N\int_0^\epsilon R_n d\rho=NR_n\epsilon=N\frac{(p\mathfrak{I})^n}{(1+p\mathfrak{I})^{n+1}} \tag{270}$$

where $n=0, 1, 2, 3, \ldots\ldots$

CHAPTER IV

THE LAW OF THE NORMAL DISTRIBUTION OF ENERGY. ELEMENTARY QUANTA OF MATTER AND ELECTRICITY

155. In the preceding chapter we have made ourselves familiar with all the details of a system of oscillators exposed to uniform radiation. We may now develop the idea put forth at the end of Sec. 144. That is to say, we may identify the stationary state of the oscillators just found with the state of maximum entropy of the system of oscillators which was derived directly from the hypothesis of quanta in the preceding part, and we may then equate the temperature of the radiation to the temperature of the oscillators. It is, in fact, possible to obtain perfect agreement of the two states by a suitable coordination of their corresponding quantities.

According to Sec. 139, the "distribution density" w of the oscillators in the state of statistical equilibrium changes abruptly from one region element to another, while, according to Sec. 138, the distribution within a single region element is uniform. The region elements of the state plane $(f\psi)$ are bounded by concentric similar and similarly situated ellipses which correspond to those values of the energy U of an oscillator which are integral multiples of $h\nu$. We have found exactly the same thing for the stationary state of the oscillators when they are exposed to uniform radiation, and the distribution density w_n in the nth region element may be found from (270), if we remember that the nth region element contains the energies between $(n-1)h\nu$ and $nh\nu$. Hence:

$$w_n=\frac{(p\mathsf{l})^{n-1}}{(1+p\mathsf{l})^n}=\frac{1}{p\mathsf{l}}\left(\frac{p\mathsf{l}}{1+p\mathsf{l}}\right)^n. \tag{271}$$

This is in perfect agreement with the previous value (220) of w_n if we put

$$\alpha=\frac{1}{p\mathsf{l}} \text{ and } \gamma=\frac{p\mathsf{l}}{1+p\mathsf{l}},$$

and each of these two equations leads, according to (221), to the following relation between the intensity of the exciting vibration I and the total energy E of the N oscillators:

$$p\mathsf{I}=\frac{E}{Nh\nu}-\frac{1}{2}\ . \tag{272}$$

156. If we finally introduce the temperature T from (223), we get from the last equation, by taking account of the value (268) of the factor of proportionality p,

$$\mathsf{I}=\frac{32\pi^2h\nu^3}{3c^3}\,\frac{1}{e^{\frac{h\nu}{kT}}-1} \tag{273}$$

Moreover the specific intensity K of a monochromatic plane polarized ray of frequency ν is, according to equation (160),

$$\mathsf{K}=\frac{h\nu^3}{c^2}\,\frac{1}{e^{\frac{h\nu}{kT}}-1} \tag{274}$$

and the space density of energy of uniform monochromatic unpolarized radiation of frequency ν is, from (159),

$$\mathsf{u}=\frac{8\pi h\nu^3}{c^3}\,\frac{1}{e^{\frac{h\nu}{kT}}-1} \tag{275}$$

Since, among all the forms of radiation of differing constitutions, black radiation is distinguished by the fact that all monochromatic rays contained in it have the same temperature (Sec. 93) these equations also give the law of distribution of energy in the normal spectrum, *i.e.*, in the emission spectrum of a body which is black with respect to the vacuum.

If we refer the specific intensity of a monochromatic ray not to the frequency ν but, as is usually done in experimental physics, to the wave length λ, by making use of (15) and (16) we obtain the expression

$$E_\lambda=\frac{c^2h}{\lambda^5}\,\frac{1}{e^{\frac{ch}{k\lambda T}}-1}=\frac{c_1}{\lambda^5}\,\frac{1}{e^{\frac{c_2}{\lambda T}}-1} \tag{276}$$

This is the specific intensity of a monochromatic plane polarized ray of the wave length λ which is emitted from a black body at the temperature T into a vacuum in a direction perpendicular to the surface. The corresponding space density of unpolarized radiation is obtained by multiplying E_λ by $\frac{8\pi}{c}$.

Experimental tests have so far confirmed equation (276).[1] According to the most recent measurements made in the Physikalisch-technische Reichsanstalt[2] the value of the second radiation constant c_2 is approximately

$$c_2 = \frac{ch}{k} = 1.436 \text{ cm degree.}$$

More detailed information regarding the history of the equation of radiation is to be found in the original papers and in the first edition of this book. At this point it may merely be added that equation (276) was not simply extrapolated from radiation measurements, but was originally found in a search after a connection between the entropy and the energy of an oscillator vibrating in a field, a connection which would be as simple as possible and consistent with known measurements.

157. The entropy of a ray is, of course, also determined by its temperature. In fact, by combining equations (138) and (274) we readily obtain as an expression for the entropy radiation L of a monochromatic plane polarized ray of the specific intensity of radiation K and the frequency ν,

$$\mathsf{L} = \frac{k\nu^2}{c^2}\left\{\left(1+\frac{c^2\mathsf{K}}{h\nu^3}\right)\log\left(1+\frac{c^2\mathsf{K}}{h\nu^3}\right) - \frac{c^2\mathsf{K}}{h\nu^3}\log\frac{c^2\mathsf{K}}{h\nu^3}\right\} \qquad (278)$$

which is a more definite statement of equation (134) for *Wien's* displacement law.

Moreover it follows from (135), by taking account of (273), that the space density of the entropy s of uniform monochromatic unpolarized radiation as a function of the space density of energy u is

$$\mathsf{s} = \frac{8\pi k\nu^2}{c^3}\left\{\left(1+\frac{c^3\mathsf{u}}{8\pi h\nu^3}\right)\log\left(1+\frac{c^3\mathsf{u}}{8\pi h\nu^3}\right) - \frac{c^3\mathsf{u}}{8\pi h\nu^3}\log\frac{c^3\mathsf{u}}{8\pi h\nu^3}\right\} \qquad (279)$$

This is a more definite statement of equation (119).

[1] See among others *H. Rubens* und *F. Kurlbaum*, Sitz. Ber. d. Akad. d. Wiss. zu Berlin vom 25. Okt., 1900, p. 929. Ann. d. Phys. **4**, p. 649, 1901. *F. Paschen*, Ann. d. Phys. **4**, p. 277, 1901. *O. Lummer* und *E. Pringsheim*, Ann. d. Phys. **6**, p. 210, 1901. Tätigkeitsbericht der Phys.-Techn. Reichsanstalt vom J. 1911, Zeitschr. f. Instrumentenkunde, 1912, April, p. 134 ff.

[2] According to private information kindly furnished by the president, *Mr. Warburg*.

158. For *small* values of λT (*i.e.*, small compared with the constant $\left.\frac{ch}{k}\right)$ equation (276) becomes

$$E_\lambda = \frac{c^2 h}{\lambda^5} e^{-\frac{ch}{k\lambda T}} \tag{280}$$

an equation which expresses *Wien's*[1] law of energy distribution.

The specific intensity of radiation K then becomes, according to (274),

$$\mathsf{K} = \frac{h\nu^3}{c^2} e^{-\frac{h\nu}{kT}} \tag{281}$$

and the space density of energy u is, from (275),

$$\mathsf{u} = \frac{8\pi h\nu^3}{c^3} e^{-\frac{h\nu}{kT}} \tag{282}$$

159. On the other hand, for *large* values of λT (276) becomes

$$E_\lambda = \frac{ckT}{\lambda^4} \tag{283}$$

a relation which was established first by *Lord Rayleigh*[2] and which we may, therefore, call "*Rayleigh's* law of radiation."

We then find for the specific intensity of radiation K from (274)

$$\mathsf{K} = \frac{k\nu^2 T}{c^2} \tag{284}$$

and from (275) for the space density of monochromatic radiation we get

$$\mathsf{u} = \frac{8\pi k\nu^2 T}{c^3} \tag{285}$$

Rayleigh's law of radiation is of very great theoretical interest, since it represents that distribution of energy which is obtained for radiation in statistical equilibrium with material molecules by means of the classical dynamics, and without introducing the hypothesis of quanta.[3] This may also be seen from the fact that for a vanishingly small value of the quantity element of action, h, the general formula (276) degenerates into *Rayleigh's* formula (283). See also below, Sec. 168 *et seq.*

[1] *W. Wien*, Wied. Ann. **58,** p. 662, 1896.

[2] *Lord Rayleigh*, Phil. Mag. **49,** p. 539, 1900.

[3] *J. H. Jeans*, Phil. Mag. Febr., 1909, p. 229, *H. A. Lorentz*, Nuovo Cimento V, vol. **16,** 1908.

160. For the total space density, u, of black radiation at any temperature T we obtain, from (275),

$$u=\int_o^\infty \mathfrak{u}\,d\nu=\frac{8\pi h}{c^3}\int_o^\infty \frac{\nu^3 d\nu}{e^{\frac{h\nu}{kT}}-1}$$

or

$$u=\frac{8\pi h}{c^3}\int_o^\infty\left(e^{-\frac{h\nu}{kT}}+e^{-\frac{2h\nu}{kT}}+e^{-\frac{3h\nu}{kT}}+\ .\ .\ .\ .\ .\ \right)\nu^3 d\nu$$

and, integrating term by term,

$$u=\frac{48\pi h}{c^3}\left(\frac{kT}{h}\right)^4\alpha \qquad (286)$$

where α is an abbreviation for

$$\alpha=1+\frac{1}{2^4}+\frac{1}{3^4}+\frac{1}{4^4}+\ .\ .\ .\ .\ .\ =1.0823. \qquad (287)$$

This relation expresses the *Stefan-Boltzmann* law (75) and it also tells us that the constant of this law is given by

$$a=\frac{48\pi\alpha k^4}{c^3h^3}. \qquad (288)$$

161. For that wave length λ_m to which the maximum of the intensity of radiation corresponds in the spectrum of black radiation, we find from (276)

$$\left(\frac{dE_\lambda}{d\lambda}\right)_{\lambda=\lambda_m}=0.$$

On performing the differentiation and putting as an abbreviation

$$\frac{ch}{k\lambda_m T}=\beta,$$

we get

$$e^{-\beta}+\frac{\beta}{5}-1=0.$$

The root of this transcendental equation is

$$\beta=4.9651, \qquad (289)$$

and accordingly $\lambda_m T=\frac{ch}{\beta k}$, and this is a constant, as demanded by *Wien's* displacement law. By comparison with (109) we find the meaning of the constant b, namely,

$$b=\frac{ch}{\beta k}, \tag{290}$$

and, from (277),

$$b=\frac{c_2}{\beta}=\frac{1.436}{4.9651}=0.289 \text{ cm} \cdot \text{degree}, \tag{291}$$

while *Lummer* and *Pringsheim* found by measurements 0.294 and *Paschen* 0.292.

162. By means of the measured values[1] of a and c_2 the universal constants h and k may be readily calculated. For it follows from equations (277) and (288) that

$$h=\frac{ac_2^4}{48\pi\alpha c} \qquad k=\frac{ac_2^3}{48\pi\alpha} \tag{292}$$

Substituting the values of the constants a, c_2, α, c, we get

$$h=6.415 \cdot 10^{-27} \text{ erg sec.}, \qquad k=1.34 \cdot 10^{-16} \frac{\text{erg}}{\text{degree}} \tag{293}$$

163. To ascertain the full physical significance of the quantity element of action, h, much further research work will be required. On the other hand, the value obtained for k enables us readily to state numerically in the C. G. S. system the general connection between the entropy S and the thermodynamic probability W as expressed by the universal equation (164). The general expression for the entropy of a physical system is

$$S=1.34 \cdot 10^{-16} \log W \frac{\text{erg}}{\text{degree}} \tag{294}$$

This equation may be regarded as the most general definition of entropy. Herein the thermodynamic probability W is an integral number, which is completely defined by the macroscopic state of the system. Applying the result expressed in (293) to the kinetic

[1] Here as well as later on the value given above (79) has been replaced by $a=7.39 \cdot 10^{-15}$, obtained from $\sigma=a\,c/4=5.54 \cdot 10^{-5}$. This is the final result of the newest measurements made by *W. Westphal*, according to information kindly furnished by him and Mr. *H. Rubens*. (Nov., 1912). [Compare p. 64, footnote. Tr.]

theory of gases, we obtain from equation (194) for the ratio of the mass of a molecule to that of a mol,

$$\omega=\frac{k}{R}=\frac{1.34\times10^{-16}}{831\times10^{5}}=1.61\times10^{-24}, \tag{295}$$

that is to say, there are in one mol

$$\frac{1}{\omega}=6.20\times10^{23}$$

molecules, where the mol of oxygen, O_2, is always assumed as 32 gr. Hence, for example, the absolute mass of a hydrogen atom ($\frac{1}{2}H_2=1.008$) equals 1.62×10^{-24} gr. With these numerical values the number of molecules contained in 1 cm.³ of an ideal gas at 0° C. and 1 atmosphere pressure becomes

$$\mathbf{N}=\frac{76\cdot13.6\cdot981}{831\cdot10^{5}\cdot273\omega}=2.77\cdot10^{19}. \tag{296}$$

The mean kinetic energy of translatory motion of a molecule at the absolute temperature $T=1$ is, in the absolute C. G. S. system, according to (200),

$$\frac{3}{2}k=2.01\cdot10^{-16} \tag{297}$$

In general the mean kinetic energy of translatory motion of a molecule is expressed by the product of this number and the absolute temperature T.

The elementary quantity of electricity or the free charge of a monovalent ion or electron is, in electrostatic units,

$$e=\omega\cdot9654\cdot3\cdot10^{10}=4.67\cdot10^{-10}. \tag{298}$$

Since absolute accuracy is claimed for the formulæ here employed, the degree of approximation to which these numbers represent the corresponding physical constants depends only on the accuracy of the measurements of the two radiation constants a and c_2.

164. Natural Units.—All the systems of units which have hitherto been employed, including the so-called absolute C. G. S. system, owe their origin to the coincidence of accidental circumstances, inasmuch as the choice of the units lying at the base of every system has been made, not according to general points of view which would necessarily retain their importance for all places and all times, but essentially with reference to the special needs of our terrestrial civilization.

Thus the units of length and time were derived from the present dimensions and motion of our planet, and the units of mass and temperature from the density and the most important temperature points of water, as being the liquid which plays the most important part on the surface of the earth, under a pressure which corresponds to the mean properties of the atmosphere surrounding us. It would be no less arbitrary if, let us say, the invariable wave length of Na-light were taken as unit of length. For, again, the particular choice of Na from among the many chemical elements could be justified only, perhaps, by its common occurrence on the earth, or by its double line, which is in the range of our vision, but is by no means the only one of its kind. Hence it is quite conceivable that at some other time, under changed external conditions, every one of the systems of units which have so far been adopted for use might lose, in part or wholly, its original natural significance.

In contrast with this it might be of interest to note that, with the aid of the two constants h and k which appear in the universal law of radiation, we have the means of establishing units of length, mass, time, and temperature, which are independent of special bodies or substances, which necessarily retain their significance for all times and for all environments, terrestrial and human or otherwise, and which may, therefore, be described as "natural units."

The means of determining the four units of length, mass, time, and temperature, are given by the two constants h and k mentioned, together with the magnitude of the velocity of propagation of light in a vacuum, c, and that of the constant of gravitation, f. Referred to centimeter, gram, second, and degrees Centigrade, the numerical values of these four constants are as follows:

$$h = 6.415 \cdot 10^{-27} \frac{g\,cm^2}{sec}$$

$$k = 1.34 \cdot 10^{-16} \frac{g\,\text{cm}^2}{\text{sec}^2\text{degree}}$$

$$c = 3 \cdot 10^{10} \frac{\text{cm}}{\text{sec}}$$

$$f = 6.685 \cdot 10^{-8} \frac{\text{cm}^3}{g\,\text{sec}^2}$$ [1]

If we now choose the natural units so that in the new system of measurement each of the four preceding constants assumes the value 1, we obtain, as unit of length, the quantity

$$\sqrt{\frac{fh}{c^3}} = 3.99 \cdot 10^{-33} \text{ cm},$$

as unit of mass

$$\sqrt{\frac{ch}{f}} = 5.37 \cdot 10^{-5}\, g,$$

as unit of time

$$\sqrt{\frac{fh}{c^5}} = 1.33 \cdot 10^{-43} \text{ sec},$$

as unit of temperature

$$\frac{1}{k}\sqrt{\frac{c^5 h}{f}} = 3.60 \cdot 10^{32} \text{ degree}.$$

These quantities retain their natural significance as long as the law of gravitation and that of the propagation of light in a vacuum and the two principles of thermodynamics remain valid; they therefore must be found always the same, when measured by the most widely differing intelligences according to the most widely differing methods.

165. The relations between the intensity of radiation and the temperature expressed in Sec. 156 hold for radiation in a pure vacuum. If the radiation is in a medium of refractive index n, the way in which the intensity of radiation depends on the frequency and the temperature is given by the proposition of Sec. 39, namely, the product of the specific intensity of radiation K_ν and the square of the velocity of propagation of the radiation

[1] *F. Richarz* and *O. Krigar-Menzel*, Wied. Ann. **66**, p. 190, 1898.

has the same value for all substances. The form of this universal function (42) follows directly from (274)

$$\mathsf{K}q^2 = \frac{\epsilon_\nu}{\alpha_\nu} q^2 = \frac{h\nu^3}{e^{\frac{h\nu}{kT}} - 1} \tag{299}$$

Now, since the refractive index n is inversely proportional to the velocity of propagation, equation (274) is, in the case of a medium with the index of refraction n, replaced by the more general relation

$$\mathsf{K}_\nu = \frac{h\nu^3 n^2}{c^2} \frac{1}{e^{\frac{h\nu}{kT}} - 1} \tag{300}$$

and, similarly, in place of (275) we have the more general relation

$$\mathsf{u} = \frac{8\pi h\nu^3 n^3}{c^3} \frac{1}{e^{\frac{h\nu}{kT}} - 1} \tag{301}$$

These expressions hold, of course, also for the emission of a body which is black with respect to a medium with an index of refraction n.

166. We shall now use the laws of radiation we have obtained to calculate the temperature of a monochromatic unpolarized radiation of given intensity in the following case. Let the light pass normally through a small area (slit) and let it fall on an arbitrary system of diathermanous media separated by spherical surfaces, the centers of which lie on the same line, the axis of the system. Such radiation consists of homocentric pencils and hence forms behind every refracting surface a real or virtual image of the emitting surface, the image being likewise normal to the axis. To begin with, we assume the last as well as the first medium to be a pure vacuum. Then, for the determination of the temperature of the radiation according to equation (274), we need calculate only the specific intensity of radiation K_ν in the last medium, and this is given by the total intensity of the monochromatic radiation I_ν, the size of the area of the image F, and the solid angle Ω of the cone of rays passing through a point of the image. For the specific intensity of radiation K_ν is, according to (13), determined by the fact that an amount

$$2\mathsf{K}_\nu \, d\sigma \, d\Omega \, d\nu \, dt$$

of energy of unpolarized light corresponding to the interval of frequencies from ν to $\nu+d\nu$ is, in the time dt, radiated in a normal direction through an element of area $d\sigma$ within the conical element $d\Omega$. If now $d\sigma$ denotes an element of the area of the surface image in the last medium, then the total monochromatic radiation falling on the image has the intensity

$$I_\nu = 2\mathsf{K}_\nu \int d\sigma \int d\Omega.$$

I_ν is of the dimensions of energy, since the product $d\nu\, dt$ is a mere number. The first integral is the whole area, F, of the image, the second is the solid angle, Ω, of the cone of rays passing through a point of the surface of the image. Hence we get

$$I_\nu = 2\mathsf{K}_\nu F\Omega, \tag{302}$$

and, by making use of (274), for the temperature of the radiation

$$T = \frac{h\nu}{k} \cdot \frac{1}{\log\left(\frac{2h\nu^3 F\Omega}{c^2 I_\nu}+1\right)} \tag{303}$$

If the diathermanous medium considered is not a vacuum but has an index of refraction n, (274) is replaced by the more general relation (300), and, instead of the last equation, we obtain

$$T = \frac{h\nu}{k} \frac{1}{\log\left(\frac{2h\nu^3 F\Omega n^2}{c^2 I_\nu}+1\right)} \tag{304}$$

or, on substituting the numerical values of c, h, and k,

$$T = \frac{0.479 \cdot 10^{-10}\nu}{\log\left(\frac{1.43 \cdot 10^{-47}\nu^3 F\Omega n^2}{I_\nu}+1\right)} \text{ degree Centigrade.}$$

In this formula the natural logarithm is to be taken, and I_ν is to be expressed in ergs, ν in "reciprocal seconds," *i.e.*, (seconds)$^{-1}$, F in square centimeters. In the case of visible rays the second term, 1, in the denominator may usually be omitted.

The temperature thus calculated is retained by the radiation considered, so long as it is propagated without any disturbing

12

influence in the diathermanous medium, however great the distance to which it is propagated or the space in which it spreads. For, while at larger distances an ever decreasing amount of energy is radiated through an element of area of given size, this is contained in a cone of rays starting from the element, the angle of the cone continually decreasing in such a way that the value of K remains entirely unchanged. Hence the free expansion of radiation is a perfectly reversible process. (Compare above, Sec. 144.) It may actually be reversed by the aid of a suitable concave mirror or a converging lens.

Let us next consider the temperature of the radiation in the other media, which lie between the separate refracting or reflecting spherical surfaces. In every one of these media the radiation has a definite temperature, which is given by the last formula when referred to the real or virtual image formed by the radiation in that medium.

The frequency ν of the monochromatic radiation is, of course, the same in all media; moreover, according to the laws of geometrical optics, the product $n^2F\Omega$ is the same for all media. Hence, if, in addition, the total intensity of radiation I_ν remains constant on refraction (or reflection), T also remains constant, or in other words: The temperature of a homocentric pencil is not changed by regular refraction or reflection, unless a loss in energy of radiation occurs. Any weakening, however, of the total intensity I_ν by a subdivision of the radiation, whether into two or into many different directions, as in the case of diffuse reflection, leads to a lowering of the temperature of the pencil. In fact, a certain loss of energy by refraction or reflection does occur, in general, on a refraction or reflection, and hence also a lowering of the temperature takes place. In these cases a fundamental difference appears, depending on whether the radiation is weakened merely by free expansion or by subdivision or absorption. In the first case the temperature remains constant, in the second it decreases.[1]

167. The laws of emission of a black body having been deter-

[1] Nevertheless regular refraction and reflection are not irreversible processes; for the refracted and the reflected rays are coherent and the entropy of two coherent rays is not equal to the sum of the entropies of the separate rays. (Compare above, Sec. 104.) On the other hand, diffraction is an irreversible process. *M. Laue*, Ann. d. Phys. **31,** p. 547, 1910.

mined, it is possible to calculate, with the aid of *Kirchhoff's* law (48), the emissive power E of any body whatever, when its absorbing power A or its reflecting power $1-A$ is known. In the case of metals this calculation becomes especially simple for long waves, since *E. Hagen* and *H. Rubens*[1] have shown experimentally that the reflecting power and, in fact, the entire optical behavior of the metals in the spectral region mentioned is represented by the simple equations of *Maxwell* for an electromagnetic field with homogeneous conductors and hence depends only on the specific conductivity for steady electric currents. Accordingly, it is possible to express completely the emissive power of a metal for long waves by its electric conductivity combined with the formulæ for black radiation.[2]

168. There is, however, also a method, applicable to the case of long waves, for the direct theoretical determination of the electric conductivity and, with it, of the absorbing power, A, as well as the emissive power, E, of metals. This is based on the ideas of the electron theory, as they have been developed for the thermal and electrical processes in metals by *E. Riecke*[3] and especially by *P. Drude.*[4] According to these, all such processes are based on the rapid irregular motions of the negative electrons, which fly back and forth between the positively charged molecules of matter (here of the metal) and rebound on impact with them as well as with one another, like gas molecules when they strike a rigid obstacle or one another. The velocity of the heat motions of the material molecules may be neglected compared with that of the electrons, since in the stationary state the mean kinetic energy of motion of a material molecule is equal to that of an electron, and since the mass of a material molecule is more than a thousand times as large as that of an electron. Now, if there is an electric field in the interior of the metal, the oppositely charged particles are driven in opposite directions with average velocities depending on the mean free path, among other factors, and this explains the conductivity of the metal for the electric current. On the other hand, the emissive power of the metal for the radiant heat follows from the calculation of the impacts of the electrons. For,

[1] *E. Hagen* und *H. Rubens*, Ann. d. Phs.y**11**, p. 873, 1903.
[2] *E. Aschkinass*, Ann. d. Phys. **17**, p. 960, 1905.
[3] *E. Riecke*, Wied. Ann. **66**, p. 353, 1898.
[4] *P. Drude*, Ann. d. Phys. **1**, p. 566, 1900.

so long as an electron flies with constant speed in a constant direction, its kinetic energy remains constant and there is no radiation of energy; but, whenever it suffers by impact a change of its velocity components, a certain amount of energy, which may be calculated from electrodynamics and which may always be represented in the form of a *Fourier's* series, is radiated into the surrounding space, just as we think of *Roentgen rays* as being caused by the impact on the anticathode of the electrons ejected from the cathode. From the standpoint of the hypothesis of quanta this calculation cannot, for the present, be carried out without ambiguity except under the assumption that, during the time of a partial vibration of the *Fourier* series, a large number of impacts of electrons occurs, *i.e.*, for comparatively long waves, for then the fundamental law of impact does not essentially matter.

Now this method may evidently be used to derive the laws of black radiation in a new way, entirely independent of that previously employed. For if the emissive power, E, of the metal, thus calculated, is divided by the absorbing power, A, of the same metal, determined by means of its electric conductivity, then, according to *Kirchhoff's* law (48), the result must be the emissive power of a black body, irrespective of the special substance used in the determination. In this manner *H. A. Lorentz*[1] has, in a profound investigation, derived the law of radiation of a black body and has obtained a result the contents of which agree exactly with equation (283), and where also the constant k is related to the gas constant R by equation (193). It is true that this method of establishing the laws of radiation is, as already said, restricted to the range of long waves, but it affords a deeper and very important insight into the mechanism of the motions of the electrons and the radiation phenomena in metals caused by them. At the same time the point of view described above in Sec. 111, according to which the normal spectrum may be regarded as consisting of a large number of quite irregular processes as elements, is expressly confirmed.

169. A further interesting confirmation of the law of radiation of black bodies for long waves and of the connection of the radiation constant k with the absolute mass of the material

[1] *H. A. Lorentz*, Proc. Kon. Akad. v. Wet. Amsterdam, 1903, p. 666.

molecules was found by *J. H. Jeans*[1] by a method previously used by *Lord Rayleigh*,[2] which differs essentially from the one pursued here, in the fact that it entirely avoids making use of any special mutual action between matter (molecules, oscillators) and the ether and considers essentially only the processes in the vacuum through which the radiation passes. The starting point for this method of treatment is given by the following proposition of statistical mechanics. (Compare above, Sec. 140.) When irreversible processes take place in a system, which satisfies *Hamilton's* equations of motion, and whose state is determined by a large number of independent variables and whose total energy is found by addition of different parts depending on the squares of the variables of state, they do so, on the average, in such a sense that the partial energies corresponding to the separate independent variables of state tend to equality, so that finally, on reaching statistical equilibrium, their mean values have become equal. From this proposition the stationary distribution of energy in such a system may be found, when the independent variables which determine the state are known.

Let us now imagine a perfect vacuum, cubical in form, of edge l, and with metallically reflecting sides. If we take the origin of coordinates at one corner of the cube and let the axes of coordinates coincide with the adjoining edges, an electromagnetic process which may occur in this cavity is represented by the following system of equations:

$$\mathsf{E}_x = \cos\frac{\mathsf{a}\pi x}{l}\sin\frac{\mathsf{b}\pi y}{l}\sin\frac{\mathsf{c}\pi z}{l}(e_1\cos 2\pi\nu t + e'_1\sin 2\pi\nu t),$$

$$\mathsf{E}_y = \sin\frac{\mathsf{a}\pi x}{l}\cos\frac{\mathsf{b}\pi y}{l}\sin\frac{\mathsf{c}\pi z}{l}(e_2\cos 2\pi\nu t + e'_2\sin 2\pi\nu t),$$

$$\mathsf{E}_z = \sin\frac{\mathsf{a}\pi x}{l}\sin\frac{\mathsf{b}\pi y}{l}\cos\frac{\mathsf{c}\pi z}{l}(e_3\cos 2\pi\nu t + e'_3\sin 2\pi\nu t),$$

$$\mathsf{H}_x = \sin\frac{\mathsf{a}\pi x}{l}\cos\frac{\mathsf{b}\pi y}{l}\cos\frac{\mathsf{c}\pi z}{l}(h_1\sin 2\pi\nu t - h'_1\cos 2\pi\nu t),$$

[1] *J. H. Jeans*, Phil. Mag. **10**, p. 91, 1905.
[2] *Lord Rayleigh*, Nature **72**, p. 54 and p. 243, 1905.

$$\mathsf{H}_y = \cos\frac{\mathsf{a}\pi x}{l}\sin\frac{\mathsf{b}\pi y}{l}\cos\frac{\mathsf{c}\pi z}{l}(h_2\sin 2\pi\nu t - h'_2\cos 2\pi\nu t),$$

$$\mathsf{H}_z = \cos\frac{\mathsf{a}\pi x}{l}\cos\frac{\mathsf{b}\pi y}{l}\sin\frac{\mathsf{c}\pi z}{l}(h_3\sin 2\pi\nu t - h'_3\cos 2\pi\nu t),$$

where a, b, c represent any three positive integral numbers. The boundary conditions in these expressions are satisfied by the fact that for the six bounding surfaces $x=0$, $x=l$, $y=0$, $y=l$, $z=0$, $z=l$ the tangential components of the electric field-strength E vanish. Maxwell's equations of the field (52) are also satisfied, as may be seen on substitution, provided there exist certain conditions between the constants which may be stated in a single proposition as follows: Let a be a certain positive constant, then there exist between the nine quantities written in the following square:

$$\begin{array}{ccc} \frac{\mathsf{a}c}{2l\nu} & \frac{\mathsf{b}c}{2l\nu} & \frac{\mathsf{c}c}{2l\nu} \\ \\ \frac{h_1}{a} & \frac{h_2}{a} & \frac{h_3}{a} \\ \\ \frac{e_1}{a} & \frac{e_2}{a} & \frac{e_3}{a} \end{array}$$

all the relations which are satisfied by the nine so-called "direction cosines" of two orthogonal right-handed coordinate systems, *i.e.*, the cosines of the angles of any two axes of the systems.

Hence the sum of the squares of the terms of any horizontal or vertical row equals 1, for example,

$$\frac{c^2}{4l^2\nu^2}(\mathsf{a}^2+\mathsf{b}^2+\mathsf{c}^2) = 1 \qquad (306)$$

$$h_1^2+h_2^2+h_3^2 = a^2 = e_1^2+e_2^2+e_3^2.$$

Moreover the sum of the products of corresponding terms in any two parallel rows is equal to zero, for example,

$$\begin{aligned} \mathsf{a}e_1+\mathsf{b}e_2+\mathsf{c}e_3 &= 0 \\ \mathsf{a}h_1+\mathsf{b}h_2+\mathsf{c}h_3 &= 0. \end{aligned} \qquad (307)$$

Moreover there are relations of the following form:

$$\frac{h_1}{a}=\frac{e_2}{a}\cdot\frac{\mathsf{c}c}{2l\nu}-\frac{e_3}{a}\frac{\mathsf{b}c}{2l\nu},$$

and hence

$$h_1=\frac{c}{2l\nu}(\mathsf{c}e_2-\mathsf{b}e_3),\text{ etc.} \tag{308}$$

If the integral numbers a, b, c are given, then the frequency ν is immediately determined by means of (306). Then among the six quantities e_1, e_2, e_3, h_1, h_2, h_3, only two may be chosen arbitrarily, the others then being uniquely determined by them by linear homogeneous relations. If, for example, we assume e_1 and e_2 arbitrarily, e_3 follows from (307) and the values of h_1, h_2, h_3 are then found by relations of the form (308). Between the quantities with accent e_1', e_2', e_3', h_1', h_2', h_3' there exist exactly the same relations as between those without accent, of which they are entirely independent. Hence two also of them, say h_1' and h_2', may be chosen arbitrarily so that in the equations given above for given values of a, b, c four constants remain undetermined. If we now form, for all values of a b c whatever, expressions of the type (305) and add the corresponding field components, we again obtain a solution for *Maxwell's* equations of the field and the boundary conditions, which, however, is now so general that it is capable of representing any electromagnetic process possible in the hollow cube considered. For it is always possible to dispose of the constants e_1, e_2, h_1', h_2' which have remained undetermined in the separate particular solutions in such a way that the process may be adapted to any initial state ($t=0$) whatever.

If now, as we have assumed so far, the cavity is entirely void of matter, the process of radiation with a given initial state is uniquely determined in all its details. It consists of a set of stationary vibrations, every one of which is represented by one of the particular solutions considered, and which take place entirely independent of one another. Hence in this case there can be no question of irreversibility and hence also none of any tendency to equality of the partial energies corresponding to the separate partial vibrations. As soon, however, as we assume the presence in the cavity of only the slightest trace of matter which can influence the electrodynamic vibrations, *e.g.*, a few gas molecules, which emit or absorb radiation, the process becomes chaotic and a passage from less to more probable states will take place, though perhaps slowly. Without considering any further details of the electromagnetic constitution of the molecules, we may from the law of statistical mechanics quoted above draw the conclusion that, among all possible processes, that one in which the energy is distributed uniformly among all the independent variables of the state has the stationary character.

From this let us determine these independent variables. In the first place there are the velocity components of the gas molecules. In the stationary state to every one of the three mutually independent velocity components of a molecule there corresponds on the average the energy $\frac{1}{3}\bar{L}$ where $\bar{L}$ represents the mean energy of a molecule and is given by (200). Hence the partial energy, which on the average corresponds to any one of the independent variables of the electromagnetic system, is just as large.

Now, according to the above discussion, the electro-magnetic state of the whole cavity for every stationary vibration corresponding to any one system of values of the numbers a b c is determined, at any instant, by four mutually independent quantities. Hence for the radiation processes the number of independent variables of state is four times as large as the number of the possible systems of values of the positive integers a, b, c.

We shall now calculate the number of the possible systems of values a, b, c, which correspond to the vibrations within a certain small range of the spectrum, say between the frequencies ν and $\nu+d\nu$. According to (306), these systems of values satisfy the inequalities

$$\left(\frac{2l\nu}{c}\right)^2 < \mathsf{a}^2+\mathsf{b}^2+\mathsf{c}^2 < \left(\frac{2l(\nu+d\nu)}{c}\right)^2, \tag{309}$$

where not only $\frac{2l\nu}{c}$ but also $\frac{2ld\nu}{c}$ is to be thought of as a large number. If we now represent every system of values of a, b, c graphically by a point, taking a, b, c as coordinates in an orthogonal coordinate system, the points thus obtained occupy one octant of the space of infinite extent, and condition (309) is equivalent to requiring that the distance of any one of these points from the origin of the coordinates shall lie between $\frac{2l\nu}{c}$ and $\frac{2l(\nu+d\nu)}{c}$. Hence the required number is equal to the number of points which lie between the two spherical surface-octants corresponding to the radii $\frac{2l\nu}{c}$ and $\frac{2l(\nu+d\nu)}{c}$. Now since to every point there corresponds a cube of volume 1 and *vice versa*, that number is simply equal to the space between the two spheres mentioned, and hence equal to

$$\frac{1}{8}4\pi\left(\frac{2l\nu}{c}\right)^2\frac{2ld\nu}{c},$$

and the number of the independent variables of state is four times as large or

$$\frac{16\pi l^3\nu^2\,d\nu}{c^3}$$

Since, moreover, the partial energy $\frac{\overline{L}}{3}$ corresponds on the average to every independent variable of state in the state of equilibrium, the total energy falling in the interval from ν to $\nu+d\nu$ becomes

$$\frac{16\pi l^3\nu^2 d\nu}{3c^3}\overline{L}.$$

Since the volume of the cavity is l^3, this gives for the space density of the energy of frequency ν

$$\mathsf{u}d\nu=\frac{16\pi\nu^2 d\nu}{3c^3}\overline{L},$$

and, by substitution of the value of $\overline{L}=\frac{L}{N}$ from (200),

$$\mathsf{u}=\frac{8\pi\nu^2 kT}{c^3}, \tag{310}$$

which is in perfect agreement with *Rayleigh's* formula (285).

If the law of the equipartition of energy held true in all

cases, *Rayleigh's* law of radiation would, in consequence, hold for all wave lengths and temperatures. But since this possibility is excluded by the measurements at hand, the only possible conclusion is that the law of the equipartition of energy and, with it, the system of *Hamilton's* equations of motion does not possess the general importance attributed to it in classical dynamics. Therein lies the strongest proof of the necessity of a fundamental modification of the latter.

PART V

IRREVERSIBLE RADIATION PROCESSES

CHAPTER I

FIELDS OF RADIATION IN GENERAL

170. According to the theory developed in the preceding section, the nature of heat radiation within an isotropic medium, when the state is one of stable thermodynamic equilibrium, may be regarded as known in every respect. The intensity of the radiation, uniform in all directions, depends for all wave lengths only on the temperature and the velocity of propagation, according to equation (300), which applies to black radiation in any medium whatever. But there remains another problem to be solved by the theory. It is still necessary to explain how and by what processes the radiation which is originally present in the medium and which may be assigned in any way whatever, passes gradually, when the medium is bounded by walls impermeable to heat, into the stable state of black radiation, corresponding to the maximum of entropy, just as a gas which is enclosed in a rigid vessel and in which there are originally currents and temperature differences assigned in any way whatever gradually passes into the state of rest and of uniform distribution of temperature.

To this much more difficult question only a partial answer can, at present, be given. In the first place, it is evident from the extensive discussion in the first chapter of the third part that, since irreversible processes are to be dealt with, the principles of pure electrodynamics alone will not suffice. For the second principle of thermodynamics or the principle of increase of entropy is foreign to the contents of pure electrodynamics as well as of pure mechanics. This is most immediately shown by the fact that the fundamental equations of mechanics as well as those of electrodynamics allow the direct reversal of every process as regards time, which contradicts the principle of increase of entropy. Of course all kinds of friction and of electric conduction of currents must be assumed to be excluded; for these processes, since they are always connected with the production of heat, do not belong to mechanics or electrodynamics proper.

This assumption being made, the time t occurs in the fundamental equations of mechanics only in the components of acceleration; that is, in the form of the square of its differential. Hence, if instead of t the quantity $-t$ is introduced as time variable in the equations of motion, they retain their form without change, and hence it follows that if in any motion of a system of material points whatever the velocity components of all points are suddenly reversed at any instant, the process must take place in the reverse direction. For the electrodynamic processes in a homogeneous non-conducting medium a similar statement holds. If in *Maxwell's* equations of the electrodynamic field $-t$ is written everywhere instead of t, and if, moreover, the sign of the magnetic field-strength H is reversed, the equations remain unchanged, as can be readily seen, and hence it follows that if in any electrodynamic process whatever the magnetic field-strength is everywhere suddenly reversed at a certain instant, while the electric field-strength keeps its value, the whole process must take place in the opposite sense.

If we now consider any radiation processes whatever, taking place in a perfect vacuum enclosed by reflecting walls, it is found that, since they are completely determined by the principles of classical electrodynamics, there can be in their case no question of irreversibility of any kind. This is seen most clearly by considering the perfectly general formulæ (305), which hold for a cubical cavity and which evidently have a periodic, *i.e.*, reversible character. Accordingly we have frequently (Sec. 144 and 166) pointed out that the simple propagation of free radiation represents a reversible process. An irreversible element is introduced by the addition of emitting and absorbing substance.

171. Let us now try to define for the general case the state of radiation in the thermodynamic-macroscopic sense as we did above in Sec. 107, *et seq.*, for a stationary radiation. Every one of the three components of the electric field-strength, *e.g.*, E_z may, for the long time interval from $t=0$ to $t=\mathsf{T}$, be represented at every point, *e.g.*, at the origin of coordinates, by a *Fourier's* integral, which in the present case is somewhat more convenient than the *Fourier's* series (149):

$$\mathsf{E}_z = \int_0^\infty d\nu C_\nu \cos (2\pi \nu t - \theta_\nu), \tag{311}$$

where C_ν (positive) and θ_ν denote certain functions of the positive variable of integration ν. The values of these functions are not wholly determined by the behavior of E_z in the time interval mentioned, but depend also on the manner in which E_z varies as a function of the time beyond both ends of that interval. Hence the quantities C_ν and θ_ν possess separately no definite physical significance, and it would be quite incorrect to think of the vibration E_z as, say, a continuous spectrum of periodic vibrations with the constant amplitudes C_ν. This may, by the way, be seen at once from the fact that the character of the vibration E_z may vary with the time in any way whatever. How the spectral resolution of the vibration E_z is to be performed and to what results it leads will be shown below (Sec. 174).

172. We shall, as heretofore (158), define J, the "intensity of the exciting vibration,"[1] as a function of the time to be the mean value of E_z^2 in the time interval from t to $t+\tau$, where τ is taken as large compared with the time $1/\nu$, which is the duration of one of the periodic partial vibrations contained in the radiation, but as small as possible compared with the time T. In this statement there is a certain indefiniteness, from which results the fact that J will, in general, depend not only on t but also on τ. If this is the case one cannot speak of the intensity of the exciting vibration at all. For it is an essential feature of the conception of the intensity of a vibration that its value should change but unappreciably within the time required for a single vibration. (Compare above, Sec. 3.) Hence we shall consider in future only those processes for which, under the conditions mentioned, there exists a mean value of E_z^2 depending only on t. We are then obliged to assume that the quantities C_ν in (311) are negligible for all values of ν which are of the same order of magnitude as $\frac{1}{\tau}$ or smaller, *i.e.*,

$$\nu\tau \text{ is large.} \tag{312}$$

[1] Not to be confused with the "field intensity" (field-strength) $\mathbf{E}_z$ of the exciting vibration.

In order to calculate J we now form from (311) the value of E_z^2 and determine the mean value $\overline{\mathsf{E}_z^2}$ of this quantity by integrating with respect to t from t to $t+\tau$, then dividing by τ and passing to the limit by decreasing τ sufficiently. Thus we get

$$\mathsf{E}_z^2 = \int_0^\infty \int_0^\infty d\nu' \, d\nu \, C_{\nu'} \, C_\nu \cos(2\pi\nu' t - \theta_{\nu'}) \cos(2\pi\nu t - \theta_\nu).$$

If we now exchange the values of ν and ν', the function under the sign of integration does not change; hence we assume

$$\nu' > \nu$$

and write:

$$\mathsf{E}_z^2 = 2\int\int d\nu' \, d\nu \, C_{\nu'} \, C_\nu \cos(2\pi\nu' t - \theta_{\nu'}) \cos(2\pi\nu t - \theta_\nu),$$

or

$$\mathsf{E}_z^2 = \int\int d\nu' \, d\nu \, C_{\nu'} \, C_\nu \{\cos[2\pi(\nu'-\nu)t - \theta_{\nu'} + \theta_\nu] + \cos[2\pi(\nu'+\nu)t - \theta_{\nu'} - \theta_\nu]\}.$$

And hence

$$J = \overline{\mathsf{E}_z^2} = \frac{1}{\tau}\int_t^{t+\tau} \mathsf{E}_z^2 \, dt$$

$$= \int\int d\nu' \, d\nu \, C_{\nu'} C_\nu \left\{ \frac{\sin\pi(\nu'-\nu)\tau \cdot \cos[\pi(\nu'-\nu)(2t+\tau) - \theta_{\nu'} + \theta_\nu]}{\pi(\nu'-\nu)\tau} + \frac{\sin\pi(\nu'+\nu)\tau \cdot \cos[\pi(\nu'+\nu)(2t+\tau) - \theta_{\nu'} - \theta_\nu]}{\pi(\nu'+\nu)\tau} \right\}.$$

If we now let τ become smaller and smaller, since $\nu\tau$ remains large, the denominator $(\nu'+\nu)\tau$ of the second fraction remains large under all circumstances, while that of the first fraction $(\nu'-\nu)\tau$ may decrease with decreasing value of τ to less than any finite value. Hence for sufficiently small values of $\nu'-\nu$ the integral reduces to

$$\int\int d\nu' \, d\nu \, C_{\nu'} \, C_\nu \cos[2\pi(\nu'-\nu)t - \theta_{\nu'} + \theta_\nu]$$

which is in fact independent of τ. The remaining terms of the double integral, which correspond to larger values of $\nu'-\nu$, *i.e.*, to more rapid changes with the time, depend in general on τ and therefore must vanish, if the intensity J is not to depend on τ. Hence in our case on introducing as a second variable of integration instead of ν

$$\mu = \nu' - \nu (> 0)$$

we have

$$J = \int\int d\mu \, d\nu \, C_{\nu+\mu} C_\nu \cos (2\pi\mu t - \theta_{\nu+\mu} + \theta_\nu) \qquad (313)$$

or

$$J = \int d\mu (A_\mu \cos 2\pi\mu t + B_\mu \sin 2\pi\mu t)$$

$$\text{where } A_\mu = \int d\nu C_{\nu+\mu} C_\nu \cos (\theta_{\nu+\mu} - \theta_\nu) \qquad (314)$$

$$B_\mu = \int d\nu C_{\nu+\mu} C_\nu \sin (\theta_{\nu+\mu} - \theta_\nu)$$

By this expression the intensity J of the exciting vibration, if it exists at all, is expressed by a function of the time in the form of a *Fourier's* integral.

173. The conception of the intensity of vibration J necessarily contains the assumption that this quantity varies much more slowly with the time t than the vibration E_z itself. The same follows from the calculation of J in the preceding paragraph. For there, according to (312), $\nu\tau$ and $\nu'\tau$ are large, but $(\nu' - \nu)\tau$ is small for all pairs of values C_ν and $C_{\nu'}$ that come into consideration; hence, *a fortiori*,

$$\frac{\nu' - \nu}{\nu} = \frac{\mu}{\nu} \text{ is small,} \qquad (315)$$

and accordingly the *Fourier's* integrals E_z in (311) and J in (314) vary with the time in entirely different ways. Hence in the following we shall have to distinguish, as regards dependence on time, two kinds of quantities, which vary in different ways: Rapidly varying quantities, as E_z, and slowly varying quantities as J and I the spectral intensity of the exciting vibration, whose value we shall calculate in the next paragraph. Nevertheless this difference in the variability with respect to time of the quanti-

13

ties named is only relative, since the absolute value of the differential coefficient of J with respect to time depends on the value of the unit of time and may, by a suitable choice of this unit, be made as large as we please. It is, therefore, not proper to speak of $J(t)$ simply as a slowly varying function of t. If, in the following, we nevertheless employ this mode of expression for the sake of brevity, it will always be in the relative sense, namely, with respect to the different behavior of the function $\mathsf{E}_z(t)$.

On the other hand, as regards the dependence of the phase constant θ_ν on its index ν it necessarily possesses the property of rapid variability in the *absolute* sense. For, although μ is small compared with ν, nevertheless the difference $\theta_{\nu+\mu}-\theta_\nu$ is in general not small, for if it were, the quantities A_μ and B_μ in (314) would have too special values and hence it follows that $(\partial\theta_\nu/\partial\nu)\cdot\nu$ must be large. This is not essentially modified by changing the unit of time or by shifting the origin of time.

Hence the rapid variability of the quantities θ_ν and also C_ν with ν is, in the absolute sense, a necessary condition for the existence of a definite intensity of vibration J, or, in other words, for the possibility of dividing the quantities depending on the time into those which vary rapidly and those which vary slowly—a distinction which is also made in other physical theories and upon which all the following investigations are based.

174. The distinction between rapidly variable and slowly variable quantities introduced in the preceding section has, at the present stage, an important physical aspect, because in the following we shall assume that only slow variability with time is capable of direct measurement. On this assumption we approach conditions as they actually exist in optics and heat radiation. Our problem will then be to establish relations between slowly variable quantities exclusively; for these only can be compared with the results of experience. Hence we shall now determine the most important one of the slowly variable quantities to be considered here, namely, the "spectral intensity" I of the exciting vibration. This is effected as in (158) by means of the equation

$$J=\int_0^\infty \mathsf{I}\,d\nu.$$

By comparison with 313 we obtain:

$$\left\{\begin{aligned} &\mathsf{I} = \int d\mu (\mathsf{A}_\mu \cos 2\pi\mu t + \mathsf{B}_\mu \sin 2\pi\mu t) \\ &\text{where} \\ &\mathsf{A}_\mu = \overline{C_{\nu+\mu} C_\nu \cos\,(\theta_{\nu+\mu} - \theta_\nu)} \\ &\mathsf{B}_\mu = \overline{C_{\nu+\mu} C_\nu \sin\,(\theta_{\nu+\mu} - \theta_\nu)}\;. \end{aligned}\right. \qquad (316)$$

By this expression the spectral intensity, I, of the exciting vibration at a point in the spectrum is expressed as a slowly variable function of the time t in the form of a *Fourier's* integral. The dashes over the expressions on the right side denote the mean values extended over a narrow spectral range for a given value of μ. If such mean values do not exist, there is no definite spectral intensity.

CHAPTER II

ONE OSCILLATOR IN THE FIELD OF RADIATION

175. If in any field of radiation whatever we have an ideal oscillator of the kind assumed above (Sec. 135), there will take place between it and the radiation falling on it certain mutual actions, for which we shall again assume the validity of the elementary dynamical law introduced in the preceding section. The question is then, how the processes of emission and absorption will take place in the case now under consideration.

In the first place, as regards the emission of radiant energy by the oscillator, this takes place, as before, according to the hypothesis of emission of quanta (Sec. 147), where the probability quantity η again depends on the corresponding spectral intensity I through the relation (265).

On the other hand, the absorption is calculated, exactly as above, from (234), where the vibrations of the oscillator also take place according to the equation (233). In this way, by calculations analogous to those performed in the second chapter of the preceding part, with the difference only that instead of the Fourier's series (235) the Fourier's integral (311) is used, we obtain for the energy absorbed by the oscillator in the time τ the expression

$$\frac{\tau}{4L}\int d\mu(\mathsf{A}_\mu \cos 2\pi\mu t + \mathsf{B}_\mu \sin 2\pi\mu t),$$

where the constants A_μ and B_μ denote the mean values expressed in (316), taken for the spectral region in the neighborhood of the natural frequency ν_o of the oscillator. Hence the law of absorption will again be given by equation (249), which now holds also for an intensity of vibration I varying with the time.

176. There now remains the problem of deriving the expression for I, the spectral intensity of the vibration exciting the oscillator, when the thermodynamic state of the field of radiation at the oscillator is given in accordance with the statements made in Sec. 17.

Let us first calculate the total intensity $J=\overline{\mathsf{E}_z^2}$ of the vibration exciting an oscillator, from the intensities of the heat rays striking the oscillator from all directions.

For this purpose we must also allow for the polarization of the monochromatic rays which strike the oscillator. Let us begin by considering a pencil which strikes the oscillator within a conical element whose vertex lies in the oscillator and whose solid angle, $d\Omega$, is given by (5), where the angles θ and ϕ, polar coordinates, designate the direction of the propagation of the rays. The whole pencil consists of a set of monochromatic pencils, one of which may have the principal values of intensity K and K' (Sec. 17). If we now denote the angle which the plane of vibration belonging to the principal intensity K makes with the plane through the direction of the ray and the z-axis (the axis of the oscillator) by ψ, no matter in which quadrant it lies, then, according to (8), the specific intensity of the monochromatic pencil may be resolved into the two plane polarized components at right angles with each other,

$$\mathsf{K}\cos^2\psi+\mathsf{K}'\sin^2\psi$$
$$\mathsf{K}\sin^2\psi+\mathsf{K}'\cos^2\psi,$$

the first of which vibrates in a plane passing through the z-axis and the second in a plane perpendicular thereto.

The latter component does not contribute anything to the value of E_z^2, since its electric field-strength is perpendicular to the axis of the oscillator. Hence there remains only the first component whose electric field-strength makes the angle $\frac{\pi}{2}-\theta$ with the z-axis. Now according to *Poynting's* law the intensity of a plane polarized ray in a vacuum is equal to the product of $\frac{c}{4\pi}$ and the mean square of the electric field-strength. Hence the mean square of the electric field-strength of the pencil here considered is

$$\frac{4\pi}{c}(\mathsf{K}\cos^2\psi+\mathsf{K}'\sin^2\psi)\,d\nu\,d\Omega,$$

and the mean square of its component in the direction of the z-axis is

$$\frac{4\pi}{c}(\mathsf{K}\cos^2\psi+\mathsf{K}'\sin^2\psi)\sin^2\theta\,d\nu\,d\Omega. \tag{317}$$

By integration over all frequencies and all solid angles we then obtain the value required

$$\overline{\mathsf{E}_z^2}=\frac{4\pi}{c}\int\sin^2\theta\,d\Omega\int d\nu(\mathsf{K}_\nu\cos^2\psi+\mathsf{K}_\nu'\sin^2\psi)=J. \tag{318}$$

The space density u of the electromagnetic energy at a point of the field is

$$u=\frac{1}{8\pi}(\overline{\mathsf{E}_x^2}+\overline{\mathsf{E}_y^2}+\overline{\mathsf{E}_z^2}+\overline{\mathsf{H}_x^2}+\overline{\mathsf{H}_y^2}+\overline{\mathsf{H}_z^2}),$$

where E_x^2, E_y^2, E_z^2, H_x^2, H_y^2, H_z^2 denote the squares of the field-strengths, regarded as "slowly variable" quantities, and are hence supplied with the dash to denote their mean value. Since for every separate ray the mean electric and magnetic energies are equal, we may always write

$$u=\frac{1}{4\pi}(\overline{\mathsf{E}_x^2}+\overline{\mathsf{E}_y^2}+\overline{\mathsf{E}_z^2}). \tag{319}$$

If, in particular, all rays are unpolarized and if the intensity of radiation is constant in all directions, $\mathsf{K}_\nu=\mathsf{K}_\nu'$ and, since

$$\int\sin^2\theta\,d\Omega=\int\int\sin^3\theta\,d\theta\,d\phi=\frac{8\pi}{3} \tag{319a}$$

$$\overline{\mathsf{E}_z^2}=\frac{32\pi^2}{3c}\int\mathsf{K}_\nu\,d\nu=\overline{\mathsf{E}_x^2}=\overline{\mathsf{E}_y^2}$$

and, by substitution in (319),

$$u=\frac{8\pi}{c}\int\mathsf{K}_\nu\,d\nu,$$

which agrees with (22) and (24).

177. Let us perform the spectral resolution of the intensity J according to Sec. 174; namely,

$$J=\int\mathsf{I}_\nu d\nu.$$

Then, by comparison with (318), we find for the intensity of a definite frequency ν contained in the exciting vibration the value

$$\mathsf{I}=\frac{4\pi}{c}\int \sin^2\theta\, d\Omega(\mathsf{K}_\nu \cos^2\psi+\mathsf{K}_\nu' \sin^2\psi). \tag{320}$$

For radiation which is unpolarized and uniform in all directions we obtain again, in agreement with (160),

$$\mathsf{I}=\frac{32\pi^2}{3c}\mathsf{K}.$$

178. With the value (320) obtained for I the total energy absorbed by the oscillator in an element of time dt from the radiation falling on it is found from (249) to be

$$\frac{\pi dt}{cL}\int \sin^2\theta\, d\Omega(\mathsf{K}\cos^2\psi+\mathsf{K}'\sin^2\psi).$$

Hence the oscillator absorbs in the time dt from the pencil striking it within the conical element $d\Omega$ an amount of energy equal to

$$\frac{\pi dt}{cL}\sin^2\theta(\mathsf{K}\cos^2\psi+\mathsf{K}'\sin^2\psi)d\Omega. \tag{321}$$

CHAPTER III

A SYSTEM OF OSCILLATORS

179. Let us suppose that a large number N of similar oscillators with parallel axes, acting quite independently of one another, are distributed irregularly in a volume-element of the field of radiation, the dimensions of which are so small that within it the intensities of radiation K do not vary appreciably. We shall investigate the mutual action between the oscillators and the radiation which is propagated freely in space.

As before, the state of the field of radiation may be given by the magnitude and the azimuth of vibration ψ of the principal intensities K_ν and K_ν' of the pencils which strike the system of oscillators, where K_ν and K_ν' depend in an arbitrary way on the direction angles θ and ϕ. On the other hand, let the state of the system of oscillators be given by the densities of distribution $w_1, w_2, w_3, \ldots\ldots$ (166), with which the oscillators are distributed among the different region elements, $w_1, w_2, w_3, \ldots\ldots$ being any proper fractions whose sum is 1. Herein, as always, the nth region element is supposed to contain the oscillators with energies between $(n-1)h\nu$ and $nh\nu$.

The energy absorbed by the system in the time dt within the conical element $d\Omega$ is, according to (321),

$$\frac{\pi N dt}{cL} \sin^2\theta(\mathsf{K}\cos^2\psi+\mathsf{K}'\sin^2\psi)d\Omega. \tag{322}$$

Let us now calculate also the energy emitted within the same conical element.

180. The total energy emitted in the time element dt by all N oscillators is found from the consideration that a single oscillator, according to (249), takes up an energy element $h\nu$ during the time

$$\frac{4h\nu L}{\mathsf{I}}=\tau, \tag{323}$$

and hence has a chance to emit once, the probability being η. We shall assume that the intensity I of the exciting vibration does not change appreciably in the time τ. Of the Nw_n oscillators which at the time t are in the nth region element a number $Nw_n\eta$ will emit during the time τ, the energy emitted by each being $nh\nu$. From (323) we see that the energy emitted by all oscillators during the time element dt is

$$\sum Nw_n\,\eta\,nh\nu\frac{dt}{\tau}=\frac{N\eta\mathsf{I}dt}{4L}\Sigma nw_n,$$

or, according to (265),

$$\frac{N(1-\eta)dt}{4pL}\Sigma nw_n. \tag{324}$$

From this the energy emitted within the conical element $d\Omega$ may be calculated by considering that, in the state of thermodynamic equilibrium, the energy emitted in every conical element is equal to the energy absorbed and that, in the general case, the energy emitted in a certain direction is independent of the energy simultaneously absorbed. For the stationary state we have from (160) and (265)

$$\mathsf{K}=\mathsf{K}'=\frac{3c}{32\pi^2}\mathsf{I}=\frac{3c}{32\pi^2}\,\frac{1-\eta}{p\eta} \tag{325}$$

and further from (271) and (265)

$$w_n=\frac{1}{p\mathsf{I}}\left(\frac{p\mathsf{I}}{1+p\mathsf{I}}\right)^n=\eta(1-\eta)^{n-1}, \tag{326}$$

and hence

$$\Sigma nw_n=\eta\,\Sigma n(1-\eta)^{n-1}=\frac{1}{\eta}. \tag{327}$$

Thus the energy emitted (324) becomes

$$\frac{N(1-\eta)dt}{4Lp\eta}. \tag{328}$$

This is, in fact, equal to the total energy absorbed, as may be found by integrating the expression (322) over all conical elements $d\Omega$ and taking account of (325).

Within the conical element $d\Omega$ the energy emitted or absorbed will then be

$$\frac{\pi N dt}{cL} \sin^2\theta \, \mathsf{K} d\Omega,$$

or, from (325), (327) and (268),

$$\frac{\pi h \nu^3 (1-\eta) N}{c^3 L} \sum n w_n \sin^2 \theta \, d\Omega \, dt, \tag{329}$$

and this is the general expression for the energy emitted by the system of oscillators in the time element dt within the conical element $d\Omega$, as is seen by comparison with (324).

181. Let us now, as a preparation for the following deductions, consider more closely the properties of the different pencils passing the system of oscillators. From all directions rays strike the volume-element that contains the oscillators; if we again consider those which come toward it in the direction (θ, ϕ) within the conical element $d\Omega$, the vertex of which lies in the volume-element, we may in the first place think of them as being resolved into their monochromatic constituents, and then we need consider further only that one of these constituents which corresponds to the frequency ν of the oscillators; for all other rays simply pass the oscillators without influencing them or being influenced by them. The specific intensity of a monochromatic ray of frequency ν is

$$\mathsf{K} + \mathsf{K}'$$

where K and K' represent the principal intensities which we assume as non-coherent. This ray is now resolved into two components according to the directions of its principal planes of vibration (Sec. 176).

The first component,

$$\mathsf{K} \sin^2 \psi + \mathsf{K}' \cos^2 \psi,$$

passes by the oscillators and emerges on the other side with no change whatever. Hence it gives a plane polarized ray, which starts from the system of oscillators in the direction (θ, ϕ) within the solid angle $d\Omega$ and whose vibrations are perpendicular to the axis of the oscillators and whose intensity is

$$\mathsf{K} \sin^2 \psi + \mathsf{K}' \cos^2 \psi = \mathsf{K}''. \tag{330}$$

The second component,

$$\mathsf{K}\cos^2\psi+\mathsf{K}'\sin^2\psi,$$

polarized at right angles to the first consists again, according to Sec. 176, of two parts

$$(\mathsf{K}\cos^2\psi+\mathsf{K}'\sin^2\psi)\cos^2\theta \tag{331}$$

and

$$(\mathsf{K}\cos^2\psi+\mathsf{K}'\sin^2\psi)\sin^2\theta, \tag{332}$$

of which the first passes by the system without any change, since its direction of vibration is at right angles to the axes of the oscillators, while the second is weakened by absorption, say by the small fraction β. Hence on emergence this component has only the intensity

$$(1-\beta)\ (\mathsf{K}\cos^2\psi+\mathsf{K}'\sin^2\psi)\sin^2\theta. \tag{333}$$

It is, however, strengthened by the radiation emitted by the system of oscillators (329), which has the value

$$\beta'(1-\eta)\ \Sigma n w_n \sin^2\theta, \tag{334}$$

where β' denotes a certain other constant, which depends only on the nature of the system and whose value is obtained at once from the condition that, in the state of thermodynamic equilibrium, the loss is just compensated by the gain.

For this purpose we make use of the relations (325) and (327) corresponding to the stationary state, and thus find that the sum of the expressions (333) and (334) becomes just equal to (332); and thus for the constant β' the following value is found:

$$\beta'=\beta\frac{3c}{32\pi^2 p}=\beta\frac{h\nu^3}{c^2}.$$

Then by addition of (331), (333) and (334) the total specific intensity of the radiation which emanates from the system of oscillators within the conical element $d\Omega$, and whose plane of vibration is parallel to the axes of the oscillators, is found to be

$$\begin{aligned}\mathsf{K}'''=\mathsf{K}\cos^2\psi+\mathsf{K}'\sin^2\psi+&\\ \beta\sin^2\theta(\mathsf{K}_e-(\mathsf{K}\cos^2\psi+\mathsf{K}'\sin^2\psi))&\end{aligned} \tag{335}$$

where for the sake of brevity the term referring to the emission is written

$$\frac{h\nu^3}{c^2}(1-\eta)\ \Sigma n w_n=\mathsf{K}_e. \tag{336}$$

Thus we finally have a ray starting from the system of oscillators in the direction (θ,ϕ) within the conical element $d\Omega$ and consisting of two components K'' and K''' polarized perpendicularly to each other, the first component vibrating at right angles to the axes of the oscillators.

In the state of thermodynamic equilibrium

$$\mathsf{K}=\mathsf{K}'=\mathsf{K}''=\mathsf{K}'''=\mathsf{K}_e,$$

a result which follows in several ways from the last equations.

182. The constant β introduced above, a small positive number, is determined by the spacial and spectral limits of the radiation influenced by the system of oscillators. If q denotes the cross-section at right angles to the direction of the ray, $\Delta\nu$ the spectral width of the pencil cut out of the total incident radiation by the system, the energy which is capable of absorption and which is brought to the system of oscillators within the conical element $d\Omega$ in the time dt is, according to (332) and (11),

$$q\Delta\nu(\mathsf{K}\cos^2\psi+\mathsf{K}'\sin^2\psi)\sin^2\theta\, d\Omega\, dt. \qquad (337)$$

Hence the energy actually absorbed is the fraction β of this value. Comparing this with (322) we get

$$\beta=\frac{\pi N}{q\cdot\Delta\nu\cdot cL}. \qquad (338)$$

CHAPTER IV

CONSERVATION OF ENERGY AND INCREASE OF ENTROPY. CONCLUSION

183. It is now easy to state the relation of the two principles of thermodynamics to the irreversible processes here considered. Let us consider first the *conservation of energy*. If there is no oscillator in the field, every one of the elementary pencils, infinite in number, retains, during its rectilinear propagation, both its specific intensity K and its energy without change, even though it be reflected at the surface, assumed as plane and reflecting, which bounds the field (Sec. 166). The system of oscillators, on the other hand, produces a change in the incident pencils and hence also a change in the energy of the radiation propagated in the field. To calculate this we need consider only those monochromatic rays which lie close to the natural frequency ν of the oscillators, since the rest are not altered at all by the system.

The system is struck in the direction (θ, ϕ) within the conical element $d\Omega$ which converges toward the system of oscillators by a pencil polarized in some arbitrary way, the intensity of which is given by the sum of the two principal intensities K and K'. This pencil, according to Sec. 182, conveys the energy

$$q\triangle\nu(\mathsf{K}+\mathsf{K}')d\Omega\, dt$$

to the system in the time dt; hence this energy is taken from the field of radiation on the side of the rays arriving within $d\Omega$. As a compensation there emerges from the system on the other side in the same direction (θ, ϕ) a pencil polarized in some definite way, the intensity of which is given by the sum of the two components K'' and K'''. By it an amount of energy

$$q\triangle\nu(\mathsf{K}''+\mathsf{K}''')d\Omega\, dt,$$

is added to the field of radiation. Hence, all told, the change in energy of the field of radiation in the time dt is obtained by subtracting the first expression from the second and by integrating with respect to $d\Omega$. Thus we get

$$dt \,\triangle\nu \int (\mathsf{K}''+\mathsf{K}'''-\mathsf{K}-\mathsf{K}')q\, d\Omega,$$

or by taking account of (330), (335), and (338)

$$\frac{\pi N dt}{cL}\int d\Omega \sin^2\theta\, (\mathsf{K}_c-(\mathsf{K}\cos^2\psi+\mathsf{K}'\sin^2\psi)). \qquad (339)$$

184. Let us now calculate the change in energy of the system of oscillators which has taken place in the same time dt. According to (219), this energy at the time t is

$$E=Nh\nu\sum_1^\infty (n-\frac{1}{2})w_n,$$

where the quantities w_n whose total sum is equal to 1 represent the densities of distribution characteristic of the state. Hence the energy change in the time dt is

$$dE=Nh\nu\sum_1^\infty (n-\frac{1}{2})dw_n=Nh\nu\sum_1^\infty n dw_n. \qquad (340)$$

To calculate dw_n we consider the nth region element. All of the oscillators which lie in this region at the time t have, after the lapse of time τ, given by (323), left this region; they have either passed into the $(n+1)$st region, or they have performed an emission at the boundary of the two regions. In compensation there have entered $(1-\eta)Nw_{n-1}$ oscillators during the time τ, that is, all oscillators which, at the time t, were in the $(n-1)$st region element, excepting such as have lost their energy by emission. Thus we obtain for the required change in the time dt

$$Ndw_n=\frac{dt}{\tau}N((1-\eta)w_{n-1}-w_n). \qquad (341)$$

A separate discussion is required for the first region element $n=1$. For into this region there enter in the time τ all those oscillators which have performed an emission in this time. Their number is

$$\eta(w_1+w_2+w_3+\ .\ .\ .\ .\ .\ .\ )N=\eta N.$$

Hence we have

$$N\,dw_1 = \frac{dt}{\tau} N(\eta - w_1).$$

We may include this equation in the general one (341) if we introduce as a new expression

$$w_o = \frac{\eta}{1-\eta}. \qquad (342)$$

Then (341) gives, substituting τ from (323),

$$dw_n = \frac{I\,dt}{4h\nu L}((1-\eta)w_{n-1} - w_n), \qquad (343)$$

and the energy change (340) of the system of oscillators becomes

$$dE = \frac{NI\,dt}{4L} \sum_1^\infty n((1-\eta)w_{n-1} - w_n).$$

The sum Σ may be simplified by recalling that

$$\sum_1^\infty n w_{n-1} = \sum_1^\infty (n-1) w_{n-1} + \sum_1^\infty . w_{n-1}$$

$$= \sum_1^\infty n w_n + w_o + 1 = \sum_1^\infty n w_n + \frac{1}{1-\eta}.$$

Then we have

$$dE = \frac{NI\,dt}{4L}(1 - \eta \sum_1^\infty n w_n). \qquad (344)$$

This expression may be obtained more readily by considering that dE is the difference of the total energy absorbed and the total energy emitted. The former is found from (250), the latter from (324), by taking account of (265).

The principle of the conservation of energy demands that the sum of the energy change (339) of the field of radiation and the energy change (344) of the system of oscillators shall be zero, which, in fact, is quite generally the case, as is seen from the relations (320) and (336).

185. We now turn to the discussion of the second principle, the principle of the *increase of entropy*, and follow closely the above discussion regarding the energy. When there is no oscillator in the field, every one of the elementary pencils, infinite in number, retains during rectilinear propagation both its specific intensity and its entropy without change, even when reflected at the surface, assumed as plane and reflecting, which bounds the field. The system of oscillators, however, produces a change in the incident pencils and hence also a change in the entropy of the radiation propagated in the field. For the calculation of this change we need to investigate only those monochromatic rays which lie close to the natural frequency ν of the oscillators, since the rest are not altered at all by the system.

The system of oscillators is struck in the direction (θ,ϕ) within the conical element $d\Omega$ converging toward the system by a pencil polarized in some arbitrary way, the spectral intensity of which is given by the sum of the two principal intensities K and K' with the azimuth of vibration ψ and $\frac{\pi}{2}+\psi$ respectively, which are assumed to be non-coherent. According to (141) and Sec. 182 this pencil conveys the entropy

$$q\Delta\nu[\mathsf{L}(\mathsf{K})+\mathsf{L}(\mathsf{K}')]\, d\Omega\, dt \qquad (345)$$

to the system of oscillators in the time dt, where the function $\mathsf{L}(\mathsf{K})$ is given by (278). Hence this amount of entropy is taken from the field of radiation on the side of the rays arriving within $d\Omega$. In compensation a pencil starts from the system on the other side in the same direction (θ,ϕ) within $d\Omega$ having the components K'' and K''' with the azimuth of vibration $\frac{\pi}{2}$ and 0 respectively, but its entropy radiation is not represented by $\mathsf{L}(\mathsf{K}'')+\mathsf{L}(\mathsf{K}''')$, since K'' and K''' are not non-coherent, but by

$$\mathsf{L}(\mathsf{K}_o)+\mathsf{L}(\mathsf{K}_o') \qquad (346)$$

where K_o and K_o' represent the principal intensities of the pencil.

For the calculation of K_o and K_o' we make use of the fact that, according to (330) and (335), the radiation K'' and K''', of which the component K''' vibrates in the azimuth 0, consists of the following three components, non-coherent with one another:

$$\mathsf{K}_1=\mathsf{K}\sin^2\psi+\mathsf{K}\cos^2\psi\,(1-\beta\sin^2\theta)=\mathsf{K}(1-\beta\sin^2\theta\cos^2\psi)$$

$$\text{with the azimuth of vibration } tg^2\,\psi_1=\frac{\operatorname{tg}^2\psi}{1-\beta\sin^2\theta},$$

$$\mathsf{K}_2 = \mathsf{K}' \cos^2\psi + \mathsf{K}' \sin^2\psi(1-\beta \sin^2\theta) = \mathsf{K}'(1-\beta \sin^2\theta \sin^2\psi)$$

with the azimuth of vibration $\operatorname{tg}^2 \psi_2 = \dfrac{\cot^2 \psi}{1-\beta \sin^2\theta}$,

and,

$$\mathsf{K}_3 = \beta \sin^2 \theta \, \mathsf{K}_e$$

with the azimuth of vibration $\operatorname{tg} \psi_3 = 0$.

According to (147) these values give the principal intensi ties K_o and K_o' required and hence the entropy radiation (346). Thereby the amount of entropy

$$q\Delta\nu[\mathsf{L}(\mathsf{K}_o)+\mathsf{L}(\mathsf{K}_o')]d\Omega \, dt \tag{347}$$

is added to the field of radiation in the time dt. All told, the entropy change of the field of radiation in the time dt, as given by subtraction of the expression (345) from (347) and integration with respect to $d\Omega$, is

$$dt\Delta\nu \int q d\Omega[\mathsf{L}(\mathsf{K}_o)+\mathsf{L}(\mathsf{K}_o')-\mathsf{L}(\mathsf{K})-\mathsf{L}(\mathsf{K}')]. \tag{348}$$

Let us now calculate the entropy change of the system of oscillators which has taken place in the same time dt. According to (173) the entropy at the time t is

$$S = -kN \sum_1^\infty w_n \log w_n.$$

Hence the entropy change in the time dt is

$$dS = -kN \sum_1^\infty \log w_n \, dw_n,$$

and, by taking account of (343), we have:

$$dS = \frac{Nk\mathsf{I}dt}{4h\nu L} \sum_1^\infty \left(w_n - (1-\eta)\, w_{n-1}\right) \log w_n. \tag{349}$$

186. The principle of increase of entropy requires that the sum of the entropy change (348) of the field of radiation and the entropy change (349) of the system of oscillators be always positive, or zero in the limiting case. That this condition is in fact satisfied we shall prove only for the special case when all rays falling on the oscillators are unpolarized, *i.e.*, when $\mathsf{K}' = \mathsf{K}$.

14

In this case we have from (147) and Sec. 185.

$$\left.\begin{matrix}\mathsf{K}_o\\ \mathsf{K}_o'\end{matrix}\right\} = \tfrac{1}{2}\left\{2\mathsf{K}+\beta\sin^2\theta(\mathsf{K}_e-\mathsf{K})\pm\beta\sin^2\theta(\mathsf{K}_e-\mathsf{K})\right\},$$

and hence

$$\mathsf{K}_o=\mathsf{K}+\beta\sin^2\theta(\mathsf{K}_e-\mathsf{K}),\quad \mathsf{K}_o'=\mathsf{K}.$$

The entropy change (348) of the field of radiation becomes

$$dt\Delta\nu\int q\,d\Omega\{\mathsf{L}(\mathsf{K}_o)-\mathsf{L}(\mathsf{K})\}$$

$$=dt\Delta\nu\int q\,d\Omega\ \beta\sin^2\theta(\mathsf{K}_e-\mathsf{K})\frac{d\mathsf{L}(\mathsf{K})}{d\mathsf{K}}$$

or, by (338) and (278),

$$=\frac{\pi kN\,dt}{hc\nu L}\int d\Omega\sin^2\theta(\mathsf{K}_e-\mathsf{K})\log\left(1+\frac{h\nu^3}{c^2\mathsf{K}}\right).$$

On adding to this the entropy change (349) of the system of oscillators and taking account of (320), the total increase in entropy in the time dt is found to be equal to the expression

$$\frac{\pi kN\,dt}{ch\nu L}\int d\Omega\sin^2\theta\left\{\mathsf{K}\sum_1^\infty(w_n-\zeta w_{n-1})\log w_n+(\mathsf{K}_e-\mathsf{K})\log\left(1+\frac{h\nu^3}{c^2\mathsf{K}}\right)\right\}$$

where

$$\zeta=1-\eta. \tag{350}$$

We now must prove that the expression

$$F=\int d\Omega\sin^2\theta\left\{\mathsf{K}\sum_1^\infty(w_n-\zeta w_{n-1})\log w_n+(\mathsf{K}_e-\mathsf{K})\log\left(1+\frac{h\nu^3}{c^2\mathsf{K}}\right)\right\} \tag{351}$$

is always positive and for that purpose we set down once more the meaning of the quantities involved. K is an arbitrary positive function of the polar angles θ and ϕ. The positive proper fraction ζ is according to (350), (265), and (320) given by

$$\frac{\zeta}{1-\zeta}=\frac{3c^2}{8\pi h\nu^3}\int\mathsf{K}\sin^2\theta\,d\Omega. \tag{352}$$

The quantities w_1, w_2, w_3, ... are any positive proper fractions whatever which, according to (167), satisfy the condition

$$\sum_{1}^{\infty} w_n = 1 \tag{353}$$

while, according to (342),

$$w_o = \frac{1-\zeta}{\zeta}. \tag{354}$$

Finally we have from (336)

$$\mathsf{K}_e = \frac{h\nu^3\zeta}{c^2} \sum_{1}^{\infty} n w_n. \tag{355}$$

187. To give the proof required we shall show that the least value which the function F can assume is positive or zero. For this purpose we consider first that positive function, K, of θ and ϕ, which, with fixed values of ζ, w_1, w_2, w_3, ... and K_e, will make F a minimum. The necessary condition for this is $\delta F = 0$, where according to (352)

$$\int \delta \mathsf{K} \sin^2 \theta \, d\,\Omega = 0.$$

This gives, by considering that the quantities w and ζ do not depend on θ and ϕ, as a necessary condition for the minimum,

$$\delta F = 0 = \int d\Omega \sin^2\theta \, \delta \, \mathsf{K} \left\{ -\log\left(1+\frac{h\nu^3}{c^2\mathsf{K}}\right) - \frac{\mathsf{K}_e - \mathsf{K}}{\frac{c^2\mathsf{K}}{h\nu^3}+1} \cdot \frac{1}{\mathsf{K}} \right\}$$

and it follows, therefore, that the quantity in brackets, and hence also K itself is independent of θ and ϕ. That in this case F really has a minimum value is readily seen by forming the second variation

$$\delta^2 F = \int d\Omega \sin^2 \theta \, \delta \mathsf{K} \delta \left\{ -\log\left(1+\frac{h\nu^3}{c^2\mathsf{K}}\right) - \frac{\mathsf{K}_e - \mathsf{K}}{\frac{c^2\mathsf{K}}{h\nu^3}+1} \cdot \frac{1}{\mathsf{K}} \right\}$$

which may by direct computation be seen to be positive under all circumstances.

In order to form the minimum value of F we calculate the value of K, which, from (352), is independent of θ and ϕ. Then it follows, by taking account of (319a), that

$$\mathsf{K} = \frac{h\nu^3}{c^2} \, \frac{\zeta}{1-\zeta}$$

and, by also substituting K_e from (355),

$$F=\frac{8\pi h\nu^3}{3c^2}\ \frac{\zeta}{1-\zeta}\ \sum_1^\infty (w_n-\zeta w_{n-1}) \log w_n-[(1-\zeta)n-1]\ w_n \log \zeta.$$

188. It now remains to prove that the sum

$$\Phi=\sum_1^\infty (w_n-\zeta w_{n-1}) \log w_n-[(1-\zeta)n-1]\ w_n \log \zeta, \tag{356}$$

where the quantities w_n are subject only to the restrictions that (353) and (354) can never become negative. For this purpose we determine that system of values of the w's which, with a fixed value of ζ, makes the sum Φ a minimum. In this case $\delta\Phi=0$, or

$$\sum_1^\infty (\delta w_n-\zeta\ \delta w_{n-1}) \log w_n+(w_n-\zeta w_{n-1})\ \frac{\delta w_n}{w_n} \tag{357}$$

$$-[(1-\zeta)n-1]\ \delta w_n \log \zeta=0,$$

where, according to (353) and (354),

$$\sum^\infty \delta w_n=0 \text{ and } \delta w_o=0. \tag{358}$$

If we suppose all the separate terms of the sum to be written out, the equation may be put into the following form:

$$\sum_1^\infty \delta w_n\{\log w_n-\zeta \log w_{n+1}+\frac{w_n-\zeta w_{n-1}}{w_n}-[(1-\zeta)n-1] \log \zeta\}=0. \tag{359}$$

From this, by taking account of (358), we get as the condition for a minimum, that

$$\log w_n-\zeta \log w_{n+1}+\frac{w_n-\zeta w_{n-1}}{w_n}-[(1-\zeta)\,n-1] \log \zeta \tag{360}$$

must be independent of n.

The solution of this functional equation is

$$w_n=(1-\zeta)\zeta^{n-1} \tag{361}$$

for it satisfies (360) as well as (353) and (354). With this value (356) becomes

$$\Phi=0. \tag{362}$$

189. In order to show finally that the value (362) of Φ is really the minimum value, we form from (357) the second variation

$$\delta^2\Phi = \sum_1^\infty (\delta w_n - \zeta\delta w_{n-1})\frac{\delta w_n}{w_n} - \frac{\zeta\delta w_{n-1}}{w_n}\delta w_n + \frac{\zeta w_{n-1}}{w_n^2}\delta w_n^2,$$

where all terms containing the second variation $\delta^2 w_n$ have been omitted since their coefficients are, by (360), independent of n and since

$$\sum_1^\infty \delta^2 w_n = 0.$$

This gives, taking account of (361),

$$\delta^2\Phi = \sum_1^\infty \frac{2\delta w_n^2}{(1-\zeta)\zeta^{n-1}} - \frac{2\zeta\delta w_{n-1}\delta w_n}{(1-\zeta)\zeta^{n-1}}$$

or

$$\delta^2\Phi = \frac{2\zeta}{1-\zeta}\sum_1^\infty \frac{\delta w_n^2}{\zeta^n} - \frac{\delta w_{n-1}\delta w_n}{\zeta^{n-1}}.$$

That the sum which occurs here, namely,

$$\frac{\delta w_1^2}{\zeta} - \frac{\delta w_1\delta w_2}{\zeta} + \frac{\delta w_2^2}{\zeta^2} - \frac{\delta w_2\delta w_3}{\zeta^2} + \frac{\delta w_3^2}{\zeta^3} - \frac{\delta w_3\delta w_4}{\zeta^3} + \; . \; . \; . \; . \; . \qquad (363)$$

is essentially positive may be seen by resolving it into a sum of squares. For this purpose we write it in the form

$$\sum_1^\infty \frac{1-\alpha_n}{\zeta^n}\delta w_n^2 - \frac{\delta w_n\delta w_{n+1}}{\zeta^n} + \frac{\alpha_{n+1}}{\zeta^{n+1}}\delta w_{n+1}^2,$$

which is identical with (363) provided $\alpha_1 = 0$. Now the α's may be so determined that every term of the last sum is a perfect square, *i.e.*, that

$$4\cdot\frac{1-\alpha_n}{\zeta^n}\cdot\frac{\alpha_{n+1}}{\zeta^{n+1}} = \left(\frac{1}{\zeta^n}\right)^2$$

or

$$\alpha_{n+1} = \frac{\zeta}{4(1-\alpha_n)}. \qquad (364)$$

By means of this formula the α's may be readily calculated. The first values are:

$$\alpha_1 = 0, \quad \alpha_2 = \frac{\zeta}{4}, \quad \alpha_3 = \frac{\zeta}{4-\zeta}, \; . \; . \; . \; . \; .$$

Continuing the procedure α_n remains always positive and less than $\alpha' = \frac{1}{2}\left(1 - \sqrt{1-\zeta}\right)$. To prove the correctness of this statement we show that, if it holds for α_n, it holds also for α_{n+1}. We assume, therefore, that α_n is positive and $< \alpha'$. Then from (364) α_{n+1} is positive and $< \frac{\zeta}{4(1-\alpha')}$. But $\frac{\zeta}{4(1-\alpha')} = \alpha'$. Hence $\alpha_{n+1} < \alpha'$. Now, since the assumption made does actually hold for $n=1$, it holds in general. The sum (363) is thus essentially positive and hence the value (362) of Φ really is a minimum, so that the increase of entropy is proven generally.

The limiting case (361), in which the increase of entropy vanishes, corresponds, of course, to the case of thermodynamic equilibrium between radiation and oscillators, as may also be seen directly by comparison of (361) with (271), (265), and (360).

190. Conclusion.—The theory of irreversible radiation processes here developed explains how, with an arbitrarily assumed initial state, a stationary state is, in the course of time, established in a cavity through which radiation passes and which contains oscillators of all kinds of natural vibrations, by the intensities and polarizations of all rays equalizing one another as regards magnitude and direction. But the theory is still incomplete in an important respect. For it deals only with the mutual actions of rays and vibrations of oscillators of the same period. For a definite frequency the increase of entropy in every time element until the maximum value is attained, as demanded by the second principle of thermodynamics, has been proven directly. But, for all frequencies taken together, the maximum thus attained does not yet represent the absolute maximum of the entropy of the system and the corresponding state of radiation does not, in general, represent the absolutely stable equilibrium (compare Sec. 27). For the theory gives no information as to the way in which the intensities of radiation corresponding to different frequencies equalize one another, that is to say, how from any arbitrary initial spectral distribution of energy the normal energy distribution corresponding to black radiation is, in the course of time, developed. For the oscillators on which the consideration was based influence only the intensities of rays which correspond to their natural vibration, but they are not capable of changing their frequencies, so long as they exert or suffer no other action than emitting or absorbing radiant energy.[1]

To get an insight into those processes by which the exchange of energy between rays of different frequencies takes place in nature would require also an investigation of the influence which the motion of the oscillators and of the electrons flying back and forth between them exerts on the radiation phenomena. For, if the oscillators and electrons are in motion, there will be impacts between them, and, at every impact, actions must come into play which influence the energy of vibration of the oscillators in a quite different and much more radical way than the simple emission and absorption of radiant energy. It is true that the final result of all such impact actions may be anticipated by the aid of the probability considerations discussed in the third section, but to show in detail how and in what time intervals this result is arrived at will be the problem of a future theory. It is certain that, from such a theory, further information may be expected as to the nature of the oscillators which really exist in nature, for the very reason that it must give a closer explanation of the physical significance of the universal elementary quantity of action, a significance which is certainly not second in importance to that of the elementary quantity of electricity.

[1] Compare *P. Ehrenfest*, Wien. Ber. **114** [2a], p. 1301, 1905. Ann. d. Phys. **36,** p. 91, 1911. *H. A. Lorentz*, Phys. Zeitschr. **11,** p. 1244, 1910. *H. Poincaré*, Journ. de Phys. (5) **2,** p. 5, p. 347, 1912.

AUTHOR'S BIBLIOGRAPHY

List of the papers published by the author on heat radiation and the hypothesis of quanta, with references to the sections of this book where the same subject is treated.

Absorption und Emission elektrischer Wellen durch Resonanz. Sitzungsber. d. k. preuss. Akad. d. Wissensch. vom 21. März 1895, p. 289–301. WIED. Ann. **57,** p. 1–14, 1896.

Ueber elektrische Schwingungen, welche durch Resonanz erregt und durch Strahlung gedämpft werden. Sitzungsber. d. k. preuss. Akad. d. Wissensch. vom 20. Februar 1896, p. 151–170. WIED. Ann. **60,** p. 577–599, 1897.

Ueber irreversible Strahlungsvorgänge. (Erste Mitteilung.) Sitzungsber. d. k. preuss. Akad. d. Wissensch. vom 4. Februar 1897, p. 57–68.

Ueber irreversible Strahlungsvorgänge. (Zweite Mitteilung.) Sitzungsber. d. k. preuss. Akad. d. Wissensch. vom 8. Juli 1897, p. 715–717.

Ueber irreversible Strahlungsvorgänge. (Dritte Mittelung.) Sitzungsber. d. k. preuss. Akad. d. Wissensch. vom 16. Dezember 1897, p. 1122–1145.

Ueber irreversible Strahlungsvorgänge. (Vierte Mitteilung.) Sitzungsber. d. k. preuss. Akad. d. Wissensch. vom 7. Juli 1898, p. 449–476.

Ueber irreversible Strahlungsvorgänge. (Fünfte Mitteilung.) Sitzungsber. d. k. preuss. Akad. d. Wissensch. vom 18. Mai 1899, p. 440–480. (§§ 144 bis 190. § 164.)

Ueber irreversible Strahlungsvorgänge. Ann. d. Phys. **1,** p. 69–122, 1900. (§§ 144–190. § 164.)

Entropie und Temperatur strahlender Wärme. Ann. d. Phys. **1,** p. 719 bis 737, 1900. (§ 101. § 166.)

Ueber eine Verbesserung der WIENschen Spektralgleichung. Verhandlungen der Deutschen Physikalischen Gesellschaft **2,** p. 202–204, 1900. (§ 156.)

Ein vermeintlicher Widerspruch des magneto-optischen FARADAY-Effektes mit der Thermodynamik. Verhandlungen der Deutschen Physikalischen Gesellschaft **2,** p. 206–210, 1900.

Kritik zweier Sätze des Herrn W. WIEN. Ann. d. Phys. **3,** p. 764–766, 1900.

Zur Theorie des Gesetzes der Energieverteilung im Normalspektrum. Verhandlungen der Deutschen Physikalischen Gesellschaft **2,** p. 237–245, 1900. (§§ 141–143. § 156 f. § 163.)

Ueber das Gesetz der Energieverteilung im Normalspektrum. Ann. d. Phys. **4,** p. 553–563, 1901. (§§ 141–143. §§ 156–162.)

Ueber die Elementarquanta der Materie und der Elektrizität. Ann. d. Phys. **4,** p. 564–566, 1901. (§ 163.)

Ueber irreversible Strahlungsvorgänge (Nachtrag). Sitzungsber. d. k. preuss. Akad. d. Wissensch. vom 9. Mai 1901, p. 544–555. Ann. d. Phys. **6,** p. 818–831, 1901. (§§ 185–189.)

Vereinfachte Ableitung der Schwingungsgesetze eines linearen Resonators im stationär durchstrahlten Felde. Physikalische Zeitschrift **2,** p. 530 bis p. 534, 1901.

Ueber die Natur des weissen Lichtes. Ann. d. Phys. **7,** p. 390–400, 1902. (§§ 107–112. §§ 170–174.)

Ueber die von einem elliptisch schwingenden Ion emittierte und absorbierte Energie. Archives Néerlandaises, Jubelband für H. A. LORENTZ, 1900, p. 164–174. Ann. d. Phys. **9,** p. 619–628, 1902.

Ueber die Verteilung der Energie zwischen Aether und Materie. Archives Néerlandaises, Jubelband für J. BOSSCHA, 1901, p. 55–66. Ann. d. Phys. **9,** p. 629–641, 1902. (§§ 121–132.)

Bemerkung über die Konstante des WIENschen Verschiebungsgesetzes. Verhandlungen der Deutschen Physikalischen Gesellschaft **8,** p. 695–696, 1906. (§ 161.)

Zur Theorie der Wärmestrahlung. Ann. d. Phys. **31,** p. 758–768, 1910.

Eine neue Strahlungshypothese. Verhandlungen der Deutschen Physikalischen Gesellschaft **13,** p. 138–148, 1911. (§ 147.)

Zur Hypothese der Quantenemission. Sitzungsber. d. k. preuss. Akad. d. Wissensch. vom 13. Juli 1911, p. 723–731. (§§ 150–152.)

Ueber neuere thermodynamische Theorien (NERNSTsches Wärmetheorem und Quantenhypothese). Ber. d. Deutschen Chemischen Gesellschaft **45,** p. 5–23, 1912. Physikalische Zeitschrift **13,** p. 165–175, 1912. Akademische Verlagsgesellschaft m. b. H., Leipzig 1912. (§§ 120–125.)

Ueber die Begründung des Gesetzes der schwarzen Strahlung. Ann. d. Phys. **37,** p. 642–656, 1912. (§§ 145–156.)

APPENDIX I

On Deductions from Stirling's Formula.

The formula is

(a) $$\lim_{n=\infty} \frac{n!}{n^n e^{-n}\sqrt{2\pi n}} = 1,$$

or, to an approximation quite sufficient for all practical purposes, provided that n is larger than 7

(b) $$n! = \left(\frac{n}{e}\right)^n \sqrt{2\pi n}.$$

For a proof of this relation and a discussion of its limits of accuracy a treatise on probability must be consulted.

On substitution in (170) this gives

$$W = \frac{\left(\frac{N}{e}\right)^N}{\left(\frac{N_1}{e}\right)^{N_1} \cdot \left(\frac{N_2}{e}\right)^{N_2} \cdots} \cdot \frac{\sqrt{2\pi N}}{\sqrt{2\pi N_1}\ \sqrt{2\pi N_2} \cdots}$$

On account of (165) this reduces at once to

$$\frac{N^N}{N_1^{N_1}\ N_2^{N_2} \cdots} \cdot \frac{\sqrt{2\pi N}}{\sqrt{2\pi N_1} \cdot \sqrt{2\pi N_2} \cdots}$$

Passing now to the logarithmic expression we get

$$S = k \log W = k[N \log N - N_1 \log N_1 - N_2 \log N_2 - \cdots \cdots$$
$$+ \log \sqrt{2\pi N} - \log \sqrt{2\pi N_1} - \log \sqrt{2\pi N_2} - \ \ldots\ ],$$

or,

$$S = k \log W = k[(N \log N - \log \sqrt{2\pi N}) + (N_1 \log N_1 - \log \sqrt{2\pi N_1}) +$$
$$(N_2 \log N_2 - \log \sqrt{2\pi N_2}) + \ \ldots\ ].$$

Now, for a large value of N_i, the term $N_i \log N_i$ is very much larger than $\log \sqrt{2\pi N_i}$, as is seen by writing the latter in the form $\frac{1}{2} \log 2\pi + \frac{1}{2} \log N_i$. Hence the last expression will, with a fair approximation, reduce to

$$S = k \log W = k[N \log N - N_1 \log N_1 - N_2 \log N_2 - \ \ldots\ldots\ ].$$

Introducing now the values of the densities of distribution w by means of the relation

$$N_i = w_i N$$

we obtain

$$S = k \log W = kN[\log N - w_1 \log N_1 - w_2 \log N_2 - \ . \ . \ . \ ],$$

or, since

$$w_1 + w_2 + w_3 + \ . \ . \ . \ = 1,$$

and hence

$$(w_1 + w_2 + w_3 + \ . \ . \ . \ ) \log N = \log N,$$

and

$$\log N - \log N_1 = \log \frac{N}{N_1} = \log \frac{1}{w_1} = -\log w_1,$$

we obtain by substitution, after one or two simple transformations

$$S = k \log W = -kN \Sigma\, w_1 \log w_1,$$

a relation which is identical with (173).

The statements of Sec. 143 may be proven in a similar manner. From (232) we get at once

$$S = k \log W_m = k \log \frac{(N+P-1)!}{(N-1)!\, P!}$$

Now $\log (N-1)! = \log N! - \log N,$

and, for large values of N, $\log N$ is negligible compared with $\log N!$ Applying the same reasoning to the numerator we may without appreciable error write

$$S = k \log W_m = k \log \frac{(N+P)!}{N!\, P!}$$

Substituting now for $(N+P)!$, $N!$, and $P!$ their values from (b) and omitting, as was previously shown to be approximately correct, the terms arising from the $\sqrt{2\pi(N+P)}$ *etc.*, we get, since the terms containing e cancel out

$$S = k[(N+P) \log (N+P) - N \log N - P \log P]$$

$$= k\left[(N+P) \log \frac{N+P}{N} + P \log N - P \log P\right]$$

$$= kN\left[\left(\frac{P}{N} + 1\right) \log \left(\frac{P}{N} + 1\right) - \frac{P}{N} \log \frac{P}{N}\right].$$

This is the relation of Sec. 143.

APPENDIX II

REFERENCES

Among general papers treating of the application of the theory of quanta to different parts of physics are:

1. *A. Sommerfeld*, Das Planck'sche Wirkungsquantum und seine allgemeine Bedeutung für die Molekularphysik, Phys. Zeitschr., **12**, p. 1057. Report to the Versammlung Deutscher Naturforscher und Aerzte. Deals especially with applications to the theory of specific heats and to the photoelectric effect. Numerous references are quoted.

2. Meeting of the British Association, Sept., 1913. See Nature, **92**, p. 305, Nov. 6, 1913, and Phys. Zeitschr., **14**, p. 1297. Among the principal speakers were *J. H. Jeans* and *H. A. Lorentz*.

(Also American Phys. Soc., Chicago Meeting, 1913.[1])

3. *R. A. Millikan*, Atomic Theories of Radiation, Science, **37**, p. 119, Jan. 24, 1913. A non-mathematical discussion.

4. *W. Wien*, Neuere Probleme der Theoretischen Physik, 1913. (*Wien's* Columbia Lectures, in German.) This is perhaps the most complete review of the entire theory of quanta.

H. A. Lorentz, Alte und Neue Probleme der Physik, Phys. Zeitschr., **11**, p. 1234. Address to the Versammlung Deutscher Naturforscher und Aerzte, Königsberg, 1910, contains also some discussion of the theory of quanta.

Among the papers on radiation are:

E. Bauer, Sur la théorie du rayonnement, Comptes Rendus, **153**, p. 1466. Adheres to the quantum theory in the original form, namely, that emission and absorption both take place in a discontinuous manner.

E. Buckingham, Calculation of c_2 in Planck's equation, Bull. Bur. Stand. **7**, p. 393.

E. Buckingham, On *Wien's* Displacement Law, Bull. Bur. Stand. **8**, p. 543. Contains a very simple and clear proof of the displacement law.

[1] Not yet published (Jan. 26, 1914. Tr.)

P. Ehrenfest, Strahlungshypothesen, Ann. d. Phys., **36,** p. 91.

A. Joffé, Theorie der Strahlung, Ann. d. Phys., **36,** p. 534.

Discussions of the method of derivation of the radiation formula are given in many papers on the subject. In addition to those quoted elsewhere may be mentioned:

C. Benedicks, Ueber die Herleitung von *Planck's* Energieverteilungsgesetz, Ann. d. Phys., **42,** p. 133. Derives *Planck's* law without the help of the quantum theory. The law of equipartition of energy is avoided by the assumption that solids are not always monatomic, but that, with decreasing temperature, the atoms form atomic complexes, thus changing the number of degrees of freedom. The equipartition principle applies only to the free atoms.

P. Debye, *Planck's* Strahlungsformel, Ann. d. Phys., **33,** p. 1427. This method is fully discussed by *Wien* (see 4, above). It somewhat resembles *Jeans*' method (Sec. 169) since it avoids all reference to resonators of any particular kind and merely establishes the most probable energy distribution. It differs, however, from *Jeans*' method by the assumption of discrete energy quanta $h\nu$. The physical nature of these units is not discussed at all and it is also left undecided whether it is a property of matter or of the ether or perhaps a property of the energy exchange between matter and the ether that causes their existence. (Compare also some remarks of *Lorentz* in 2.)

P. Frank, Zur Ableitung der Planckschen Strahlungsformel, Phys. Zeitschr., **13,** p. 506.

L. Natanson, Statistische Theorie der Strahlung, Phys. Zeitschr., **12,** p. 659.

W. Nernst, Zur Theorie der Specifischen Wärme und über die Anwendung, der Lehre von den Energiequanten auf Physikalisch-chemische Fragen überhaupt, Zeitschr. f. Elektochemie, **17,** p. 265.

The experimental facts on which the recent theories of specific heat (quantum theories) rely, were discovered by *W. Nernst* and his fellow workers. The results are published in a large number of papers that have appeared in different periodicals. See, *e.g.*, *W. Nernst*, Der Energieinhalt fester Substanzen, Ann. d. Phys., **36,** p. 395, where also numerous other papers are quoted. (See also references given in 1.) These experimental facts give very strong support to the heat theorem of *Nernst* (Sec. 120), according to which the entropy approaches a definite limit (perhaps the value zero, see *Planck's* Thermodynamics, 3. ed., sec. 282, *et seq.*) at the absolute zero of temperature, and which is consistent with the quantum theory. This work is in close connection with the recent attempts to develop an equation of state applicable to the solid state of matter. In addition to the papers by *Nernst* and his school there may be mentioned:

K. Eisenmann, Canonische Zustandsgleichung einatomiger fester Körper und die Quantentheorie, Verhandlungen der Deutschen Physikalischen Gesellschaft, **14,** p. 769.

W. H. Keesom, Entropy and the Equation of State, Konink. Akad. Wetensch. Amsterdam Proc., **15,** p. 240.

L. Natanson, Energy Content of Bodies, Acad. Science Cracovie Bull. Ser. A, p. 95. In *Einstein's* theory of specific heats (Sec. 140) the atoms of actual bodies in nature are apparently identified with the ideal resonators of *Planck.* In this paper it is pointed out that this is implying too special features for the atoms of real bodies, and also, that such far-reaching specializations do not seem necessary for deriving the laws of specific heat from the quantum theory.

L. S. Ornstein, Statistical Theory of the Solid State, Konink. Akad. Wetensch. Amsterdam Proc., **14,** p. 983.

S. Ratnowsky, Die Zustandsgleichung einatomiger fester Körper und die Quantentheorie, Ann. d. Phys., **38,** p. 637.

Among papers on the law of equipartition of energy (Sec. 169) are:

J. H. Jeans, Planck's Radiation Theory and *Non-Newtonian* Mechanics, Phil. Mag., **20,** p. 943.

S. B. McLaren, Partition of Energy between Matter and Radiation, Phil. Mag., **21,** p. 15.

S. B. McLaren, Complete Radiation, Phil. Mag. **23,** p. 513. This paper and the one of Jeans deal with the fact that from Newtonian Mechanics (Hamilton's Principle) the equipartition principle necessarily follows, and that hence either Planck's law or the fundamental principles of mechanics need a modification.

For the law of equipartition compare also the discussion at the meeting of the British Association (see 2).

In many of the papers cited so far deductions from the quantum theory are compared with experimental facts. This is also done by:

F. Haber, Absorptionsspectra fester Körper und die Quantentheorie, Verhandlungen der Deutschen Physikalischen Gesellschaft, **13,** p. 1117.

J. Franck und G. Hertz, Quantumhypothese und Ionisation, Ibid., **13,** p. 967.

Attempts of giving a concrete physical idea of Planck's constant h are made by:

A. Schidlof, Zur Aufklärung der universellen electrodynamischen Bedeutung der Planckschen Strahlungsconstanten h, Ann. d. Phys., **35,** p. 96.

D. A. Goldhammer, Ueber die Lichtquantenhypothese, Phys. Zeitschr., **13,** p. 535.

J. J. Thomson, On the Structure of the Atom, Phil. Mag., **26,** p. 792.

N. Bohr, On the Constitution of the Atom, Phil. Mag., **26,** p. 1.

S. B. McLaren, The Magneton and Planck's Universal Constant, Phil. Mag., **26,** p. 800.

The line of reasoning may be briefly stated thus: Find some quantity of the same dimension as h, and then construct a model of an atom where this property plays an important part and can be made, by a simple hypothesis, to vary by finite amounts instead of continuously. The simplest of these is *Bohr's*, where h is interpreted as angular momentum.

The logical reason for the quantum theory is found in the fact that the *Rayleigh-Jeans* radiation formula does not agree with experiment. Formerly *Jeans* attempted to reconcile theory and experiment by the assumption that the equilibrium of radiation and a black body observed and agreeing with *Planck's* law rather than his own, was only apparent, and that the true state of equilibrium which really corresponds to his law and the equipartition of energy among all variables, is so slowly reached that it is never actually observed. This standpoint, which was strongly objected to by authorities on the experimental side of the question (see, *e.g.*, *E. Pringsheim* in 2), he has recently abandoned. *H. Poincaré*, in a profound mathematical investigation (*H. Poincaré*, Sur. la Théorie des Quanta,

Journal de Physique (5), **2,** p. 1, 1912) reached the conclusion that whatever the law of radiation may be, it must always, if the total radiation is assumed as finite, lead to a function presenting similar discontinuites as the one obtained from the hypothesis of quanta.

While most authorities have accepted the quantum theory for good (see *J. H. Jeans* and *H. A. Lorentz* in 2), a few still entertain doubts as to the general validity of Poincaré's conclusion (see above *C. Benedicks* and *R. A Millikan* 3). Others still reject the quantum theory on account of the fact that the experimental evidence in favor of *Planck's* law is not absolutely conclusive (see *R. A. Millikan* 3); among these is *A. E. H. Love* (2), who suggests that *Korn's* (*A. Korn,* Neue Mechanische Vorstellungen uber die Schwarze Strahlung und eine sich aus denselben ergebende Modification des Planckschen Verteilungsgesetzes, Phys. Zeitschr., **14,** p. 632) radiation formula fits the facts as well as that of Planck.

H. A. Callendar, Note on Radiation and Specific Heat, Phil. Mag., **26,** p. 787, has also suggested a radiation formula that fits the data well. Both Korn's and Callendar's formulæ conform to *Wien's* displacement law and degenerate for large values of λT into the Rayleigh-Jeans, and for small values of λT into Wien's radiation law. Whether Planck's law or one of these is the correct law, and whether, if either of the others should prove to be right, it would eliminate the necessity of the adoption of the quantum theory, are questions as yet undecided. Both Korn and Callendar have promised in their papers to follow them by further ones.

ERRATA

Page 77. The last sentence of Sec. 77 should be replaced by: The corresponding additional terms may, however, be omitted here without appreciable error, since the correction caused by them would consist merely of the addition to the energy change here calculated of a comparatively infinitesimal energy change of the same kind with an external work that is infinitesimal of the second order.

Page 83. Insert at the end of Sec. 84 a:
These laws hold for any original distribution of energy whatever; hence, *e. g.*, an originally monochromatic radiation remains monochromatic during the process described, its color changing in the way stated.

VORLESUNGEN

ÜBER DIE

THEORIE DER WÄRMESTRAHLUNG

VON

DR. MAX PLANCK,
PROFESSOR DER THEORETISCHEN PHYSIK
AN DER UNIVERSITÄT BERLIN

MIT 6 ABBILDUNGEN

LEIPZIG, 1906
VERLAG VON JOHANN AMBROSIUS BARTH

Vorwort.

In dem vorliegenden Buche ist der Hauptinhalt der Vorlesungen wiedergegeben, welche ich im Wintersemester 1905/6 an der Berliner Universität gehalten habe. Ursprünglich war es nur meine Absicht gewesen, die Ergebnisse meiner eigenen, vor zehn Jahren begonnenen Untersuchungen über die Theorie der Wärmestrahlung in eine zusammenhängende Darstellung zu vereinigen; doch bald zeigte es sich als wünschenswert, auch die Grundlagen dieser Theorie, von den KIRCHHOFFschen Sätzen über das Emissions- und Absorptionsvermögen angefangen, mit in die Behandlung hineinzuziehen, und so machte ich den Versuch, ein Lehrbuch zu schreiben, welches zugleich auch zur Einführung in das Studium der gesamten Theorie der strahlenden Wärme auf einheitlicher thermodynamischer Grundlage zu dienen geeignet ist. Dementsprechend nimmt die Darstellung ihren Ausgang von den einfachen bekannten Erfahrungssätzen der Optik, um durch allmähliche Erweiterung und Hinzuziehung der Ergebnisse der Elektrodynamik und der Thermodynamik zu den Problemen der spektralen Energieverteilung und der Irreversibilität vorzudringen. Hierbei bin ich öfters, wo es mir sachliche oder didaktische Gründe nahelegten, von der sonst üblichen Art der Betrachtung abgewichen, so insbesondere bei der Ableitung der KIRCHHOFFschen Sätze, der Berechnung des MAXWELLschen Strahlungsdrucks, der Ableitung des WIENschen Verschiebungsgesetzes und seiner Verallgemeinerung auf Strahlungen von beliebiger spektraler Energieverteilung. Die Resultate meiner eigenen Untersuchungen habe ich überall an den entsprechenden Stellen mit in die Darstellung hineingearbeitet. Ein Verzeichnis derselben findet sich zur bequemeren

Vergleichung und Kontrollierung näherer Einzelheiten am Schluß des Buches zusammengestellt.

Es liegt mir aber daran, auch an dieser Stelle noch besonders hervorzuheben, was sich im letzten Paragraphen des Buches näher ausgeführt findet, daß die hier entwickelte Theorie keineswegs den Anspruch erhebt, als vollkommen abgeschlossen zu gelten, wenn sie auch, wie ich glaube, einen gangbaren Weg eröffnet, um die Vorgänge der Energiestrahlung von dem nämlichen Gesichtspunkt aus zu überblicken wie die der Molekularbewegungen.

München, Ostern 1906.

Der Verfasser.

Inhalt.

Erster Abschnitt.

Grundtatsachen und Definitionen.

Erstes Kapitel. Allgemeines.

§ 1. Wärme kann sich in einem ruhenden Medium auf zwei gänzlich verschiedene Arten fortpflanzen: durch Leitung und durch Strahlung. Die Wärmeleitung ist bedingt durch die Temperatur des Mediums, in welchem sie stattfindet, speziell durch die Ungleichmäßigkeit der räumlichen Temperaturverteilung, welche gemessen wird durch die Größe des Temperaturgefälles oder Temperaturgradienten. In Gebieten, wo die Temperatur des Mediums sich nicht mit dem Orte ändert, verschwindet jede Spur von Wärmeleitung.

Die Wärmestrahlung dagegen ist an sich gänzlich unabhängig von der Temperatur des Mediums, durch welches sie hindurchgeht. So kann man durch eine Sammellinse von Eis hindurch, die sich auf der konstanten Temperatur von 0° C. befindet, Sonnenstrahlen in einen Brennpunkt konzentrieren und zur Entzündung eines leicht brennbaren Körpers benutzen. Im allgemeinen ist die Wärmestrahlung ein viel komplizierterer Vorgang als die Wärmeleitung, weil der Strahlungszustand in einem bestimmten Augenblicke an einer bestimmten Stelle des Mediums sich nicht, wie der Wärmeleitungsstrom, durch einen einzigen Vektor, d. h. durch eine einzige gerichtete Größe, charakterisieren läßt. Vielmehr sind die Wärmestrahlen, welche in einem bestimmten Augenblicke das Medium an einem bestimmten Punkte durchkreuzen, von vornherein gänzlich unabhängig voneinander, und man darf den Strahlungszustand nicht eher als vollkommen bekannt ansehen, als bis die Intensität der Strahlung nach jeder einzelnen der unendlich vielen von einem Punkte ausgehenden

Richtungen des Raumes gegeben ist. Dabei zählen zwei gerade entgegengesetzte Richtungen doppelt, weil die Strahlung nach der einen Seite ganz unabhängig ist von der nach der entgegengesetzten Seite.

§ 2. Ohne vorläufig auf eine speziellere Theorie der Wärmestrahlung einzugehen, werden wir stets von dem durch die mannigfaltigsten Erfahrungen bewährten Satze Gebrauch machen, daß die Wärmestrahlen, rein physikalisch betrachtet, nichts anderes sind als Lichtstrahlen von entsprechender Wellenlänge, und werden die Bezeichnung „Wärmestrahlung" ganz allgemein für alle diejenigen physikalischen Vorgänge gebrauchen, welche zur Klasse der Lichtstrahlen gehören. Jeder Lichtstrahl ist demnach zugleich auch ein Wärmestrahl. Auch werden wir gelegentlich zur Abkürzung von der „Farbe" eines Wärmestrahls sprechen, um seine Wellenlänge oder seine Schwingungsdauer zu kennzeichnen. Daher wenden wir auch alle aus der Optik bekannten Erfahrungssätze auf die Wärmestrahlung an, vor allem die der Fortpflanzung, der Spiegelung (Reflexion) und der Brechung (Refraktion). Nur die Erscheinungen der Beugung (Diffraktion), wenigstens soweit sie sich in Gebieten von größeren Dimensionen abspielen, wollen wir wegen ihres komplizierteren Charakters nicht berücksichtigen, und sind daher genötigt, von vornherein eine besondere Einschränkung hinsichtlich der von uns zu betrachtenden Räume zu machen. Es soll nämlich im folgenden überall die Voraussetzung gelten, daß die linearen Dimensionen aller betrachteten Räume und auch die Krümmungsradien aller betrachteten Oberflächen groß sind gegen die Wellenlängen der betrachteten Strahlen. Dann können wir ohne merklichen Fehler von den Einflüssen der durch die Form der Grenzflächen bedingten Beugung ganz absehen und können überall die gewöhnlichen Gesetze der optischen Reflexion und Brechung zur Anwendung bringen. Hiermit führen wir also ein für allemal eine strenge Scheidung ein zwischen zwei Arten von Längen, die ganz getrennten Größenordnungen angehören: Körperdimensionen und Wellenlängen. Auch die Differentiale der Körperdimensionen: Längen-, Flächen- und Volumenelemente, nehmen wir immer noch groß an gegen die entsprechenden Potenzen der Wellenlängen. Je langwelligere Strahlen wir berücksichtigen wollen, um so größere Räume müssen wir daher betrachten. Da wir aber in der Wahl unserer Räume im übrigen gar nicht beschränkt sind, so wird uns aus dieser Festsetzung keine weitere Schwierigkeit erwachsen.

§ 3. Wesentlicher noch als die Unterscheidung zwischen großen und kleinen Längen ist für die gesamte Theorie der Wärmestrahlung die Unterscheidung zwischen großen und kleinen Zeiten. Denn es liegt schon in der Definition der Intensität eines Wärmestrahles als derjenigen Energie, welche von dem Strahle in der Zeiteinheit geliefert wird, die Voraussetzung mit enthalten, daß die gewählte Zeiteinheit groß ist gegenüber der Zeitdauer einer Schwingung, wie sie der Farbe des Strahles entspricht. Sonst würde nämlich offenbar der Betrag der Strahlungsintensität im allgemeinen davon abhängig sein, bei welcher Phase der Schwingung die Messung der vom Strahl gelieferten Energie begonnen wird. Nur wenn die Zeiteinheit zufällig gerade eine ganze Anzahl Schwingungen umfassen würde, wäre die Intensität eines Strahles von konstanter Periode und konstanter Amplitude unabhängig von der anfänglichen Phase. Um dieser Unzuträglichkeit zu entgehen, sind wir genötigt ganz allgemein festzusetzen, daß die Zeiteinheit, oder besser gesagt: daß die Zeit, welche der Definition einer Strahlungsintensität zugrunde gelegt wird, mag sie auch als Differential auftreten, groß ist gegen die Schwingungszeit jeder der Farben, die in dem Strahle enthalten sind.

Diese Festsetzung führt zu einer wichtigen Folgerung für Strahlungen von veränderlicher Intensität. Wenn wir z. B. bei periodisch schwankenden Strahlungsintensitäten wie in der Akustik von „Schwebungen" der Intensität sprechen, so muß selbstverständlich die zur Definition der augenblicklichen Strahlungsintensität benötigte Zeit klein sein gegen die Periode einer Schwebung. Da sie nun aber nach dem vorigen groß sein muß gegen die Zeitdauer einer Schwingung, so folgt daraus, daß die Zeitperiode einer Schwebung immer groß ist gegen die Zeitperiode einer Schwingung. Wäre diese Bedingung nicht erfüllt, so könnte man Schwingungen und Schwebungen gar nicht streng auseinanderhalten. Ebenso müssen im allgemeinen Fall, bei beliebig veränderlicher Strahlungsintensität, die Schwingungen immer sehr schnell erfolgen gegen die Intensitätsänderungen. Es versteht sich, daß in diesen Festsetzungen eine gewisse, und

1*

zwar eine sehr wesentliche Beschränkung der Allgemeinheit der zu betrachtenden Strahlungsvorgänge gelegen ist.

Eine ganz ähnliche und ebenso wesentliche Beschränkung der Allgemeinheit macht man übrigens, wie gleich hier bemerkt sein möge, in der kinetischen Gastheorie, wenn man die in einem chemisch einfachen Gase stattfindenden Bewegungen einteilt in sichtbare, grobe, molare, und in unsichtbare, feine, molekulare. Denn da die Geschwindigkeit einer einzelnen Molekel eine durchaus einheitliche Größe ist, so kann diese Einteilung nur unter der Voraussetzung durchgeführt werden, daß die Geschwindigkeitskomponenten der in hinreichend kleinen Volumina enthaltenen Molekeln gewisse, von der Größe der Volumina unabhängige Mittelwerthe besitzen, was im allgemeinen keineswegs der Fall zu sein braucht. Wenn ein derartiger Mittelwert, einschließlich des Wertes Null, nicht existiert, so kann man gar nicht zwischen der sichtbaren und der Wärmebewegung des Gases unterscheiden.

Wenn wir uns nun der Frage zuwenden, nach welchen Gesetzen sich die Strahlungsvorgänge in irgend einem Körpersystem, das wir stets als ruhend annehmen wollen, abspielen, so können wir das Problem von zwei verschiedenen Seiten angreifen: wir können nämlich entweder eine bestimmte Stelle im Raume ins Auge fassen und nach den verschiedenen Strahlen fragen, welche im Laufe der Zeit diese Stelle durchkreuzen, oder wir können einen bestimmten Strahl ins Auge fassen und nach seiner Geschichte fragen, d. h. nach seiner Entstehung, seiner Fortpflanzung und seiner Vernichtung. Für die folgende Darstellung wird es bequemer sein, die letztere Behandlungsart voranzustellen und zunächst der Reihe nach die drei genannten Vorgänge einzeln zu betrachten.

§ 4. Emission. Der Akt der Entstehung eines Wärmestrahles wird allgemein als „Emission“ bezeichnet. Nach dem Prinzip der Erhaltung der Energie erfolgt die Emission stets auf Kosten von anderweitiger Energie (Körperwärme, chemische Energie, elektrische Energie) und daraus geht hervor, daß nur substanzielle Partikel Wärmestrahlen emittieren können, nicht aber geometrische Räume oder Flächen. Man spricht zwar häufig in abkürzendem Sinne davon, daß die Oberfläche eines Körpers Wärme nach außen strahlt, aber diese Ausdrucksweise hat nicht den Sinn, daß die Oberfläche Wärmestrahlen emittiert. Die Oberfläche eines Körpers emittiert niemals im eigentlichen Sinne, sondern sie läßt die Strahlen, welche aus dem Innern des Körpers kommend die Oberfläche treffen, teils nach außen hindurch, teils reflektiert sie dieselben in das Innere zurück, und je nachdem der hindurchgehende Bruchteil größer oder kleiner ist, scheint die Oberfläche stärker oder schwächer auszustrahlen.

Betrachten wir nun das Innere einer emittierenden physikalisch homogenen Substanz und greifen dort irgend ein nicht zu kleines Volumenelement von der Größe $d\tau$ heraus. Dann wird die von allen in dem Volumenelement befindlichen Partikeln zusammengenommen in der Zeiteinheit durch Strahlung emittierte Energie proportional $d\tau$ sein. Wollten wir versuchen, näher auf die Analyse des Vorgangs der Emission einzugehen und ihn in seine elementaren Bestandteile zu zerlegen, so würden wir jedenfalls sehr komplizierte Verhältnisse antreffen. Denn es wird sich hierbei um die Betrachtung von Räumen handeln, deren Dimensionen so klein sind, daß man die Substanz nicht mehr als homogen betrachten kann, sondern auf ihre atomistische Konstitution Rücksicht nehmen muß. Deshalb ist die endliche Größe, welche man erhält, wenn man die von dem Volumenelement $d\tau$ emittierte Strahlung durch $d\tau$ dividiert, nur als ein gewisser Mittelwert anzusehen. Wir werden aber trotzdem für gewöhnlich, was für die Rechnung viel bequemer ist, den Vorgang der Emission so behandeln können, als ob alle Punkte des Volumenelementes $d\tau$ sich gleichmäßig an der Emission beteiligten, so daß jeder Punkt innerhalb $d\tau$ die Spitze eines nach allen Richtungen ausgehenden Bündels von Strahlen bildet. Ein solches elementares Punktbündel repräsentiert natürlich keine endliche Energiemenge; denn eine solche wird immer nur von den Punkten eines endlichen Volumens emittiert.

Wir wollen ferner die Substanz als isotrop annehmen. Dann wird die Strahlung von dem Volumenelement $d\tau$ nach allen Richtungen des Raumes gleichmäßig emittiert, d. h. die von einem Punkte des Elements innerhalb eines beliebigen Kegels emittierte Strahlung ist proportional der Öffnung des Kegels, wie sie gemessen wird durch die Größe der Fläche, welche der Kegel aus der mit dem Radius 1 um seine Spitze als Mittelpunkt beschriebenen Kugelfläche ausschneidet. Das gilt für beliebig große Kegelöffnungen. Nimmt man die Öffnung des Kegels unendlich klein, von der Größe $d\Omega$, so kann man von der „in einer bestimmten Richtung“ emittierten Strahlung sprechen, doch immer nur in dem Sinne, daß zur Emission einer endlichen Energiemenge eine unendliche Anzahl von Richtungen gehören, die eine endliche Kegelöffnung miteinander bilden.

§ 5. Die emittierte Strahlung wird eine gewisse, von vornherein ganz beliebige spektrale Energieverteilung besitzen, d. h. die verschiedenen Farben werden in ihr mit ganz verschiedener Intensität vertreten sein. Zur Bezeichnung der Farbe eines Strahles bedient man sich in der Experimentalphysik gewöhnlich der Angabe der Wellenlänge λ, weil dieselbe am direktesten gemessen wird. Für die theoretische Behandlung ist es aber meist bequemer, statt dessen die Anzahl der Schwingungen in der Zeiteinheit ν zu benutzen; denn für die Farbe eines bestimmten Licht- oder Wärmestrahls ist weniger seine Wellenlänge, welche sich beim Übergang des Strahles in ein anderes Medium ändert, als vielmehr seine Schwingungszahl charakteristisch, welche dem Strahl in allen Medien, wenigstens soweit sie ruhen, ungeändert erhalten bleibt. Wir bezeichnen also künftig eine bestimmte Farbe durch den entsprechenden Wert von ν, und ein bestimmtes Farbenintervall durch die Grenzen des Intervalls ν und ν', wobei $\nu' > \nu$ sein möge. Die auf ein bestimmtes Farbenintervall entfallende Strahlung, dividiert durch die Größe des Intervalls $\nu' - \nu$, nennen wir die mittlere Strahlung innerhalb des Farbenintervalls von ν bis ν'. Nehmen wir, bei festgehaltenem ν, die Differenz $\nu' - \nu$ hinreichend klein, gleich $d\nu$, so wollen wir annehmen, daß sich der Betrag der mittleren Strahlung einem bestimmten, von der Größe des Intervalls $d\nu$ unabhängigen Grenzwert nähert, den wir kurz als die „Strahlung von der Schwingungszahl ν“ bezeichnen. Zu einer endlichen Strahlung gehört dann offenbar immer ein endliches, wenn auch unter Umständen sehr kleines, Intervall von Schwingungszahlen.

Endlich haben wir noch auf den Polarisationszustand der emittierten Strahlung Rücksicht zu nehmen. Da wir das Medium als isotrop vorausgesetzt haben, so folgt, daß alle emittierten Strahlen unpolarisiert sind und daß daher jeder Strahl die doppelte Intensität besitzt wie eine seiner geradlinig polarisierten Komponenten, die man z. B. erhält, wenn man den Strahl durch ein NICOLsches Prisma hindurchschickt.

§ 6. Alles Bisherige zusammengefaßt können wir die gesamte in der Zeit dt vom Volumenelement $d\tau$ in der Richtung des Elementarkegels $d\Omega$ im Schwingungsintervall von ν bis $\nu + d\nu$ emittierte Energie gleich setzen:

$$dt \cdot d\tau \cdot d\Omega \cdot d\nu \cdot 2\,\varepsilon_\nu. \tag{1}$$

Die endliche Größe ε_ν nennen wir den „Emissionskoeffizienten" des Mediums für die Schwingungszahl ν. Er ist eine gewisse positive Funktion von ν und entspricht einem geradlinig polarisierten Strahl von bestimmter Farbe und bestimmter Richtung. Für die gesamte Emission des Volumenelements $d\tau$ erhält man hieraus durch Integration über alle Richtungen und über alle Schwingungszahlen, da ε_ν von der Richtung unabhängig ist und da das Integral über alle Kegelöffnungen $d\Omega$ den Wert 4π besitzt:

$$dt \cdot d\tau \cdot 8\pi \int_0^\infty \varepsilon_\nu\, d\nu. \tag{2}$$

§ 7. Der Emissionskoeffizient ε hängt außer von der Schwingungszahl ν noch von dem Zustand der in dem Volumenelement $d\tau$ enthaltenen emittierenden Substanz ab, und zwar im allgemeinen in sehr verwickelter Weise, je nach den physikalisch-chemischen Vorgängen, welche sich in dem betreffenden Zeitelement in dem Raume abspielen. Doch gilt ganz allgemein der Erfahrungssatz, daß die Emission eines Körperelements nur abhängt von den Vorgängen innerhalb des Körperelements. (Theorie von PREVOST.) Ein Körper A von 100^0 C. emittiert gegen einen ihm gegenüber befindlichen Körper B von 0^0 C. genau dieselbe Wärmestrahlung, wie gegen einen gleichgroßen und gleichgelegenen Körper B' von 1000^0 C., und wenn der Körper A von dem Körper B abgekühlt, von dem Körper B' aber erwärmt wird, so ist dies nur eine Folge des Umstandes, daß B schwächer, B' aber stärker emittiert als A.

Wir wollen nun weiter für das Folgende überall die vereinfachende Annahme einführen, daß die chemische Natur der emittierenden Substanz unveränderlich ist, und daß ihr physikalischer Zustand nur von einer einzigen Variabeln abhängt: der absoluten Temperatur T. Dann ergibt sich mit Notwendigkeit, daß auch der Emissionskoeffizient ε außer von der Schwingungszahl ν und von der chemischen Natur des Mediums allein von der Temperatur T abhängig ist. Damit sind eine Reihe von Strahlungsvorgängen, die als Fluoreszenz, als Phosphoreszenz, als elektrisches oder chemisches Leuchten bezeichnet werden, und die von E. Wiedemann unter dem Namen „Lumineszenzphänomene" zusammengefaßt worden sind, von der Betrachtung ausgeschlossen. Wir haben es hier vielmehr nur mit reiner „Temperaturstrahlung" zu tun. Nach dem Prinzip der Erhaltung der Energie erfolgt bei der Temperaturstrahlung die Emission vollständig auf Kosten der Körperwärme und bedingt daher, wenn nicht anderweitig Energie zugeführt wird, eine Temperaturerniedrigung der emittierenden Substanz, welche durch den Betrag der emittierten Energie, sowie durch die Wärmekapazität der Substanz bestimmt ist.

§ 8. Fortpflanzung. Die Fortpflanzung der emittierten Strahlung im Innern des als homogen, isotrop und ruhend angenommenen Mediums erfolgt, da wir von Beugungserscheinungen ganz absehen (§ 2), geradlinig und nach allen Richtungen mit der nämlichen Geschwindigkeit; doch erleidet dabei im allgemeinen jeder Strahl während seiner Fortpflanzung eine Schwächung dadurch, daß beständig ein gewisser Teil seiner Energie aus seiner Richtung abgelenkt und nach allen Richtungen des Raumes zerstreut wird. Dieser Vorgang der „Zerstreuung", der also weder Erzeugung noch Vernichtung, sondern nur eine geänderte Verteilung der strahlenden Energie bedeutet, findet prinzipiell genommen in allen Medien statt, die sich vom absoluten Vakuum unterscheiden, auch in chemisch vollkommen reinen Substanzen,[1] und wird durch den Umstand bedingt, daß die Substanz eines Mediums nicht im absoluten Sinne homogen ist, sondern in den kleinsten Räumen Unstetigkeiten besitzt, die durch ihre atomistische Struktur bedingt werden. Fremde eingelagerte kleine Partikel, wie Staubteilchen, befördern den Einfluß der Zerstreuung, ohne jedoch ihren Charakter im wesentlichen zu ändern. Denn auch solche, mit fremden Bestandteilen durchsetzten, sogenannten „trüben" Medien, können sehr wohl als optisch homogen

[1] Vgl. z. B. Lobry de Bruyn und L. K. Wolff, Rec. des Trav. Chim. des Pays-Bas **23**, p. 155, 1904.

betrachtet werden,[1] falls nur die Lineardimensionen der fremden Partikel, sowie die Abstände benachbarter Partikel, hinreichend klein sind gegen die Wellenlängen der betrachteten Strahlen. In optischer Beziehung besteht also zwischen chemisch reinen Substanzen und trüben Medien von der genannten Beschaffenheit kein prinzipieller Unterschied. Optisch leer in absolutem Sinne ist nur das reine Vakuum. Daher kann man auch jede chemisch homogene Substanz als ein durch eingelagerte Moleküle getrübtes Vakuum bezeichnen.

Ein typisches Beispiel für das Phänomen der Zerstreuung bietet das Verhalten des Sonnenlichts in der Atmosphäre. Wenn bei heiterem Himmel die Sonne im Zenit steht, erreicht nur etwa $^2/_3$ der direkten Sonnenstrahlung die Erdoberfläche; der Rest wird in der Atmosphäre aufgehalten, und zwar zum Teil absorbiert und in Luftwärme verwandelt, zum Teil aber zerstreut und in diffuses Himmelslicht verwandelt. Ob und inwieweit hierbei nur die Luftmoleküle selber oder auch die in der Atmosphäre suspendierten Partikel eine Rolle spielen, ist noch nicht mit voller Sicherheit entschieden.

Auf welchen physikalischen Vorgängen der Akt der Zerstreuung beruht: ob auf Reflexion, Beugung, Resonanzwirkung an den Molekülen oder Partikeln, können wir hier ganz dahingestellt sein lassen. Wir bringen nur zum Ausdruck, daß ein jeder im Innern eines Mediums fortschreitender Strahl auf der sehr kleinen Strecke s seiner Bahn durch Zerstreuung um den Bruchteil

$$\beta_\nu \cdot s \tag{3}$$

seiner Intensität geschwächt wird, und nennen die positive, von der Strahlungsintensität unabhängige Größe β_ν den „Zerstreuungskoeffizienten" des Mediums. Da das Medium als isotrop angenommen ist, so ist β_ν auch unabhängig von der Richtung und von der Polarisation des Strahles; dagegen hängt β_ν außer von der physikalischen und chemischen Beschaffenheit des Mediums in beträchtlichem Maße von der Schwingungszahl ab, wie schon durch den Index ν angedeutet ist. Für gewisse Werte von ν kann β_ν so große Beträge annehmen, daß von einer geradlinigen

[1] Wollte man das Wort „homogen" nur in absolutem Sinne gebrauchen, so dürfte man es auf keine einzige ponderable Substanz anwenden.

Fortpflanzung der Strahlung gar nicht mehr die Rede ist. Für andere Werte von ν kann β wieder so klein werden, daß die Zerstreuung gänzlich vernachlässigt werden kann. Wir werden der Allgemeinheit halber β als mittelgroß annehmen. In den wichtigsten Fällen nimmt β mit wachsendem ν zu, und zwar ziemlich stark, d. h. die Zerstreuung ist für Strahlen von kürzerer Wellenlänge beträchtlich größer.[1] Daher auch die blaue Farbe des diffusen Himmelslichts.

Die zerstreute Strahlungsenergie pflanzt sich von der Zerstreuungsstelle ebenso wie die emittierte Strahlung von der Emissionsstelle nach allen Seiten des Raumes, vorwärts, seitwärts und rückwärts, fort, doch nicht nach allen Richtungen mit gleicher Intensität. Auch ist sie nicht unpolarisiert, sondern es zeigen sich gewisse Vorzugsrichtungen, wobei natürlich die Richtung des ursprünglichen Strahles eine Rolle spielt. Doch brauchen wir diese Fragen hier nicht weiter zu verfolgen.

§ 9. Während das Phänomen der Zerstreuung eine stetig wirkende Modifikation der fortschreitenden Strahlung im Innern des Mediums bedeutet, tritt eine plötzliche Änderung sowohl der Intensität als auch der Richtung eines Strahles ein, wenn er an die Grenze des Mediums gelangt und dort auf die Oberfläche eines anderen Mediums trifft, dessen Substanz wir ebenfalls als homogen und isotrop voraussetzen wollen. In diesem Falle wird im allgemeinen ein merklicher Teil des Strahles reflektiert, der andere Teil durchgelassen. Reflexion und Brechung erfolgen entweder „regulär", indem ein einziger reflektierter und ein einziger gebrochener Strahl auftritt, gemäß dem einfachen Reflexionsgesetz und dem Snelliusschen Brechungsgesetz, oder sie erfolgen „diffus", indem die Strahlung von der Oberfläche sich nach verschiedenen Richtungen mit verschiedener Intensität in beide Medien hinein ausbreitet. Im ersten Falle nennen wir die Oberfläche des zweiten Mediums „glatt", im zweiten Falle „rauh". Von der diffusen Reflexion, die an einer rauhen Fläche eintritt, wohl zu unterscheiden ist die Reflexion eines Strahles an der glatten Oberfläche eines trüben Mediums. In beiden Fällen gelangt ein Teil des einfallenden Strahles als diffuse Strahlung in das erste Medium zurück. Aber im ersten Falle findet die

---

[1] Lord Rayleigh, Phil. Mag. **47**, p. 379, 1899.

Zerstreuung an der Oberfläche statt, im zweiten dagegen ausschließlich im Innern des zweiten Mediums, in mehr oder weniger tiefen Schichten desselben.

§ 10. Wenn eine glatte Fläche alle auf sie fallenden Strahlen vollständig reflektiert, wie das z. B. viele Metalloberflächen mit großer Annäherung tun, so nennen wir sie „spiegelnd". Wenn aber eine rauhe Fläche alle auf sie fallenden Strahlen vollständig und nach allen Richtungen gleichmäßig reflektiert, so nennen wir sie „weiß". Der entgegengesetzte Grenzfall, daß die Oberfläche eines Mediums alle auf sie fallenden Strahlen vollständig hindurchläßt, kommt bei glatten Flächen nicht vor, falls die beiden aneinander grenzenden Medien überhaupt optisch verschieden sind. Eine rauhe Fläche, welche die Eigenschaft besitzt, alle auffallenden Strahlen durchzulassen, keinen zu reflektieren, nennen wir „schwarz".

Außer von schwarzen Flächen sprechen wir auch von schwarzen Körpern, und nennen im Anschluß an G. KIRCHHOFF[1] einen Körper schwarz, wenn er die Eigenschaft besitzt, alle auf seine Oberfläche fallenden Strahlen ohne jede Reflexion in sich aufzunehmen und keinen derselben wieder herauszulassen. Damit ein Körper schwarz ist, müssen mithin drei verschiedene, voneinander ganz unabhängige Bedingungen erfüllt sein. Erstens muß der Körper eine schwarze Oberfläche besitzen, damit alle auffallenden Strahlen ohne Reflexion eindringen. Da die Eigenschaften einer Oberfläche im allgemeinen durch beide an sie grenzenden Substanzen beeinflußt werden, so zeigt diese Bedingung, daß die Eigenschaft eines Körpers, schwarz zu sein, nicht nur von seiner eigenen Natur abhängt, sondern auch von der Natur des angrenzenden Mediums. Ein Körper, welcher gegen Luft schwarz ist, braucht es gegen Glas nicht zu sein, und umgekehrt. Zweitens muß der schwarze Körper mindestens eine gewisse, je nach dem Grade seiner Absorptionsfähigkeit verschieden zu wählende Dicke besitzen, damit die von ihm aufgenommenen Strahlen nicht an irgend einer anderen Stelle

---

[1] G. KIRCHHOFF, POGG. Ann. **109**, p. 275, 1860. Gesammelte Abhandlungen, J. A. Barth, Leipzig 1882, p. 573. KIRCHHOFF setzt bei der Definition eines schwarzen Körpers auch voraus, daß die Absorption der auffallenden Strahlen innerhalb einer Schicht von „unendlich kleiner Dicke" erfolgt, was wir hier nicht tun.

der Oberfläche wieder austreten können. Je kräftiger der Körper absorbiert, um so geringer darf seine Dicke sein; Körper mit verschwindend kleinem Absorptionsvermögen müssen dagegen unendlich dick angenommen werden, damit sie als schwarz gelten können. Endlich drittens muß der schwarze Körper einen verschwindend kleinen Zerstreuungskoeffizienten (§ 8) besitzen. Denn sonst würden die von ihm aufgenommenen Strahlen in seinem Innern teilweise zerstreut werden und wieder durch die Oberfläche hinausgelangen.[1]

§ 11. Alle in den beiden vorigen Paragraphen genannten Unterscheidungen und Definitionen beziehen sich zunächst immer nur auf Strahlen einer bestimmten Farbe. Eine Fläche z. B., die für eine gewisse Strahlengattung rauh ist, kann für eine andere Strahlengattung als glatt betrachtet werden. Im allgemeinen verliert eine Fläche für Strahlen von wachsender Wellenlänge immer mehr von ihrer Rauhigkeit, wie leicht zu verstehen. Da nun glatte nichtreflektierende Flächen nicht existieren (§ 10), so zeigen alle herstellbaren nahezu schwarzen Flächen (Lampenruß, Platinmoor) für Strahlen hinreichend großer Wellenlänge merkliche Reflexion.

§ 12. Absorption. Die Vernichtung eines Wärmestrahles erfolgt durch den Akt der „Absorption". Nach dem Prinzip der Erhaltung der Energie wird dabei die Energie der Wärmestrahlung in anderweitige Energie (Körperwärme, chemische Energie) verwandelt, und daraus folgt, daß nur substanzielle Partikel Wärmestrahlen absorbieren können, nicht aber Flächenelemente, wenn man auch manchmal der Kürze halber von absorbierenden Oberflächen spricht. Da wir uns schon oben (§ 7) ausdrücklich auf solche Substanzen beschränkt haben, deren Zustand nur von der Temperatur abhängt, so kommt die absorbierte Strahlungsenergie hier lediglich der Körperwärme zugute, dient also zur Temperaturerhöhung der Substanz, entsprechend ihrer spezifischen Wärme und ihrer Dichtigkeit.

Der Vorgang der Absorption äußert sich darin, daß jeder in dem betrachteten Medium fortschreitende Wärmestrahl auf

---

[1] Vgl. hierüber namentlich A. Schuster, Astrophysical Journal, **21**, p. 1, 1905, welcher besonders darauf hingewiesen hat, daß eine unendlich dicke Gasschicht mit schwarzer Oberfläche noch keineswegs ein schwarzer Körper zu sein braucht.

einer gewissen Strecke seiner Bahn um einen gewissen Bruchteil seiner Intensität geschwächt wird, und zwar ist für eine hinreichend kleine Strecke s dieser Bruchteil proportional der Länge der Strecke; wir setzen ihn also gleich:

$$\alpha_\nu \cdot s \tag{4}$$

und nennen α_ν den „Absorptionskoeffizienten" des Mediums für einen Strahl von der Schwingungszahl ν. Von der Intensität der Strahlung setzen wir den Absorptionskoeffizienten als unabhängig voraus; dagegen wird α_ν im allgemeinen, für inhomogene und anisotrope Medien, vom Orte, von der Richtung und außerdem auch von der Art der Polarisation des Strahles abhängen so z. B. beim Thurmalin). Da wir aber hier nur homogene und isotrope Substanzen betrachten, so dürfen wir α_ν für alle Stellen des Mediums und für alle Richtungen gleich groß und nur von der Schwingungszahl ν, von der Temperatur T und von der chemischen Beschaffenheit des Mediums abhängig annehmen.

Wenn α_ν nur für einen beschränkten Spektralbezirk von Null verschieden ist, so besitzt das Medium „auswählende" (selektive) Absorption. Für diejenigen Farben, für welche $\alpha_\nu = 0$, und auch der Zerstreuungskoeffizient $\beta_\nu = 0$, ist das Medium „vollkommen durchsichtig" oder „diatherman". Die Eigenschaften der selektiven Absorption und der Diathermansie können sich aber für ein bestimmtes Medium mit der Temperatur stark ändern. Im allgemeinen nehmen wir α_ν als von mittlerer Größe an, worin enthalten ist, daß die Absorption längs einer einzigen Wellenlänge sehr schwach ist. Denn die Strecke s enthält, obwohl klein, immer noch viele Wellenlängen (§ 2).

§ 13. Die im vorhergehenden in bezug auf die Emission, die Fortpflanzung und die Absorption eines Wärmestrahles angestellten Überlegungen genügen, um, falls die nötigen Konstanten bekannt sind, bei gegebenen Anfangs- und Grenzbedingungen den gesamten zeitlichen Verlauf eines Strahlungsvorgangs, einschließlich der durch ihn bewirkten Temperaturänderungen, in einem oder mehreren aneinander grenzenden Medien der betrachteten Art mathematisch zu verfolgen, — eine allerdings meist sehr verwickelte Aufgabe. Wir wollen aber, ehe wir zur Behandlung spezieller Fälle übergehen, die allgemeinen Strahlungsvorgänge zunächst noch von einer anderen

Seite betrachten, indem wir nämlich nicht mehr einen bestimmten Strahl, sondern eine bestimmte Stelle im Raume ins Auge fassen.

§ 14. Denken wir uns irgendwo im Innern eines beliebig durchstrahlten Mediums ein unendlich kleines Flächenelement $d\sigma$ herausgegriffen, so wird dieses Element in einem bestimmten Augenblick nach den verschiedensten Richtungen von Strahlen durchkreuzt werden, und die in dem Zeitelement dt durch das Element $d\sigma$ in einer bestimmten Richtung hindurchgestrahlte Energie wird proportional dt und $d\sigma$ sein, und außerdem proportional dem cos des Winkels ϑ, welchen die Normale von $d\sigma$ mit der Richtung der Strahlung bildet. Denn wenn $d\sigma$ hinreichend klein genommen wird, so können wir uns vorstellen, obgleich das den tatsächlichen Verhältnissen nur angenähert entsprechen wird, daß alle Punkte von $d\sigma$ in vollkommen gleicher Weise von der Strahlung betroffen werden. Dann muß die durch $d\sigma$ in einer bestimmten Richtung hindurchgestrahlte Energie proportional der Größe der Öffnung sein, welche das Element $d\sigma$ jener Strahlung darbietet, und diese Öffnung wird gemessen durch die Größe $d\sigma \cdot \cos\vartheta$. Wenn das Element $d\sigma$ gegen die Strahlung gedreht wird, so verschwindet für $\vartheta = \frac{\pi}{2}$ die hindurchgestrahlte Energie vollständig, wie leicht einzusehen.

Von jedem Punkt des Flächenelementes $d\sigma$ aus pflanzt sich nun im allgemeinen ein Bündel von Strahlen nach allen Richtungen des Raumes fort, und zwar nach verschiedenen Richtungen mit verschiedener Intensität, und alle diese Strahlenbündel sind bis auf kleine Abweichungen von höherer Ordnung identisch. Doch kommt einem einzelnen dieser Punktbündel niemals eine endliche Energie zu, da eine endliche Energie nur durch eine endliche Fläche gestrahlt wird. Dies gilt auch für den Durchgang von Strahlen durch einen sogenannten Brennpunkt. Wenn z. B. Sonnenlicht durch eine Sammellinse in deren Brennebene konzentriert wird, so vereinigen sich die Sonnenstrahlen nicht etwa alle in einem einzigen Punkt, sondern jedes Bündel paralleler Strahlen liefert einen besonderen Brennpunkt, und alle diese Brennpunkte bilden zusammen eine Fläche, welche ein zwar kleines, aber doch endlich ausgedehntes Bild der Sonne darstellt. Eine endliche Energie geht nur durch einen endlichen Teil dieser Fläche.

§ 15. Betrachten wir nun für den allgemeinen Fall das Strahlenbündel, welches von einem Punkte des Flächenelementes $d\sigma$ als Spitze nach allen Richtungen des Raumes, zu beiden Seiten von $d\sigma$, sich fortpflanzt. Die einer gewissen, durch den schon oben benutzten Winkel ϑ (zwischen 0 und π) und durch das Azimut φ (zwischen 0 und 2π) bestimmten Richtung entsprechende Strahlungsintensität wird gemessen durch die Energie, welche sich innerhalb eines unendlich dünnen, durch die Werte der Winkel ϑ und $\vartheta + d\vartheta$, φ und $\varphi + d\varphi$, begrenzten Kegels fortpflanzt. Die Öffnung dieses Kegels ist:

$$d\Omega = \sin\vartheta \cdot d\vartheta \cdot d\varphi\,. \tag{5}$$

Auf diese Weise erhalten wir für die Energie, welche in der Zeit dt durch das Flächenelement $d\sigma$ in der Richtung des Kegels $d\Omega$ hindurchgestrahlt wird, den Ausdruck:

$$dt\, d\sigma \cos\vartheta\, d\Omega\, K = K \sin\vartheta \cos\vartheta\, d\vartheta\, d\varphi\, d\sigma\, dt\,. \tag{6}$$

Die endliche Größe K nennen wir die „spezifische Intensität“ oder auch die „Helligkeit“, $d\Omega$ den „Öffnungswinkel“ des von einem Punkte des Elementes $d\sigma$ in der Richtung (ϑ, φ) ausgehenden Strahlenbündels. K ist eine positive Funktion des Ortes, der Zeit und der beiden Winkel ϑ und φ. Die spezifischen Strahlungsintensitäten nach verschiedenen Richtungen sind im allgemeinen gänzlich unabhängig voneinander. Setzt man z. B. in der Funktion K für ϑ den Wert $\pi - \vartheta$, und für φ den Wert $\pi + \varphi$, so erhält man die spezifische Strahlungsintensität in der gerade entgegengesetzten Richtung, eine im allgemeinen von der vorigen ganz verschiedene Größe.

Die Gesamtstrahlung durch das Flächenelement $d\sigma$ nach einer Seite, etwa derjenigen, für welche der Winkel ϑ ein spitzer ist, ergibt sich durch Integration über φ von 0 bis 2π, und über ϑ von 0 bis $\frac{\pi}{2}$:

$$\int_0^{2\pi} d\varphi \int_0^{\pi/2} d\vartheta\, K \sin\vartheta \cos\vartheta\, d\sigma\, dt\,.$$

Ist die Strahlung nach allen Richtungen gleichmäßig, also K konstant, so folgt hieraus für die Gesamtstrahlung durch $d\sigma$ nach einer Seite:

$$\pi\, K\, d\sigma\, dt\,. \tag{7}$$

§ 16. Wenn man von der Strahlung in einer bestimmten Richtung (ϑ, φ) spricht, so ist dabei doch immer zu bedenken, daß eine endliche Energiestrahlung stets nur innerhalb eines Kegels von endlicher Öffnung stattfindet. Es gibt keine endliche Licht- und Wärmestrahlung, die sich in einer einzigen ganz bestimmten Richtung fortpflanzt, oder, was dasselbe ist: es gibt in der Natur kein absolut paralleles Licht, keine absolut ebenen Lichtwellen. Aus einem sogenannten parallelen Strahlenbündel ist eine endliche Strahlungsenergie nur dann zu gewinnen, wenn die Strahlen oder die Wellennormalen des Bündels innerhalb eines endlichen, wenn auch unter Umständen sehr schmalen Kegels divergieren.

§ 17. Die spezifische Intensität K der Energiestrahlung nach jeder Richtung zerfällt weiter in die Intensitäten der einzelnen, den verschiedenen Gebieten des Spektrums angehörigen Strahlen, die sich unabhängig voneinander fortpflanzen. Hierfür ist maßgebend die Strahlungsintensität innerhalb eines Intervalls von Schwingungszahlen, etwa von ν bis ν'. Ist das Intervall $\nu' - \nu$ hinreichend klein, gleich $d\nu$, so ist die Strahlungsintensität innerhalb des Intervalls proportional $d\nu$; die Strahlung heißt dann homogen oder monochromatisch.

Endlich ist bei einem Strahl von bestimmter Richtung, Intensität und Farbe noch die Art der Polarisation charakteristisch. Zerlegt man einen in bestimmter Richtung fortschreitenden Strahl von bestimmter Schwingungszahl ν und beliebigem Polarisationszustand in zwei geradlinig polarisierte Komponenten, deren Polarisationsebenen senkrecht aufeinander stehen, im übrigen aber beliebig sind, so ist die Summe der Intensitäten der beiden Komponenten stets gleich der Intensität des ganzen Strahles, unabhängig von der Orientierung des Ebenenpaares, und zwar kann die Größe der beiden Komponenten stets dargestellt werden durch zwei Ausdrücke von der Form:

$$(8) \quad \left\{ \begin{array}{ll} & \mathfrak{K}_\nu \cos^2\omega + \mathfrak{K}_\nu' \sin^2\omega \\ \text{und} & \mathfrak{K}_\nu \sin^2\omega + \mathfrak{K}_\nu' \cos^2\omega, \end{array} \right.$$

wobei ω das Azimut der Polarisationsebene einer Komponente bedeutet. Die Summe dieser beiden Ausdrücke, welche wir die „Komponenten der spezifischen Strahlungsintensität von der Schwingungszahl ν“ nennen, ergibt in der Tat die Intensität des ganzen Strahls $\mathfrak{K}_\nu + \mathfrak{K}_\nu'$ unabhängig von ω. $\mathfrak{K}_\nu$ und $\mathfrak{K}_\nu'$ repräsentieren zugleich den größten und den kleinsten Wert der Intensität, den eine Komponente überhaupt annehmen kann $\left(\text{für } \omega = 0 \text{ und } \omega = \frac{\pi}{2}\right)$. Daher nennen wir diese Werte die „Hauptwerte der Intensität" oder die „Hauptintensitäten", und die entsprechenden Polarisationsebenen die „Hauptpolarisationsebenen" des Strahles. Beide sind natürlich im allgemeinen mit der Zeit veränderlich. Somit können wir allgemein setzen:

$$K = \int_0^\infty d\nu (\mathfrak{K}_\nu + \mathfrak{K}_\nu'), \tag{9}$$

wobei die positiven Größen $\mathfrak{K}_\nu$ und $\mathfrak{K}_\nu'$, die beiden Hauptwerte der spezifischen Strahlungsintensität (Helligkeit) von der Schwingungszahl ν, außer von ν noch vom Ort, von der Zeit und von den Winkeln ϑ und φ abhängen. Durch Substitution in (6) erhält man hieraus für die Energie, welche in der Zeit dt durch das Flächenelement $d\sigma$ in der Richtung des Elementarkegels $d\Omega$ hindurchgestrahlt wird, den Ausdruck:

$$dt\, d\sigma \cos\vartheta\, d\Omega \int_0^\infty d\nu (\mathfrak{K}_\nu + \mathfrak{K}_\nu') \tag{10}$$

und für monochromatische geradlinig polarisierte Strahlung von der Helligkeit $\mathfrak{K}_\nu$:

$$dt\, d\sigma \cos\vartheta\, d\Omega\, \mathfrak{K}_\nu\, d\nu = dt\, d\sigma \sin\vartheta \cos\vartheta\, d\vartheta\, d\varphi\, \mathfrak{K}_\nu\, d\nu. \tag{11}$$

Für unpolarisierte Strahlen ist $\mathfrak{K}_\nu = \mathfrak{K}_\nu'$; folglich:

$$K = 2\int_0^\infty d\nu\, \mathfrak{K}_\nu \tag{12}$$

und die Energie eines monochromatischen Strahles von der Schwingungszahl ν wird:

$$2\,dt\, d\sigma \cos\vartheta\, d\Omega\, \mathfrak{K}_\nu\, d\nu = 2\,dt\, d\sigma \sin\vartheta \cos\vartheta\, d\vartheta\, d\varphi\, \mathfrak{K}_\nu\, d\nu. \tag{13}$$

Ist außerdem die Strahlung nach allen Richtungen gleichmäßig, so ergibt sich für die Gesamtstrahlung durch $d\sigma$ nach einer Seite aus (7) und (12):

$$2\pi\, d\sigma\, dt \int_0^\infty \mathfrak{K}_\nu\, d\nu. \tag{14}$$

§ 18. Da $\mathfrak{K}_\nu$ in der Natur nicht unendlich groß werden kann, so ist ein endlicher Wert von K nur dann möglich, wenn $\mathfrak{K}_\nu$ in einem endlichen Intervall von Schwingungszahlen ν von Null verschieden ist. Daher gibt es in der Natur keine in absolutem Sinne homogene oder monochromatische Licht- oder Wärmestrahlung. Eine endliche Strahlung umfaßt immer auch ein endliches, wenn auch unter Umständen sehr schmales Spektralgebiet. Hierin liegt ein prinzipieller Unterschied gegenüber den entsprechenden Erscheinungen in der Akustik, wo eine endliche Schallintensität auf eine ganz bestimmte Schwingungszahl treffen kann. Auf diesen Unterschied gründet sich u. a., wie wir später sehen werden, der Umstand, daß der zweite Hauptsatz der Wärmetheorie nur für Licht- und Wärmestrahlen, nicht aber für Schallwellen Bedeutung hat.

§ 19. Wie man aus der Gleichung (9) ersieht, ist die Größe $\mathfrak{K}_\nu$, die Strahlungsintensität der Schwingungszahl ν, von anderer Dimension als die Größe K, die Strahlungsintensität des ganzen Spektrums. Ferner ist zu beachten, daß, wenn man die spektrale Zerlegung nicht nach Schwingungszahlen ν, sondern nach Wellenlängen λ vornimmt, die Strahlungsintensität E_λ der der Schwingungszahl ν entsprechenden Wellenlänge λ nicht einfach dadurch erhalten wird, daß man in dem Ausdruck von $\mathfrak{K}_\nu$ ν durch den entsprechenden Wert von λ ersetzt, also:

$$\nu = \frac{q}{\lambda}, \tag{15}$$

wenn q die Fortpflanzungsgeschwindigkeit bedeutet. Denn es ist nicht E_λ gleich $\mathfrak{K}_\nu$, sondern es ist $E_\lambda \, d\lambda = \mathfrak{K}_\nu \, d\nu$, wenn sich $d\lambda$ und $d\nu$ auf dasselbe Spektralintervall beziehen. Nun ist, wenn $d\lambda$ und $d\nu$ beide positiv genommen werden:

$$d\nu = \frac{q \cdot d\lambda}{\lambda^2},$$

folglich durch Substitution:

$$E_\lambda = \frac{q\,\mathfrak{K}_\nu}{\lambda^2}. \tag{16}$$

Hieraus geht u. a. hervor, daß in einem bestimmten Spektrum die Maxima von E_λ und von $\mathfrak{K}_\nu$ an verschiedenen Stellen des Spektrums liegen!

§ 20. Wenn die Hauptintensitäten $\mathfrak{K}_\nu$ und $\mathfrak{K}_\nu'$ aller monochromatischen Strahlen nach allen Richtungen in allen Punkten des Mediums gegeben sind, so ist damit der Strahlungszustand in allen Einzelheiten bestimmt, und es lassen sich sämtliche darauf bezügliche Fragen beantworten. Wir wollen dies noch an einigen speziellen Anwendungen zeigen. Fragen wir zunächst nach der Energiemenge, welche durch irgend ein Flächenelement $d\sigma$ einem beliebigen anderen Flächenelement $d\sigma'$ zugestrahlt wird. Die Entfernung r der beiden Flächenelemente sei groß gegen die Lineardimensionen jedes der Elemente, aber doch so klein, daß auf der Strecke r keine merkliche Absorption oder Zerstreuung der Strahlung stattfindet. Für diathermane Medien ist natürlich die letzte Bedingung gegenstandslos.

Nun gehen durch irgend einen bestimmten Punkt von $d\sigma$ Strahlen nach allen Punkten von $d\sigma'$. Diese Strahlen bilden einen Kegel, dessen Spitze in $d\sigma$ liegt und dessen Öffnung gegeben ist durch:

$$d\Omega = \frac{d\sigma' \cos(\nu', r)}{r^2},$$

wobei ν' die Normale von $d\sigma'$ bedeutet und der Winkel (ν', r) spitz zu nehmen ist. Dieser Wert von $d\Omega$ ist bis auf kleine Größen höherer Ordnung unabhängig von der speziellen Lage der Spitze des Kegels auf $d\sigma$.

Bezeichnen wir weiter mit ν die entsprechend gerichtete Normale von $d\sigma$, so ergibt sich aus der Gleichung (6), da hier $\vartheta = (\nu, r)$ zu setzen ist, die gesuchte Strahlungsenergie:

$$K \cdot \frac{d\sigma \cdot d\sigma' \cdot \cos(\nu, r) \cdot \cos(\nu', r)}{r^2} \cdot dt \tag{17}$$

und für monochromatische geradlinig polarisierte Strahlung von der Schwingungszahl ν nach Gleichung (11):

$$\mathfrak{K}_\nu \, d\nu \cdot \frac{d\sigma \, d\sigma' \cos(\nu, r) \cos(\nu', r)}{r^2} \cdot dt. \tag{18}$$

Das Größenverhältnis der Flächenelemente $d\sigma$ und $d\sigma'$ zueinander ist dabei ganz beliebig, sie können von gleicher oder auch von verschiedener Größenordnung genommen werden, wenn nur r groß ist gegen die Lineardimensionen jedes der beiden Elemente. Nimmt man $d\sigma$ unendlich klein gegen $d\sigma'$, so divergieren die Strahlen von $d\sigma$ gegen $d\sigma'$; nimmt man aber $d\sigma$ unendlich groß gegen $d\sigma'$, so konvergieren sie von $d\sigma$ gegen $d\sigma'$.

§ 21. Da jeder Punkt von $d\sigma$ die Spitze eines nach $d\sigma'$ gehenden Strahlenkegels bildet, so besteht das ganze hier be-

2*

trachtete, durch die Flächen $d\sigma$ und $d\sigma'$ bestimmte Strahlenbündel aus zweifach unendlich vielen Punktbündeln oder aus vierfach unendlich vielen Strahlen, welche alle in gleicher Weise für die Energiestrahlung in Betracht kommen. Ebensogut kann man sich das Strahlenbündel auch zusammengesetzt denken aus den Kegeln, welche von allen Punkten des Elementes $d\sigma$ ausgehend in je einem Punkte von $d\sigma'$ als Spitze konvergieren. Schneiden wir nun das ganze Strahlenbündel durch irgend eine Ebene in beliebiger Entfernung von den Elementen $d\sigma$ und $d\sigma'$, sei es zwischen ihnen oder außerhalb, so werden die Querschnitte der einzelnen Punktbündel im allgemeinen nicht dieselben sein, auch nicht annähernd, sondern sie werden sich teilweise überdecken, teilweise aber auseinanderfallen, so daß man von einem bestimmten Querschnitt des ganzen Strahlenbündels im Sinne einer gleichförmigen Bestrahlung desselben gar nicht reden kann Nur wenn die Schnittebene mit $d\sigma$ oder mit $d\sigma'$ zusammenfällt, hat das Strahlenbündel einen bestimmten Querschnitt. Diese beiden Flächen spielen also in ihm eine ausgezeichnete Rolle; wir wollen sie die beiden „Brennflächen" des Bündels nennen.

In dem schon oben erwähnten speziellen Falle, daß eine der beiden Brennflächen unendlich klein ist gegen die andere, nimmt das ganze Strahlenbündel den Charakter eines Punktbündels an, insofern seine Gestalt nahezu die eines Kegels wird, der seine Spitze in der gegen die andere unendlich kleinen Brennfläche hat, und man kann dann auch in bestimmtem Sinne von einem Querschnitt des Bündels an irgend einer Stelle im Raume sprechen. Ein solches, einem Kegel ähnlich sehendes Strahlenbündel nennen wir ein Elementarbündel, und die unendlich kleine Brennfläche die erste Brennfläche des Elementarbündels. Die Strahlung erfolgt entweder konvergierend, auf die erste Brennfläche zu, oder divergierend, von der ersten Brennfläche fort. Alle in einem Medium fortschreitenden Strahlenbündel lassen sich auffassen als zusammengesetzt aus solchen Elementarbündeln, und wir können daher unseren künftigen Betrachtungen stets lauter Elementarbündel zugrunde legen, was wegen ihrer einfacheren Beschaffenheit viel bequemer ist.

Die Begrenzung eines Elementarbündels bei gegebener erster Brennfläche $d\sigma$ kann, außer durch die zweite Brennfläche $d\sigma'$, auch durch die Größe des Öffnungswinkels $d\,\Omega$ festgelegt werden, unter welchem $d\,\sigma'$ von $d\,\sigma$ aus gesehen wird; während dagegen bei einem beliebigen Bündel, d. h. wenn die beiden Brennflächen von gleicher Größenordnung sind, die zweite Brennfläche nicht allgemein durch den Öffnungswinkel $d\,\Omega$ ersetzt werden kann, ohne daß das Bündel seinen Charakter wesentlich ändert. Denn wenn statt $d\,\sigma'$ die Größe und Richtung von $d\,\Omega$ (konstant für alle Punkte von $d\,\sigma$) gegeben ist, so bilden die von $d\,\sigma$ ausgehenden Strahlen nicht mehr das vorige Bündel, sondern vielmehr ein Elementarbündel, dessen erste Brennfläche $d\,\sigma$ ist und dessen zweite Brennfläche im Unendlichen liegt.

§ 22. Da die Energiestrahlung sich in dem Medium mit endlicher Geschwindigkeit q fortpflanzt, so befindet sich in einem endlichen Raumteile desselben ein endlicher Betrag von Energie; wir sprechen daher von der „räumlichen Strahlungsdichte" als dem Verhältnis der gesamten in einem Volumenelement enthaltenen Strahlungsenergie zu der Größe des Volumenelements. Berechnen wir nun die räumliche Strahlungsdichte u an irgend einer Stelle des Mediums. Wenn wir an der betreffenden Stelle ein unendlich kleines Volumen v von beliebiger Form betrachten, so haben wir alle Strahlen zu berücksichtigen, welche das Volumenelement v durchkreuzen. Zu diesem Zwecke legen wir um irgend einen Punkt O des Volumenelements als Mittelpunkt eine Kugelfläche vom Radius r, der groß ist gegen die Lineardimensionen von v, aber doch so klein, daß auf der Strecke r keine merkliche Schwächung der Strahlung durch Absorption oder Zerstreuung stattfindet (Fig. 1). Jeder Strahl, der das Volumen v trifft, kommt von einem Punkte der Kugelfläche her. Wenn wir also zunächst die Strahlen ins Auge fassen, welche von den Punkten eines bestimmten unendlich kleinen Elements $d\,\sigma$ der Kugelfläche ausgehend das Volumen v treffen, so erhalten wir daraus durch Summation über alle Elemente der Kugelfläche sämtliche in Betracht kommenden Strahlen, und jeden nur einmal.

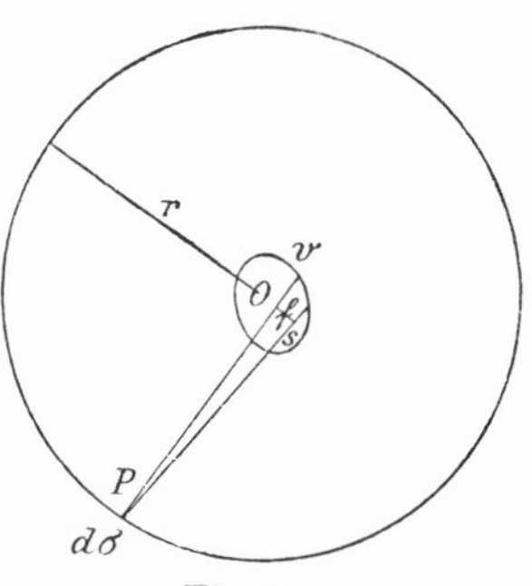

Fig. 1.

Berechnen wir daher zunächst den Beitrag, welchen das Flächenelement $d\sigma$ durch seine gegen das Volumen v gerichtete Strahlung zu der in v enthaltenen Strahlungsenergie liefert. Wir nehmen die Lineardimensionen von $d\sigma$ unendlich klein gegen die von v und betrachten den Strahlenkegel, der von einem in $d\sigma$ gelegenen Punkt P ausgehend das Volumen v trifft. Dieser Kegel zerfällt in unendlich viele unendlich dünne Elementarkegel, alle mit P als Spitze, deren jeder ein bestimmtes Stück von der Länge s aus dem Volumen v ausschneidet. Die Öffnung eines solchen Elementarkegels beträgt $\frac{f}{r^2}$, wenn f den senkrechten Querschnitt des Kegels in der Entfernung r von der Spitze bezeichnet. Nun braucht die Strahlung, um die Strecke s zurückzulegen, die Zeit:

$$\tau = \frac{s}{q}.$$

Während dieser Zeit τ gelangt nach der Gleichung (6), da $d\Omega$ hier gleich $\frac{f}{r^2}$ und ϑ gleich Null zu setzen ist, die Energiemenge:

$$\tau\, d\sigma \frac{f}{r^2} K = \frac{fs}{r^2 q} \cdot K\, d\sigma \tag{19}$$

innerhalb des betrachteten Elementarkegels in das Volumen v hinein und verteilt sich hier auf den von dem Elementarkegel ausgeschnittenen Raum, dessen Volumen fs beträgt. Summiert man über alle von dem Flächenelement $d\sigma$ ausgehenden Elementarkegel, welche v treffen, so erhält man

$$\frac{K\, d\sigma}{r^2 q} \cdot \sum fs = \frac{K\, d\sigma}{r^2 q} \cdot v$$

für die ganze in dem Volumen v befindliche strahlende Energie, soweit sie von der Strahlung durch das Flächenelement $d\sigma$ herrührt. Um die gesamte in v enthaltene Strahlungsenergie zu erhalten, hat man schließlich noch über alle Elemente $d\sigma$ der Kugelfläche zu integrieren. Dies liefert, wenn man den Öffnungswinkel $\frac{d\sigma}{r^2}$ des Kegels, dessen Spitze in O liegt, und der das Element $d\sigma$ aus der Kugelfläche ausschneidet, mit $d\Omega$ bezeichnet:

$$\frac{v}{q} \cdot \int K\, d\Omega$$

und als gesuchte räumliche Strahlungsdichte, durch Division mit v:

$$u = \frac{1}{q} \cdot \int K \, d\,\Omega. \tag{20}$$

Da r hier ganz fortgefallen ist, so kann man unter K einfach die Strahlungsintensität in dem Punkte O selber verstehen. Bei der Integration ist zu beachten, daß K im allgemeinen von der Richtung (ϑ, φ) abhängt.

Für gleichmäßige Strahlung nach allen Richtungen ist K konstant, und man erhält:

$$u = \frac{4\pi K}{q}. \tag{21}$$

§ 23. Wie von der räumlichen Dichte der Gesamtstrahlung u, so spricht man auch von der räumlichen Dichte der Strahlung einer bestimmten Schwingungszahl $\mathfrak{u}_\nu$, indem man die spektrale Zerlegung vornimmt:

$$u = \int_0^\infty \mathfrak{u}_\nu \, d\nu. \tag{22}$$

Dann ergibt sich durch Kombination der Gleichungen (20) und (9):

$$\mathfrak{u}_\nu = \frac{1}{q} \int (\mathfrak{K}_\nu + \mathfrak{K}_\nu') \, d\,\Omega \tag{23}$$

und hieraus für unpolarisierte und nach allen Richtungen gleichmäßige Strahlung:

$$\mathfrak{u}_\nu = \frac{8\pi\,\mathfrak{K}_\nu}{q}. \tag{24}$$

Zweites Kapitel. Strahlung beim thermodynamischen Gleichgewicht. Kirchhoffsches Gesetz. Schwarze Strahlung.

§ 24. Wir wollen jetzt die im vorigen Kapitel aufgestellten Sätze auf den speziellen Fall des thermodynamischen Gleichgewichts anwenden und stellen daher an die Spitze der folgenden Überlegungen die aus dem zweiten Hauptsatz der Thermodynamik fließende Folgerung: Ein System ruhender Körper von beliebiger Natur, Form und Lage, welches von einer festen, für Wärme undurchlässigen Hülle umschlossen ist, geht, bei beliebig gewähltem Anfangszustand, im Laufe der Zeit in einen

Dauerzustand über, bei welchem die Temperatur in allen Körpern des Systems die nämliche ist. Dies ist der thermodynamische Gleichgewichtszustand, in welchem die Entropie des Systems unter allen Werten, die sie vermöge der durch die Anfangsbedingungen gegebenen Gesamtenergie anzunehmen vermag, einen Maximalwert besitzt, von welchem aus daher keine weitere Vermehrung der Entropie mehr möglich ist.

Es kann in gewissen Fällen vorkommen, daß unter den gegebenen Bedingungen die Entropie nicht ein einziges, sondern mehrere verschiedene Maxima annehmen kann, von denen dann eins das absolute ist, während die übrigen nur relative Bedeutung haben.[1] In diesen Fällen stellt jeder Zustand, der einem Maximalwerte der Entropie entspricht, einen thermodynamischen Gleichgewichtszustand des Systems dar. Aber nur derjenige unter ihnen, der dem absolut größten Wert der Entropie entspricht, bezeichnet das absolut stabile Gleichgewicht. Die übrigen sind alle in gewissem Sinne labil, insofern eine geeignete, wenn auch minimale Störung des Gleichgewichts eine dauernde Veränderung des Systems in Richtung nach dem absolut stabilen Gleichgewicht hin veranlassen kann. Ein Beispiel hierfür bietet ein in ein festes Gefäß eingeschlossener Dampf im Zustand der Übersättigung, oder irgend eine explosible Substanz. Auch bei den Strahlungsvorgängen werden wir Beispiele solcher labilen Gleichgewichte antreffen (§ 52).

§ 25. Wir setzen nun wieder, wie im vorigen Kapitel, homogene isotrope Medien voraus, deren Zustand nur von der Temperatur abhängt, und fragen nach den Gesetzen, denen die Strahlungsvorgänge in ihnen gehorchen müssen, wenn sie mit der im vorigen Paragraph angeführten Folgerung aus dem zweiten Hauptsatz der Thermodynamik im Einklang sein sollen. Das Mittel zur Beantwortung dieser Frage gibt uns die Untersuchung des thermodynamischen Gleichgewichtszustandes eines oder mehrerer solcher Medien, unter Benutzung der im vorigen Kapitel aufgestellten Begriffe und Sätze.

Wir beginnen mit dem einfachsten Fall: einem einzigen Medium, welches nach allen Richtungen des Raumes sehr weit

---

[1] Vgl. z. B. M. PLANCK, Vorlesungen über Thermodynamik, Leipzig, Veit & Comp., 1905, § 165 und § 189 ff.

ausgedehnt ist, und, wie alle hier betrachteten Systeme, von einer festen, für Wärme undurchlässigen Hülle umschlossen wird. Das Medium besitze, wie wir vorläufig annehmen wollen, einen endlichen Absorptionskoeffizienten, einen endlichen Emissionskoeffizienten und einen endlichen Zerstreuungskoeffizienten.

Betrachten wir nun zunächst solche Stellen des Mediums, welche von der Oberfläche sehr weit entfernt liegen. Hier wird der Einfluß der Oberfläche jedenfalls verschwindend klein sein und wir werden wegen der Homogenität und Isotropie des Mediums schließen müssen, daß im thermodynamischen Gleichgewichtszustand die Wärmestrahlung überall und nach allen Richtungen von gleicher Beschaffenheit ist, oder daß $\mathfrak{K}_\nu$, die spezifische Strahlungsintensität eines geradlinig polarisierten Strahles von der Schwingungszahl ν (§ 17), unabhängig ist vom Azimut der Polarisation, von der Richtung des Strahles und vom Orte. Daher entspricht auch jedem von einem Flächenelement $d\sigma$ ausgehenden, innerhalb eines Elementarkegels $d\Omega$ divergierenden Strahlenbündel ein genau gleiches und entgegengesetzt gerichtetes innerhalb des nämlichen Elementarkegels gegen das Flächenelement hin konvergierendes Bündel.

Nun erfordert die Bedingung des thermodynamischen Gleichgewichts, daß die Temperatur überall gleich und unveränderlich ist, daß also in jedem Volumenelement des Mediums während einer beliebigen Zeit ebensoviel strahlende Wärme absorbiert wie emittiert wird. Denn da wegen der Gleichmäßigkeit der Temperatur keinerlei Wärmeleitung stattfindet, wird die Körperwärme lediglich durch die Wärmestrahlung beeinflußt. Das Phänomen der Zerstreuung spielt bei dieser Bedingung keine Rolle; denn die Zerstreuung betrifft nur eine Richtungsänderung der gestrahlten Energie, nicht aber Erzeugung oder Vernichtung derselben. Wir berechnen daher die in der Zeit dt von einem Volumenelement v emittierte und absorbierte Energie.

Die emittierte Energie beträgt nach Gleichung (2):

$$dt \cdot v \cdot 8\pi \int_0^\infty \varepsilon_\nu \, d\nu,$$

wobei der Emissionskoeffizient ε_ν des Mediums außer von seiner chemischen Natur nur von der Schwingungszahl ν und von der Temperatur abhängt.

§ 26. Zur Berechnung der absorbierten Energie bedienen wir uns der nämlichen Betrachtung, wie die, welche durch die Fig. 1 (§ 22) illustriert wurde, und behalten auch die dortigen Bezeichnungen bei. Die von dem Volumenelement v in der Zeit dt absorbierte Energiestrahlung ergibt sich, wenn wir die Intensitäten aller das Element v durchkreuzenden Strahlen betrachten und von jedem dieser Strahlen den in v absorbierten Bruchteil berücksichtigen. Nun besitzt der von $d\sigma$ ausgehende, aus dem Volumen v den Teil fs ausschneidende Elementarstrahlenkegel nach (19) die Intensität (Energiestrahlung in der Zeiteinheit)

$$d\sigma \cdot \frac{f}{r^2} \cdot K$$

oder in spektraler Zerlegung nach (12):

$$2\, d\sigma \cdot \frac{f}{r^2} \cdot \int_0^\infty \mathfrak{K}_\nu \, d\nu \, .$$

Die Intensität eines monochromatischen Strahles ist daher:

$$2\, d\sigma \cdot \frac{f}{r^2} \cdot \mathfrak{K}_\nu \, d\nu \, .$$

Der Betrag der auf der Strecke s in der Zeit dt absorbierten Energie dieses Strahles ist daher nach (4):

$$dt \cdot \alpha_\nu \, s \cdot 2\, d\sigma \, \frac{f}{r^2} \, \mathfrak{K}_\nu \, d\nu$$

und die ganze aus dem Elementarstrahlenkegel absorbierte Energie beträgt, durch Integration über alle Schwingungszahlen:

$$dt \cdot 2\, d\sigma \, \frac{fs}{r^2} \int_0^\infty \alpha_\nu \, \mathfrak{K}_\nu \, d\nu \, .$$

Summiert man diesen Ausdruck erstens über die verschiedenen Querschnitte f der von $d\sigma$ ausgehenden, das Volumen v treffenden Elementarstrahlenkegel, wobei zu beachten, daß $\sum fs = v$, und zweitens über alle Elemente $d\sigma$ der Kugelfläche vom Radius r, wobei $\int \frac{d\sigma}{r^2} = 4\pi$, so erhält man als Ausdruck für die gesamte in der Zeit dt vom Volumenelement v absorbierte strahlende Energie:

$$dt \cdot v \cdot 8\pi \int_0^\infty \alpha_\nu \, \mathfrak{K}_\nu \, d\nu \tag{25}$$

und durch Gleichsetzen mit der emittierten Energie:

$$\int_0^\infty \varepsilon_\nu \, d\nu = \int_0^\infty \alpha_\nu \, \mathfrak{K}_\nu \, d\nu .$$

Diese Beziehung läßt sich noch spektral zerlegen. Denn die Gleichheit der emittierten und der absorbierten Energie beim thermodynamischen Gleichgewicht gilt nicht nur für die Gesamtstrahlung des ganzen Spektrums, sondern auch, wie sich leicht einsehen läßt, für jede monochromatische Strahlung. Da nämlich die Größen ε_ν, α_ν und $\mathfrak{K}_\nu$ unabhängig vom Orte sind, so würde, wenn für eine einzelne Farbe die absorbierte Energie der emittierten nicht gleich wäre, überall im ganzen Medium eine fortwährende Zunahme oder Abnahme der Energiestrahlung der betreffenden Farbe, auf Kosten anderer Farben, stattfinden, was der Bedingung widerspricht, daß $\mathfrak{K}_\nu$ für jede einzelne Schwingungszahl sich mit der Zeit nicht ändert. Es gilt also auch für jede Schwingungszahl die Beziehung:

$$\varepsilon_\nu = \alpha_\nu \, \mathfrak{K}_\nu \tag{26}$$

oder:

$$\mathfrak{K}_\nu = \frac{\varepsilon_\nu}{\alpha_\nu}, \tag{27}$$

d. h. im Innern eines im thermodynamischen Gleichgewicht befindlichen Mediums ist die spezifische Strahlungsintensität einer bestimmten Schwingungszahl gleich dem Quotienten aus dem Emissionskoeffizienten und dem Absorptionskoeffizienten des Mediums für diese Schwingungszahl.

§ 27. Da ε_ν und α_ν außer von der Natur des Mediums nur von der Temperatur und der Schwingungszahl ν abhängen, so ist mithin auch die Strahlungsintensität einer bestimmten Farbe beim thermodynamischen Gleichgewicht durch die Natur des Mediums und durch die Temperatur vollständig bestimmt. Eine Ausnahme bildet jedoch der Fall, daß $\alpha_\nu = 0$, d. h. daß das Medium die betreffende Farbe gar nicht absorbiert. Da $\mathfrak{K}_\nu$ nicht unendlich groß werden kann, so folgt zunächst, daß dann auch $\varepsilon_\nu = 0$, d. h. ein Medium emittiert keine Farbe, welche es nicht absorbiert. Ferner aber erkennt man, daß, wenn sowohl ε als auch α verschwinden, die Gleichung (26) durch jeden Wert von $\mathfrak{K}$ befriedigt wird. In einem für eine bestimmte

Farbe diathermanen Medium kann bei jeder beliebigen Strahlungsintensität der betreffenden Farbe thermodynamisches Gleichgewicht bestehen.

Hier haben wir schon ein Beispiel für die oben (§ 24) besprochenen Fälle, in denen bei gegebener Gesamtenergie eines von einer festen adiabatischen Hülle umschlossenen Systems mehrere Gleichgewichtszustände möglich sind, entsprechend mehreren relativen Maxima der Entropie. Denn da die Strahlungsintensität der betreffenden Farbe im thermodynamischen Gleichgewicht ganz unabhängig von der Temperatur des für sie diathermanen Mediums ist, so läßt sich die gegebene Gesamtenergie ganz beliebig auf die Strahlung jener Farbe und auf die Körperwärme verteilen, ohne daß das thermodynamische Gleichgewicht unmöglich wird. Unter allen diesen Verteilungen gibt es aber eine ganz bestimmte, dem absoluten Maximum der Entropie entsprechende, welche das absolut stabile Gleichgewicht bezeichnet und welche, im Gegensatz zu den übrigen, in gewissem Sinne labilen Zuständen, die Eigenschaft besitzt, durch keinerlei minimale Störung eine merkliche Änderung zu erleiden. In der Tat werden wir unten (§ 48) sehen, daß unter den unendlich vielen Werten, deren der Quotient $\frac{\varepsilon_\nu}{\alpha_\nu}$ fähig ist, wenn Zähler und Nenner beide verschwinden, ein ausgezeichneter, in bestimmter Weise von der Natur des Mediums, der Schwingungszahl ν und der Temperatur abhängiger Wert existiert, den man als die stabile Strahlungsintensität $\mathfrak{K}_\nu$ in dem für die Schwingungszahl ν diathermanen Medium bei der betreffenden Temperatur zu bezeichnen hat.

Was hier von einem für eine bestimmte Strahlenart diathermanen Medium gesagt ist, das gilt ebenso vom absoluten Vakuum, als einem für sämtliche Strahlenarten diathermanen Medium, nur daß man hier nicht mehr von der Körperwärme und von der Temperatur des Mediums sprechen kann.

Fürs erste wollen wir aber von dem speziellen Falle der Diathermansie wieder ganz absehen und voraussetzen, daß alle betrachteten Medien einen endlichen Absorptionskoeffizienten besitzen.

§ 28. Widmen wir nun auch noch dem Vorgang der Zerstreuung beim thermodynamischen Gleichgewicht eine kurze

Betrachtung. Jeder Strahl, der das Volumenelement v trifft, erleidet dortselbst eine gewisse Schwächung seiner Intensität dadurch, daß ein gewisser Bruchteil seiner Energie nach anderen Richtungen abgelenkt wird. Der Betrag der gesamten in der Zeit dt vom Volumenelement v von allen Richtungen des Raumes durch Zerstreuung aufgefangenen Energiestrahlung berechnet sich auf Grund des Ausdruckes (3) genau in derselben Weise wie der der absorbierten Energiestrahlung in § 26, und wir erhalten daher für ihn, wie in (25):

$$dt \cdot v \cdot 8\pi \int_0^\infty \beta_\nu \mathfrak{K}_\nu \, d\nu. \tag{28}$$

Die Frage nach dem Verbleib dieser Energie läßt sich ebenfalls leicht beantworten. Denn wegen der Isotropie des Mediums muß die vom Volumenelement v ausgehende Strahlung der dortselbst zerstreuten Energie (28), ebenso wie die Einstrahlung, nach allen Richtungen gleichmäßig erfolgen. Dies ergibt für denjenigen Teil der vom Element v durch Zerstreuung aufgefangenen Energie, welcher durch den Öffnungswinkel $d\Omega$ wieder ausgestrahlt wird, durch Multiplikation mit $\frac{d\Omega}{4\pi}$:

$$2\, dt\, dv\, d\Omega \int_0^\infty \beta_\nu \mathfrak{K}_\nu \, d\nu$$

und für monochromatische geradlinig polarisierte Strahlung:

$$dt\, dv\, d\Omega \cdot \beta_\nu \mathfrak{K}_\nu \, d\nu. \tag{29}$$

Dabei ist allerdings wohl zu beachten, daß diese Gleichmäßigkeit der Ausstrahlung nach allen Richtungen nur für alle das Element v treffenden Strahlen zusammengenommen gilt; denn ein einzelner Strahl wird, auch in einem isotropen Medium, nach verschiedenen Richtungen mit verschiedener Intensität und Polarisation zerstreut (vgl. § 8 am Schluß).

Auf diese Weise ergibt sich, daß beim thermodynamischen Strahlungsgleichgewicht im Innern des Mediums der Vorgang der Zerstreuung im ganzen überhaupt keinen Effekt hervorbringt. Die von allen Seiten auf ein Volumenelement auffallende und dort wieder nach allen Seiten zerstreute Strahlung verhält sich genau ebenso, als ob sie ohne jede Modifikation durch das Volumenelement direkt hindurchgegangen wäre. Was ein Strahl durch Zerstreuung an Energie verliert, das gewinnt er wieder durch Zerstreuung anderer Strahlen.

§ 29. Wir wollen nun die Strahlungsvorgänge im Innern eines sehr weit ausgedehnten, im thermodynamischen Gleichgewicht befindlichen homogenen isotropen Mediums noch von einem anderen Standpunkt aus betrachten, indem wir nicht mehr ein bestimmtes Volumenelement, sondern ein bestimmtes Strahlenbündel, und zwar ein Elementarbündel (§ 21) ins Auge fassen. Dasselbe sei charakterisiert durch die unendlich kleine Brennfläche $d\sigma$ beim Punkte O (Fig. 2), senkrecht zur Achse des Bündels, und durch den Öffnungswinkel $d\Omega$, und die Strahlung erfolge gegen die Brennfläche hin in der Richtung des Pfeiles. Wir betrachten ausschließlich nur solche Strahlen, welche diesem Bündel angehören.

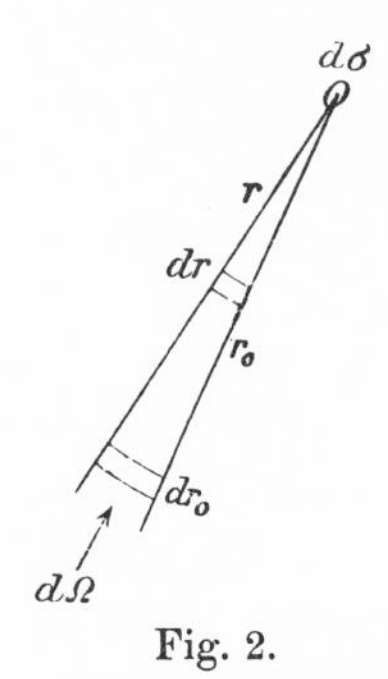

Fig. 2.

Die in der Zeiteinheit durch $d\sigma$ hindurchtretende Energie monochromatischer geradlinig polarisierter Strahlung ist nach (11), da hier $dt = 1$, und $\vartheta = 0$ zu setzen ist:

$$d\sigma \cdot d\Omega \cdot \mathfrak{K}_\nu \, d\nu \tag{30}$$

und dieser nämliche Wert gilt auch für jeden anderen Querschnitt des Bündels. Denn erstens ist $\mathfrak{K}_\nu \, d\nu$ überall gleich groß (§ 25), und zweitens besitzt auch das Produkt aus irgend einem senkrechten Querschnitt des Bündels und dem Öffnungswinkel, unter dem die Brennfläche $d\sigma$ von diesem Querschnitte aus gesehen wird, den konstanten Wert $d\sigma \cdot d\Omega$, da die Größe des Querschnittes sich mit der Entfernung von der Spitze O des Bündels in demselben Maße vergrößert, wie die jenes Öffnungswinkels sich verkleinert. Die Strahlung innerhalb des Bündels erfolgt also genau ebenso, als ob das Medium vollkommen diatherman wäre.

Andererseits modifiziert sich aber die Strahlung auf ihrer Bahn fortwährend durch die Einflüsse der Emission, der Absorption und der Zerstreuung. Wir wollen uns von dem Betrage dieser Wirkungen im einzelnen Rechenschaft geben.

§ 30. Ein Raumelement des Strahlenbündels, welches begrenzt ist durch zwei Querschnitte in den Entfernungen r_0

(beliebig groß) und $r_0 + d\,r_0$ von der Spitze O, und welches daher das Volumen $dr_0 \cdot r_0^{\,2}\,d\Omega$ besitzt, emittiert gegen die bei O gelegene Brennfläche $d\sigma$ in der Zeiteinheit eine Energiemenge E monochromatischer geradlinig polarisierter Strahlung, welche man aus (1) erhält, wenn man darin setzt:

$$d\,t = 1\,, \qquad d\,\tau = d\,r_0 \cdot r_0^{\,2}\,d\,\Omega\,, \qquad d\,\Omega = \frac{d\,\sigma}{r_0^{\,2}}$$

und den Zahlenfaktor 2 unterdrückt; also die Energie:

$$E = d\,r_0 \cdot d\,\Omega \cdot d\,\sigma \cdot \varepsilon_\nu\,d\,\nu\,. \tag{31}$$

Aber von dieser Energie E kommt nur ein Teilbetrag E_0 in O an, da auf jeder unendlich kleinen Wegstrecke s, welche sie bis O zurückzulegen hat, durch Absorption und Zerstreuung der Bruchteil $(\alpha_\nu + \beta_\nu)\,s$ verloren geht. Ist nämlich E_r derjenige Teil der Energie E, welcher in dem um die Strecke $r\,(< r_0)$ von O entfernten Querschnitt ankommt, so haben wir, für $s = d\,r$:

$$E_{r+d\,r} - E_r = E_r \cdot (\alpha_\nu + \beta_\nu)\,d\,r$$

oder:

$$\frac{d\,E_r}{d\,r} = E_r\,(\alpha_\nu + \beta_\nu)\,.$$

Integriert:

$$E_r = E\,e^{(\alpha_\nu + \beta_\nu)\,(r - r_0)}\,,$$

da für $r = r_0$ $E_r = E$ durch die Gleichung (31) gegeben ist.

Hieraus folgt für $r = 0$ die in O ankommende, von dem Raumelement bei r_0 emittierte Energie:

$$E_0 = E\,e^{-(\alpha_\nu + \beta_\nu)\,r_0} = d\,r_0 \cdot d\,\Omega \cdot d\,\sigma\,\varepsilon_\nu\,e^{-(\alpha_\nu + \beta_\nu)\,r_0}\,d\,\nu. \tag{32}$$

Alle Raumelemente des Strahlenbündels zusammen liefern also durch ihre Emission die in $d\,\sigma$ ankommende Energie:

$$d\,\Omega \cdot d\,\sigma \cdot d\,\nu\,\varepsilon_\nu \cdot \int\limits_0^\infty d\,r_0 \cdot e^{-(\alpha_\nu + \beta_\nu)\,r_0} = d\,\Omega \cdot d\,\sigma \cdot \frac{\varepsilon_\nu}{\alpha_\nu + \beta_\nu}\,d\,\nu. \tag{33}$$

§ 31. Wäre gar keine Zerstreuung der Strahlung wirksam, so müßte die gesamte in $d\sigma$ ankommende Energie sich zusammensetzen aus den von den einzelnen Raumelementen des Strahlenbündels emittierten Energiebeträgen, unter Berücksichtigung der Verluste, die unterwegs durch Absorption eintreten, und in der Tat sind für $\beta_\nu = 0$ die Ausdrücke (33) und (30) identisch, wie ein Vergleich mit (27) zeigt. Im allgemeinen ist aber (30) größer als (33), weil die in $d\sigma$ ankommende Energie auch noch Strahlen enthält, welche gar nicht innerhalb des Strahlenbündels, sondern irgendwo anders emittiert, und später durch Zerstreuung in das betrachtete Strahlenbündel hineingeraten sind. In der Tat: die Raumelemente des Strahlenbündels zerstreuen nicht nur die innerhalb des Bündels fortschreitende Strahlung nach außen, sondern sie sammeln auch Strahlen, die von außen kommen, in das Bündel hinein, und zwar ergibt sich die von dem Raumelement bei r_0 auf diese Weise gesammelte Strahlung E', wenn man in dem Ausdruck (29) setzt:

$$dt = 1, \qquad v = dr_0 \cdot d\Omega\, r_0^2, \qquad d\Omega = \frac{d\sigma}{r_0^2},$$

$$E' = dr_0\, d\Omega\, d\sigma\, \beta_\nu\, \mathfrak{K}_\nu\, d\nu.$$

Diese Energie kommt zu der von dem Raumelement emittierten oben in (31) berechneten Energie E hinzu, so daß man für die gesamte in dem Raumelement bei r_0 in das Strahlenbündel neu eintretende Energie erhält:

$$E + E' = dr_0\, d\Omega\, d\sigma (\varepsilon_\nu + \beta_\nu\, \mathfrak{K}_\nu)\, d\nu.$$

Von dieser Energie kommt, analog (32), in O der Betrag an:

$$dr_0\, d\Omega\, d\sigma (\varepsilon_\nu + \beta_\nu \mathfrak{K}_\nu)\, d\nu \cdot e^{-r_0(\alpha_\nu + \beta_\nu)}$$

und alle Raumelemente des Strahlenbündels zusammen liefern durch Emission und Sammlung zerstreuter Strahlung unter Berücksichtigung der unterwegs durch Absorption und Zerstreuung eintretenden Verluste, die in $d\sigma$ ankommende Energie:

$$d\Omega\, d\sigma (\varepsilon_\nu + \beta_\nu \mathfrak{K}_\nu)\, d\nu \cdot \int_0^\infty dr_0 \cdot e^{-r_0(\alpha_\nu + \beta_\nu)} = d\Omega\, d\sigma \cdot \frac{\varepsilon_\nu + \beta_\nu \mathfrak{K}_\nu}{\alpha_\nu + \beta_\nu}\, d\nu,$$

welche nun in der Tat genau gleich dem Ausdruck (30) ist, wie man durch Vergleichung mit (26) erkennt.

§ 32. Die im vorhergehenden abgeleiteten Sätze über den Strahlungszustand beim thermodynamischen Gleichgewicht eines homogenen isotropen Mediums gelten zunächst nur für solche Stellen des Mediums, welche von der Oberfläche sehr weit entfernt liegen, weil nur für diese aus Symmetriegründen die Strahlung von vornherein als unabhängig vom Orte und von der Richtung angenommen werden darf. Indessen zeigt eine einfache Überlegung, daß der in (27) berechnete, nur von der Temperatur und von der Natur des Mediums abhängige Wert von $\mathfrak{K}_\nu$ auch bis unmittelbar an die Oberfläche des Mediums den Betrag der Strahlungsintensität der betreffenden Schwingungszahl nach jeder beliebigen Richtung angibt. Denn beim thermodynamischen Gleichgewicht muß jeder Strahl genau die nämliche Intensität besitzen wie der gerade entgegengesetzte Strahl, weil sonst durch die Strahlung ein einseitiger Transport von Energie bedingt werden würde. Fassen wir also einen von der Oberfläche des Mediums herkommenden, in das Innere hinein gerichteten Strahl ins Auge, so muß derselbe die nämliche Intensität besitzen wie der gerade entgegengesetzte, aus dem Innern kommende Strahl, und daraus folgt ohne weiteres, daß der gesamte Strahlungszustand des Mediums an der Oberfläche der nämliche ist wie im Innern.

§ 33. Während also die von einem Element der Oberfläche ausgehende, nach dem Innern des Mediums gerichtete Strahlung in jeder Beziehung gleich ist der von irgend einem gleichgroßen und gleichgerichteten im Innern gelegenen Flächenelement ausgehenden Strahlung, so hat sie doch eine andere Vorgeschichte als diese, sie rührt nämlich, da die Oberfläche des Mediums als für Wärme undurchlässig vorausgesetzt ist, lediglich her von der Reflexion der aus dem Innern kommenden Strahlung an der Oberfläche. Im einzelnen kann dies in sehr verschiedener Weise geschehen, je nachdem die Oberfläche als glatt, also in diesem Falle als spiegelnd, oder als rauh, etwa als weiß (§ 10) vorausgesetzt ist. Im ersteren Falle entspricht jedem auf die Oberfläche auftreffenden Strahlenbündel ein ganz bestimmtes symmetrisch dazu gelegenes von der nämlichen Intensität, im zweiten Falle aber zersplittert sich jedes einzelne auftreffende Strahlenbündel in unendlich viele reflektierte Strahlenbündel von verschiedener Richtung, Intensität und Polarisation, doch immer so, daß die von allen Seiten mit gleicher Intensität $\mathfrak{K}_\nu$ auf ein Oberflächenelement auftreffenden Bündel in ihrer Gesamtheit wieder eine gleichmäßige von der Oberfläche in das Innere des Mediums gerichtete Strahlung von der nämlichen Helligkeit $\mathfrak{K}_\nu$ liefern.

§ 34. Nun bietet es nicht die geringste Schwierigkeit mehr, die im § 25 gemachte Voraussetzung aufzuheben, daß das betrachtete Medium nach allen Richtungen des Raumes sehr weit ausgedehnt ist. Denn wenn in unserem Medium der thermodynamische Gleichgewichtszustand allenthalben eingetreten ist, so wird nach den Ergebnissen des letzten Paragraphen das Gleichgewicht in keiner Weise gestört, wenn man in dem Medium beliebig viele feste für Wärme undurchlässige, glatte oder rauhe, Flächen angebracht denkt. Hierdurch wird das ganze System in eine beliebig große Anzahl vollkommen abgeschlossener Einzelsysteme zerlegt, deren jedes so klein gewählt werden kann, als es die allgemeinen in § 2 ausgesprochenen Beschränkungen überhaupt gestatten. Daraus geht hervor, daß der in (27) gegebene Wert der spezifischen Strahlungsintensität $\mathfrak{K}_\nu$ auch für das thermodynamische Gleichgewicht einer in einem beliebig kleinen und beliebig geformten Raume eingeschlossenen Substanz Gültigkeit besitzt.

§ 35. Von dem aus einer einzigen homogenen isotropen Substanz bestehenden System gehen wir jetzt über zu einem aus zwei verschiedenen aneinander grenzenden homogenen isotropen Substanzen bestehenden System, das wiederum von einer festen für Wärme undurchlässigen Hülle begrenzt ist, und betrachten den Strahlungszustand beim thermodynamischen Gleichgewicht, zunächst wieder unter der Voraussetzung, daß beide Medien räumlich sehr weit ausgedehnt sind. Da das Gleichgewicht in Nichts gestört wird, wenn man die Trennungsfläche der beiden Substanzen sich einen Augenblick durch eine für Wärmestrahlung ganz undurchlässige Fläche ersetzt denkt, so gelten für eine jede der beiden Substanzen einzeln alle Sätze der letzten Paragraphen. Die spezifische Strahlungsintensität der Schwingungszahl ν, nach einer beliebigen Ebene polarisiert, im Innern der ersten Substanz (der oberen in Fig. 3) sei $\mathfrak{K}_\nu$, die im Innern der zweiten Substanz $\mathfrak{K}_\nu'$, wie wir überhaupt die auf die zweite Substanz bezüglichen Größen durch einen hinzugefügten Strich markieren wollen. Beide Größen $\mathfrak{K}_\nu$ und $\mathfrak{K}_\nu'$ hängen gemäß Gleichung (27) außer von der Temperatur und der Schwingungszahl ν nur von der Natur der beiden Substanzen ab, und zwar gelten diese Werte der Strahlungsintensität bis unmittelbar an die Grenzfläche der Substanzen, sind also ganz unabhängig von der Beschaffenheit dieser Fläche.

§ 36. Wir nehmen nun zunächst die Grenzfläche der beiden Medien als glatt (§ 9) an. Dann spaltet sich jeder aus dem ersten Medium kommende auf die Grenzfläche treffende Strahl in zwei Strahlen: den reflektierten und den durchgelassenen. Die Richtungen dieser beiden Strahlen variieren nach Maßgabe des Einfallswinkels und der Farbe des einfallenden Strahles, die Intensität außerdem nach Maßgabe seiner Polarisation. Bezeichnen wir mit ϱ (Reflexionskoeffizient) den Betrag der reflektierten, und infolgedessen mit $1 - \varrho$ den Betrag der durchgelassenen Strahlungsenergie im Verhältnis zur auffallenden Energie, so ist ϱ vom Einfallswinkel, von der Schwingungszahl und von der Polarisation des auffallenden Strahles abhängig. Entsprechendes gilt von ϱ', dem Reflexionskoeffizienten für einen aus dem zweiten Medium kommenden auf die Grenzfläche treffenden Strahl.

Nun ist die Energie der monochromatischen geradlinig polarisierten Strahlung von der Schwingungszahl ν von einem

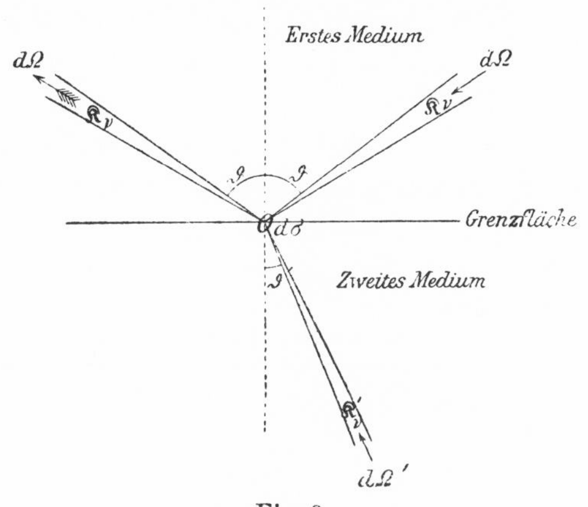

Fig. 3.

Element $d\sigma$ der Grenzfläche aus innerhalb des Elementarkegels $d\Omega$ in der Richtung nach dem ersten Medium (s. den gefiederten Pfeil oben links in Fig. 3) für die Zeit dt nach (11):

$$dt\, d\sigma \cos\vartheta\, d\Omega\, \mathfrak{K}_\nu\, d\nu\,, \qquad (34)$$

wobei:

$$d\Omega = \sin\vartheta\, d\vartheta\, d\varphi\,. \qquad (35)$$

Diese Energie wird geliefert durch die beiden Strahlen, welche

3*

aus dem ersten bez. zweiten Medium kommend von dem Flächenelement $d\sigma$ in entsprechender Richtung reflektiert bez. durchgelassen werden (s. die ungefiederten Pfeile. Von dem Flächenelement $d\sigma$ ist nur ein Punkt O gezeichnet.) Der erstere Strahl verläuft, gemäß dem Reflexionsgesetz, innerhalb des symmetrisch gelegenen Elementarkegels $d\Omega$, der zweite innerhalb des Elementarkegels:

$$d\Omega' = \sin\vartheta'\, d\vartheta'\, d\varphi', \tag{36}$$

wobei nach dem Brechungsgesetz:

$$\varphi' = \varphi \quad \text{und} \quad \frac{\sin\vartheta}{\sin\vartheta'} = \frac{q}{q'}. \tag{37}$$

Nehmen wir nun an, die Strahlung (34) sei entweder in der Einfallsebene oder senkrecht zur Einfallsebene polarisiert, so gilt das Entsprechende für die beiden Strahlungen, aus deren Energien sie sich zusammensetzt, und zwar liefert die aus dem ersten Medium kommende, von $d\sigma$ reflektierte Strahlung den Beitrag:

$$\varrho \cdot dt \cdot d\sigma \cos\vartheta \cdot d\Omega\, \mathfrak{K}_\nu\, d\nu \tag{38}$$

und die aus dem zweiten Medium kommende, von $d\sigma$ durchgelassene Strahlung den Beitrag:

$$(1-\varrho') \cdot dt \cdot d\sigma \cos\vartheta' \cdot d\Omega'\, \mathfrak{K}_\nu'\, d\nu. \tag{39}$$

Die Größen dt, $d\sigma$, ν und $d\nu$ sind hier ohne Strich hingeschrieben, weil sie in beiden Medien die nämlichen Werte besitzen.

Addiert man die Ausdrücke (38) und (39) und setzt die Summe gleich dem Ausdruck (34), so ergibt sich:

$$\varrho \cos\vartheta \cdot d\Omega\, \mathfrak{K}_\nu + (1-\varrho') \cdot \cos\vartheta'\, d\Omega'\, \mathfrak{K}_\nu' = \cos\vartheta \cdot d\Omega\, \mathfrak{K}_\nu.$$

Nun ist nach (37)

$$\frac{\cos\vartheta\, d\vartheta}{q} = \frac{\cos\vartheta'\, d\vartheta'}{q'}$$

und weiter durch Berücksichtigung von (35) und (36):

$$d\Omega' \cos\vartheta' = \frac{d\Omega \cos\vartheta \cdot q'^2}{q^2}.$$

Folglich ergibt sich:

$$\varrho\, \mathfrak{K}_\nu + (1-\varrho') \frac{q'^2}{q^2} \mathfrak{K}_\nu' = \mathfrak{K}_\nu$$

oder:

$$\frac{\mathfrak{K}_\nu}{\mathfrak{K}_\nu'} \cdot \frac{q^2}{q'^2} = \frac{1-\varrho'}{1-\varrho}.$$

§ 37. In der letzten Gleichung ist die Größe auf der linken Seite unabhängig vom Einfallswinkel ϑ und von der Art der Polarisation; folglich muß es auch die Größe auf der rechten Seite sein. Kennt man also den Wert dieser Größe für einen einzigen Einfallswinkel und eine bestimmte Art der Polarisation, so besitzt dieser Wert für alle Einfallswinkel und alle Polarisationen Gültigkeit. Nun ist in dem speziellen Falle, daß die Strahlen rechtwinklig zur Einfallsebene polarisiert sind und unter dem Polarisationswinkel auf die Grenzfläche auffallen, $\varrho = 0$ und $\varrho' = 0$. Dann wird der Ausdruck rechts gleich 1; also ist er allgemein gleich 1, und wir haben stets:

$$\varrho = \varrho' \qquad (40)$$

und:

$$q^2 \mathfrak{K}_\nu = q'^2 \mathfrak{K}_\nu' . \qquad (41)$$

§ 38. Die erste dieser beiden Beziehungen, welche besagt, daß der Reflexionskoeffizient der Grenzfläche nach beiden Seiten hin der nämliche ist, bildet den speziellen Ausdruck eines allgemeinen zuerst von HELMHOLTZ[1] ausgesprochenen Reziprozitätssatzes, wonach der Intensitätsverlust, welchen ein Strahl von bestimmter Farbe und Polarisation auf seinem Wege durch irgend welche Medien infolge von Reflexion, Brechung, Absorption, Zerstreuung erleidet, genau gleich ist dem Intensitätsverlust, welchen ein Strahl von entsprechender Intensität, Farbe und Polarisation auf dem gerade entgegengesetzten Wege erleidet. Daraus folgt unmittelbar, daß die auf die Grenzfläche zweier Medien auffallende Strahlung stets nach beiden Seiten hin gleich gut hindurchgelassen bez. reflektiert wird, für jede Farbe, Richtung und Polarisation.

§ 39. Die zweite Beziehung (41) bringt die in den beiden Substanzen bestehenden Strahlungsintensitäten miteinander in Zusammenhang, sie besagt nämlich, daß beim thermodynamischen Gleichgewicht die spezifischen Strahlungsintensitäten einer bestimmten Schwingungszahl in beiden Medien

---

[1] H. v. HELMHOLTZ, Handbuch der Physiologischen Optik. 1. Lieferung. Leipzig, Leop. Voss, 1856, p. 169. Vgl. auch HELMHOLTZ' Vorlesungen über die Theorie der Wärme, herausgegeben von F. RICHARZ, Leipzig, J. A. Barth, 1903, p. 161. Die dort für besondere Fälle gemachten Einschränkungen des Reziprozitätssatzes sind hier gegenstandslos, da es sich hier nur um Temperaturstrahlung (§ 7) handelt.

sich umgekehrt verhalten wie die Quadrate der Fortpflanzungsgeschwindigkeiten oder direkt wie die Quadrate der Brechungsexponenten.[1]

Substituiert man für $\mathfrak{K}_\nu$ seinen Wert aus (27), so kann man auch sagen: Die Größe:

$$q^2 \mathfrak{K}_\nu = q^2 \frac{\varepsilon_\nu}{\alpha_\nu} \tag{42}$$

hängt nicht ab von der Natur der Substanz, ist also eine universelle Funktion der Temperatur T und der Schwingungszahl ν.

Der hohe Wert dieses Satzes beruht offenbar darauf, daß er eine Eigenschaft der Strahlung angibt, die für alle Körper in der Natur gleichmäßig gilt, und die daher nur an einem ein zigen ganz beliebig ausgewählten Körper bekannt zu sein braucht, um sogleich vollständig allgemein ausgesprochen werden zu können. Die hierdurch gebotene Möglichkeit werden wir später benutzen, um jene universelle Funktion wirklich zu berechnen, in § 161.

§ 40. Nun fassen wir den weiteren Fall ins Auge, daß die Grenzfläche der beiden Medien rauh ist. Dieser Fall ist insofern viel allgemeiner als der vorher betrachtete, als hier die Energie eines von einem Element der Grenzfläche in das Innere des ersten Mediums hinein gerichteten Strahlenbündels nicht mehr von zwei bestimmten, sondern von beliebig vielen Strahlenbündeln, die aus beiden Medien kommend die Grenzfläche treffen, geliefert wird. Es können hier im einzelnen sehr komplizierte Verhältnisse eintreten, je nach der Beschaffenheit der Grenzfläche, die überdies von Element zu Element beliebig variieren kann. Immer bleiben dabei natürlich nach § 35 die Werte der spezifischen Strahlungsintensitäten $\mathfrak{K}_\nu$ und $\mathfrak{K}_\nu'$ in beiden Medien nach allen Richtungen die nämlichen wie im Fall einer glatten Grenzfläche. Die Erfüllung dieser für das thermodynamische Gleichgewicht notwendigen Bedingung wird begreiflich durch den HELMHOLTZschen Reziprozitätssatz, wonach bei der stationären Strahlung jedem Strahl, der auf die Grenzfläche trifft und von dieser diffus nach beiden Seiten derselben zerstreut

---

[1] G. KIRCHHOFF, Gesammelte Abhandlungen, Leipzig, J. A. Barth, 1882, p. 594. R. CLAUSIUS, POGG. Ann. **121**, p. 1, 1864.

wird, an demselben Orte ein gleich intensiver gerade entgegengesetzt gerichteter entspricht, der durch den umgekehrten Vorgang an derselben Stelle der Grenzfläche: die Sammlung diffus auftreffender Strahlung in eine bestimmte Richtung, zustande kommt, ebenso wie dies im Innern jedes der beiden Medien der Fall ist.

§ 41. Die erhaltenen Sätze wollen wir noch etwas weiter verallgemeinern. Zunächst kann, ebenso wie im § 34, die von uns gemachte Voraussetzung, daß die beiden Medien räumlich weit ausgedehnt sind, ohne weiteres aufgehoben werden, da man beliebig viele Trennungsflächen einführen kann, ohne daß das thermodynamische Gleichgewicht gestört wird. Dadurch sind wir dann auch in den Stand gesetzt, sogleich zu dem Falle beliebig vieler, beliebig großer und beliebig geformter Substanzen übergehen zu können. Denn wenn ein System von beliebig vielen sich gegenseitig berührenden Substanzen sich im thermodynamischen Gleichgewicht befindet, so wird das Gleichgewicht in keiner Weise gestört, wenn man eine oder mehrere der Berührungsflächen zum Teil oder ganz als für Wärme undurchlässig voraussetzt. Hierdurch können wir immer den Fall beliebig vieler Substanzen zurückführen auf den zweier in eine für Wärme undurchlässige Hülle eingeschlossener Substanzen, und daher den Satz ganz allgemein aussprechen, daß beim thermodynamischen Gleichgewicht eines beliebigen Systems die spezifische Strahlungsintensität $\mathfrak{K}_\nu$ in jeder einzelnen Substanz durch die universelle Funktion (42) bestimmt wird.

§ 42. Wir betrachten nun ein in eine für Wärme undurchlässige Hülle eingeschlossenes System von n nebeneinander gelagerten emittierenden und absorbierenden Körpern beliebiger Größe und Form im Zustand des thermodynamischen Gleichgewichtes, und fassen wieder, wie in § 36, ein monochromatisches geradlinig polarisiertes Strahlenbündel ins Auge, welches von einem Element $d\sigma$ der Grenzfläche zweier Medien innerhalb des Elementarkegels $d\Omega$ in der Richtung nach dem ersten Medium fortschreitet (Fig. 3, s. den gefiederten Pfeil). Dann ist die von dem Bündel gelieferte Energie für die Zeiteinheit, wie in (34):

$$d\sigma \cos\vartheta \cdot d\Omega\, \mathfrak{K}_\nu\, d\nu = I. \tag{43}$$

Diese Strahlungsenergie I setzt sich zusammen aus einem

Teil, der mittels regulärer oder diffuser Reflexion an der Grenzfläche aus dem ersten Medium kommt, und aus einem anderen Teil, der durch die Grenzfläche hindurch aus dem zweiten Medium kommt. Wir wollen aber jetzt bei dieser Art Einteilung nicht stehen bleiben, sondern wollen die Einteilung darnach einrichten, in welchem der n Medien die einzelnen Teile der Strahlung I emittiert worden sind. Dieser Gesichtspunkt ist ein von dem vorigen wesentlich verschiedener. Denn die Strahlen, welche z. B. aus dem zweiten Medium durch die Grenzfläche hindurch in das betrachtete Bündel hineingelangen, brauchen keineswegs alle im zweiten Medium emittiert worden zu sein, sondern können unter Umständen einen langen, sehr komplizierten Weg durch verschiedene Medien zurückgelegt haben, in dessen Verlauf sie den Einflüssen der Brechung, Reflexion, Zerstreuung und teilweisen Absorption beliebig oft unterworfen waren. Ebenso brauchen die Strahlen des Bündels I, welche aus dem ersten Medium kommend an $d\sigma$ reflektiert worden sind, durchaus nicht alle im ersten Medium emittiert worden zu sein. Es kann auch vorkommen, daß ein Strahl, der in einem Medium emittiert ist, auf seinem Wege durch andere Medien hindurch wieder in das ursprüngliche Medium zurückgelangt und dort entweder absorbiert wird oder zum zweiten Male aus dem Medium austritt.

Wir wollen nun, unter Berücksichtigung aller dieser Möglichkeiten, den Teil von I, der von Volumenelementen des ersten Mediums emittiert worden ist, ganz gleichgültig, welche Wege seine einzelnen Bestandteile eingeschlagen haben, mit I_1, den, der von Volumenelementen des zweiten Mediums emittiert worden ist, mit I_2 bezeichnen usw. Dann muß sein:

$$I = I_1 + I_2 + I_3 + \dots + I_n, \tag{44}$$

denn in irgend einem Körperelement muß jeder Bestandteil von I emittiert worden sein.

§ 43. Um nun näheres über die Herkunft und die Bahnen der einzelnen Strahlen zu erfahren, aus denen sich die Strahlungen I_1, I_2, ... I_n zusammensetzen, ist es am zweckmäßigsten, den umgekehrten Weg zu gehen und nach dem künftigen Schicksal desjenigen Strahlenbündels zu fragen, welches dem Bündel I gerade entgegengesetzt gerichtet ist, also vom ersten Medium kommend innerhalb des Kegels $d\Omega$ auf das Oberflächenelement $d\sigma$ des zweiten Mediums trifft. Denn da jeder optische Weg auch in umgekehrter Richtung gangbar ist, so erhält man durch diese Betrachtung sämtliche Bahnen, auf denen Strahlen in das Bündel I hineingelangen können, so kompliziert sie auch im übrigen sein mögen. Ist J die Intensität dieses umgekehrten, auf die Grenzfläche zu gerichteten, ebenso polarisierten Bündels, so ist nach § 40:

$$J = I. \tag{45}$$

Die Strahlen des Bündels J werden an der Grenzfläche $d\sigma$ teils reflektiert, teils durchgelassen, regulär oder diffus, hierauf in beiden Medien teils absorbiert, teils zerstreut, teils wiederum reflektiert oder in andere Medien durchgelassen usw., je nach der Konfiguration des Systems. Schließlich aber wird das ganze Bündel J, nachdem es sich in viele einzelne Strahlen verzweigt hat, in den n Medien vollständig absorbiert werden. Bezeichnen wir denjenigen Teil von J, welcher schließlich im ersten Medium absorbiert wird, mit J_1, denjenigen, welcher schließlich im zweiten Medium absorbiert wird, mit J_2 usw., so ist mithin:

$$J = J_1 + J_2 + J_3 + \ldots + J_n.$$

Nun sind die Volumenelemente der n Medien, in denen die Absorption der Strahlen des Bündels J stattfindet, genau dieselben, wie die, in welchen die Emission der Strahlen stattfindet, aus denen sich das oben zuerst betrachtete Bündel I zusammensetzt. Denn nach dem Helmholtzschen Reziprozitätssatz kann keine merkliche Strahlung aus dem Bündel J in ein Volumenelement dringen, aus welchem keine merkliche Strahlung in das Bündel I hineingelangt, und umgekehrt.

Bedenkt man ferner, daß die Absorption eines jeden Volumenelements nach (42) proportional ist seiner Emission, und daß nach dem Helmholtzschen Reziprozitätssatz die Schwächung, welche die Energie eines Strahles auf irgend einem Wege erleidet, immer gleich ist derjenigen Schwächung, welche die Energie des Strahles auf dem umgekehrten Wege erleidet, so erhellt, daß die betrachteten Volumenelemente die Strahlen des Bündels J gerade in demselben Verhältnis absorbieren, wie sie durch ihre Emission zur Energie des entgegengesetzten Bündels I beitragen; und da überdies die Summe I der von allen Volumenelementen durch Emission gelieferten Energien gleich ist der Summe J der von allen Elementen absorbierten Energien, so muß auch der von jedem einzelnen Element aus dem Bündel J absorbierte Energiebetrag gleich sein dem von demselben Element in das Bündel I emittierten Energiebetrag. Mit anderen Worten: Der Teil eines Strahlenbündels I, welcher aus einem bestimmten Volumen irgend eines Mediums emittiert worden ist, ist gleich demjenigen Teile des entgegengesetzt gerichteten Strahlenbündels $J(= I)$, welcher in demselben Volumen absorbiert wird.

Es sind also nicht nur die Summen I und J einander gleich, sondern auch ihre Bestandteile:

$$J_1 = I_1, \quad J_2 = I_2, \ \ldots \ J_n = I_n. \tag{46}$$

§ 44. Die Größe I_2, d. h. die Intensität des vom zweiten Medium in das erste Medium emittierten Strahlenbündels, nennen wir nach G. Kirchhoff[1] das Emissionsvermögen E des zweiten Mediums, während wir als Absorptionsvermögen A desselben Mediums das Verhältnis von J_2 zu J bezeichnen, d. h. denjenigen Bruchteil des auf das zweite Medium fallenden Strahlenbündels, welcher in diesem Medium absorbiert wird. Also:

$$E = I_2 (\leqq I), \qquad A = \frac{J_2}{J} (\leqq 1). \tag{47}$$

Die Größen E und A hängen ab von der Natur beider Medien und der Temperatur, von der Schwingungszahl ν und von der Polarisationsrichtung der betrachteten Strahlung, ferner von der Beschaffenheit der Grenzfläche, von der Größe des Flächenelements $d\sigma$ und des Öffnungswinkels $d\Omega$, endlich von der geometrischen Ausdehnung und der Form der gesamten Oberfläche beider Medien, sowie von der Natur und Form sämtlicher anderer im System vorhandener Körper. Denn wenn z. B. ein aus dem ersten in das zweite Medium eingedrungener Strahl von letzterem hindurchgelassen wird, kann er möglicherweise irgendwo anders reflektiert werden, dadurch in das zweite Medium zurückgelangen und dort absorbiert werden.

Bei diesen Festsetzungen gilt gemäß den Gleichungen (46), (45) und (43) der Kirchhoffsche Satz:

---

[1] G. Kirchhoff, Gesammelte Abhandlungen, 1882, p. 574.

$$\frac{E}{A} = I = d\sigma \cos\vartheta \cdot d\Omega \cdot \mathfrak{K}_\nu \, d\nu\,, \tag{48}$$

d. h. das Verhältnis des Emissionsvermögens zum Absorptionsvermögen eines Körpers ist unabhängig von der Beschaffenheit des Körpers. Denn dies Verhältnis ist gleich der Intensität des im ersten Medium fortschreitenden Strahlenbündels, welche nach der Gleichung (27) von dem zweiten Medium gar nicht abhängt. Von der Beschaffenheit des ersten Mediums ist aber der Wert jenes Verhältnisses abhängig, insofern als nach (42) nicht die Größe $\mathfrak{K}_\nu$, sondern die Größe $q^2\mathfrak{K}_\nu$ eine universelle Funktion der Temperatur und der Schwingungszahl ist. G. Kirchhoff hat den Beweis seines Satzes nur unter der Voraussetzung geführt, daß im ersten Medium weder Absorption noch Zerstreuung der Strahlung stattfindet. Dasselbe gilt von dem neueren sehr vereinfachten Beweise von E. Pringsheim.[1]

§ 45. Wenn speziell das zweite Medium ein schwarzer Körper ist (§ 10), so absorbiert es die ganze auffallende Strahlung. Daher ist dann $J_2 = J$, $A = 1$ und $E = I$, d. h. das Emissionsvermögen eines schwarzen Körpers ist von seiner Beschaffenheit unabhängig. Es ist größer als das Emissionsvermögen irgend eines anderen Körpers von derselben Temperatur, und direkt gleich der Intensität der Strahlung im angrenzenden Medium.

§ 46. Wir fügen hier, ohne näheren Beweis, noch einen allgemeinen Reziprozitätssatz an, der sich dem am Schluß des § 43 ausgesprochenen eng anschließt, und der folgendermaßen lautet: Beim thermodynamischen Gleichgewicht beliebiger emittierender und absorbierender Körper ist derjenige Teil der von einem Körper A emittierten Energie einer bestimmten Farbe, welcher von irgend einem anderen Körper B absorbiert wird, gleich demjenigen Teile der von B emittierten Energie derselben Farbe, welcher von A absorbiert wird. Bedenkt man, daß jeder emittierte Energiebetrag eine Verminderung der Körperwärme, jeder absorbierte Energiebetrag eine Vermehrung der

---

[1] E. Pringsheim, Verhandlungen der Deutschen Physikalischen Gesellschaft, 3, p. 81, 1901.

Körperwärme bedingt, so erhellt daraus, daß beim thermodynamischen Gleichgewicht je zwei beliebig herausgegriffene Körper (oder Körperelemente) vermittelst der Strahlung gegenseitig gleichviel Körperwärme austauschen. Dabei ist natürlich wohl zu unterscheiden zwischen der emittierten Strahlung und der gesamten Strahlung, die von einem Körper zu einem anderen hingelangt.

§ 47. Das für die Größe (42) gültige Gesetz läßt sich auch noch in einer anderen Form aussprechen, wenn man statt der spezifischen Strahlungsintensität $\mathfrak{K}_\nu$ die räumliche Dichte $\mathfrak{u}_\nu$ der monochromatischen Strahlung aus (24) einführt. Man erhält dann den Satz, daß bei der Strahlung im thermodynamischen Gleichgewicht die Größe:

$$\mathfrak{u}_\nu\, q^3 \tag{49}$$

eine für alle Substanzen identische Funktion der Temperatur T und der Schwingungszahl ν ist.[1] Eine anschaulichere Form gewinnt dieser Satz noch, wenn man bedenkt, daß auch die Größe

$$\mathfrak{u}_\nu\, d\nu \cdot \frac{q^3}{\nu^3} \tag{50}$$

eine universelle Funktion von T, ν und $d\nu$ ist, und daß das Produkt $\mathfrak{u}_\nu\, d\nu$ nach (22) die räumliche Strahlungsdichte derjenigen Strahlung ist, deren Schwingungszahl zwischen ν und $\nu + d\nu$ liegt, während der Quotient $\frac{q}{\nu}$ die Wellenlänge eines Strahles von der Schwingungszahl ν in dem betrachteten Medium darstellt. Dann erhält der Satz folgende einfache Fassung: Beim thermodynamischen Gleichgewicht beliebiger Körper ist die in dem Kubus einer Wellenlänge enthaltene Energie der monochromatischen Strahlung für eine bestimmte Schwingungszahl in allen Körpern die nämliche.

§ 48. Wir wollen schließlich noch auf den bisher unberücksichtigt gebliebenen Fall der diathermanen Medien (§ 12) eingehen. Im § 27 sahen wir, daß in einem von einer adiabatischen Hülle umschlossenen Medium, welches für eine bestimmte

[1] Bei der Anwendung auf stark dispergierende Substanzen ist zu beachten, daß in diesem Satze die Identität der Größe q in (24) und der Größe q in (37) vorausgesetzt ist.

Farbe diatherman ist, bei jeder beliebigen Strahlungsintensität dieser Farbe thermodynamisches Gleichgewicht bestehen kann, daß aber unter allen möglichen Strahlungsintensitäten eine bestimmte, dem absoluten Maximum der Gesamtentropie des Systems entsprechende, existieren muß, welche das absolut stabile Strahlungsgleichgewicht bezeichnet. In der Tat nimmt in der Gleichung (27) die Strahlungsintensität $\mathfrak{K}_\nu$ für $\alpha_\nu = 0$ und $\varepsilon_\nu = 0$ den Wert $\frac{0}{0}$ an, und kann daher aus dieser Gleichung nicht berechnet werden. Aber man sieht auch sogleich weiter, daß die nötige Ergänzung zu dieser Unbestimmtheit geliefert wird von der Gleichung (41), welche besagt, daß beim thermodynamischen Gleichgewicht das Produkt $q^2 \mathfrak{K}_\nu$ für alle Substanzen den nämlichen Wert besitzt. Daraus ergibt sich unmittelbar auch für jedes diathermane Medium ein bestimmter Wert von $\mathfrak{K}_\nu$, der hierdurch vor allen anderen Werten ausgezeichnet ist. Die physikalische Bedeutung dieses Wertes erkennt man ebenfalls unmittelbar aus der Betrachtung des Weges, auf dem jene Gleichung hergeleitet wurde: es ist diejenige Strahlungsintensität, welche in dem diathermanen Medium besteht, wenn es sich bei der Berührung mit einem beliebig absorbierenden und emittierenden Medium im thermodynamischen Gleichgewicht befindet. Auf das Volumen und die Form des zweiten Mediums kommt es dabei gar nicht an; insbesondere kann das Volumen beliebig klein genommen werden. Somit läßt sich folgender Satz aussprechen: Obwohl in einem diathermanen Medium von vornherein bei jeder beliebigen Strahlungsintensität thermodynamisches Gleichgewicht bestehen kann, so gibt es doch in jedem diathermanen Medium für eine bestimmte Schwingungszahl bei einer bestimmten Temperatur eine durch die universelle Funktion (42) bestimmte Strahlungsintensität, welche insofern die stabile zu nennen ist, als sie sich immer dann einstellt, wenn das Medium sich mit einer beliebigen emittierenden und absorbierenden Substanz im stationären Strahlungsaustausch befindet.

§ 49. Nach dem im § 45 ausgesprochenen Satze ist bei der stabilen Wärmestrahlung in einem diathermanen Medium die Intensität eines Strahlenbündels gleich dem Emissionsvermögen E eines an das Medium grenzenden schwarzen Körpers. Hierauf beruht die Möglichkeit, das Emissionsvermögen eines schwarzen Körpers zu messen, da es doch absolut schwarze Körper in der Natur nicht gibt.[1] Man stellt einen von stark emittierenden Wänden[2] begrenzten diathermanen Hohlraum her, und erhält die Wände auf einer bestimmten konstanten Temperatur T. Dann nimmt die Strahlung in dem Hohlraum zugleich mit dem Eintritt des thermodynamischen Gleichgewichtszustandes für jede Schwingungszahl ν die aus der universellen Funktion (42) durch die Fortpflanzungsgeschwindigkeit q des diathermanen Mediums bedingte Intensität an. Von jedem Flächenelement einer Wand geht dann eine Strahlung in den Hohlraum, die ebenso beschaffen ist, als ob die Wand ein schwarzer Körper von der Temperatur T wäre. Was den von den Wänden wirklich emittierten Strahlen im Vergleich mit der Emission eines schwarzen Körpers an Intensität noch fehlt, wird ersetzt durch solche Strahlen, die auf die Wand auffallen und dort zurückgeworfen werden. Ebenso wird jedes Flächenelement einer Wand von der nämlichen Strahlung getroffen.

Bohrt man nun in eine der Wände ein Loch von der Größe $d\sigma$, welches so klein ist, daß dadurch die Intensität der auf das Loch zu gerichteten Strahlung nicht geändert wird, so dringt durch das Loch nach außen, wo sich das nämliche diathermane Medium befinden möge wie im Innern, eine Strahlung, die genau die nämlichen Eigenschaften besitzt, als ob $d\sigma$ die Oberfläche eines schwarzen Körpers wäre, und diese Strahlung kann für jede Farbe zugleich mit der Temperatur T gemessen werden.

§ 50. Alle im vorstehenden für diathermane Medien abgeleiteten Sätze gelten zunächst für eine bestimmte Schwingungszahl, wobei zu bedenken ist, daß eine Substanz für eine Farbe diatherman, für eine andere adiatherman sein kann. Daher ist im thermodynamischen Gleichgewichtszustand eines rings von absolut reflektierenden Wänden umschlossenen Mediums die Strahlung für alle Farben, für welche das Medium einen endlichen Absorptionskoeffizienten besitzt, immer die der Temperatur

[1] W. Wien und O. Lummer, Wied. Ann. **56**, p. 451, 1895.

[2] Die Stärke der Emission beeinflußt nur die Zeit bis zur Herstellung der stationären Strahlung, nicht aber deren Charakter.

des Mediums entsprechende stabile, durch die Emission eines schwarzen Körpers dargestellte und daher auch kurz als „schwarz" bezeichnete[1]; dagegen ist die Strahlungsintensität für alle Farben, bezüglich derer das Medium diatherman ist, nur dann notwendig die stabile, schwarze, wenn das Medium sich mit eine absorbierenden Substanz in stationärem Strahlungsaustausch befindet.

Von Medien, die für alle Strahlenarten diatherman sind, existiert nur ein einziges: das absolute Vakuum, welches allerdings in der Natur nur annähernd herzustellen ist. Doch besitzen auch die Gase, z. B. die atmosphärische Luft, bei nicht zu großer Dichtigkeit und für nicht zu kurze Wellen, mit großer, in den meisten Fällen praktisch vollkommen hinreichender Annäherung die optischen Eigenschaften des Vakuums. Insofern dies der Fall ist, kann die Fortpflanzungsgeschwindigeit q für alle Schwingungszahlen als die gleiche:

$$c = 3 \cdot 10^{10} \frac{\text{cm}}{\text{sec}} \qquad (51)$$

angenommen werden.

§ 51. In einem von total reflektierenden Wänden umschlossenen Vakuum kann daher von vornherein jeder beliebige Strahlungszustand stationär sein. Sobald man aber in das Vakuum eine beliebig kleine Menge einer ponderablen Substanz hineinbringt, so stellt sich mit der Zeit ein stationärer Strahlungszustand her, in welchem die Strahlung einer jeden Farbe, die von der Substanz in merklichem Betrage absorbiert wird, die der Temperatur der Substanz entsprechende, durch die universelle Funktion (42) für $q = c$ bestimmte Intensität $\mathfrak{K}_\nu$ besitzt, während die Strahlungsintensität der übrigen Farben unbestimmt bleibt. Ist die eingebrachte Substanz für keine Farbe diatherman, z. B. ein beliebig kleines Stückchen Kohle, so besteht beim stationären Strahlungszustand im ganzen Vakuum für alle Farben die der Temperatur der Substanz entsprechende Intensität $\mathfrak{K}_\nu$ der schwarzen Strahlung. Die Größe $\mathfrak{K}_\nu$ als Funktion von ν betrachtet ergibt die spektrale Verteilung der schwarzen Strahlung im Vakuum oder das sogenannte normale Energiespektrum, welches ausschließlich von der Temperatur abhängt. Im Normalspektrum, als dem Spektrum der Emission eines schwarzen

---

[1] M. Thiesen, Verhandlungen der Deutschen Physikalischen Gesellschaft, **2**, p. 65, 1900.

Körpers, ist die Strahlungsintensität einer jeden Farbe die größte, welche ein Körper bei der betreffenden Temperatur überhaupt emittieren kann.

§ 52. Man kann also eine ganz beliebige Strahlung, die anfangs in dem betrachteten evakuierten Hohlraume mit total reflektierenden Wänden herrscht, durch Einbringung eines winzigen Kohlestäubchens in schwarze Strahlung verwandeln. Charakteristisch für diesen Vorgang ist der Umstand, daß die Körperwärme des Kohlestäubchens beliebig klein sein kann gegen die Strahlungsenergie, die in dem beliebig groß zu nehmenden Hohlraume vorhanden ist, und daß daher in diesem Falle nach dem Prinzip der Erhaltung der Energie die gesamte Strahlungsenergie auch bei der eintretenden Umwandlung wesentlich konstant bleibt, da die Änderungen der Körperwärme des Stäubchens selbst bei endlichen Temperaturänderungen desselben gar nicht in Betracht kommen. Das Kohlestäubchen spielt dann lediglich die Rolle einer auslösenden Wirkung, indem es den Anstoß dazu gibt, daß in der ursprünglich vorhandenen Strahlung die Intensitäten der verschieden gerichteten, verschieden polarisierten Strahlenbündel der verschiedenen Schwingungszahlen sich auf gegenseitige Kosten verändern, entsprechend dem Übergang des Systems aus einem minder stabilen in einen stabileren Strahlungszustand, oder aus einem Zustand kleinerer in einen Zustand größerer Entropie. Vom thermodynamischen Standpunkt aus ist dieser Vorgang ganz analog der Verwandlung, die in einem Quantum Knallgas durch einen minimalen Funken, oder die in einem Quantum übersättigten Dampfes durch ein winziges Flüssigkeitströpfchen hervorgerufen wird; denn auf die Zeit kommt es hier nicht an. In allen diesen Fällen ist die Größe der Störung eine minimale und steht in gar keiner Beziehung zu der Größe der an der Verwandlung beteiligten Energiemengen, so daß man bei der Anwendung der beiden Hauptsätze der Thermodynamik die Ursache der Gleichgewichtsstörung: das Kohlestäubchen, den Funken, das Tröpfchen gar nicht zu berücksichtigen braucht. Es handelt sich jedesmal um den Übergang eines Systems aus einem mehr oder minder labilen in einen stabileren Zustand, wobei nach dem ersten Hauptsatz die Energie des Systems konstant bleibt und nach dem zweiten Hauptsatz die Entropie des Systems zunimmt.

Zweiter Abschnitt.

Folgerungen aus der Elektrodynamik und der Thermodynamik.

Erstes Kapitel. Maxwellscher Strahlungsdruck.

§ 53. Während wir im vorigen Abschnitt für die Darstellung der Strahlungsvorgänge lediglich die aus der elementaren Optik bekannten, im § 2 zusammengefaßten Sätze benutzt haben, welche allen optischen Theorien gemeinsam sind, wollen wir von jetzt an die elektromagnetische Theorie des Lichtes benutzen, und beginnen damit, indem wir eine Folgerung ableiten, welche dieser Theorie eigentümlich ist. Wir wollen nämlich die Größe der mechanischen Kraft berechnen, welche ein im Vakuum fortschreitender Licht- oder Wärmestrahl beim Auftreffen auf eine ruhend gedachte spiegelnde (§ 10) Fläche ausübt.

Zu diesem Zwecke stellen wir zunächst die allgemeinen Maxwellschen Gleichungen für einen elektromagnetischen Vorgang im Vakuum auf. Sie lauten, wenn der Vektor $\mathfrak{E}$ die elektrische Feldstärke (Intensität des elektrischen Feldes) im elektrischen Maße, der Vektor $\mathfrak{H}$ die magnetische Feldstärke im magnetischen Maße bedeutet, in der abgekürzten Bezeichnung der Vektorrechnung:

$$\left.\begin{array}{ll} \dot{\mathfrak{E}} = c \operatorname{curl} \mathfrak{H} & \dot{\mathfrak{H}} = - c \operatorname{curl} \mathfrak{E} \\ \operatorname{div} \mathfrak{E} = 0 & \operatorname{div} \mathfrak{H} = 0 . \end{array}\right\} \qquad (52)$$

Wer mit den hier benutzten Symbolen nicht vertraut ist, kann sich deren Bedeutung leicht aus den folgenden Gleichungen (53) rückwärts ergänzen.

§ 54. Um zu dem Fall einer ebenen beliebig gerichteten Welle überzugehen, setzen wir voraus, daß alle Zustandsgrößen außer von der Zeit t nur von einer einzigen der drei Koordinaten x', y', z' eines orthogonalen rechtshändigen Koordinatensystems abhängen, z. B. von x'. Dann reduzieren sich die Gleichungen (52) auf:

$$(53)\quad \left\{ \begin{array}{ll} \dfrac{\partial \mathfrak{E}_{x'}}{\partial t} = 0 & \dfrac{\partial \mathfrak{H}_{x'}}{\partial t} = 0 \\[2ex] \dfrac{\partial \mathfrak{E}_{y'}}{\partial t} = -c \dfrac{\partial \mathfrak{H}_{z'}}{\partial x'} & \dfrac{\partial \mathfrak{H}_{y'}}{\partial t} = c \dfrac{\partial \mathfrak{E}_{z'}}{\partial x'} \\[2ex] \dfrac{\partial \mathfrak{E}_{z'}}{\partial t} = c \dfrac{\partial \mathfrak{H}_{y'}}{\partial x'} & \dfrac{\partial \mathfrak{H}_{z'}}{\partial t} = -c \dfrac{\partial \mathfrak{E}_{y'}}{\partial x'} \\[2ex] \dfrac{\partial \mathfrak{E}_{x'}}{\partial x'} = 0 & \dfrac{\partial \mathfrak{H}_{x'}}{\partial x'} = 0 \end{array} \right.$$

Hieraus ergibt sich als allgemeinster Ausdruck für eine ebene, in der Richtung der positiven x'-Achse im Vakuum fortschreitende Welle:

$$(54)\quad \left\{ \begin{array}{ll} \mathfrak{E}_{x'} = 0 & \mathfrak{H}_{x'} = 0 \\[1ex] \mathfrak{E}_{y'} = f\left(t - \dfrac{x'}{c}\right) & \mathfrak{H}_{y'} = -g\left(t - \dfrac{x'}{c}\right) \\[2ex] \mathfrak{E}_{z'} = g\left(t - \dfrac{x'}{c}\right) & \mathfrak{H}_{z'} = f\left(t - \dfrac{x'}{c}\right), \end{array} \right.$$

wo f und g zwei beliebige Funktionen eines einzigen Arguments vorstellen.

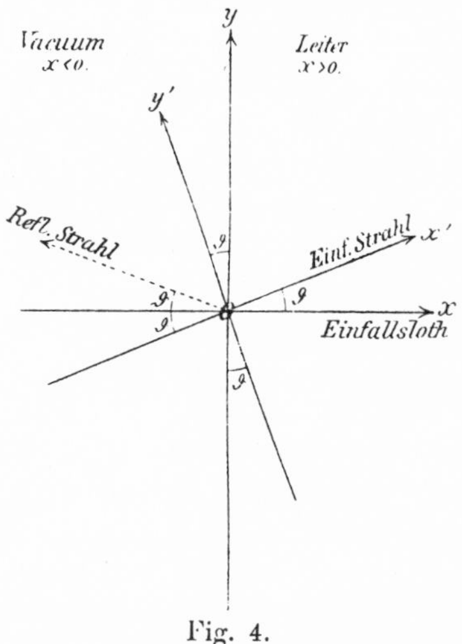

Fig. 4.

§ 55. Nun treffe diese Welle auf eine spiegelnde Fläche, z. B. auf die Oberfläche eines absoluten Leiters, d. h. einer Substanz (Metall) von unendlich großer Leitungsfähigkeit. In einem solchen Leiter bewirkt schon eine unendlich kleine elektrische Feldstärke einen endlichen Leitungsstrom; daher ist in ihr die elektrische Feldstärke $\mathfrak{E}$ stets und überall unendlich klein. Der Einfachheit halber setzen wir außerdem den Leiter als unmagnetisierbar voraus, d. h. wir nehmen die magnetische Induktion $\mathfrak{B}$ in ihm gleich der magnetischen Feldstärke $\mathfrak{H}$ an, wie im Vakuum.

Legen wir die x-Achse des rechtshändigen Koordinatensystems (x, y, z) in die nach dem Innern des Leiters gerichtete Normale seiner Oberfläche, so ist die x-Achse das Einfallslot. Die ($x'y'$)-Ebene legen wir in die Einfallsebene und machen sie zur Bildebene (Fig. 4). Ferner können wir, ohne die Allgemeinheit zu beschränken, auch die y-Achse in die Bildebene legen, so daß die z-Achse mit der z'-Achse zusammenfällt (in der Figur vom Bilde zum Beschauer gerichtet). Der beiden Koordinatensystemen gemeinsame Anfangspunkt O liege in der Oberfläche. Ist endlich ϑ der Einfallswinkel, so sind die gestrichenen und die ungestrichenen Koordinaten durch die folgenden Gleichungen miteinander verknüpft:

$$\begin{aligned} x &= x'\cos\vartheta - y'\sin\vartheta \qquad & x' &= x\cos\vartheta + y\sin\vartheta \\ y &= x'\sin\vartheta + y'\cos\vartheta & y' &= -x\sin\vartheta + y\cos\vartheta \\ z &= z' & z' &= z\,. \end{aligned}$$

Ganz dieselben Transformationsgleichungen gelten, wenn man die Koordinaten durch die Komponenten der elektrischen oder der magnetischen Feldstärke in beiden Koordinatensystemen ersetzt. Dadurch erhalten wir für die Komponenten der elektrischen und der magnetischen Feldstärke der einfallenden Welle in bezug auf das ungestrichene Koordinatensystem nach (54) die Werte:

$$\left.\begin{aligned} \mathfrak{E}_x &= -\sin\vartheta\cdot f \qquad & \mathfrak{H}_x &= \sin\vartheta\cdot g \\ \mathfrak{E}_y &= \cos\vartheta\cdot f & \mathfrak{H}_y &= -\cos\vartheta\cdot g \\ \mathfrak{E}_z &= g & \mathfrak{H}_z &= f, \end{aligned}\right\} \tag{55}$$

wo in die Funktionen f und g das Argument

$$t - \frac{x'}{c} = t - \frac{x\cos\vartheta + y\sin\vartheta}{c} \tag{56}$$

eingesetzt zu denken ist.

§ 56. In der Grenzfläche der beiden Medien ist $x = 0$. Für diesen Wert müssen also nach den allgemeinen elektromagnetischen Grenzbedingungen die in die Grenzfläche fallenden Komponenten der Feldstärken, d. h. hier die vier Größen $\mathfrak{E}_y$, $\mathfrak{E}_z$, $\mathfrak{H}_y$, $\mathfrak{H}_z$ auf beiden Seiten der Grenzfläche einander gleich sein. Nun ist im Leiter nach der oben gemachten Voraussetzung die elektrische Feldstärke $\mathfrak{E}$ unendlich klein; folglich müssen $\mathfrak{E}_y$ und $\mathfrak{E}_z$ auch im Vakuum für $x = 0$ verschwinden. Diese Bedingung kann nur

4*

erfüllt werden, wenn man im Vakuum außer der einfallenden noch eine reflektierte Welle annimmt, die sich der einfallenden Welle superponiert, und zwar in der Weise, daß die in die y- und z-Richtung fallenden elektrischen Feldkomponenten der beiden Wellen sich in allen Punkten der Grenzfläche in jedem Augenblick gegenseitig gerade aufheben. Hierdurch und durch die Bedingung, daß die reflektierte Welle eine ebene ist und sich nach rückwärts in das Innere des Vakuums hinein fortpflanzt, sind auch die übrigen vier Komponenten der reflektierten Welle vollkommen bestimmt; sie sind alle Funktionen des einen Arguments:

$$t - \frac{-x\cos\vartheta + y\sin\vartheta}{c}. \tag{57}$$

Die Ausführung der Rechnung ergibt als Komponenten des gesamten durch die Superposition der beiden Wellen im Vakuum gebildeten elektromagnetischen Feldes für die Punkte der Grenzfläche ($x = 0$) die Ausdrücke:

$$\left\{\begin{aligned}
\mathfrak{E}_x &= -\sin\vartheta\cdot f - \sin\vartheta\cdot f = -2\sin\vartheta\cdot f\\
\mathfrak{E}_y &= \cos\vartheta\cdot f - \cos\vartheta\cdot f = 0\\
\mathfrak{E}_z &= g - g = 0\\
\mathfrak{H}_x &= \sin\vartheta\cdot g - \sin\vartheta\cdot g = 0\\
\mathfrak{H}_y &= -\cos\vartheta\cdot g - \cos\vartheta\cdot g = -2\cos\vartheta\cdot g\\
\mathfrak{H}_z &= f + f = 2f
\end{aligned}\right. \tag{58}$$

wo nach (56) und (57) in die Funktionen f und g überall das Argument:

$$t - \frac{y\sin\vartheta}{c}$$

eingesetzt zu denken ist.

Mit Hilfe dieser Werte ergibt sich dann auch die elektrische und die magnetische Feldstärke innerhalb des Leiters unmittelbar an der Grenzfläche $x = 0$:

$$\left\{\begin{aligned}
\mathfrak{E}_x &= 0 \qquad & \mathfrak{H}_x &= 0\\
\mathfrak{E}_y &= 0 \qquad & \mathfrak{H}_y &= -2\cos\vartheta\cdot g\\
\mathfrak{E}_z &= 0 \qquad & \mathfrak{H}_z &= 2f
\end{aligned}\right. \tag{59}$$

wo wieder in die Funktionen f und g das Argument $t - \frac{y\sin\vartheta}{c}$ eingesetzt zu denken ist. Denn die Komponenten von $\mathfrak{E}$ verschwinden im absoluten Leiter alle, und die Komponenten $\mathfrak{H}_x$, $\mathfrak{H}_y$, $\mathfrak{H}_z$ sind an der Grenzfläche alle stetig: die letzten beiden als tangentielle Komponenten der Feldstärke, die erste als Normalkomponente der magnetischen Induktion $\mathfrak{B}$ (§ 55), welche ebenfalls durch jede Grenzfläche stetig hindurchgeht.

Dagegen ist, wie man sieht, die Normalkomponente der elektrischen Feldstärke: $\mathfrak{E}_x$ unstetig; ihr Sprung ergibt das Vorhandensein einer elektrischen Ladung an der Grenzfläche, deren Flächendichte nach Größe und Vorzeichen beträgt:

$$\frac{1}{4\pi}\cdot 2\sin\vartheta\cdot f = \frac{1}{2\pi}\sin\vartheta\cdot f. \tag{60}$$

Im Innern des Leiters, in endlicher Entfernung von der Grenzfläche, d. h. für $x > 0$, sind alle sechs Feldkomponenten unendlich klein. Daher fallen die für $x = 0$ endlichen Werte von $\mathfrak{H}_y$ und $\mathfrak{H}_z$ mit wachsendem x unendlich schnell gegen Null ab.

§ 57. Durch das im Vakuum vorhandene elektromagnetische Feld wird eine gewisse mechanische Kraft auf die Leitersubstanz ausgeübt, deren Komponente normal zur Oberfläche wir berechnen wollen. Dieselbe ist teils elektrischen, teils magnetischen Ursprungs. Betrachten wir zunächst die erstere: $\mathfrak{F}_e$. Da die an der Leiteroberfläche befindliche elektrische Ladung sich in einem elektrischen Felde befindet, so wirkt auf sie eine mechanische Kraft, die gleich ist dem Produkt der Ladung und der Feldstärke. Da aber die Feldstärke unstetig ist, nämlich auf der Seite des Vakuums: $-2\sin\vartheta\cdot f$, auf der Seite des Leiters: 0, so erhält man die Größe der auf das Flächenelement $d\sigma$ der Oberfläche des Leiters wirkenden mechanischen Kraft $\mathfrak{F}_e$ nach einem bekannten Satz der Elektrostatik durch Multiplikation der in (60) berechneten elektrischen Ladung des Flächenelements mit dem arithmetischen Mittel der elektrischen Feldstärke auf beiden Seiten, mithin:

$$\mathfrak{F}_e = \frac{\sin\vartheta}{2\pi} f\,d\sigma\cdot(-\sin\vartheta\cdot f) = -\frac{\sin^2\vartheta}{2\pi} f^2\,d\sigma.$$

Diese Kraft wirkt in der Richtung nach dem Vakuum zu, äußert sich also als Zug.

§ 58. Jetzt berechnen wir die mechanische Kraft magnetischen Ursprungs: $\mathfrak{F}_m$. Im Innern der Leitersubstanz fließen gewisse Leitungsströme, deren Intensität und Richtung durch den Vektor $\mathfrak{J}$ der Stromdichte:

$$\mathfrak{J} = \frac{c}{4\pi} \cdot \operatorname{curl} \mathfrak{H} \tag{61}$$

bestimmt ist. Nun wirkt auf jedes von einem Leitungsstrom durchflossene Raumelement $d\tau$ des Leiters eine mechanische Kraft, die gegeben ist durch das Vektorprodukt:

$$\frac{d\tau}{c} \cdot [\mathfrak{J}, \mathfrak{H}]. \tag{62}$$

Die Komponente dieser Kraft normal zur Leiteroberfläche ($x = 0$) ist daher:

$$\frac{d\tau}{c} \cdot (\mathfrak{J}_y \mathfrak{H}_z - \mathfrak{J}_z \mathfrak{H}_y),$$

und wenn man die Werte von $\mathfrak{J}_y$ und $\mathfrak{J}_z$ aus (61) einsetzt:

$$\frac{d\tau}{4\pi} \cdot \left[\mathfrak{H}_z \left(\frac{\partial \mathfrak{H}_x}{\partial z} - \frac{\partial \mathfrak{H}_z}{\partial x}\right) - \mathfrak{H}_y \left(\frac{\partial \mathfrak{H}_y}{\partial x} - \frac{\partial \mathfrak{H}_x}{\partial y}\right)\right].$$

Die in diesem Ausdruck vorkommenden Differentialquotienten nach y und nach z sind nach der Bemerkung am Schluß von § 56 gegen die nach x verschwindend klein; daher reduziert er sich auf:

$$-\frac{d\tau}{4\pi} \cdot \left(\mathfrak{H}_y \frac{\partial \mathfrak{H}_y}{\partial x} + \mathfrak{H}_z \frac{\partial \mathfrak{H}_z}{\partial x}\right).$$

Wir betrachten nun einen aus dem Leiter ausgeschnittenen, auf seiner Oberfläche senkrechten Zylinder mit dem Querschnitt $d\sigma$, der von $x = 0$ bis $x = \infty$ reicht. Die gesamte auf diesen Zylinder in der Richtung der x-Achse wirkende mechanische Kraft magnetischen Ursprungs ist dann, da $d\tau = d\sigma \cdot dx$:

$$\mathfrak{F}_m = -\frac{d\sigma}{4\pi} \int_0^\infty dx \cdot \left(\mathfrak{H}_y \frac{\partial \mathfrak{H}_y}{\partial x} + \mathfrak{H}_z \frac{\partial \mathfrak{H}_z}{\partial x}\right)$$

und durch Integration, da für $x = \infty$ $\mathfrak{H}$ verschwindet:

$$\mathfrak{F}_m = \frac{d\sigma}{8\pi} (\mathfrak{H}_y^2 + \mathfrak{H}_z^2)_{x=0}$$

oder nach den Gleichungen (59):

$$\mathfrak{F}_m = \frac{d\sigma}{2\pi} \cdot (\cos^2 \vartheta \cdot g^2 + f^2).$$

Durch Addition von $\mathfrak{F}_e$ und $\mathfrak{F}_m$ ergibt sich die ganze auf den betrachteten Zylinder in der Richtung der x-Achse wirkende mechanische Kraft:

$$\mathfrak{F} = \frac{d\sigma}{2\pi} \cos^2 \vartheta (f^2 + g^2), \tag{63}$$

welche sich als ein in normaler Richtung auf die Oberfläche des Leiters nach dem Innern desselben wirkender Druck äußert, der als „MAXWELLscher Strahlungsdruck" bezeichnet wird. Die Existenz und auch die Größe des Strahlungsdruckes wurde zuerst von P. LEBEDEW[1] durch subtile Messungen mit dem Radiometer als mit der Theorie übereinstimmend gefunden.

§ 59. Wir wollen den Strahlungsdruck in Beziehung bringen zu der auf das Oberflächenelement $d\sigma$ des Leiters im Zeitelement dt auffallenden Strahlungsenergie $I\,dt$. Dieselbe beträgt nach dem POYNTINGschen Energieströmungsgesetz:

$$I\,dt = \frac{c}{4\pi}(\mathfrak{E}_y \mathfrak{H}_z - \mathfrak{E}_z \mathfrak{H}_y)\,d\sigma\,dt,$$

also nach (55):

$$I\,dt = \frac{c}{4\pi} \cos\vartheta (f^2 + g^2)\,d\sigma\,dt.$$

Durch Vergleich mit (63) ergibt sich:

$$\mathfrak{F} = \frac{2\cos\vartheta}{c} \cdot I. \tag{64}$$

Hieraus berechnen wir endlich den gesamten Druck p, d. h. diejenige mechanische Kraft, welche eine beliebige, aus dem Vakuum kommende, den Leiter treffende und von ihm vollständig reflektierte Strahlung auf die Oberflächeneinheit des Leiters in normaler Richtung ausübt. Die Energie, welche innerhalb des Elementarkegels:

$$d\Omega = \sin\vartheta\,d\vartheta\,d\varphi$$

in der Zeit dt auf das Flächenelement $d\sigma$ gestrahlt wird, beträgt nach (6):

$$I\,dt = K\cos\vartheta \cdot d\Omega\,d\sigma\,dt,$$

wo K die spezifische Intensität der Strahlung in der Richtung von $d\Omega$ auf den Spiegel zu bedeutet. Dies in (64) eingesetzt und über $d\Omega$ integriert, ergibt für den Gesamtdruck aller auf die Oberfläche fallender und dort reflektierter Strahlenbündel:

---

[1] P. LEBEDEW, DRUDES Ann. **6**, p. 433, 1901. Vgl. ferner: E. F. NICHOLS und G. F. HULL, DRUDES Ann. **12**, p. 225, 1903.

(65) $$p = \frac{2}{c} \int K \cos^2 \vartheta \, d\Omega,$$

wo die Integration in bezug auf φ von 0 bis 2π, in bezug auf ϑ von 0 bis $\frac{\pi}{2}$ auszuführen ist.

Wenn speziell K von der Richtung unabhängig ist, wie bei der schwarzen Strahlung, so erhält man für den Druck derselben:

$$p = \frac{2K}{c} \int_0^{2\pi} d\varphi \cdot \int_0^{\pi/2} d\vartheta \cos^2 \vartheta \sin \vartheta = \frac{4\pi K}{3c}$$

oder, wenn statt K die räumliche Strahlungsdichte u aus (21) eingeführt wird:

(66) $$p = \frac{u}{3}.$$

Dieser Wert des Strahlungsdruckes gilt zunächst nur für den Fall, daß die Reflexion der Strahlung an der Oberfläche eines absoluten unmagnetisierbaren Leiters erfolgt, und wir werden ihn daher bei den thermodynamischen Deduktionen im nächsten Kapitel auch nur für diesen Fall benutzen. Indessen wird sich später (§ 66) zeigen, daß die Gleichung (66) den Druck der gleichmäßigen Strahlung auch gegen eine ganz beliebige vollständig reflektierende Fläche ergibt, gleichgültig ob sie regelmäßig oder diffus reflektiert.

§ 60. Angesichts der überaus einfachen und nahen Beziehung zwischen dem Strahlungsdrucke und der Strahlungsenergie könnte man die Frage aufwerfen, ob diese Beziehung wirklich eine spezielle Folgerung der elektromagnetischen Theorie ist, oder ob sie sich vielleicht auch auf allgemeinere energetische bezw. thermodynamische Überlegungen gründen läßt. Um diese Frage zu entscheiden, wollen wir denjenigen Strahlungsdruck berechnen, der sich aus der NEWTONschen Emanationstheorie des Lichtes ergeben würde, welche Theorie ja mit dem Energieprinzip an sich wohl verträglich ist. Nach ihr ist die durch einen im Vakuum fortschreitenden Lichtstrahl einer Fläche zugestrahlte Energie gleich der lebendigen Kraft der auf die Fläche treffenden Lichtpartikel, die sich alle mit der konstanten Geschwindigkeit c bewegen. Die Abnahme der Intensität der Energiestrahlung mit der Entfernung erklärt sich dann einfach aus der Abnahme der räumlichen Verteilungsdichte der Lichtpartikel.

Nennen wir also n die Anzahl der in der Volumeneinheit enthaltenen Lichtpartikel, m die Masse einer Partikel, so ist zunächst für ein paralleles Lichtbündel die Zahl der in der Zeiteinheit auf das Element $d\sigma$ einer spiegelnden Oberfläche unter dem Einfallswinkel ϑ treffenden Partikel:

$$n \cdot c \cdot \cos\vartheta \cdot d\sigma \tag{67}$$

und ihre lebendige Kraft:

$$I = n c \cos\vartheta \; d\sigma \cdot \frac{m c^2}{2} = n m \cos\vartheta \cdot \frac{c^3}{2} \cdot d\sigma . \tag{68}$$

Um andererseits den Normaldruck dieser Partikel auf die Oberfläche zu bestimmen, beachten wir, daß die Normalkomponente der Geschwindigkeit $c \cdot \cos\vartheta$ einer jeden Partikel bei der Reflexion in die entgegengesetzte verwandelt wird. Daher wird die Normalkomponente der Bewegungsgröße (Impulskoordinate) einer jeden Partikel bei der Reflexion um $-2 m c \cdot \cos\vartheta$ geändert. Dies ergibt für alle betrachteten Partikel nach (67) die Änderung der Bewegungsgröße:

$$-2 n m \cos^2\vartheta \cdot c^2 d\sigma . \tag{69}$$

Ist nun der spiegelnde Körper in der Richtung der Spiegelnormalen frei beweglich, und es wirkt außer dem Stoße der Lichtpartikel keine Kraft auf ihn, so wird er durch die Stöße in Bewegung gesetzt werden, und zwar nach dem Gesetz von Wirkung und Gegenwirkung in der Weise, daß die Bewegungsgröße, welche er in einem gewissen Zeitintervall annimmt, gleich und entgegengesetzt ist der Änderung der Bewegungsgrößen aller in demselben Zeitintervall an ihm reflektierten Lichtpartikel. Läßt man aber noch eine besondere konstante Kraft von außen auf den Spiegel wirken, so kommt zu der Bewegungsgrößenänderung noch die von dieser Kraft gelieferte Bewegungsgröße hinzu, welche gleich ist dem Impuls der Kraft, d. h. dem Produkt der Kraft mal dem betrachteten Zeitintervall.

Daher wird der Spiegel dauernd in Ruhe bleiben, wenn die konstante von außen auf ihn wirkende Kraft so gewählt wird, daß ihr Impuls für irgend eine Zeit gerade gleich ist der in derselben Zeit eintretenden Änderung der Bewegungsgrößen der an dem Spiegel reflektierten Partikel, und daraus folgt, daß die Kraft $\mathfrak{F}$ selbst, welche die Partikel durch ihren Anprall gegen das Flächenelement $d\sigma$ ausüben, gleich und entgegengesetzt ist der Änderung ihrer Bewegungsgröße für die Zeiteinheit, wie sie durch (69) ausgedrückt ist, nämlich:

$$\mathfrak{F} = 2\,n\,m\cos^2\vartheta \cdot c^2\,d\sigma$$

und mit Benutzung von (68):

$$\mathfrak{F} = \frac{4\cos\vartheta}{c} \cdot I.$$

Vergleicht man diese Beziehung mit der Gleichung (64), in welcher alle Zeichen die nämliche physikalische Bedeutung haben, so erkennt man, daß der NEWTONsche Strahlungsdruck doppelt so groß ist als der MAXWELLsche bei gleicher Energiestrahlung, und daraus folgt mit Notwendigkeit, daß die Größe des MAXWELLschen Strahlungsdruckes nicht aus allgemeinen energetischen Überlegungen abgeleitet werden kann, sondern daß sie der elektromagnetischen Theorie eigentümlich ist. Daher sind auch alle aus dem MAXWELLschen Strahlungsdrucke abgeleiteten Folgerungen als Folgerungen der elektromagnetischen Lichttheorie, und alle Bestätigungen derselben als Bestätigungen dieser speziellen Theorie anzusehen.

Zweites Kapitel. Stefan-Boltzmannsches Strahlungsgesetz.

§ 61. Wir denken uns im folgenden einen vollständig evakuierten Hohlzylinder mit einem absolut dicht schließenden, in vertikaler Richtung ohne Reibung frei beweglichen Kolben. Ein Teil der Wandung des Zylinders, etwa der feste Boden, bestehe aus einem schwarzen Körper, dessen Temperatur T willkürlich von außen reguliert werden kann. Die übrige Wand, auch die innere Kolbenfläche, sei vollständig reflektierend. Dann wird, bei ruhendem Kolben und bei konstant gehaltener Temperatur T, die Strahlung im Vakuum nach einiger Zeit den Charakter der schwarzen, nach allen Richtungen gleichmäßigen Strahlung (§ 50) annehmen, deren spezifische Intensität K und räumliche Dichte u nur von der Temperatur T abhängt, insbesondere auch unabhängig ist von dem Volumen V des Vakuums, also von der Stellung des Kolbens.

Bewegt man nun den Kolben nach unten, so wird die Strahlung auf einen kleineren Raum zusammengedrängt, bewegt man ihn nach oben, so dehnt sie sich auf einen größeren Raum aus. Gleichzeitig kann man auch die Temperatur T des schwarzen Bodenkörpers durch Zuleitung oder Ableitung äußerer Wärme willkürlich verändern. Dadurch treten jedesmal gewisse Störungen des stationären Zustandes ein; es läßt sich aber durch gehörige Verlangsamung der willkürlich vorgenommenen Änderungen von V und T immer erreichen, daß die Abweichungen von den Bedingungen des stationären Zustandes beliebig klein bleiben, und daß man daher, ohne einen merklichen Fehler zu begehen, den Strahlungszustand im Vakuum immer als einen thermodynamischen Gleichgewichtszustand betrachten kann, ganz ähnlich, wie es in der Thermodynamik ponderabler Substanzen bei sogenannten unendlich langsamen Prozessen geschieht, in denen die jeweiligen Abweichungen vom Gleichgewichtszustand zu vernachlässigen sind gegenüber den Änderungen, die das behandelte System schließlich durch den ganzen Prozeß erleidet.

Hält man z. B. die Temperatur T des schwarzen Bodenkörpers konstant, was durch geeignete Verbindung desselben mit einem Wärmereservoir von großer Kapazität geschehen kann, so wird bei Hebung des Kolbens der schwarze Körper so lange stärker emittieren als absorbieren, bis der neu geschaffene Raum mit der nämlichen Strahlungsdichte wie früher angefüllt ist. Umgekehrt wird bei Senkung des Kolbens der schwarze Körper die überschüssige Strahlung absorbieren, bis wieder die ursprüngliche, der Temperatur T entsprechende Strahlung hergestellt ist. Ebenso wird bei Erhöhung der Temperatur T des schwarzen Körpers, die durch Wärmezuleitung aus einem um ein äußerst Geringes wärmeren Reservoir bewirkt werden kann, die Strahlungsdichte im Vakuum durch Mehremission entsprechend erhöht werden, usw. Zur größeren Beschleunigung der Herstellung des Strahlungsgleichgewichtes kann man den reflektierenden Mantel des Hohlzylinders als weiß (§ 10) voraussetzen, da durch die diffuse Reflexion die durch die Bewegungsrichtung des Kolbens etwa entstehenden Vorzugsrichtungen der Strahlung schneller ausgeglichen werden. Als reflektierende Kolbenfläche wollen wir aber bis auf weiteres einen vollkommenen Metallspiegel wählen, um des Maxwellschen Strahlungsdruckes (66) auf den

Kolben sicher zu sein. Dann muß, um das mechanische Gleichgewicht herzustellen, der Kolben mit einem Gewicht belastet werden, welches gleich ist dem Produkt des Strahlungsdruckes p und dem Querschnitt des Kolbens. Eine minimale Abweichung des belastenden Gewichtes von diesem Wert bringt dann eine entsprechend langsame Bewegung des Kolbens nach der einen oder der anderen Seite hervor.

Da die Einwirkungen, welche bei den hier ins Auge gefaßten Prozessen von außen auf das betrachtete System, den durchstrahlten Hohlraum, stattfinden, teils mechanischer Natur (Verschiebung des beschwerten Kolbens), teils thermischer Natur (Wärmeleitung vom und zum Reservoir) sind, so haben sie eine gewisse Ähnlichkeit mit den in der Thermodynamik gewöhnlich betrachteten Vorgängen, nur daß hier das zugrunde gelegte System kein materielles ist, wie z. B. ein Gas, sondern ein rein energetisches. Wenn aber die Hauptsätze der Thermodynamik in der Natur universelle Gültigkeit besitzen, was wir hier überall voraussetzen, so müssen sie auch für das hier betrachtete System Bedeutung haben. Es muß nämlich bei irgend einer in der Natur eintretenden Veränderung die Energie aller an der Änderung beteiligten Systeme konstant bleiben (erster Hauptsatz), und es muß ferner die Entropie aller an der Änderung beteiligten Systeme größer werden, im Grenzfall, bei reversibeln Prozessen, ungeändert bleiben (zweiter Hauptsatz).

§ 62. Bilden wir zunächst die Gleichung des ersten Hauptsatzes für eine unendlich kleine Änderung des betrachteten Systems. Daß dem durchstrahlten Hohlraum eine bestimmte Energie zukommt, haben wir schon früher (§ 22) aus dem Umstand abgeleitet, daß die Energiestrahlung sich mit endlicher Geschwindigkeit fortpflanzt. Wir bezeichnen sie mit U. Dann ist:

$$U = V \cdot u, \tag{70}$$

wobei u, die räumliche Strahlungsdichte, allein von der Temperatur T des schwarzen Bodenkörpers abhängt.

Die bei einer Vergrößerung des Volumens V des Hohlraumes um dV von dem System gegen die äußeren Druckkräfte (Gewicht des belasteten Kolbens) geleistete Arbeit ist $p \cdot dV$, wobei p den MAXWELLschen Strahlungsdruck (66) darstellt. Dieser Betrag von mechanischer Energie wird also außerhalb des Systems gewonnen, indem das Gewicht gehoben wird. Der Fehler, den wir dadurch begehen, daß wir hier den Strahlungsdruck auf eine ruhende Fläche benutzen, während doch die reflektierende Fläche während der Volumenänderung sich bewegt, ist offenbar zu vernachlässigen, da man sich die Bewegung mit beliebig kleiner Geschwindigkeit erfolgend denken kann.

Bezeichnet ferner Q die unendlich kleine Wärmemenge im mechanischen Maße, welche von dem schwarzen Bodenkörper an den durchstrahlten Raum durch Mehremission abgegeben wird, so verliert der Bodenkörper bez. das mit ihm in Verbindung stehende Wärmereservoir diese Wärme Q, wodurch seine innere Energie sich um diesen Betrag vermindert. Folglich ist nach dem ersten Hauptsatz der Thermodynamik, da die Summe der Strahlungsenergie und der Energie der materiellen Körper konstant bleibt:

$$dU + p\,dV - Q = o. \tag{71}$$

Nach dem zweiten Hauptsatz der Thermodynamik kommt dem durchstrahlten Vakuum auch eine bestimmte Entropie zu. Denn wenn die Wärme Q von dem Wärmereservoir an den Hohlraum abgegeben wird, so verkleinert sich die Entropie des Reservoirs, und zwar verändert sie sich um:

$$-\frac{Q}{T}.$$

Infolgedessen muß, da in anderen Körpern keine Änderungen eintreten — denn der starre und absolut reflektierende Kolben mit dem darauf lastenden Gewicht ändert auch bei der Bewegung seinen inneren Zustand nicht — als Kompensation eine Entropieänderung mindestens im Betrage $\frac{Q}{T}$ in der Natur eintreten, durch welche jene Verkleinerung kompensiert wird, und hierfür kann nur die Entropie des durchstrahlten Hohlraumes in Anspruch genommen werden, die wir mit S bezeichnen wollen.

Da nun aber die hier beschriebenen Prozesse aus lauter Gleichgewichtszuständen bestehen, so sind sie vollkommen reversibel, es findet also keine Entropievermehrung statt, sondern wir haben:

$$dS - \frac{Q}{T} = o, \tag{72}$$

oder aus (71):

$$dS = \frac{dU + p\,dV}{T}. \tag{73}$$

In dieser Gleichung stellen die Größen U, p, V und S gewisse Eigenschaften der Wärmestrahlung vor, die durch den augenblicklichen Zustand der Strahlung vollkommen bestimmt sind. Folglich ist auch die Größe T eine gewisse Eigenschaft des Zustandes der Strahlung, d. h. die schwarze Strahlung im Hohlraum besitzt eine gewisse Temperatur T, und diese Temperatur ist diejenige eines mit der Strahlung im Wärmegleichgewicht stehenden Körpers.

§ 63. Wir wollen nun aus der letzten Gleichung diejenige Folgerung ableiten, die aus dem Umstand entspringt, daß der Zustand des betrachteten Systems und mithin auch seine Entropie durch die Werte zweier unabhängiger Variablen bestimmt ist. Als erste Variable wählen wir V, als zweite können wir entweder T, oder u, oder p wählen, von welchen drei Größen zwei durch die dritte allein bestimmt sind. Wir wollen die Temperatur T neben dem Volumen V als unabhängige Variable nehmen. Dann ergibt die Substitution von (66) und (70) in (73):

$$dS = \frac{V}{T}\frac{du}{dT}\,dT + \frac{4u}{3T}\,dV. \tag{74}$$

Daraus: $\left(\frac{\partial S}{\partial T}\right)_V = \frac{V}{T}\frac{du}{dT}$, und $\left(\frac{\partial S}{\partial V}\right)_T = \frac{4u}{3T}$.

Differentiiert man die erste dieser Gleichungen partiell nach V, die zweite partiell nach T, so ergibt sich:

$$\frac{\partial^2 S}{\partial T\,\partial V} = \frac{1}{T}\frac{du}{dT} = \frac{4}{3T}\frac{du}{dT} - \frac{4u}{3T^2},$$

oder:

$$\frac{du}{dT} = \frac{4u}{T}.$$

Integriert:

$$u = aT^4 \tag{75}$$

und nach (21) als spezifische Intensität der schwarzen Strahlung:

$$K = \frac{c}{4\pi}u = \frac{ac}{4\pi}\cdot T^4. \tag{76}$$

Ferner als Druck der schwarzen Strahlung:

$$p = \frac{a}{3}T^4 \tag{77}$$

und als Gesamtenergie der Strahlung:

$$U = a T^4 \cdot V. \tag{78}$$

Dieses Gesetz, welches ausspricht, daß die räumliche Dichte und die spezifische Intensität der schwarzen Strahlung der vierten Potenz der absoluten Temperatur proportional sind, ist zuerst von J. STEFAN[1] auf Grund ziemlich roher Messungen aufgestellt, später von L. BOLTZMANN[2] auf thermodynamischer Grundlage aus dem MAXWELLschen Strahlungsdruck abgeleitet, und in neuerer Zeit durch exakte Messungen von O. LUMMER und E. PRINGSHEIM[3] zwischen 100° und 1300° C., wobei die Temperatur durch das Gasthermometer definiert wurde, bestätigt worden. In Temperaturgebieten und bei Genauigkeitsanforderungen, für welche die Angaben der verschiedenen Gasthermometer nicht mehr genügend miteinander übereinstimmen oder überhaupt nicht zu ermitteln sind, kann das STEFAN-BOLTZMANNsche Strahlungsgesetz zu einer absoluten, von jeder Substanz unabhängigen Definition der Temperatur verwendet werden.

§ 64. Der Zahlenwert der Konstanten a ergibt sich aus Messungen von F. KURLBAUM.[4] Hiernach ist, wenn man mit S_t die gesamte Energie bezeichnet, die von 1 qcm eines auf t^0 C. befindlichen schwarzen Körpers in 1 sec in die Luft ausgestrahlt wird:

$$S_{100} - S_0 = 0{,}0731 \frac{\text{Watt}}{\text{cm}^2} = 7{,}31 \cdot 10^5 \frac{\text{erg}}{\text{cm}^2\,\text{sec}}.$$

Da nun die Strahlung in Luft nahezu identisch ist mit der Strahlung ins Vakuum, so kann nach (7) und (76) gesetzt werden:

$$S_t = \pi K = \frac{a\,c}{4} \cdot (273 + t)^4$$

und wir erhalten:

$$S_{100} - S_0 = \frac{a\,c}{4} \cdot (373^4 - 273^4).$$

Also: $$a = \frac{4 \cdot 7{,}31 \cdot 10^5}{3 \cdot 10^{10} \cdot (373^4 - 273^4)} = 7{,}061 \cdot 10^{-15} \frac{\text{erg}}{\text{cm}^3\,\text{grad}^4}. \tag{79}$$

---

[1] J. STEFAN, Wien. Ber. **79**, p. 391, 1879.

[2] L. BOLTZMANN, WIED. Ann. **22**, p. 291, 1884.

[3] O. LUMMER und E. PRINGSHEIM, WIED. Ann. **63**, p. 395, 1897. DRUDES Ann. **3**, p. 159, 1900.

[4] F. KURLBAUM, WIED. Ann. **65**, p. 759, 1898.

§ 65. Die Größe der Entropie S der schwarzen Strahlung ergibt sich durch Integration der Differentialgleichung (73) zu:

$$S = \frac{4}{3} a T^3 \cdot V, \tag{80}$$

wenn man eine belanglose additive Konstante fortläßt. Daraus die Entropie der Volumeneinheit, oder die räumliche Entropiedichte der schwarzen Strahlung:

$$\frac{S}{V} = s = \frac{4}{3} a T^3. \tag{81}$$

§ 66. Wir wollen uns zunächst noch von einer beschränkenden Voraussetzung befreien, die wir machen mußten, um den von uns im vorigen Kapitel berechneten Wert des MAXWELLschen Strahlungsdruckes anwenden zu können. Bisher hatten wir den Zylinder als fest und nur den Kolben als frei beweglich angenommen. Jetzt wollen wir uns das ganze Gefäß, bestehend aus dem Zylinder, dem schwarzen Boden und dem Kolben, der in einer bestimmten Höhe über dem Boden an der Wandung des Zylinders befestigt sei, uns im Raume frei beweglich denken. Dann muß das Gefäß als Ganzes nach dem Prinzip von Wirkung und Gegenwirkung, da gar keine Kraft von außen darauf wirkt, dauernd in Ruhe bleiben. Dies würde übrigens auch dann gefolgert werden müssen, wenn man das Gegenwirkungsprinzip nicht von vornherein für diesen Fall als gültig anerkennen wollte. Denn würde das Gefäß in Bewegung geraten, so könnte die lebendige Kraft dieser Bewegung nur auf Kosten der Wärme des Bodenkörpers oder der Strahlungsenergie entstehen, da sonst keine andere disponible Energie in dem von einer starren Hülle umschlossenen System vorhanden ist, und es müßte zugleich mit der Energie auch die Entropie des Körpers oder der Strahlung abnehmen, was dem zweiten Hauptsatz der Thermodynamik widersprechen würde, da sonst keine Entropieänderungen in der Natur eintreten. Das Gefäß befindet sich also als Ganzes im mechanischen Gleichgewicht. Daraus folgt sogleich, daß der Druck der Strahlung auf den schwarzen Boden ebensogroß ist wie der entgegengesetzt gerichtete auf den spiegelnden Kolben, daß also der Druck der schwarzen Strahlung auf einen schwarzen Körper von der nämlichen Temperatur ebensogroß ist wie der auf einen spiegelnden

Körper, und das nämliche läßt sich leicht für eine beliebige vollständig reflektierende Fläche beweisen, die man am Boden des Zylinders befindlich annehmen kann, ohne den stationären Strahlungszustand irgendwie zu stören. Daher läßt sich bei allen vorhergehenden Betrachtungen das Spiegelmetall durch einen beliebigen vollständig reflektierenden oder auch durch einen schwarzen Körper von der Temperatur des Bodenkörpers ersetzen, und man kann allgemein den Satz aussprechen, daß der Strahlungsdruck nur von der Beschaffenheit der hin- und hergehenden Strahlung, nicht aber von der Beschaffenheit der angrenzenden Substanz abhängt.

§ 67. Wenn bei der Hebung des Kolbens die Temperatur des schwarzen Bodenkörpers durch entsprechende Wärmezufuhr aus dem Reservoir konstant erhalten wird, so verläuft der Vorgang isotherm. Dann bleibt mit der Temperatur T auch die Energiedichte u, der Strahlungsdruck p und die Entropiedichte s konstant; infolgedessen wächst die Gesamtenergie der Strahlung von $U = u\,V$ auf $U' = u\,V'$, die Entropie von $S = s\,V$ auf $S' = s\,V'$, und für die aus dem Wärmereservoir zugeführte Wärme erhält man durch Integration von (72) bei konstantem T:

$$Q = T\cdot(S' - S) = T s\cdot(V' - V)$$

oder nach (81) und (75):

$$Q = \frac{4}{3} a\,T^4 (V' - V) = \frac{4}{3}(U' - U).$$

Wie man sieht, übersteigt die von außen zugeführte Wärme den Betrag der Vermehrung der Strahlungsenergie $(U' - U)$ um $\frac{1}{3}(U' - U)$. Diese Mehrzufuhr von Wärme ist nötig, um die mit der Vergrößerung des Strahlungsvolumens verbundene äußere Arbeit zu leisten.

§ 68. Betrachten wir auch einen reversibeln adiabatischen Prozeß. Hierfür ist notwendig, daß nicht nur der Kolben und die Mantelfläche, sondern auch der Boden des Zylinders als vollständig reflektierend, etwa als weiß, vorausgesetzt wird. Dann ist bei der Kompression oder Ausdehnung des Strahlungsraumes die von außen zugeführte Wärme $Q = 0$, und die Energie der Strahlung ändert sich nur um den Betrag der äußeren Arbeit $p \cdot d\,V$. Um indessen sicher zu sein, daß bei einem endlichen adiabatischen Prozeß die Strahlung in jedem Augenblick vollständig stabil ist, d. h. den Charakter der schwarzen Strahlung besitzt, wollen wir innerhalb des evakuierten Hohlraumes noch ein minimales Kohlenstäubchen als vorhanden voraussetzen. Dieses Körperchen, von dem wir annehmen können, daß es für sämtliche Strahlenarten ein von Null verschiedenes Absorptionsvermögen besitzt, dient nur dazu, um das stabile Gleichgewicht der Strahlung im Hohlraum herzustellen (§ 51 f.) und dadurch die Reversibilität des Vorgangs zu verbürgen, während seine Körperwärme gegen die Strahlungsenergie U so klein angenommen werden kann, daß die zu einer merklichen Temperaturänderung des Stäubchens erforderliche Wärmezufuhr ganz zu vernachlässigen ist. Dann herrscht in jedem Augenblick des Prozesses absolut stabiles Strahlungsgleichgewicht, und die Strahlung besitzt die Temperatur des in dem Hohlraum befindlichen Stäubchens. Volumen, Energie und Entropie des Stäubchens können ganz vernachlässigt werden.

Bei der reversibeln adiabatischen Änderung bleibt nach (72) die Entropie S des Systems konstant. Es folgt also als Bedingung dieses Prozesses aus (80):

$$T^3 \cdot V = \text{const}$$

oder auch nach (77):

$$p \cdot V^{\frac{4}{3}} = \text{const},$$

d. h. bei adiabatischer Kompression steigt die Strahlungstemperatur und der Strahlungsdruck in der angegebenen Weise. Die Energie der Strahlung U ändert sich dabei nach dem Gesetz:

$$\frac{U}{T} = \tfrac{3}{4} S = \text{const},$$

d. h. sie wächst proportional der absoluten Temperatur, trotzdem das Volumen kleiner wird.

§ 69. Wir wollen schließlich noch als weiteres Beispiel einen einfachen Fall eines irreversibeln Prozesses betrachten. Der allseitig von absolut reflektierenden Wänden umschlossene Hohlraum vom Volumen V sei gleichmäßig von schwarzer Strahlung erfüllt. Nun stelle man, etwa durch Drehen eines Hahnes, an irgend einer Stelle der Wandung eine kleine Öffnung her, durch welche die Strahlung in einen anderen ebenfalls von absolut reflektierenden festen Wänden umgebenen vollständig evakuierten Raum austreten kann. Dann wird die Strahlung zunächst einen sehr unregelmäßigen Charakter annehmen, nach einiger Zeit aber wird sich ein stationärer Strahlungszustand einstellen, der beide kommunizierende Räume, deren Gesamtvolumen V' sei, gleichmäßig erfüllt. Durch die Anwesenheit eines Kohlestäubchens sei dafür gesorgt, daß auch im neuen Zustand alle Bedingungen der schwarzen Strahlung erfüllt sind. Dann ist, weil weder äußere Arbeit noch äußere Wärmezufuhr stattgefunden hat, nach dem ersten Hauptsatz die Energie im neuen Zustand gleich der im alten: $U' = U$, und daher nach (78):

$$T'^4 V' = T^4 V$$

$$\frac{T'}{T} = \sqrt[4]{\frac{V}{V'}}$$

und dadurch der neue Gleichgewichtszustand vollkommen bestimmt. Da $V' > V$, so ist die Temperatur der Strahlung durch den Vorgang erniedrigt worden.

Nach dem zweiten Hauptsatz muß die Entropie des Systems gewachsen sein, da sonst keine äußeren Veränderungen stattgefunden haben; in der Tat ist nach (80):

$$\frac{S'}{S} = \frac{T'^3 V'}{T^3 V} = \sqrt[4]{\frac{V'}{V}} > 1 . \tag{82}$$

§ 70. Wenn der Vorgang der irreversibeln adiabatischen Ausdehnung der Strahlung vom Volumen V auf das Volumen V' genau ebenso erfolgt, wie vorhin beschrieben, nur mit dem einzigen Unterschied, daß kein Kohlenstäubchen in das Vakuum eingelagert ist, so wird nach Herstellung des gleichmäßigen Strahlungszustandes, welche infolge der diffusen Reflexion an den Wänden des Hohlraums nach gehöriger Zeit eintreten wird, im neuen Volumen V' die Strahlung nicht mehr den Charakter der schwarzen Strahlung haben, also auch keine bestimmte Temperatur besitzen. Wohl aber besitzt auch dann die Strahlung, wie überhaupt jedes in einem bestimmten Zustand befindliche physikalische System, eine bestimmte Entropie, die nach dem zweiten Hauptsatz größer ist als die anfängliche S, aber nicht so groß als die oben in (82) ausgedrückte S'. Ihre Berechnung kann erst auf Grund späterer Sätze erfolgen (vgl. § 103). Bringt man dann nachträglich in das Vakuum ein Kohlestäubchen, so stellt sich durch einen zweiten irreversibeln Prozeß das absolut stabile Strahlungsgleichgewicht her, indem die Strahlung bei

5*

konstanter Gesamtenergie die normale Energieverteilung der schwarzen Strahlung annimmt, und die Entropie steigt dabei auf den durch (82) gegebenen Maximalwert S'.

Drittes Kapitel. Wiensches Verschiebungsgesetz.

§ 71. Wenn durch das STEFAN-BOLTZMANNsche Gesetz die Abhängigkeit der räumlichen Dichte u und der spezifischen Intensität K der schwarzen Strahlung von der Temperatur bestimmt ist, so ist dadurch für die Kenntnis der auf eine bestimmte Schwingungszahl ν bezogenen räumlichen Strahlungsdichte $\mathfrak{u}_\nu$, und der spezifischen Strahlungsintensität $\mathfrak{K}_\nu$ der monochromatischen Strahlung, welche miteinander durch die Gleichung (24), und mit u und K durch die Gleichungen (22) und (12) verknüpft sind, noch verhältnismäßig wenig gewonnen, und es bleibt als ein Hauptproblem der Theorie der Wärmestrahlung die Aufgabe bestehen, die Größen $\mathfrak{u}_\nu$ und $\mathfrak{K}_\nu$ für die schwarze Strahlung im Vakuum, und dadurch nach (42) auch in jedem beliebigen Medium, als Funktionen von ν und T zu bestimmen, oder mit anderen Worten: die Verteilung der Energie im Normalspektrum für jede beliebige Temperatur anzugeben. Ein wesentlicher Schritt zur Lösung dieser Aufgabe ist in dem von W. WIEN aufgestellten sogenannten „Verschiebungsgesetz“[1] enthalten, dessen Bedeutung darin besteht, daß es die Funktionen $\mathfrak{u}_\nu$ und $\mathfrak{K}_\nu$ der beiden Argumente ν und T auf eine Funktion eines einzigen Arguments zurückführt.

Den Ausgangspunkt des WIENschen Verschiebungsgesetzes bildet folgender Satz. Wenn die in einem vollständig evakuierten Hohlraum mit absolut reflektierenden Wänden enthaltene schwarze Strahlung adiabatisch und unendlich langsam komprimiert oder dilatiert wird, wie im § 68 beschrieben wurde, so behält die Strahlung, auch ohne daß ein Kohlestäubchen sich im Vakuum befindet, stets den Charakter der schwarzen

[1] W. WIEN, Sitzungsber. d. Akad. d. Wissensch. Berlin vom 9. Febr. 1893, p. 55. WIED. Ann. **52**, p. 132, 1894. Vgl. ferner u. a.: M. THIESEN, Verhandlungen der Deutschen Physikalischen Gesellschaft, **2**, p. 65, 1900. H. A. LORENTZ, Akad. d. Wissensch. Amsterdam, 18. Mai 1901, p. 607. M. ABRAHAM, DRUDES Ann. **14**, p. 236, 1904.

Strahlung bei. Der Prozeß verläuft also auch im absoluten Vakuum genau so wie im § 68 berechnet wurde, und die dort als Vorsichtsmaßregel angewandte Einführung des Kohlestäubchens zeigt sich als überflüssig, allerdings nur in diesem speziellen Falle, nicht etwa auch in dem § 70 beschriebenen Falle.

Die Richtigkeit des ausgesprochenen Satzes ergibt sich aus folgendem. Man komprimiere den anfänglich mit schwarzer Strahlung erfüllten vollständig evakuierten Hohlzylinder adiabatisch und unendlich langsam auf einen endlichen Bruchteil seines ursprünglichen Volumens. Wäre nun, nach Vollendung der Kompression, die Strahlung nicht mehr schwarz, so bestände kein stabiles thermodynamisches Gleichgewicht der Strahlung (§ 51). Dann könnte man durch Einbringung eines Kohlestäubchens, dem im Vergleich zur Strahlungsenergie keine merkliche Körperwärme zukommt, bei konstantem Volumen und konstanter Gesamtenergie der Strahlung eine endliche Umwandlung, nämlich den Übergang zum absolut stabilen Strahlungszustand, und damit eine endliche Entropievermehrung des Systems herbeiführen. Diese Veränderung würde natürlich nur die spektrale Strahlungsdichte u_ν betreffen, während dagegen die gesamte Energiedichte u konstant bleibt. Nachdem dies geschehen, könnte man, das Kohlestäubchen in dem Raume belassend, den Hohlzylinder wieder adiabatisch und unendlich langsam auf sein ursprüngliches Volumen vergrößern und hierauf das Kohlestäubchen entfernen. Dann hat das System einen Kreisprozeß durchgemacht, ohne daß irgendwelche äußere Veränderungen zurückgeblieben sind. Denn Wärme ist überhaupt weder zu- noch abgeleitet worden, und die auf die Kompression verwendete mechanische Arbeit ist bei der Ausdehnung genau wieder gewonnen worden; denn diese hängt, ebenso wie der Strahlungsdruck, nur von der Gesamtdichte u der Strahlungsenergie, nicht von ihrer spektralen Verteilung ab. Infolgedessen ist nach dem ersten Hauptsatz der Thermodynamik auch die gesamte Strahlungsenergie am Schluß wieder dieselbe wie am Anfang und daher auch die Temperatur der schwarzen Strahlung wieder die nämliche. Das Kohlestäubchen und seine Veränderungen zählt nicht mit; denn seine Energie und Entropie sind verschwindend klein gegen die betreffenden Größen des Systems. Der Prozeß ist also in allen Einzelheiten rückgängig gemacht, man kann ihn beliebig oft hintereinander wiederholen, ohne daß irgend eine dauernde Veränderung in der Natur eintritt. Dies widerspricht der oben gemachten Voraussetzung einer endlichen Entropievermehrung; denn eine solche läßt sich, wenn einmal eingetreten, auf keinerlei Weise vollständig rückgängig machen. Es kann also durch die Einbringung des Kohlestäubchens in den Strahlungsraum keine endliche Entropievermehrung herbeigeführt worden sein, sondern die Strahlung befand sich schon vorher und jederzeit im stabilen Gleichgewichtszustand.

§ 72. Damit das Wesentliche dieses wichtigen Beweises noch klarer hervortritt, sei auf eine analoge einigermaßen naheliegende Betrachtung hingewiesen. Ein Hohlraum, in dem sich anfänglich ein Dampf gerade im Zustand der Sättigung befindet, werde adiabatisch und unendlich langsam komprimiert.

„Dann verbleibt der Dampf bei beliebiger endlicher adiabatischer Kompression immer gerade im Zustand der Sättigung. Denn würde er z. B. bei der Kompression übersättigt werden, so könnte man, nachdem die Kompression auf einen merklichen Bruchteil des ursprünglichen Volumens stattgefunden hat, durch Einbringung eines winzigen Flüssigkeitströpfchens, dem keine merkliche Masse und Wärmekapazität zukommt, bei konstantem Volumen und konstanter Gesamtenergie die Kondensation einer endlichen Menge Dampf, und damit den endlichen Übergang in einen stabileren Zustand, also eine endliche Entropievermehrung des Systems herbeiführen. Nachdem dies geschehen, könnte man das Volumen wieder adiabatisch und unendlich langsam vergrößern, bis alle Flüssigkeit verdampft ist, und dadurch den Prozeß vollständig rückgängig machen, was der angenommenen Entropievermehrung widerspricht.“

Ein solches Beweisverfahren würde deshalb fehlerhaft sein, weil durch den beschriebenen Prozeß die eingetretene Veränderung keineswegs vollständig rückgängig gemacht ist. Denn da die bei der Kompression des übersättigten Dampfes aufgewendete mechanische Arbeit nicht gleich ist der bei der Ausdehnung des gesättigten Dampfes wiedergewonnenen, so entspricht einem bestimmten Volumen des Systems bei der Kompression eine andere Energie als bei der Ausdehnung, und deshalb kann auch das Volumen, bei dem alle Flüssigkeit gerade wieder verdampft ist, nicht gleich dem ursprünglichen Anfangsvolumen sein. Die gemutmaßte Analogie ist also hinfällig, und die oben in Anführungszeichen gesetzte Behauptung unrichtig.

§ 73. Wir wollen uns nun wieder den in § 68 beschriebenen reversibeln adiabatischen Prozeß mit der in dem evakuierten Hohlzylinder mit weißen Wänden und weißem Boden befindlichen schwarzen Strahlung ausgeführt denken, indem wir den aus absolut spiegelndem Metall bestehenden Kolben unendlich langsam herabsinken lassen, nur mit dem Unterschied, daß sich diesmal kein Kohlestäubchen in dem Zylinder befindet. Dann verläuft der Prozeß, wie wir jetzt wissen, genau so wie dort. Wir können uns aber nun, da jetzt überhaupt keine Emission und Absorption der Strahlung stattfindet, Rechenschaft geben von den Änderungen, welche die einzelnen Strahlenbündel des Systems an Farbe und Intensität erleiden. Solche Änderungen treten natürlich nur bei der Reflexion an dem bewegten Metallspiegel, nicht bei der Reflexion an den ruhenden Wänden und an dem ruhenden Boden des Zylinders ein.

Wenn der spiegelnde Kolben mit der konstanten unendlich kleinen Geschwindigkeit v sich herabsenkt, so werden die ihn währenddem treffenden monochromatischen Strahlenbündel bei der Reflexion eine Änderung ihrer Farbe, ihrer Intensität und ihrer Richtung erleiden. Betrachten wir diese verschiedenen Einflüsse der Reihe nach hintereinander.[1]

§ 74. Zunächst fragen wir nach der Farbenänderung, die ein monochromatischer Strahl durch Reflexion an dem unendlich langsam bewegten Spiegel erleidet, und betrachten zu diesem Zweck erst den Fall eines in normaler Richtung, von unten nach oben, auf den Spiegel fallenden und daher auch in normaler Richtung, von oben nach unten, reflektierten Strahles. Die Ebene A (Fig. 5) bedeute die Lage des Spiegels zur Zeit t, die Ebene A' die zur Zeit $t + \delta t$, wobei die Entfernung $AA' = v \cdot \delta t$, wenn v die Geschwindigkeit des Spiegels bedeutet. Denken wir uns nun durch das durchstrahlte Vakuum in gehöriger Entfernung

---

[1] Eine vollständige Lösung des Problems der Reflexion eines Strahlenbündels an einer bewegten absolut spiegelnden Fläche, auch bei beliebig großer Geschwindigkeit derselben, findet sich in der § 71 zitierten Abhandlung von M. Abraham. Hierbei sind die Gesetze der Elektrodynamik bewegter Körper zugrunde gelegt. Vgl. auch desselben Autors Lehrbuch: Elektromagnetische Theorie der Strahlung, 1905 (Leipzig, B. G. Teubner).

vom Spiegel eine dem Spiegel parallele ruhende Ebene B gelegt, und nennen wir λ die Wellenlänge des auf den Spiegel treffenden, λ' die Wellenlänge des vom Spiegel reflektierten Strahles, so liegen zur Zeit t auf der Strecke AB des durchstrahlten Vakuums $\frac{AB}{\lambda}$ Wellen des einfallenden, und $\frac{AB}{\lambda'}$ Wellen des reflektierten Strahles, was man sich etwa dadurch versinnlichen kann, daß man die elektrische Feldstärke in den verschiedenen Punkten je eines der beiden Strahlen zur Zeit t sich in Form einer Sinuskurve aufgezeichnet denkt. Im ganzen liegen also zur Zeit t in dem Zwischenraum zwischen A und B

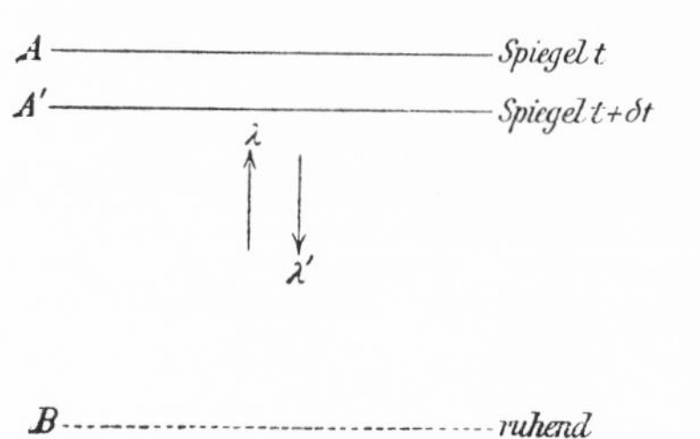

Fig. 5.

$$AB \cdot \left(\frac{1}{\lambda} + \frac{1}{\lambda'}\right)$$

Wellen, einfallender und reflektierter Strahl zusammengenommen. Da diese Anzahl sehr groß ist, so kommt es nicht darauf an, ob es eine ganze Zahl ist oder nicht.

Ebenso liegen zur Zeit $t + \delta t$, wenn sich der Spiegel in A' befindet, in dem Zwischenraum zwischen A' und B im ganzen

$$A'B \cdot \left(\frac{1}{\lambda} + \frac{1}{\lambda'}\right) \text{ Wellen.}$$

Die letzte Zahl wird kleiner sein als die erste, da in dem kleineren Zwischenraum $A'B$ eine geringere Anzahl Wellen von beiden Arten Platz finden, als vorher in dem größeren Zwischenraum AB. Der Rest der Wellen muß während der Zeit δt aus dem Zwischenraum zwischen dem bewegten Spiegel und der ruhenden Ebene B hinausgedrängt worden sein, und zwar durch die Ebene B hindurch nach unten; denn auf andere Weise kann keine Welle aus dem betrachteten Raum verschwinden.

Nun gehen durch die ruhende Ebene B in der Zeit δt in der Richtung nach oben: $\nu \cdot \delta t$ Wellen, in der Richtung nach unten: $\nu' \cdot \delta t$ Wellen; folglich ist die Differenz:

$$(\nu' - \nu)\,\delta t = (AB - A'B) \cdot \left(\frac{1}{\lambda} + \frac{1}{\lambda'}\right)$$

oder, da:

$$AB - A'B = v \cdot \delta t$$

und
$$\lambda = \frac{c}{\nu}, \quad \lambda' = \frac{c}{\nu'}$$

$$\nu' = \frac{c+v}{c-v} \cdot \nu,$$

oder, da v unendlich klein gegen c,

$$\nu' = \nu\left(1 + \frac{2v}{c}\right).$$

§ 75. Wenn die Strahlung nicht in normaler Richtung, sondern unter dem spitzen Einfallswinkel ϑ auf den Spiegel fällt, so kann man eine ganz ähnliche Betrachtung anstellen,

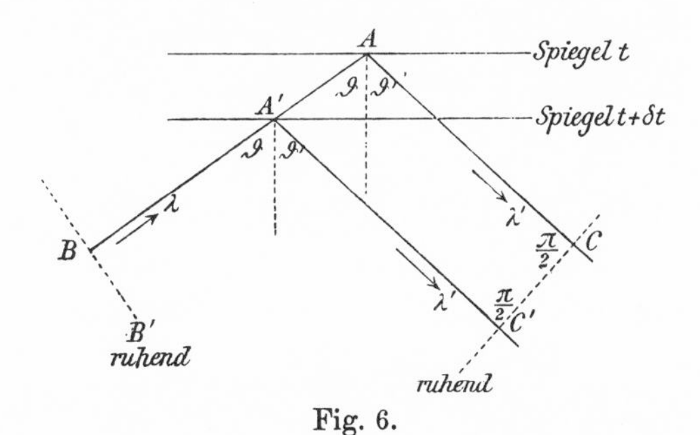

Fig. 6.

nur mit dem Unterschied, daß dann der Schnittpunkt A eines bestimmten ins Auge gefaßten Strahles BA mit dem Spiegel zur Zeit t eine andere Lage auf dem Spiegel hat als der Schnittpunkt A' desselben Strahles mit dem Spiegel zur Zeit $t + \delta t$ (Fig. 6). Die Anzahl der Wellen, welche zur Zeit t auf der Strecke BA liegen, ist $\frac{BA}{\lambda}$. Ebenso ist zu der nämlichen Zeit t die Anzahl der Wellen auf der Strecke AC, welche die Entfernung des Punktes A von einer im Vakuum ruhenden Wellenebene CC' des reflektierten Strahles darstellt, $\frac{AC}{\lambda'}$. Im ganzen liegen also zur Zeit t auf der Strecke BAC des betrachteten Strahles

$$\frac{BA}{\lambda} + \frac{AC}{\lambda'}$$

Wellen. Hierbei sei noch bemerkt, daß der Reflexionswinkel ϑ'

nicht genau gleich dem Einfallswinkel ist, sondern, wie sich durch eine einfache geometrische Überlegung zeigen läßt, etwas kleiner. Die Differenz zwischen ϑ und ϑ' wird sich aber für unsere Berechnung als unwesentlich erweisen.

Ferner liegen zur Zeit $t + \delta t$, wenn der Spiegel durch A' geht, auf der Strecke $BA'C'$

$$\frac{BA'}{\lambda} + \frac{A'C'}{\lambda'}$$

Wellen. Die letzte Zahl ist kleiner als die erste, und zwar muß die Differenz gleich der Anzahl der Wellen sein, welche während der Zeit δt aus dem Raume, der durch die ruhende Ebene BB' und durch die ruhende Ebene CC' begrenzt wird, im ganzen hinausgedrängt worden sind.

Nun gehen durch die Ebene BB' in der Zeit δt in den Raum hinein $\nu \cdot \delta t$ Wellen, durch die Ebene CC' aus dem Raum hinaus $\nu' \cdot \delta t$ Wellen. Folglich ist

$$(\nu' - \nu) \cdot \delta t = \left(\frac{BA}{\lambda} + \frac{AC}{\lambda'}\right) - \left(\frac{BA'}{\lambda} + \frac{A'C'}{\lambda'}\right).$$

Es ist aber:

$$BA - BA' = AA' = \frac{v \cdot \delta t}{\cos\vartheta}$$

$$AC - A'C' = AA' \cdot \cos(\vartheta + \vartheta')$$

$$\lambda = \frac{c}{\nu}, \qquad \lambda' = \frac{c}{\nu'}.$$

Folglich: $$\nu' = \frac{c\cos\vartheta + v}{c\cos\vartheta - v\cos(\vartheta + \vartheta')} \cdot \nu.$$

Diese Beziehung gilt für eine beliebig große Geschwindigkeit v des bewegten Spiegels. Da nun in unserem Fall v unendlich klein ist gegen c, so wird einfacher:

$$\nu' = \nu \cdot \left(1 + \frac{v}{c\cos\vartheta} \cdot [1 + \cos(\vartheta + \vartheta')]\right).$$

Die Differenz der Winkel ϑ und ϑ' ist jedenfalls von der Größenordnung $\frac{v}{c}$; daher kann man hier ohne merklichen Fehler ϑ' durch ϑ ersetzen, und erhält so:

$$\nu' = \nu \cdot \left(1 + \frac{2v\cos\vartheta}{c}\right) \tag{83}$$

als Schwingungszahl des reflektierten Strahles für schiefe Inzidenz.

§ 76. Aus dem Vorstehenden erhellt, daß die Schwingungszahlen aller auf den bewegten Spiegel treffenden Strahlen durch die Reflexion vergrößert werden, wenn sich der Spiegel gegen die Strahlung bewegt, dagegen verkleinert werden, wenn der Spiegel sich in Richtung der auffallenden Strahlung bewegt ($v < o$). Dabei wird aber die gesamte auf den bewegten Spiegel fallende Strahlung einer bestimmten Schwingungszahl ν keineswegs wieder als monochromatische Strahlung reflektiert, sondern die Farbenänderung bei der Reflexion hängt wesentlich mit von dem Einfallswinkel ϑ ab. Daher kann man von einer bestimmten spektralen „Verschiebung" der Farbe nur bei einem einzelnen bestimmt gerichteten Strahlenbündel, dagegen bei der gesamten monochromatischen Strahlung höchstens von einer spektralen „Zersplitterung" reden. Die Farbenänderung ist am größten für normale Inzidenz, sie verschwindet ganz für streifende Inzidenz.

§ 77. Berechnen wir zweitens die Energieänderung, welche der bewegte Spiegel der auftreffenden Strahlung erteilt, und zwar gleich für den allgemeinen Fall der schiefen Inzidenz. Ein monochromatisches unendlich dünnes unpolarisiertes Strahlenbündel, welches unter dem Einfallswinkel ϑ auf ein Flächenelement des Spiegels trifft, möge in der Zeit δt die Energie $I \cdot \delta t$ auf den Spiegel fallen lassen. Dann beträgt die mechanische Druckkraft des Strahlenbündels normal auf den Spiegel nach Gleichung (64) bis auf verschwindend kleine Größen:

$$\mathfrak{F} = \frac{2 \cos \vartheta}{c} \cdot I$$

und die bei der Bewegung des Spiegels in der Zeit δt von außen gegen die auftreffende Strahlung geleistete Arbeit ist mit demselben Grade der Annäherung:

$$\mathfrak{F} \, v \, \delta t = \frac{2 \, v \cos \vartheta}{c} \cdot I \, \delta t. \tag{84}$$

Nach dem Prinzip der Erhaltung der Energie muß sich dieser Arbeitsbetrag in der Energie der reflektierten Strahlung wiederfinden. Daher besitzt das reflektierte Strahlenbündel eine größere Intensität als das auffallende, es liefert nämlich in der Zeit δt die Energie:[1]

[1] Es versteht sich, daß die durch die Bewegung des Spiegels verursachte Intensitätsänderung bei der Reflexion sich auch rein elektro-

$$(85) \qquad I\,\delta t + \mathfrak{F}\, v\, \delta t = I\left(1 + \frac{2\,v \cos\vartheta}{c}\right)\delta t = I'\,\delta t\,.$$

Man kann also zusammenfassend sagen: Durch die Reflexion eines unter dem Einfallswinkel ϑ auftreffenden monochromatischen unpolarisierten Elementarstrahlenbündels an dem gegen die Strahlung mit der unendlich kleinen Geschwindigkeit v bewegten Spiegel wird während der Zeit δt die Strahlungsenergie $I\,\delta t$, deren Schwingungszahlen von ν bis $\nu + d\,\nu$ reichen, in die Strahlungsenergie $I'\,\delta t$ mit dem Schwingungsintervall $(\nu', \nu' + d\,\nu')$ verwandelt, wobei I' durch (85), ν' durch (83) und dementsprechend $d\,\nu'$, die Spektralbreite des reflektierten Bündels, durch:

$$(86) \qquad d\nu' = d\nu\left(1 + \frac{2\,v \cos\vartheta}{c}\right)$$

gegeben ist. Ein Vergleich dieser Werte zeigt, daß

$$(87) \qquad \frac{I'}{I} = \frac{\nu'}{\nu} = \frac{d\,\nu'}{d\,\nu}\,.$$

Der absolute Betrag der bei dieser Verwandlung verschwundenen Strahlungsenergie ist nach Gleichung (13):

$$(88) \qquad I \cdot \delta t = 2\,\mathfrak{K}_\nu\, d\sigma \cos\vartheta\; d\,\Omega\; d\nu\; \delta t$$

und daher der absolute Betrag der dabei neu entstandenen Strahlungsenergie nach (85):

$$(89) \qquad I' \cdot \delta t = 2\,\mathfrak{K}_\nu\, d\sigma \cos\vartheta\; d\,\Omega\; d\nu\left(1 + \frac{2\,v \cos\vartheta}{c}\right)\delta t\,.$$

In diesen beiden Ausdrücken wäre streng genommen noch eine unendlich kleine Korrektur anzubringen, da I die Energiestrahlung auf ein ruhendes Flächenelement $d\,\sigma$ darstellt, während doch durch die Bewegung von $d\,\sigma$ gegen das auftreffende Strahlenbündel die auffallende Strahlung etwas vermehrt wird. Indessen können die entsprechenden Zusatzglieder hier ohne merklichen Fehler fortgelassen werden, weil die Differenz der beiden Ausdrücke: $(I' - I)\,\delta t$, welche durch (84) dargestellt wird, von jener Korrektur offenbar nicht merklich betroffen wird.

---

dynamisch ableiten läßt, da ja die Elektrodynamik mit dem Energieprinzip in Übereinstimmung ist. Dieser Weg ist etwas umständlicher, dafür gewährt er aber einen tieferen Einblick in die Einzelheiten des Vorgangs der Reflexion.

§ 78. Was endlich die Richtungsänderungen betrifft, welche den auftreffenden Strahlen durch die Reflexion an dem bewegten Spiegel erteilt werden, so brauchen wir uns der Berechnung derselben hier gar nicht zu unterziehen. Denn wenn die Bewegung des Spiegels nur hinreichend langsam erfolgt, so werden alle Ungleichmäßigkeiten in der Richtung der Strahlung sogleich wieder durch die weitere Reflexion an den Gefäßwänden ausgeglichen werden. Wir können uns ja den ganzen Prozeß in sehr vielen kleinen Intervallen ausgeführt denken, in der Weise, daß der Kolben, nachdem er eine sehr kleine Wegstrecke mit sehr kleiner Geschwindigkeit zurückgelegt hat, eine Zeitlang in Ruhe gehalten wird, und zwar so lange, bis alle etwa entstandenen Ungleichmäßigkeiten in den Strahlungsrichtungen durch die Reflexion an den weißen Wänden des Hohlzylinders wieder zum Verschwinden gebracht sind. Wenn man dies Verfahren genügend lange fortsetzt, so kann man die Kompression der Strahlung bis zu einem beliebig kleinen Bruchteil des ursprünglichen Volumens fortsetzen, und dabei stets die Strahlung als nach allen Richtungen gleichmäßig betrachten. Dieser stetig wirkende Ausgleichungsprozeß betrifft natürlich nur die Verschiedenheit der Strahlungsrichtungen; denn Farben- und Intensitätsänderungen der Strahlung, die einmal eingetreten sind, wenn auch noch so minimaler Größe, können offenbar durch Reflexion an total reflektierenden ruhenden Wänden mit der Zeit niemals ausgeglichen werden, sondern bleiben konstant weiter bestehen.

§ 79. Mit Hilfe der gewonnenen Sätze sind wir nun imstande, für den Fall der unendlich langsamen adiabatischen Kompression des von gleichmäßiger Strahlung erfüllten vollkommen evakuierten Hohlzylinders die Änderung der Strahlungsdichte für jede einzelne Schwingungszahl zu berechnen. Wir fassen zu diesem Zwecke die Strahlung innerhalb eines bestimmten unendlich kleinen Intervalls von Schwingungszahlen, nämlich von ν bis $\nu + d\nu$, zur Zeit t ins Auge, und fragen nach der Änderung, welche die gesamte Strahlungsenergie, die in dieses bestimmte unveränderliche Intervall fällt, während der Zeit δt erleidet.

Zur Zeit t ist diese Strahlungsenergie nach (22) $V\mathfrak{u} \cdot d\nu$, zur Zeit $t + \delta t$ ist sie $\left(V\mathfrak{u} + \delta(V\mathfrak{u})\right) \cdot d\nu$, also die zu berechnende Änderung:

(90) $$\delta(V\mathfrak{u})\cdot d\nu.$$

Die monochromatische Strahlungsdichte $\mathfrak{u}$ ist hierbei als Funktion der beiden voneinander unabhängigen Variabeln ν und t zu betrachten, deren Differentiale durch die Zeichen d und δ unterschieden sind.

Die Änderung der monochromatischen Strahlungsenergie kommt lediglich bei der Reflexion an dem bewegten Spiegel, und zwar dadurch zustande, daß erstens gewisse Strahlen, welche zur Zeit t dem Intervall $(\nu, d\nu)$ angehören, durch die bei der Reflexion erlittene Farbenänderung aus diesem Intervall austreten, und daß zweitens gewisse Strahlen, welche zur Zeit t nicht dem Intervall $(\nu, d\nu)$ angehören, durch die bei der Reflexion erlittene Farbenänderung in dieses Intervall eintreten. Wir berechnen beide Einflüsse nacheinander. Die Berechnung wird wesentlich vereinfacht, wenn wir die Breite dieses Intervalls: $d\nu$ so klein nehmen, daß:

(91) $$d\nu \text{ klein gegen } \frac{v}{c}\cdot\nu,$$

was deshalb möglich ist, weil $d\nu$ und v gar nicht voneinander abhängen.

§ 80. Die Strahlen, welche zur Zeit t dem Intervall $(\nu, d\nu)$ angehören und in der Zeit δt infolge der Reflexion am bewegten Spiegel aus diesem Intervall austreten, sind einfach alle diejenigen, welche während der Zeit δt den bewegten Spiegel treffen. Denn die Farbenänderung, die ein solcher Strahl erleidet, ist nach (83) und (91) groß gegen die Breite $d\nu$ des ganzen Intervalls. Wir haben hier also nur die Energie zu berechnen, welche während der Zeit δt durch die Strahlen des Intervalls $(\nu, d\nu)$ auf den Spiegel geworfen wird.

Für ein Elementarstrahlenbündel, das unter dem Einfallswinkel ϑ auf das Element $d\sigma$ der Spiegelfläche fällt, ist diese Energie nach (88) und (5):

$$I\,\delta t = 2\mathfrak{K}_\nu\, d\sigma \cos\vartheta\; d\Omega\; d\nu\; \delta t = 2\mathfrak{K}_\nu\, d\sigma \sin\vartheta \cos\vartheta\; d\vartheta\; d\varphi\; d\nu\; \delta t,$$

also für die gesamte monochromatische Strahlung, die auf die ganze Spiegelfläche F auffällt, durch Integration über φ von 0 bis 2π, über ϑ von 0 bis $\frac{\pi}{2}$, und über $d\sigma$ von 0 bis F:

(92) $$2\pi\, F\, \mathfrak{K}_\nu\, d\nu\, \delta t.$$

Diese Strahlungsenergie tritt also während der Zeit δt aus dem betrachteten Schwingungszahlintervall $(\nu, d\nu)$ heraus.

§ 81. Bei der Berechnung derjenigen Strahlungsenergie, welche während der Zeit δt durch die Reflexion an dem bewegten Spiegel in das Intervall $(\nu, d\nu)$ eintritt, müssen wir die unter verschiedenen Einfallswinkeln auf den Spiegel treffenden Strahlen gesondert betrachten. Da bei positivem v die Schwingungszahl durch die Reflexion vergrößert wird, so besitzen die hier zu betrachtenden Strahlen zur Zeit t eine Schwingungszahl $\nu_1 < \nu$. Nehmen wir nun zur Zeit t ein monochromatisches Strahlenbündel vom Schwingungsintervall $(\nu_1, d\nu_1)$, welches unter dem Einfallswinkel ϑ auf den Spiegel trifft, so wird es durch die Reflexion immer und nur dann in das Intervall $(\nu, d\nu)$ eintreten, wenn

$$\nu = \nu_1\left(1 + \frac{2\,v\cos\vartheta}{c}\right) \quad \text{und} \quad d\nu = d\nu_1\left(1 + \frac{2\,v\cos\vartheta}{c}\right).$$

Diese Beziehungen ergeben sich, wenn man in die Gleichungen (83) und (86) an die Stelle von ν und ν', der Schwingungszahlen vor der Reflexion und nach der Reflexion, beziehungsweise ν_1 und ν setzt.

Die Energie, welche dieses Strahlenbündel während der Zeit δt in das Intervall $(\nu, d\nu)$ hineinbringt, ergibt sich aus (89), wenn man darin ebenfalls ν_1 an die Stelle von ν setzt, zu:

$$2\,\mathfrak{K}_{\nu_1}\,d\sigma\cos\vartheta\,d\Omega\,d\nu_1\left(1 + \frac{2\,v\cos\vartheta}{c}\right)\delta t = 2\,\mathfrak{K}_{\nu_1}\,d\sigma\cos\vartheta\,d\Omega\,d\nu\,\delta t.$$

Nun ist:

$$\mathfrak{K}_{\nu_1} = \mathfrak{K}_\nu + (\nu_1 - \nu)\cdot\frac{\partial\mathfrak{K}}{\partial\nu} + \ldots,$$

wobei wir voraussetzen, daß $\frac{\partial\mathfrak{K}}{\partial\nu}$ endlich ist. Also bis auf unendlich kleine Größen höherer Ordnung:

$$\mathfrak{K}_{\nu_1} = \mathfrak{K}_\nu - \frac{2\,\nu\,v\cos\vartheta}{c}\,\frac{\partial\mathfrak{K}}{\partial\nu}.$$

Dadurch wird die gesuchte Energie:

$$2\,d\sigma\left(\mathfrak{K}_\nu - \frac{2\,\nu\,v\cos\vartheta}{c}\,\frac{\partial\mathfrak{K}}{\partial\nu}\right)\sin\vartheta\cos\vartheta\,d\vartheta\,d\varphi\,d\nu\,\delta t,$$

und durch Integration dieses Ausdrucks über $d\sigma$, φ und ϑ, wie oben, ergibt sich die gesamte Strahlungsenergie, welche während der Zeit δt in das Schwingungszahlintervall (ν, $d\nu$) neu eintritt:

$$2\pi F\left(\mathfrak{K}_\nu - \frac{4}{3}\frac{\nu v}{c}\frac{\partial \mathfrak{K}}{\partial \nu}\right) d\nu\, \delta t. \tag{93}$$

§ 82. Die Differenz der Ausdrücke (93) und (92) ergibt die gesamte Änderung (90), also:

$$-\frac{8\pi}{3} F \frac{\nu v}{c} \frac{\partial \mathfrak{K}}{\partial \nu} \delta t = \delta(V\mathfrak{u})$$

oder nach (24):

$$-\frac{1}{3} F \nu v \frac{\partial \mathfrak{u}}{\partial \nu} \delta t = \delta(V\mathfrak{u})$$

oder endlich, da $F v \delta t$ gleich ist der Abnahme des Volumens V:

$$\frac{1}{3}\nu \frac{\partial \mathfrak{u}}{\partial \nu} \delta V = \delta(V\mathfrak{u}) = \mathfrak{u}\,\delta V + V\delta\mathfrak{u}, \tag{94}$$

woraus folgt:

$$\delta\mathfrak{u} = \left(\frac{\nu}{3}\frac{\partial \mathfrak{u}}{\partial \nu} - \mathfrak{u}\right) \cdot \frac{\delta V}{V}. \tag{95}$$

Diese Gleichung ergibt die bei einer unendlich langsamen adiabatischen Kompression der Strahlung eintretende Änderung der räumlichen Energiedichte irgend einer bestimmten Schwingungszahl ν. Sie gilt übrigens, wie die Art ihrer Ableitung zeigt, nicht allein für schwarze Strahlung, sondern für eine Strahlung von anfänglich ganz beliebiger Energieverteilung.

Da die während der Zeit δt im Strahlungszustand eintretenden Änderungen der unendlich kleinen Geschwindigkeit v proportional sind und mit deren Vorzeichen sich umkehren, so gilt die Gleichung für jedes Vorzeichen von δV, der Vorgang ist also reversibel.

§ 83. Bevor wir zur allgemeinen Integration der Gleichung (95) schreiten, wollen wir sie einer naheliegenden Prüfung unterziehen. Nach dem Energieprinzip muß nämlich die bei der adiabatischen Kompression eintretende Änderung der gesamten Strahlungsenergie:

$$U = V \cdot u = V \cdot \int_0^\infty \mathfrak{u}\, d\nu$$

gleich sein der bei der Kompression von außen gegen den Strahlungsdruck geleisteten Arbeit:

$$-p\,\delta V = -\frac{u}{3}\delta V = -\frac{\delta V}{3}\int_0^\infty \mathfrak{u}\, d\nu. \tag{96}$$

Nun ergibt sich mit Benutzung von (94) für die Änderung der Gesamtenergie:

$$\delta U = \int_0^\infty d\nu \cdot \delta(V\mathfrak{u}) = \frac{\delta V}{3} \cdot \int_0^\infty \nu \frac{\partial \mathfrak{u}}{\partial \nu} d\nu$$

oder durch partielle Integration:

$$\delta U = \frac{\delta V}{3} \cdot \left([\nu \mathfrak{u}]_0^\infty - \int_0^\infty \mathfrak{u}\, d\nu \right)$$

und dieser Ausdruck ist in der Tat mit (96) identisch, da das Produkt $\nu\mathfrak{u}$ sowohl für $\nu = 0$ als auch für $\nu = \infty$ verschwindet. Letzteres könnte einen Augenblick zweifelhaft erscheinen; man ersieht aber leicht, daß, wenn $\nu\mathfrak{u}$ für $\nu = \infty$ einen von Null verschiedenen Wert annähme, dann das Integral von $\mathfrak{u}$ nach ν, von 0 bis ∞ genommen, keinen endlichen Wert besitzen könnte, was doch sicher der Fall ist.

§ 84. Wir haben schon oben in § 79 hervorgehoben, daß $\mathfrak{u}$ als Funktion zweier unabhängiger Variabeln anzusehen ist, von denen wir als erste die Schwingungszahl ν, als zweite die Zeit t genommen haben. Da nun die Zeit t in der Gleichung (95) explicite gar nicht vorkommt, so ist es sachgemäßer, als zweite unabhängige Variable statt t direkt das Volumen V einzuführen, welches ja von t allein abhängig ist. Dann schreibt sich die Gleichung (95) folgendermaßen als partielle Differentialgleichung:

$$V \frac{\partial \mathfrak{u}}{\partial V} = \frac{V}{3} \frac{\partial \mathfrak{u}}{\partial \nu} - \mathfrak{u}, \tag{97}$$

aus welcher $\mathfrak{u}$, wenn es für ein bestimmtes V als Funktion von ν bekannt ist, für alle anderen Werte von V als Funktion von ν berechnet werden kann. Das allgemeine Integral dieser Differentialgleichung lautet, wie man sich leicht durch nachträgliche Substitution überzeugen kann:

$$\mathfrak{u} = \frac{1}{V} \varphi(\nu^3 V), \tag{98}$$

wobei φ eine beliebige Funktion eines einzigen Arguments $\nu^3 V$ bedeutet. Statt dessen kann man auch schreiben, indem man $\nu^3 V \cdot \varphi(\nu^3 V)$ statt $\varphi(\nu^3 V)$ einsetzt:

$$\mathfrak{u} = \nu^3 \varphi(\nu^3 V). \tag{99}$$

Jede der beiden letzten Gleichungen ist der allgemeine Ausdruck des WIENschen Verschiebungsgesetzes.

Wenn also für ein bestimmtes gegebenes Volumen V die spektrale Energieverteilung, d. h. $\mathfrak{u}$ als Funktion von ν, bekannt ist, so läßt sich daraus die Abhängigkeit der Funktion φ von ihrem Argument ableiten, und dadurch ergibt sich dann unmittelbar die Energieverteilung für jedes beliebige andere Volumen V, in welches die den Hohlzylinder erfüllende Strahlung durch einen reversibeln adiabatischen Prozeß gebracht wird.

§ 85. Nun führen wir, zu dem Gedankengang des § 73 zurückkehrend, die Voraussetzung ein, daß am Anfang die spektrale Energieverteilung die normale, der schwarzen Strahlung entsprechende ist. Dann behält nach dem damals bewiesenen Satze die Strahlung bei der reversibeln adiabatischen Volumenänderung diese Eigenschaft unverändert bei, und es gelten für den Prozeß alle in § 68 abgeleiteten Gesetzmäßigkeiten. Die Strahlung besitzt also dann in jedem Zustand eine bestimmte Temperatur T, welche mit dem Volumen V durch die dort abgeleitete Gleichung:

$$T^3 \cdot V = \text{const} \tag{100}$$

zusammenhängt. Daher kann man nun die Gleichung (99) auch so schreiben:

$$\mathfrak{u} = \nu^3 \varphi\left(\frac{\nu^3}{T^3}\right)$$

oder auch:

$$\mathfrak{u} = \nu^3 \varphi\left(\frac{T}{\nu}\right).$$

Ist also für eine einzige Temperatur die spektrale Energieverteilung der schwarzen Strahlung, d. h. $\mathfrak{u}$ als Funktion von ν, bekannt, so ergibt sich daraus die Abhängigkeit der Funktion φ von ihrem Argument, und dadurch die spektrale Energieverteilung für jede andere Temperatur.

Nimmt man noch den in § 47 bewiesenen Satz hinzu, daß bei der schwarzen Strahlung einer bestimmten Temperatur das Produkt $\mathfrak{u} q^3$ für alle Medien den nämlichen Wert hat, so kann man auch schreiben:

$$\mathfrak{u} = \frac{\nu^3}{q^3} F\left(\frac{T}{\nu}\right), \tag{101}$$

wo nun die Funktion F die Fortpflanzungsgeschwindigkeit nicht mehr enthält,

§ 86. Für die gesamte räumliche Strahlungsdichte der schwarzen Strahlung im Vakuum ergibt sich:

$$u = \int_0^\infty \mathfrak{u}\, d\nu = \frac{1}{c^3}\int_0^\infty \nu^3 F\left(\frac{T}{\nu}\right) d\nu$$

oder, wenn man statt ν die Größe $\frac{T}{\nu} = x$ als Integrationsvariable einführt:

$$u = \frac{T^4}{c^3}\int_0^\infty \frac{F(x)}{x^5}\, dx\,.$$

Setzt man die absolute Konstante:

$$\frac{1}{c^3}\int_0^\infty \frac{F(x)}{x^5}\, dx = a\,,$$

so kehrt man damit zu der in Gleichung (75) ausgesprochenen Form des STEFAN-BOLTZMANNschen Strahlungsgesetzes zurück.

§ 87. Eine anschauliche Fassung gewinnt die Gleichung (101), wenn man sie in folgender Form schreibt:

$$\frac{\mathfrak{u}\,c^3}{\nu^3}\, d\nu = F\left(\frac{T}{\nu}\right) d\nu = \eta \qquad (102)$$

und bedenkt, daß der Ausdruck links die in dem Kubus einer Wellenlänge λ enthaltene Strahlungsenergie zwischen den Schwingungszahlen ν und $\nu + d\nu$ darstellt, die wir vorübergehend mit η bezeichnen wollen. Wenn nun durch reversible adiabatische Kompression die Strahlung auf die höhere Temperatur T' übergeht, so ist für ein beliebiges anderes, von ν' bis $\nu' + d\nu'$ reichendes Spektralgebiet:

$$\eta' = F\left(\frac{T'}{\nu'}\right) d\nu'\,. \qquad (103)$$

Setzen wir nun $$\nu' = \frac{T'}{T}\nu \qquad (104)$$

und entsprechend: $$d\nu' = \frac{T'}{T}\nu\,,$$

d. h. fassen wir im zweiten Zustand solche Schwingungszahlen ν' ins Auge, die sich im Verhältnis der Temperaturen von den ursprünglichen Schwingungszahlen ν unterscheiden, so ergibt sich durch Division von (103) und (102):

$$\frac{\eta'}{\eta} = \frac{d\nu'}{d\nu} = \frac{T'}{T}\,. \qquad (105$$

6*

Ferner durch Berücksichtigung der Bedeutung von η:

$$\mathfrak{u}' : \mathfrak{u} = T'^3 : T^3$$

und:
$$\mathfrak{u}' d\nu' : \mathfrak{u}\, d\nu = T'^4 : T^4 .$$

Man kann also folgende Sätze aussprechen. Das Spektrum der schwarzen Strahlung verändert sich beim Übergang von der Temperatur T zu einer höheren Temperatur T' derartig, als ob alle Schwingungszahlen ν sich im Verhältnis der Temperaturen T' und T vergrößern, und dabei die im Kubus einer Wellenlänge λ enthaltene Energie η der Strahlung eines unendlich kleinen Spektralbezirks sich ebenfalls in demselben Verhältnis vergrößert. Die monochromatische Strahlungsdichte $\mathfrak{u}$ vergrößert sich dann im Verhältnis der dritten Potenz, und die Strahlungsdichte $\mathfrak{u}\, d\nu$ eines unendlich kleinen Spektralbezirks im Verhältnis der vierten Potenz der Temperaturen. Für die gesamte räumliche Strahlungsdichte u, als der Summe der Strahlungsdichten aller Spektralbezirke, ergibt sich dann wieder das STEFAN-BOLTZMANNsche Gesetz.

Da ferner nach (104) und (100):

$$\frac{\nu^3 V}{q^3} = \frac{\nu'^3 V'}{q^3},$$

so wird bei dieser Veränderung die im ganzen Volumen V der Strahlung enthaltene Anzahl von Wellenkuben einer jeden Schwingungszahl durch die Kompression nicht geändert.

Diese Sätze haben natürlich nur zusammenfassende Bedeutung; denn wie wir oben gesehen haben, verändern sich in Wirklichkeit bei der adiabatischen Kompression die Schwingungszahlen ν der einzelnen Strahlen durchaus nicht alle, sondern nur insofern sie während der Kompression an dem bewegten Kolben reflektiert werden, und dann auch nicht gleichmäßig, sondern je nach der Größe des Einfallswinkels in verschiedener Weise.

§ 88. Wie für die räumliche Strahlungsdichte $\mathfrak{u}$, so läßt sich das WIENsche Verschiebungsgesetz auch für die spezifische Strahlungsintensität $\mathfrak{K}_\nu$ eines geradlinig polarisierten monochromatischen Strahles bei der schwarzen Strahlung aussprechen, und lautet nach (24) in dieser Form:

$$\mathfrak{K}_\nu = \frac{\nu^3}{q^2} F\left(\frac{T}{\nu}\right). \tag{106}$$

Bezieht man die Strahlungsintensität, wie es in der Experimentalphysik meistens geschieht, statt auf Schwingungszahlen ν, auf die Wellenlänge λ, setzt also nach (16):

$$E_\lambda = \frac{\mathfrak{c} \cdot \mathfrak{K}_\nu}{\lambda^2},$$

so nimmt die letzte Gleichung die Form an:

$$E_\lambda = \frac{\mathfrak{c}^2}{\lambda^5} \cdot F\left(\frac{\lambda T}{\mathfrak{c}}\right). \tag{107}$$

Diese Form des WIENschen Verschiebungsgesetzes hat meistens den Ausgangspunkt zur experimentellen Prüfung gebildet, welche in allen Fällen zu einer merklichen Bestätigung des Gesetzes geführt hat.[1]

§ 89. Da E_λ sowohl für $\lambda = 0$ als auch für $\lambda = \infty$ verschwindet, so besitzt E_λ in bezug auf λ ein Maximum, welches sich aus der Gleichung ergibt:

$$\frac{d E_\lambda}{d \lambda} = 0 = -\frac{5}{\lambda^6} F\left(\frac{\lambda T}{\mathfrak{c}}\right) + \frac{1}{\lambda^5} \frac{T}{\mathfrak{c}} \dot{F}\left(\frac{\lambda T}{\mathfrak{c}}\right),$$

wobei $\dot{F}$ den Differentialquotienten von F nach seinem Argument bedeutet. Oder:

$$\frac{\lambda T}{\mathfrak{c}} \dot{F}\left(\frac{\lambda T}{\mathfrak{c}}\right) - 5 F\left(\frac{\lambda T}{\mathfrak{c}}\right) = 0. \tag{108}$$

Diese Gleichung ergibt für das Argument $\frac{\lambda T}{c}$ einen ganz bestimmten Wert, so daß für die Wellenlänge λ_m des Maximums der Strahlungsintensität E_λ die Beziehung gilt:

$$\lambda_m T = b. \tag{109}$$

Das Strahlungsmaximum verschiebt sich also bei Erhöhung der Temperatur nach der Seite der kürzeren Wellenlängen.

Der Zahlenwert der Konstante b ist von LUMMER und PRINGSHEIM[2] gemessen worden zu:

$$b = 0{,}294 \text{ cm} \cdot \text{grad}. \tag{110}$$

PASCHEN[3] hat einen etwas kleineren Wert gefunden, etwa 0,292.

---

[1] F. PASCHEN, Sitzungsber. d. Akad. d. Wissensch. Berlin, p. 405 u. 959, 1899. O. LUMMER und E. PRINGSHEIM, Verhandlungen der Deutschen Physikalischen Gesellschaft **1**, p. 23 u. 215, 1899. DRUDES Ann. **6**, p. 192, 1901.

[2] O. LUMMER und E. PRINGSHEIM, a. a. O.

[3] F. PASCHEN, DRUDES Ann. **6**, p. 657, 1901.

Es sei hier übrigens noch einmal ausdrücklich darauf hingewiesen, daß nach § 19 das Maximum von E_λ keineswegs auf dieselbe Stelle im Spektrum fällt wie das Maximum von $\mathfrak{K}_\nu$, und daß daher die Bedeutung der Konstante b wesentlich mit dadurch bedingt ist, daß die Intensität der monochromatischen Strahlung auf Wellenlängen und nicht auf Schwingungszahlen bezogen wird.

§ 90. Auch der Betrag des Maximums von E_λ ergibt sich aus (107), wenn man darin $\lambda = \lambda_m$ einsetzt. Dann erhält man unter Berücksichtigung von (109):

$$E_{\max} = \text{const} \cdot T^5, \tag{111}$$

d. h. der Betrag des Strahlungsmaximums im Spektrum der schwarzen Strahlung ist proportional der fünften Potenz der absoluten Temperatur.

Würde man die Intensität der monochromatischen Strahlung nicht durch E_λ, sondern durch $\mathfrak{K}_\nu$ messen, so erhielte man für den Betrag des Strahlungsmaximums ein ganz anderes Gesetz, nämlich:

$$\mathfrak{K}_{\max} = \text{const} \cdot T^3. \tag{112}$$

Viertes Kapitel. Strahlung von beliebiger spektraler Energieverteilung. Entropie und Temperatur monochromatischer Strahlung.

§ 91. Wir hatten das Wiensche Verschiebungsgesetz bisher nur auf den Fall der schwarzen Strahlung angewendet; dasselbe besitzt aber eine noch viel allgemeinere Bedeutung. Denn die Gleichung (95) gibt, wie schon dort bemerkt wurde, für jede beliebige anfängliche spektrale Verteilung der im evakuierten Hohlraum befindlichen nach allen Richtungen gleichmäßigen Energiestrahlung die Änderung dieser Energieverteilung bei einer reversibeln adiabatischen Änderung des Gesamtvolumens. Jeder durch einen derartigen Prozeß herbeigeführte Strahlungszustand ist vollkommen stationär und kann unbegrenzte Zeiten lang fortbestehen, allerdings nur unter der Bedingung, daß keine Spur emittierender und absorbierender Substanz in dem Strahlungsraum vorhanden ist. Denn sonst würde sich nach § 51 durch den auslösenden Einfluß der Substanz mit der Zeit die Energieverteilung auf irreversible Weise, d. h. unter Vermehrung der Gesamtentropie, in die stabile, der schwarzen Strahlung entsprechende Verteilung verwandeln.

Der Unterschied dieses allgemeineren Falles gegen den im vorigen Kapitel behandelten speziellen ist der, daß man hier nicht mehr, wie bei der schwarzen Strahlung, von einer bestimmten Temperatur der Strahlung reden kann. Wohl aber besitzt, da der zweite Hauptsatz der Thermodynamik als allgemein gültig vorausgesetzt wird, die Strahlung, wie überhaupt jedes in einem bestimmten Zustand befindliche physikalische System, eine bestimmte Entropie $S = V \cdot s$, und diese Entropie setzt sich, da die einzelnen Strahlengattungen unabhängig voneinander sind, durch Addition aus den Entropien der monochromatischen Strahlungen zusammen, also:

$$s = \int_0^\infty \mathfrak{s}\, d\nu, \qquad S = V \cdot \int_0^\infty \mathfrak{s}\, d\nu, \tag{113}$$

wobei $\mathfrak{s}\, d\nu$ die Entropie der in der Volumeneinheit enthaltenen Strahlung zwischen den Schwingungszahlen ν und $\nu + d\nu$ bezeichnet. $\mathfrak{s}$ ist eine bestimmte Funktion der beiden unabhängigen Variabeln ν und $\mathfrak{u}$, und wird im folgenden stets als solche behandelt werden.

§ 92. Würde der analytische Ausdruck der Funktion $\mathfrak{s}$ bekannt sein, so könnte man daraus unmittelbar das Gesetz der Energieverteilung im Normalspektrum ableiten; denn unter allen spektralen Energieverteilungen ist ja die normale, oder die der schwarzen Strahlung, dadurch ausgezeichnet, daß sie das Maximum der Strahlungsentropie S aufweist.

Nehmen wir also einmal $\mathfrak{s}$ als bekannte Funktion von ν und $\mathfrak{u}$ an, so ergibt sich als Bedingung der schwarzen Strahlung:

$$\delta S = 0 \tag{114}$$

für alle beliebigen Variationen der Energieverteilung, welche bei konstantem Gesamtvolumen V und konstanter Gesamtenergie U der Strahlung möglich sind. Die Variation der Energieverteilung denken wir uns dadurch charakterisiert, daß die Energie $\mathfrak{u}$ jeder einzelnen bestimmten Schwingungszahl ν eine unendlich kleine Änderung $\delta\mathfrak{u}$ erleidet. Dann haben wir als feste Bedingungen:

$$\delta V = 0 \quad \text{und} \quad \int_0^\infty \delta\mathfrak{u} \cdot d\nu = 0. \tag{115}$$

Die Änderungen d und δ sind natürlich ganz unabhängig voneinander.

Nun ist nach (114) und (113), da $\delta V = 0$:

$$\int_0^\infty \delta \mathfrak{s} \cdot d\nu = 0$$

oder, da ν unvariiert bleibt:

$$\int_0^\infty \frac{\partial \mathfrak{s}}{\partial \mathfrak{u}} \delta \mathfrak{u} \cdot d\nu = 0$$

und die Gültigkeit dieser Gleichung für alle beliebigen Werte von $\delta \mathfrak{u}$ erfordert mit Rücksicht auf (115), daß

$$\frac{\partial \mathfrak{s}}{\partial \mathfrak{u}} = \text{const} \tag{116}$$

für alle verschiedenen Schwingungszahlen. Diese Gleichung spricht das Gesetz der Energieverteilung bei der schwarzen Strahlung aus.

§ 93. Die Konstante der Gleichung (116) steht in einfachem Zusammenhang mit der Temperatur der schwarzen Strahlung. Denn wenn die schwarze Strahlung bei konstantem Volumen V durch Zuleitung einer gewissen Wärmemenge eine unendlich kleine Energieänderung δU erfährt, so ist nach (73) die Änderung ihrer Entropie:

$$\delta S = \frac{\delta U}{T}.$$

Nun ist aber nach (113) und (116):

$$\delta S = V \int_0^\infty \frac{\partial \mathfrak{s}}{\partial \mathfrak{u}} \delta \mathfrak{u}\, d\nu = \frac{\partial \mathfrak{s}}{\partial \mathfrak{u}} V \int_0^\infty \delta \mathfrak{u}\, d\nu = \frac{\partial \mathfrak{s}}{\partial \mathfrak{u}} \cdot \delta U,$$

folglich ist:

$$\frac{\partial \mathfrak{s}}{\partial \mathfrak{u}} = \frac{1}{T} \tag{117}$$

und die obige Größe, welche bei der schwarzen Strahlung als für alle Schwingungszahlen gleich gefunden wurde, erweist sich als die reziproke Temperatur der schwarzen Strahlung.

Durch diesen Satz erhält der Begriff der Temperatur eine Bedeutung auch für Strahlungen von ganz beliebiger Energieverteilung. Denn da $\mathfrak{s}$ nur von $\mathfrak{u}$ und ν abhängt, so besitzt eine nach allen Richtungen gleichmäßige monochromatische Strahlung, welche eine bestimmte Energiedichte $\mathfrak{u}$ hat, auch eine ganz bestimmte, durch (117) gegebene Temperatur, und unter allen denkbaren Energieverteilungen ist die normale dadurch charakterisiert, daß die Strahlungen aller Schwingungszahlen die nämliche Temperatur haben.

Jede Änderung der Energieverteilung besteht in einem Energieübergang von einer monochromatischen Strahlung auf eine andere, und je nachdem die Temperatur der ersten oder die der zweiten Strahlung höher ist, bedingt der Energieübergang eine Vermehrung oder eine Verminderung der Gesamtentropie, ist also in der Natur ohne Kompensation möglich oder nicht ohne Kompensation möglich, gerade wie das bei dem Wärmeübergang zwischen zwei verschieden temperierten Körpern zutrifft.

§ 94. Wir wollen nun sehen, was das WIENsche Verschiebungsgesetz über die Abhängigkeit der Größe $\mathfrak{s}$ von den Variabeln $\mathfrak{u}$ und ν aussagt. Aus der Gleichung (101) folgt, wenn man sie nach T auflöst und dafür den in (117) gegebenen Wert einsetzt:

$$\frac{1}{T} = \frac{1}{\nu} F\left(\frac{c^3 \mathfrak{u}}{\nu^3}\right) = \frac{\partial \mathfrak{s}}{\partial \mathfrak{u}}, \tag{118}$$

wo F wieder eine gewisse Funktion eines einzigen Arguments darstellt, deren Konstante die Fortpflanzungsgeschwindigkeit c nicht enthalten. Nach dem Argument integriert ergibt dies bei analoger Bezeichnung:

$$\mathfrak{s} = \frac{\nu^2}{c^3} F\left(\frac{c^3 \mathfrak{u}}{\nu^3}\right). \tag{119}$$

In dieser Form besitzt das WIENsche Verschiebungsgesetz für jede monochromatische Strahlung einzeln, und dadurch auch für Strahlungen von beliebiger Energieverteilung, Bedeutung.

§ 95. Nach dem zweiten Hauptsatz der Thermodynamik muß die Gesamtentropie einer Strahlung von ganz beliebiger Energieverteilung bei reversibler adiabatischer Kompression konstant bleiben. Den direkten Nachweis dieses Satzes können wir jetzt in der Tat auf Grund der Gleichung (119) führen. Es ist nämlich für einen solchen Vorgang nach Gleichung (113):

$$\delta S = \int_0^\infty d\nu (V\,\delta\mathfrak{s} + \mathfrak{s}\,\delta V)$$

$$(120) \qquad = \int_0^\infty d\nu \left(V \frac{\partial \mathfrak{s}}{\partial \mathfrak{u}} \delta \mathfrak{u} + \mathfrak{s}\,\delta V\right).$$

Hier ist $\mathfrak{s}$, wie stets, als Funktion von $\mathfrak{u}$ und ν zu betrachten.

Nun gilt für eine reversible adiabatische Zustandsänderung die Beziehung (95), aus welcher wir den Wert von $\delta\mathfrak{u}$ entnehmen, so daß sich ergibt:

$$\delta S = \delta V \cdot \int_0^\infty d\nu \left\{\frac{\partial \mathfrak{s}}{\partial \mathfrak{u}} \left(\frac{\nu}{3} \frac{d\mathfrak{u}}{d\nu} - \mathfrak{u}\right) + \mathfrak{s}\right\}.$$

Dabei bezieht sich der Differentialquotient von $\mathfrak{u}$ nach ν auf die in beliebiger Weise von vornherein gegebene spektrale Energieverteilung der Strahlung, er ist daher, im Gegensatz zu den partiellen Differentialquotienten, mit dem Buchstaben d bezeichnet.

Nun ist das vollständige Differential:

$$\frac{d\mathfrak{s}}{d\nu} = \frac{\partial \mathfrak{s}}{\partial \mathfrak{u}} \frac{d\mathfrak{u}}{d\nu} + \frac{\partial \mathfrak{s}}{\partial \nu},$$

folglich, durch Substitution:

$$(121) \qquad \delta S = \delta V \cdot \int_0^\infty d\nu \left\{\frac{\nu}{3}\left(\frac{d\mathfrak{s}}{d\nu} - \frac{\partial \mathfrak{s}}{\partial \nu}\right) - \mathfrak{u}\frac{\partial \mathfrak{s}}{\partial \mathfrak{u}} + \mathfrak{s}\right\}.$$

Aus Gleichung (119) folgt aber durch Differentiation:

$$\frac{\partial \mathfrak{s}}{\partial \mathfrak{u}} = \frac{1}{\nu} \dot{F}\left(\frac{c^3 \mathfrak{u}}{\nu^3}\right) \quad \text{und} \quad \frac{\partial \mathfrak{s}}{\partial \nu} = \frac{2\nu}{c^3} F\left(\frac{c^3 \mathfrak{u}}{\nu^3}\right) - \frac{3\mathfrak{u}}{\nu^2} \dot{F}\left(\frac{c^3 \mathfrak{u}}{\nu^3}\right).$$

Mithin:

$$\nu \frac{\partial \mathfrak{s}}{\partial \nu} = 2\mathfrak{s} - 3\mathfrak{u}\frac{\partial \mathfrak{s}}{\partial \mathfrak{u}}.$$

Dies in (121) eingesetzt, ergibt:

$$\delta S = \delta V \cdot \int_0^\infty d\nu \left(\frac{\nu}{3} \frac{d\mathfrak{s}}{d\nu} + \frac{1}{3}\mathfrak{s}\right)$$

oder:

$$\delta S = \frac{\delta V}{3} \cdot [\nu \mathfrak{s}]_0^\infty = 0,$$

wie es sein muß. Daß das Produkt $\nu\mathfrak{s}$ auch für $\nu = \infty$ verschwindet, läßt sich ebenso wie in § 83 bei dem Produkte $\nu\mathfrak{u}$ beweisen.

§ 96. Mittels der Gleichung (119) kann man den Gesetzen der reversibeln adiabatischen Kompression eine anschauliche

Fassung geben, welche die Verallgemeinerung der in § 87 für die schwarze Strahlung ausgesprochenen Sätze auf eine Strahlung von beliebiger Energieverteilung bildet. Zu diesem Zwecke führen wir wieder die in dem Kubus einer Wellenlänge λ enthaltene Strahlungsenergie zwischen den Schwingungszahlen ν und $\nu + d\nu$ ein:

$$\frac{\mathfrak{u}\, c^3}{\nu^3}\, d\nu = \eta\,,$$

ferner die entsprechende Strahlungsentropie: (122)

$$\frac{\mathfrak{s}\, c^3}{\nu^3}\, d\nu = \sigma\,,$$

endlich die in dem ganzen Strahlungsvolumen V enthaltene Anzahl von Wellenlängenkuben:

$$\frac{V\nu^3}{c^3} = N. \tag{123}$$

Wenn nun die Strahlung, deren Energieverteilung eine ganz beliebige sein mag, durch reversible adiabatische Kompression auf ein kleineres Volumen V' gebracht wird, so ist für ein beliebiges anderes Intervall von Schwingungszahlen ν' und $\nu' + d\nu'$:

$$\eta' = \frac{\mathfrak{u}' c^3}{\nu'^3}\, d\nu', \qquad \sigma' = \frac{\mathfrak{s}' c^3}{\nu'^3}\, d\nu', \tag{124}$$

$$N' = \frac{V'\nu'^3}{c^3}. \tag{125}$$

Setzt man nun $N = N'$, also nach (123) und (125):

$$V \cdot \nu^3 = V'\nu'^3$$

und dementsprechend für die beiden Intervalle $d\nu$ und $d\nu'$:

$$V\nu^2\, d\nu = V'\nu'^2\, d\nu'$$

oder durch Division:

$$\frac{d\nu}{\nu} = \frac{d\nu'}{\nu'}\,,$$

so ergibt sich zunächst aus (99):

$$\frac{\mathfrak{u}'}{\nu'^3} = \frac{\mathfrak{u}}{\nu^3}$$

folglich, nach (122) und (124):

$$\frac{\eta'}{\eta} = \frac{d\nu'}{d\nu} = \frac{\nu'}{\nu} \quad \text{und} \quad \sigma' = \sigma\,, \tag{126}$$

da nach (119): $\dfrac{\mathfrak{s}'}{\nu'^2} = \dfrac{\mathfrak{s}}{\nu^2}\,.$

Daher kann man sagen: die reversible adiabatische Kompression einer Strahlung von beliebiger Energieverteilung erfolgt ebenso, als ob die Schwingungszahlen ν der einzelnen Farben sich in der Weise verändern, daß für jede Farbe die Anzahl der im ganzen Volumen enthaltenen Wellenlängenkuben bei der Kompression ungeändert bleibt. Die in einem solchen Kubus enthaltene monochromatische Strahlungsenergie η wächst dann proportional der Schwingungszahl ν, und ebenso wächst nach (118) die Temperatur T der betreffenden Strahlung, während dagegen die Entropie derselben σ konstant bleibt. Hiermit ist zugleich wiederum der Beweis geliefert, daß die Gesamtentropie der Strahlung, als die Summe der Entropien aller darin enthaltenen monochromatischen Strahlungen, konstant bleibt.

§ 97. Wir können noch einen Schritt weiter gehen, und von der Entropie $\mathfrak{s}$ und der Temperatur T einer nach allen Richtungen gleichmäßigen unpolarisierten monochromatischen Strahlung auf die Entropie und die Temperatur eines einzelnen geradlinig polarisierten monochromatischen Strahlenbündels schließen. Daß auch jedem einzelnen Strahlenbündel eine bestimmte Entropie zukommt, folgt nach dem zweiten Hauptsatz der Thermodynamik schon aus dem Phänomen der Emission. Denn da durch den Akt der Emission Körperwärme in Strahlungswärme verwandelt wird, so nimmt hierbei die Entropie des emittierenden Körpers ab, und dafür muß nach dem Satz der Vermehrung der Gesamtentropie als Kompensation eine andere Form der Entropie auftreten, welche durch nichts anderes bedingt sein kann als durch die Energie der emittierten Strahlung. Jedes einzelne geradlinig polarisierte monochromatische Strahlenbündel besitzt also seine bestimmte Entropie, die nur von seiner Energie und seiner Schwingungszahl abhängen kann, und sich mit ihm im Raume fortpflanzt und ausbreitet. Dadurch erhalten wir den Begriff der Entropiestrahlung, welche ganz analog der Energiestrahlung gemessen wird durch den Betrag der Entropie, die in der Zeiteinheit in einer bestimmten Richtung durch die Flächeneinheit hindurchgeht. Es gelten daher für die Entropiestrahlung genau dieselben Betrachtungen, wie die, welche wir vom § 14 an für die Energiestrahlung angestellt haben, indem jedes Strahlenbündel außer seiner Energie auch seine Entropie besitzt und befördert. Wir wollen, unter Hinweis auf die dortigen Ausführungen, hier nur die wichtigsten Sätze für den späteren Gebrauch zusammenstellen.

§ 98. In einem von beliebiger Strahlung erfüllten Raume ist die Entropie, welche in der Zeit dt durch ein Flächenelement $d\sigma$ in der Richtung des Elementarkegels $d\Omega$ hindurchgestrahlt wird, gleich einem Ausdruck von der Form:

$$dt\, d\sigma \cos\vartheta\, d\Omega \cdot L = L \sin\vartheta \cos\vartheta\, d\vartheta\, d\varphi\, d\sigma\, dt. \qquad (127)$$

Die positive Größe L nennen wir die „spezifische Intensität der Entropiestrahlung“ am Orte des Flächenelements $d\sigma$ in der Richtung des Öffnungswinkels $d\Omega$. L ist im allgemeinen eine Funktion des Ortes, der Zeit und der Richtung.

Die Gesamtstrahlung der Entropie durch das Flächenelement $d\sigma$ nach einer Seite, etwa derjenigen, für welche der Winkel ϑ ein spitzer ist, ergibt sich durch Integration über φ von 0 bis 2π, und über ϑ von 0 bis $\frac{\pi}{2}$ zu:

$$d\sigma\, dt \cdot \int_0^{2\pi} d\varphi \cdot \int_0^{\pi/2} d\vartheta\, L \sin\vartheta \cos\vartheta.$$

Ist die Strahlung nach allen Richtungen gleichmäßig, also L konstant, so ist die Entropiestrahlung durch $d\sigma$ nach einer Seite:

$$\pi L\, d\sigma\, dt. \qquad (128)$$

Die spezifische Intensität L der Entropiestrahlung nach jeder Richtung zerfällt weiter in die Intensitäten der einzelnen, den verschiedenen Gebieten des Spektrums angehörigen Strahlen, die sich unabhängig voneinander fortpflanzen. Endlich ist bei einem Strahle von bestimmter Farbe und Intensität noch die Art seiner Polarisation charakteristisch. Wenn ein monochromatischer Strahl von der Schwingungszahl ν aus zwei voneinander unabhängigen[1], senkrecht aufeinander polarisierten Komponenten mit den Intensitäten $\mathfrak{K}_\nu$ und $\mathfrak{K}_\nu'$, den „Hauptintensitäten“ der Energie-

---

1) „unabhängig“ im Sinne von „inkohärent“. Ist z. B. der Strahl mit den Hauptintensitäten $\mathfrak{K}$ und $\mathfrak{K}'$ elliptisch polarisiert, so ist seine Entropie nicht gleich $\mathfrak{L} + \mathfrak{L}'$, sondern gleich der Entropie eines geradlinig polarisierten Strahls von der Intensität $\mathfrak{K} + \mathfrak{K}'$. Denn ein elliptisch polarisierter Strahl läßt sich ohne weiteres in einen geradlinig polarisierten verwandeln und umgekehrt, z. B. mittels totaler Reflexion.

strahlung (§ 17) besteht, so ist die spezifische Intensität der Entropiestrahlung aller Schwingungszahlen von der Form:

$$L = \int_0^\infty d\nu (\mathfrak{L}_\nu + \mathfrak{L}'_\nu). \tag{129}$$

Hierbei sind die positiven Größen $\mathfrak{L}_\nu$ und $\mathfrak{L}_\nu'$, die „Hauptintensitäten" der Entropiestrahlung von der Schwingungszahl ν, durch die Werte von $\mathfrak{K}_\nu$ und $\mathfrak{K}_\nu'$ bestimmt. Durch Substitution in (127) erhält man hieraus für die Entropie, welche in der Zeit dt durch das Flächenelement $d\sigma$ in der Richtung des Elementarkegels $d\Omega$ hindurchgestrahlt wird, den Ausdruck

$$dt\, d\sigma \cos\vartheta\, d\Omega \int_0^\infty d\nu (\mathfrak{L}_\nu + \mathfrak{L}_\nu')$$

und für monochromatische geradlinig polarisierte Strahlung:

$$dt\, d\sigma \cos\vartheta\, d\Omega\, \mathfrak{L}_\nu\, d\nu = \mathfrak{L}_\nu\, d\nu \cdot \sin\vartheta \cos\vartheta\, d\vartheta\, d\varphi\, d\sigma\, dt. \tag{130}$$

Für unpolarisierte Strahlen ist $\mathfrak{L}_\nu = \mathfrak{L}_\nu'$, und aus (129) wird:

$$L = 2\int_0^\infty \mathfrak{L}_\nu\, d\nu.$$

Bei gleichmäßiger Strahlung nach allen Richtungen ergibt sich dann für die gesamte Entropiestrahlung nach einer Seite, gemäß (128)

$$2\pi\, d\sigma\, dt \cdot \int_0^\infty \mathfrak{L}_\nu\, d\nu.$$

§ 99. Aus der Intensität der fortschreitenden Entropiestrahlung ergibt sich auch der Ausdruck für die räumliche Dichte der Strahlungsentropie, ganz ebenso wie die räumliche Dichte der Strahlungsenergie aus der Intensität der fortschreitenden Energiestrahlung folgt. (Vgl. § 22.) Es ist nämlich, analog der Gleichung (20), die räumliche Dichte s der Strahlungsentropie im Vakuum in irgendeinem Punkte:

$$s = \frac{1}{c}\int L\, d\Omega, \tag{131}$$

wobei die Integration über die von dem Punkte nach allen Richtungen des Raumes ausgehenden Elementarkegel zu erstrecken ist. Für gleichmäßige Strahlung ist L konstant, und man erhält:

$$s = \frac{4\pi L}{c}. \tag{132}$$

Durch spektrale Zerlegung der Größe L nach der Gleichung (129) ergibt sich aus (131) auch die räumliche Dichte der monochromatischen Strahlungsentropie:

$$\mathfrak{s} = \frac{1}{c}\int (\mathfrak{L} + \mathfrak{L}')\cdot d\Omega$$

und für unpolarisierte und nach allen Richtungen gleichmäßige Strahlung:

$$\mathfrak{s} = \frac{8\pi\mathfrak{L}}{c}. \tag{133}$$

§ 100. Über die Art der Abhängigkeit der Entropiestrahlung $\mathfrak{L}$ von der Energiestrahlung $\mathfrak{K}$ gibt das WIENsche Verschiebungsgesetz in der Form (119) sogleich Auskunft. Es folgt nämlich daraus mit Berücksichtigung von (133) und (24):

$$\mathfrak{L} = \frac{\nu^2}{c^2} F\left(\frac{c^2\mathfrak{K}}{\nu^3}\right) \tag{134}$$

und ferner, mit Berücksichtigung von (118):

$$\frac{\partial\mathfrak{L}}{\partial\mathfrak{K}} = \frac{1}{T}. \tag{135}$$

Daher auch:

$$T = \nu F\left(\frac{c^2\mathfrak{K}}{\nu^3}\right), \tag{136}$$

oder:

$$\mathfrak{K} = \frac{\nu^3}{c^2} F\left(\frac{T}{\nu}\right). \tag{137}$$

Diese Beziehungen sind zwar, wie die Gleichungen (118) und (119), zunächst nur für unpolarisierte und nach allen Richtungen gleichmäßige Strahlung abgeleitet, sie besitzen aber auch im Falle beliebiger Strahlung allgemeine Gültigkeit für jeden einzelnen geradlinig polarisierten Strahl. Denn da sich die einzelnen Strahlen gänzlich unabhängig voneinander verhalten und fortpflanzen, so kann die Intensität $\mathfrak{L}$ der Entropiestrahlung eines Strahles auch nur von der Intensität der Energiestrahlung $\mathfrak{K}$ des nämlichen Strahles abhängen. Jeder einzelne monochromatische Strahl besitzt also außer seiner Energie auch seine durch (134) bestimmte Entropie und seine durch (136) bestimmte Temperatur.

§ 101. Die hier vorgenommene Erweiterung des Begriffs der Temperatur auf einen einzelnen monochromatischen Strahl bringt es mit sich, daß in einem von beliebigen Strahlen durchsetzten Medium an einer und derselben Stelle des Mediums im allgemeinen unendlich viele verschiedene Temperaturen bestehen, indem jeder einzelne Strahl, der diese Stelle trifft, seine besondere Temperatur besitzt, ja daß sogar die in der nämlichen Richtung fortschreitenden verschieden gefärbten Strahlen je nach der spektralen Energieverteilung verschiedene Temperaturen aufweisen. Zu allen diesen Temperaturen kommt schließlich noch die Temperatur des Mediums selber, welche auch ihrerseits von vornherein ganz unabhängig von der Temperatur der Strahlung ist. Diese Kompliziertheit der Betrachtungsweise liegt aber ganz in der Natur der Sache, und entspricht der Kompliziertheit der physikalischen Vorgänge in einem solcherweise durchstrahlten Medium. Nur im Falle des stabilen thermodynamischen Gleichgewichts gibt es nur eine einzige Temperatur, die dann dem Medium selber und allen dasselbe durchkreuzenden Strahlen verschiedener Richtung und verschiedener Farbe gemeinsam ist.

Auch in der praktischen Physik hat sich die Notwendigkeit, den Begriff der Strahlungstemperatur von dem der Körpertemperatur zu trennen, schon seit längerer Zeit geltend gemacht. So hat man es für vorteilhaft gefunden, neben der wirklichen Temperatur der Sonne von einer „scheinbaren“ oder „Effektiv“-temperatur der Sonne zu sprechen, d. h. von derjenigen Temperatur, welche die Sonne haben müßte, um der Erde die tatsächlich zu beobachtende Wärmestrahlung zuzusenden, wenn sie wie ein schwarzer Körper strahlen würde. Die scheinbare Temperatur der Sonne ist nun offenbar nichts anderes als die wirkliche Temperatur der Sonnenstrahlen[1], sie hängt lediglich ab von der Beschaffenheit der Strahlen, ist also eine Eigenschaft der Strahlen und nicht eine Eigenschaft der Sonne. Daher wäre es wohl nicht nur bequemer, sondern auch exakter, diese Bezeichnung auch direkt anzuwenden, statt von einer doch nur fingierten und nur durch Einführung einer in Wirklichkeit unzutreffenden Voraussetzung verständlichen Temperatur der Sonne zu reden.

---

[1] im Durchschnitt genommen, da die Sonnenstrahlen verschiedener Farbe nicht genau die nämliche Temperatur besitzen.

In neuerer Zeit haben Helligkeitsmessungen an monochromatischem Licht L. HOLBORN und F. KURLBAUM[1] zu der Einführung des Begriffs der „schwarzen" Temperatur einer strahlenden Oberfläche geführt. Die schwarze Temperatur einer strahlenden Oberfläche wird gemessen durch die Helligkeit der Strahlen, welche sie emittiert, sie ist im allgemeinen für jeden einzelnen Strahl von bestimmter Farbe, Richtung und Polarisation, den die Oberfläche emittiert, eine besondere, und stellt eben einfach die Temperatur dieses Strahles dar. Statt der unendlich vielen „schwarzen" Temperaturen der emittierenden Oberfläche haben wir hier also eine ganz bestimmte wirkliche Temperatur eines jeden emittierten Strahles, welche durch seine Helligkeit (spezifische Intensität) $\mathfrak{K}$ und durch seine Schwingungszahl ν nach Gleichung (136) gegeben ist, ohne jede Rücksicht auf seine Abstammung und auf seine Vorgeschichte. Die numerisch bestimmte Form dieser Gleichung wird unten im § 162 angegeben werden. Da ein schwarzer Körper das maximale Emissionsvermögen besitzt, so kann die Temperatur eines emittierten Strahles niemals höher sein als die des emittierenden Körpers.

§ 102. Machen wir noch eine einfache Anwendung der zuletzt gewonnenen Sätze auf den speziellen Fall der schwarzen Strahlung. Für diese ist nach (81) die gesamte räumliche Entropiedichte:

$$s = \frac{4}{3} a T^3 . \tag{138}$$

Also nach (132) die spezifische Intensität der gesamten Entropiestrahlung nach irgend einer Richtung:

$$L = \frac{c}{3\pi} a T^3 \tag{139}$$

und die gesamte Entropiestrahlung durch ein Flächenelement $d\sigma$ nach einer Seite, gemäß (128):

$$\frac{c}{3} a T^3 \, d\sigma \, dt . \tag{140}$$

Wir wollen jetzt als spezielles Beispiel die beiden Hauptsätze der Thermodynamik anwenden auf den Fall, daß die Oberfläche eines schwarzen Körpers von der Temperatur T allseitig getroffen wird von schwarzer Strahlung der Temperatur T'. Dann emittiert

[1] L. HOLBORN und F. KURLBAUM, DRUDES Ann. **10**, p. 229, 1903.

der schwarze Körper pro Flächeneinheit und Zeiteinheit nach (7) und (76) die Energie:

$$\pi K = \frac{a\,c}{4} T^4$$

und nach (140) die Entropie:

$$\frac{a\,c}{3} T^3 .$$

Dagegen absorbiert er die Energie: $\frac{a\,c}{4} T'^4$ und die Entropie:

$$\frac{a\,c}{3} T'^3 .$$

Nach dem ersten Hauptsatz ist also die dem Körper im ganzen zugeführte Wärme, positiv oder negativ, je nachdem T' größer oder kleiner als T ist:

$$Q = \frac{a\,c}{4} T'^4 - \frac{a\,c}{4} T^4 = \frac{a\,c}{4} (T'^4 - T^4),$$

und nach dem zweiten Hauptsatz ist die Änderung der Gesamtentropie positiv oder Null. Nun ändert sich die Entropie des Körpers um $\frac{Q}{T}$, die Entropie der Strahlung im Vakuum dagegen um:

$$\frac{a\,c}{3} (T^3 - T'^3).$$

Folglich ist die Änderung der Gesamtentropie des betrachteten Systems pro Zeiteinheit und Flächeneinheit:

$$\frac{a\,c}{4} \cdot \frac{T'^4 - T^4}{T} + \frac{a\,c}{3} (T^3 - T'^3) \geqq 0 .$$

Diese Beziehung ist in der Tat für alle Werte von T und T' erfüllt. Denn der Minimalwert des Ausdrucks auf der linken Seite ist Null; derselbe wird erreicht für $T = T'$. Dann ist der Vorgang reversibel. Sobald aber T von T' verschieden ist, haben wir merkliche Entropievermehrung, der Prozeß ist also irreversibel. Im besonderen ergibt sich für $T = 0$ die Entropievermehrung gleich ∞, d. h. die Absorption von Wärmestrahlung durch einen schwarzen Körper von der absoluten Temperatur Null ist mit unendlich großer Entropievermehrung verknüpft, kann also durch keine endliche Kompensation rückgängig gemacht werden. Dagegen ist für $T' = 0$ die Entropievermehrung nur gleich $\frac{a\,c}{12} T^3$, d. h. die Emission eines schwarzen Körpers von der Temperatur T, ohne gleichzeitige Absorption von Wärmestrahlung ist irreversibel, läßt sich aber durch eine Kompensation von mindestens dem angegebenen endlichen Betrage rückgängig machen. In der Tat: Läßt man die vom Körper emittierten Strahlen wieder auf ihn zurückfallen, etwa durch geeignete Spiegelung, so wird der Körper diese Strahlen zwar wieder absorbieren, aber notwendigerweise gleichzeitig neue Strahlen emittieren, und hierin liegt die vom zweiten Hauptsatz geforderte Kompensation.

Allgemein kann man sagen: Emission ohne gleichzeitige Absorption ist irreversibel, während dagegen der umgekehrte Vorgang: Absorption ohne gleichzeitige Emission, in der Natur unmöglich ist.

§ 103. Ein weiteres Beispiel der Anwendung der beiden Hauptsätze der Thermodynamik liefert die oben im § 70 betrachtete irreversible Ausdehnung einer ursprünglich schwarzen Strahlung von dem Volumen V und der Temperatur T auf das größere Volumen V', aber diesmal bei Abwesenheit jeglicher absorbierenden und emittierenden Substanz. Dann bleibt nicht nur die Gesamtenergie, sondern auch die Energie jeder einzelnen Schwingungszahl ν erhalten, also, wenn infolge diffuser Reflexion an den Wänden die Strahlung wieder nach allen Richtungen gleichförmig geworden ist: $\mathfrak{u}_\nu V = \mathfrak{u}_\nu' V'$, und hierdurch ist nach (118) auch T_ν', die Temperatur der monochromatischen Strahlung von der Schwingungszahl ν, im Endzustand bestimmt. Die Ausführung der Berechnung kann allerdings erst mit Hilfe der späteren Gleichung (233) erfolgen. Die Gesamtentropie der Strahlung, d. h. die Summe der Entropien der Strahlungen aller Schwingungszahlen:

$$V' \cdot \int_0^\infty \mathfrak{s}_\nu' \, d\nu$$

muß nach dem zweiten Hauptsatz im Endzustand größer sein als im Anfangszustand. Da T_ν' für die verschiedenen Schwingungszahlen ν verschiedene Werte besitzt, so ist die Endstrahlung nicht mehr schwarz. Daher erhält man durch nachträgliche Einbringung eines Kohlestäubchens in den Hohlraum eine endliche Änderung der Energieverteilung, und die Entropie steigt dabei weiter bis auf den in (82) berechneten Wert S'.

7*

Dritter Abschnitt.

Emission und Absorption elektromagnetischer Wellen durch einen linearen Oszillator.

Erstes Kapitel.
Einleitung. Schwingungsgleichung eines linearen Oszillators.

§ 104. Das Hauptproblem der Theorie der Wärmestrahlung, dessen Lösung in diesem Abschnitt vorbereitet werden soll, ist die Bestimmung der Energieverteilung in dem von der schwarzen Strahlung gelieferten Normalspektrum, oder, was auf dasselbe hinauskommt, die Auffindung der im allgemeinen Ausdruck des WIENschen Verschiebungsgesetzes (119) noch unbestimmt gelassenen Funktion F. Zur Behandlung dieser Aufgabe wird es nötig sein, näher als bisher auf diejenigen Vorgänge einzugehen, welche die Entstehung und die Vernichtung der Wärmestrahlen bedingen, also auf den Akt der Emission und den der Absorption. Bei der Kompliziertheit dieser Vorgänge und der Schwierigkeit, darüber irgendwelche bestimmte Einzelheiten in Erfahrung zu bringen, wäre es freilich gänzlich aussichtslos, auf diesem Wege irgendwelche sichere Resultate zu gewinnen, wenn nicht als zuverlässiger Führer in diesem dunkeln Gebiete der im § 51 abgeleitete KIRCHHOFFsche Satz benutzt werden könnte, welcher besagt, daß ein rings durch spiegelnde Wände abgeschlossenes Vakuum, in welchem beliebige emittierende und absorbierende Körper in beliebiger Anordnung verstreut sind, im Laufe der Zeit den stationären Zustand der schwarzen Strahlung annimmt, der vollständig bestimmt ist durch einen einzigen Parameter:

die Temperatur, und insbesondere nicht abhängt von der Anzahl, der Beschaffenheit und der Anordnung der ponderablen Körper. Es ist also zur Untersuchung der Eigenschaften des Zustandes der schwarzen Strahlung ganz gleichgültig, welcher Art die Körper sind, welche man im Vakuum befindlich voraussetzt, ja es kommt nicht einmal darauf an, ob solche Körper in der Natur wirklich irgendwo vorkommen, sondern nur darauf, ob ihre Existenz und ihre Eigenschaften mit den Gesetzen der Elektrodynamik und der Thermodynamik überhaupt verträglich sind. Sobald es nur gelingt, für irgend eine beliebig herausgegriffene spezielle Art und Anordnung emittierender und absorbierender Systeme einen Strahlungszustand im umgebenden Vakuum nachzuweisen, der sich durch absolute Stabilität auszeichnet, so kann dieser Zustand kein anderer sein als der der schwarzen Strahlung.

Von der durch diesen Satz gewährleisteten Freiheit Gebrauch machend, wählen wir uns nun unter allen emittierenden und absorbierenden Systemen das denkbar einfachste aus, nämlich einen einzigen ruhenden Oszillator, bestehend aus zwei mit gleichen Elektrizitätsmengen von entgegengesetztem Vorzeichen geladenen Polen, welche auf einer festen gerade gerichteten Linie, der Achse des Oszillators, gegeneinander beweglich sind.

Allgemeiner und den natürlichen Verhältnissen näher angepaßt wäre es freilich, die Schwingungen des Oszillators, wie sie durch die Bewegungen der beiden Pole bedingt werden, statt mit einem einzigen, mit drei Graden von Bewegungsfreiheit auszustatten, d. h. dieselben nicht geradlinig, sondern räumlich vorauszusetzen. Diese Annahme läßt sich vollkommen analog der hier gemachten einfacheren durchführen, entsprechend der Zerlegung einer räumlichen Bewegung in ihre drei geradlinigen aufeinander senkrechten Komponenten. Indessen können wir uns nach der oben dargelegten prinzipiellen Überlegung, ohne eine wesentliche Einbuße in der Allgemeinheit unserer Betrachtungen befürchten zu müssen, von vornherein auf die Behandlung einer einzigen Komponente beschränken.

Dagegen könnte es prinzipielle Bedenken erregen, daß wir uns den ganzen Oszillator als ruhend vorstellen, da doch nach der kinetischen Gastheorie in Substanzen von endlicher Temperatur alle darin enthaltenen frei beweglichen materiellen Partikel im Mittel eine bestimmte endliche kinetische Energie der fortschreitenden Bewegung besitzen. Indessen läßt sich auch dies Bedenken heben durch die Erwägung, daß mit der kinetischen Energie noch nicht die Geschwindigkeit festgelegt ist. Man braucht nur den Oszillator, etwa in seinem positiven Pole, mit einer verhältnismäßig bedeutenden, elektrodynamisch vollkommen unwirksamen trägen Masse belastet zu denken, um seine Geschwindigkeit, bei bestimmter kinetischer Energie, unter jeden beliebigen Betrag herabzudrücken. Diese Überlegung hält natürlich auch dann noch Stand, wenn man, wie es jetzt mehrfach geschieht, alle Trägheit auf elektrodynamische Wirkungen zurückführt. Denn diese Wirkungen sind jedenfalls von ganz anderer Art als die im folgenden zu betrachtenden, und können sie daher nicht beeinflussen.

Der Zustand des angenommenen Oszillators sei vollständig bestimmt durch sein „Moment“ $f(t)$, d. h. durch das Produkt aus der elektrischen Ladung des auf der positiven Seite der Achse gelegenen Poles in den Polabstand, und durch dessen Differentialquotienten nach der Zeit:

(141) $$\frac{d\,f(t)}{d\,t} = \dot{f}(t).$$

Die Energie des Oszillators sei von der folgenden einfachen Form:

(142) $$U = \tfrac{1}{2} K f^2 + \tfrac{1}{2} L \dot{f}^2,$$

wobei K und L positive Konstante bezeichnen, die von der Natur des Oszillators in irgend einer, hier nicht näher zu erörternden Weise abhängen.

§ 105. Würde bei den Schwingungen des Oszillators die Energie U genau konstant bleiben, so hätte man:

$$d\,U = K\,f\,d\,f + L\,\dot{f}\,d\,\dot{f} = 0$$

oder mit Rücksicht auf (141):

(143) $$K\,f(t) + L\,\ddot{f}(t) = 0$$

und daraus ergäbe sich als allgemeine Lösung dieser Differentialgleichung eine rein periodische Schwingung:

$$f = C \cos(2\,\pi\,\nu_0\,t - \vartheta),$$

wobei C und ϑ die Integrationskonstanten, und ν_0 die Schwingungszahl pro Zeiteinheit bedeutet:

$$\nu_0 = \frac{1}{2\pi}\sqrt{\frac{K}{L}}. \tag{144}$$

Ein solcher mit konstanter Energie periodisch schwingender Oszillator würde weder von dem umgebenden elektromagnetischen Felde beeinflußt werden, noch würde er irgendwelche Strahlungswirkungen nach außen hin ausüben, er könnte also für die Wärmestrahlung im umgebenden Vakuum von keinerlei Bedeutung sein.

Nach der MAXWELLschen Theorie bleibt nun aber die Schwingungsenergie U des Oszillators im allgemeinen keineswegs konstant, sondern der Oszillator entsendet vermöge seiner Schwingungen nach allen Richtungen Kugelwellen in das umgebende Feld hinaus, und hiermit muß nach dem Prinzip der Erhaltung der Energie, wenn nicht andererseits von außen her Wirkungen auf den Oszillator ausgeübt werden, notwendig ein Verlust von Schwingungsenergie, also eine Dämpfung der Schwingungsamplitude verbunden sein. Wir wollen zunächst den Betrag dieser Dämpfung berechnen.

§ 106. Zu diesem Zweck gehen wir zunächst von folgender partikulären Lösung der MAXWELLschen Feldgleichungen (52) aus:

$$\left.\begin{aligned}
\mathfrak{E}_x &= \frac{\partial^2 F}{\partial x\,\partial z} & \mathfrak{H}_x &= \frac{1}{c}\frac{\partial^2 F}{\partial y\,\partial t}\\
\mathfrak{E}_y &= \frac{\partial^2 F}{\partial y\,\partial z} & \mathfrak{H}_y &= -\frac{1}{c}\frac{\partial^2 F}{\partial x\,\partial t}\\
\mathfrak{E}_z &= \frac{\partial^2 F}{\partial z^2} - \frac{1}{c^2}\frac{\partial^2 F}{\partial t^2} & \mathfrak{H}_z &= 0,
\end{aligned}\right\} \tag{145}$$

wobei die Funktion F von x, y, z, t der Bedingung genügt:

$$\frac{\partial^2 F}{\partial t^2} = c^2 \cdot \triangle F. \tag{146}$$

Daß diese Größen wirklich allen Bedingungsgleichungen des Feldes genügen, erkennt man direkt durch Substitution in (52).

Nehmen wir spezieller an, daß die Funktion F außer von der Zeit t nur von der Entfernung r des Feldpunktes x, y, z vom Anfangspunkt der Koordinaten abhängt, so verwandelt sich die Gleichung (146) in:

$$\frac{\partial^2 F}{\partial t^2} = \frac{c^2}{r^2}\frac{\partial}{\partial r}\left(r^2\frac{\partial F}{\partial r}\right)$$

und die allgemeine Lösung dieser Differentialgleichung ist:

$$F = \frac{1}{r} f\left(t - \frac{r}{c}\right) + \frac{1}{r} g\left(t + \frac{r}{c}\right), \tag{147}$$

wobei f und g zwei ganz beliebige Funktionen eines einzigen Arguments bedeuten. Die Funktion f entspricht einer Kugelwelle, die vom Anfangspunkt der Koordinaten nach außen, die Funktion g einer Kugelwelle, die von außen nach dem Anfangspunkt der Koordinaten fortschreitet. Die Welle g kommt in der Natur nur unter ganz besonderen Umständen zu stande (vgl. unten § 169); wir lassen sie hier fort, da wir jetzt ohnehin keinerlei von außen auf den Oszillator fallende Wellen als vorhanden annehmen wollen, und erhalten daher:

$$F = \frac{1}{r} f\left(t - \frac{r}{c}\right). \tag{148}$$

§ 107. Um nun die physikalische Bedeutung der vorliegenden partikulären Lösung der MAXWELLschen Feldgleichungen kennen zu lernen, betrachten wir solche Punkte (x, y, z) des Feldes, welche dem Anfangspunkte der Koordinaten so nahe liegen, daß für alle Zeiten $\frac{r}{c} \dot{f}$ klein ist gegen f. (Wenn f periodisch oder nahezu periodisch ist, so bedeutet dies, daß die Entfernung r vom Anfangspunkt klein ist gegen die Wellenlänge im Vakuum.) Dann ist auch $\frac{r}{c} \ddot{f}$ klein gegen $\dot{f}$, und a fortiori $\frac{r^2}{c^2} \ddot{f}$ klein gegen f, und hierdurch vereinfachen sich, wie man leicht aus dem Ausdruck (148) von F erkennt, die Gleichungen (145) folgendermaßen:

$$\left\{\begin{aligned} \mathfrak{E}_x &= \frac{\partial^2 F}{\partial x \, \partial z} & \mathfrak{H}_x &= \frac{1}{c} \frac{\partial^2 F}{\partial y \, \partial t} \\ \mathfrak{E}_y &= \frac{\partial^2 F}{\partial y \, \partial z} & \mathfrak{H}_y &= -\frac{1}{c} \frac{\partial^2 F}{\partial x \, \partial t} \\ \mathfrak{E}_z &= \frac{\partial^2 F}{\partial z^2} & \mathfrak{H}_z &= 0. \end{aligned}\right. \tag{149}$$

Die elektrischen Gleichungen besagen, daß in der Nähe des Anfangspunktes der Koordinaten das elektrische Feld den Charakter eines elektrostatischen hat:

$$\mathfrak{E} = \operatorname{grad} \frac{\partial F}{\partial z} = -\operatorname{grad} \varphi$$

mit der Potentialfunktion:

$$\varphi = -\frac{\partial F}{\partial z} = \frac{z}{r^3} \cdot f(t),$$

also entsprechend einem nach der positiven z-Achse gerichteten elektrischen Dipol vom Moment $f(t)$. Die magnetischen Gleichungen besagen, daß das magnetische Feld in der Nähe des Anfangspunktes der Koordinaten herrührt von einem in der Richtung der z-Achse fließenden Stromelement, dessen Intensität multipliziert mit der Länge den Wert $\dot{f}(t)$ besitzt. Dies ist genau dasjenige Stromelement, welches durch die Momentänderung des obigen Dipols bedingt ist.

Hierdurch ist gezeigt, daß durch die Gleichungen (145) und (148), welche im ganzen unendlichen Raume mit Ausschluß des Punktes $r = 0$ und seiner nächsten Umgebung Gültigkeit besitzen, dasjenige elektromagnetische Feld dargestellt wird, welches von einem im Koordinatenanfangspunkt befindlichen nach der z-Achse gerichteten elektrischen Dipol vom Moment $f(t)$ hervorgerufen wird. Um diesen Dipol mit dem anfangs betrachteten Oszillator identifizieren zu können, ist nur noch die Einführung der Voraussetzung nötig, daß für alle Zeiten die Lineardimensionen des Oszillators klein sind gegen die Größe $\frac{cf}{\dot{f}}$, also auch, falls der Oszillator periodisch schwingt, klein gegen die Wellenlänge dieser Schwingung im Vakuum. Denn sonst würde das elektromagnetische Feld in der Nähe des Oszillators nicht mehr durch $f(t)$ und $\dot{f}(t)$ allein bestimmt sein, vielmehr würde die Schwingung eine merkliche Zeit brauchen, um sich von einer Stelle des Oszillators zu einer anderen fortzupflanzen.

§ 108. Zur Bestimmung der vom Oszillator ausgestrahlten Energie berechnen wir die Energiemenge, welche durch eine um den Oszillator als Mittelpunkt beschriebene Kugelfläche gemäß dem Poyntingschen Satze nach außen strömt. Doch darf man nicht etwa die nach dem genannten Satze in einem unendlich kleinen Zeitintervall dt durch die Kugelfläche nach außen strömende Energie gleich der während desselben Zeitintervalls vom Oszillator ausgestrahlten Energie setzen. Denn im allgemeinen strömt die elektromagnetische Energie nicht immer in der Richtung nach außen, sondern sie strömt abwechselnd hin und her, und man würde auf diese Weise für die Größe der Ausstrahlung zu Werten gelangen, welche bald positiv, bald negativ sind und überdies noch wesentlich vom Radius der zugrunde gelegten Kugel abhängen, und zwar in der Weise, daß sie mit abnehmendem Radius ins Unbegrenzte wachsen, — was dem Begriff der ausgestrahlten Energie widerspricht. Diese wird vielmehr nur dann unabhängig von dem Kugelradius r gefunden, wenn man nicht für ein Zeitelement dt, sondern für eine endliche, hinreichend groß gewählte Zeit den Betrag der im ganzen durch die Kugelfläche nach außen strömenden Energie berechnet. Sind die Schwingungen rein periodisch, so kann man dafür die Zeit einer Periode wählen; sind sie es nicht, was wir hier der Allgemeinheit wegen annehmen müssen, so läßt sich für die mindestens notwendige Größe der Zeit von vornherein kein anderes allgemeines Kriterium angeben, als dasjenige, daß die ausgestrahlte Energie sich als unabhängig ergibt von dem Radius der zugrunde gelegten Kugel. In der Theorie der Wärmestrahlung handelt es sich immer um so schnelle Schwingungen, daß man hier praktisch stets mit einer gegen eine Sekunde sehr kleinen Zeit auskommt.

Am einfachsten gestaltet sich die Berechnung der durch die Kugelfläche strömenden Energie, wenn wir den Radius r der Kugel so groß wählen, daß für alle Zeiten $\frac{r}{c}\dot{f}$ groß gegen f. Dann ist auch $\frac{r}{c}\ddot{f}$ groß gegen $\dot{f}$, und a fortiori $\frac{r^2}{c^2}\ddot{f}$ groß gegen f. Hierdurch gehen die Feldgleichungen (145) mit Berücksichtigung von (146) und (148) über in:

$$\mathfrak{E}_x = \frac{xz}{c^2 r^3}\ddot{f}\left(t - \frac{r}{c}\right) \qquad \mathfrak{H}_x = -\frac{y}{c^2 r^2}\ddot{f}\left(t - \frac{r}{c}\right)$$

$$\mathfrak{E}_y = \frac{yz}{c^2 r^3}\ddot{f}\left(t - \frac{r}{c}\right) \qquad \mathfrak{H}_y = \frac{x}{c^2 r^2}\ddot{f}\left(t - \frac{r}{c}\right)$$

$$\mathfrak{E}_z = -\frac{x^2 + y^2}{c^2 r^3}\ddot{f}\left(t - \frac{r}{c}\right) \qquad \mathfrak{H}_z = 0.$$

Hier sind die Verhältnisse aller Komponenten unabhängig von der Zeit, also auch ihre Richtungen konstant, und die Gleichungen:

$$x\mathfrak{E}_x + y\mathfrak{E}_y + z\mathfrak{E}_z = 0,$$
$$x\mathfrak{H}_x + y\mathfrak{H}_y + z\mathfrak{H}_z = 0,$$
$$\mathfrak{E}_x\mathfrak{H}_x + \mathfrak{E}_y\mathfrak{H}_y + \mathfrak{E}_z\mathfrak{H}_z = 0$$

zeigen, daß die elektrische Feldstärke $\mathfrak{E}$, die magnetische Feldstärke $\mathfrak{H}$ und der Radiusvektor r gegenseitig senkrecht stehen. Die magnetische Feldstärke steht außerdem senkrecht auf der durch die z-Achse und r bestimmten Meridianebene, die elektrische Feldstärke liegt also in der Meridianebene. Es sind mithin reine Transversalwellen, senkrecht der Meridianebene polarisiert, die sich nach außen fortpflanzen, mit der Feldstärke:

$$\mathfrak{E} = \mathfrak{H} = \frac{\sqrt{x^2+y^2}}{c^2 r^2} \ddot{f}\left(t - \frac{r}{c}\right) = \frac{\sin\vartheta}{c^2 r} \ddot{f}\left(t - \frac{r}{c}\right),$$

wenn ϑ den Winkel des Radiusvektor r mit der z-Achse, der Achse des Oszillators, bezeichnet.

Nun ist nach dem POYNTINGschen Satz die in der Zeit dt durch das Kugelflächenelement $d\sigma = r^2 d\Omega$ nach außen strömende Energie:

$$\frac{c}{4\pi} dt\, d\sigma\, \mathfrak{E}\mathfrak{H} = \frac{\sin^2\vartheta}{4\pi c^3} \ddot{f}^2\left(t - \frac{r}{c}\right) d\Omega\, dt \tag{150}$$

$$= \frac{1}{4\pi c^3} \sin^3\vartheta\, d\vartheta\, d\varphi\, \ddot{f}^2\left(t - \frac{r}{c}\right) dt,$$

also für die ganze Kugelfläche (φ von 0 bis 2π, ϑ von 0 bis π) und für das Zeitintervall von t bis $t + T$:

$$\frac{2}{3c^3}\int_t^{t+T} \ddot{f}^2\left(t - \frac{r}{c}\right) dt.$$

In diesem Ausdruck kommt der Kugelradius r nur noch in dem Argument von $\ddot{f}$ vor, und in der Tat ist die hier berechnete durch die Kugelfläche vom Radius r in der Zeit von t bis $t + T$ nach außen strömende Energie offenbar gleich der in dem gleichlangen, aber um $\frac{r}{c}$ zurückliegenden Zeitintervall, von $t - \frac{r}{r}$ bis $t - \frac{r}{c} + T$, von dem im Mittelpunkt der Kugel befindlichen Oszillator ausgestrahlten Energie.

Daher erhalten wir für die vom Oszillator in der Zeit von t bis $t+T$ a u s g e s t r a h l t e oder e m i t t i e r t e E n e r g i e den Ausdruck:

$$\frac{2}{3c^3}\int_t^{t+T} \ddot{f}^2(t)\, dt. \tag{151}$$

Nach dem Prinzip der Erhaltung der Energie ist die in der Zeit T ausgestrahlte Energie gleich der in der nämlichen Zeit erfolgten Abnahme der Schwingungsenergie des Oszillators U, also:

$$\frac{2}{3c^3}\int_t^{t+T} \ddot{f}^2(t)\,dt = -\int_t^{t+T} dU$$

oder:

$$\int_t^{t+T}\left(\frac{dU}{dt} + \frac{2}{3c^3}\ddot{f}^2(t)\right)dt = 0\,. \tag{152}$$

§ 109. Aus dieser für ein relativ großes Zeitintervall T gültigen Beziehung läßt sich naturgemäß kein eindeutiger Schluß ziehen auf das für jeden einzelnen Zeitpunkt t gültige Schwingungsgesetz des Oszillators, und in der Tat genügen auch die hier gegebenen Daten noch durchaus nicht, um den Verlauf der Schwingungen, bei gegebenem Anfangszustand, bis ins einzelne eindeutig zu bestimmen. Damit eine exakte Lösung des Schwingungsproblems überhaupt vorhanden ist, müßte die Natur des Oszillators in allen ihren Einzelheiten, an der Oberfläche und im Innern, genau bekannt sein. Dann würde aber die Behandlung der Aufgabe mit solchen mathematischen Schwierigkeiten verknüpft sein, daß die allgemeine Durchführung der Theorie nicht wohl möglich wäre.

Nun ist zu bedenken, daß es in der ganzen Theorie der Licht- und Wärmestrahlung gar nicht auf die exakte Lösung des Schwingungsproblems, d. h. auf die absolut genaue Berechnung der Schwingungsfunktion $f(t)$ ankommt, sondern nur auf eine angenäherte Lösung von demselben Grade der Genauigkeit, wie ihn die feinsten physikalischen Messungen liefern können. Diese beziehen sich aber, wie schon in § 3 betont wurde, immer nur auf solche Zeitintervalle, welche gegen die Dauer einer einzigen Schwingungsperiode außerordentlich groß sind. Daher können die Ergebnisse auch der feinsten Strahlungsmessungen niemals über den Inhalt der Energiegleichung (152) hinausgehen, und jede Differentialgleichung für $f(t)$, welche mit dieser Gleichung verträglich ist, liefert für den Oszillator ein zulässiges Schwingungsgesetz.[1] Von dieser Überlegung Gebrauch machend wollen wir

[1] Dies gilt immer, wenn nur hinreichend große Werte von T in Betracht kommen. Für kleinere Werte von T läßt sich möglicherweise

nun aus der Gleichung (152) das denkbar einfachste Schwingungsgesetz für den Oszillator ableiten.

Würde man den in (152) mit dt multiplizierten Ausdruck direkt gleich Null setzen, so erhielte man als Schwingungsgesetz eine nichtlineare Differentialgleichung. Daher formen wir das Integral noch um, indem wir schreiben:

$$\int_t^{t+T} \left(\frac{d}{dt} \left(U + \frac{2}{3\,c^3} \dot{f} \ddot{f} \right) - \frac{2}{3\,c^3} \dot{f} \dddot{f} \right) dt = 0 . \qquad (153)$$

Um diese mit (152) ganz identische Gleichung noch weiter zu vereinfachen, führen wir eine neue naheliegende Voraussetzung ein, die wir fortan stets festhalten wollen, und deren physikalische Bedeutung im nächsten Paragraph erhellen wird. Wir setzen nämlich fest, daß für alle Zeiten:

$$\frac{1}{c^3} \dot{f} \ddot{f} \text{ klein gegen } U. \qquad (154)$$

Diese Bedingung läßt sich durch geeignete Wahl der Konstanten K und L allgemein erfüllen. Denn nach (142) ist im allgemeinen:

$$\sqrt{L}\,\dot{f} \text{ von der Größenordnung } \sqrt{K} \cdot f,$$

also auch $\quad \sqrt{L}\,\ddot{f}$ von der Größenordnung $\sqrt{K}\,\dot{f}$ oder $\dfrac{K}{\sqrt{L}} f$;

folglich, durch Substitution in (154):

$$\frac{1}{c^3} \cdot \sqrt{\frac{K}{L}} f \cdot \frac{K}{L} f \text{ klein gegen } K f^2$$

oder

$$\frac{1}{c^3} \sqrt{\frac{K}{L^3}} \text{ klein gegen } 1 . \qquad (155)$$

Wir wollen zur Abkürzung schreiben:

$$\frac{2\pi}{3\,c^3} \sqrt{\frac{K}{L^3}} = \sigma . \qquad (156)$$

Nach dieser ein für allemal gemachten Festsetzung, daß σ eine kleine Zahl ist, erhalten wir aus (153) in merklicher Annäherung:

---

an Stelle der einfachen linearen Differentialgleichung (158) ein anderes Schwingungsgesetz aufstellen, das den Vorgängen in der Natur noch besser angepaßt ist.

$$\int_{t}^{t+T}\left(\frac{dU}{dt} - \frac{2}{3c^3} \dot{f}\dddot{f}\right) dt = 0 \tag{157}$$

und daraus, indem wir den mit dt multiplizierten Ausdruck gleich Null setzen und den Wert dU aus (142) substituieren, die lineare homogene Differentialgleichung:

$$Kf + L\ddot{f} - \frac{2}{3c^3}\dddot{f} = 0 \tag{158}$$

als Schwingungsgleichung des Oszillators. Dieselbe unterscheidet sich von der Gleichung (143) für ungedämpfte Schwingungen durch das Dämpfungsglied mit $\dddot{f}$.[1]

§ 110. Um die Differentialgleichung (158) zu integrieren, setzen wir:

$$f(t) = e^{\omega t + \omega'}. \tag{159}$$

Dann wird die Differentialgleichung befriedigt, wenn

$$K + L\omega^2 - \frac{2}{3c^3}\omega^3 = 0. \tag{160}$$

Diese in ω kubische Gleichung hat eine reelle positive Wurzel und zwei komplexe Wurzeln. Die erstere hat keine physikalische Bedeutung, weil hierfür die Funktion $f(t)$ mit wachsender Zeit ungeheuer große Werte annimmt. Daher berücksichtigen wir nur die komplexen Wurzeln, indem wir setzen:

$$\omega = \alpha \pm \beta i \quad (\beta > 0) \tag{161}$$

und nach Substitution in (160) das Reelle vom Imaginären trennen:

$$K + L(\alpha^2 - \beta^2) - \frac{2}{3c^3}(\alpha^3 - 3\alpha\beta^2) = 0$$

und:
$$2L\alpha\beta - \frac{2}{3c^3}(3\alpha^2\beta - \beta^3) = 0.$$

Die zweite Gleichung ergibt:

$$\beta^2 = -3c^3L\alpha + 3\alpha^2 \tag{162}$$

und dies in die erste Gleichung substituiert:

$$K + 3c^3L^2\alpha - 8L\alpha^2 + \frac{16\alpha^3}{3c^3} = 0.$$

---

[1] Eine direkte Ableitung dieser Gleichung aus der Elektronentheorie ist kürzlich entwickelt worden von M. Abraham, Elektromagnetische Theorie der Strahlung (Leipzig, B. G. Teubner), p. 72, 1905.

Diese Gleichung in α hat nur eine einzige reelle Wurzel. Schreibt man sie mit Benutzung von (156) in der Form:

$$\frac{9\,\sigma^2 c^6 L^3}{4\,\pi^2} + 3\,c^3 L^2 \alpha - 8\,L\,\alpha^2 + \frac{16\,\alpha^3}{3\,c^3} = 0$$

und bedenkt, daß σ eine kleine Zahl ist, so erhellt, daß die reelle Wurzel α nahe gleich Null ist, und es ergibt sich mit Weglassung der Glieder mit höheren Potenzen von α als erste Annäherung:

$$\frac{9\,\sigma^2 c^6 L^3}{4\,\pi^2} + 3\,c^3 L^2 \alpha = 0\,,$$

$$\alpha = -\frac{3\,c^3 \sigma^2 L}{4\,\pi^2} = -\frac{K}{3\,c^3 L^2} \tag{163}$$

und dementsprechend nach (162):

$$\beta = \sqrt{\frac{K}{L}}\,. \tag{164}$$

Die gefundenen Werte von α und β denken wir uns in (161) und in (159) substituiert, setzen dann für ω' eine beliebige komplexe Konstante, und erhalten hierauf durch Abspaltung des reellen Teiles der Funktion $f(t)$ vom imaginären die reelle Lösung der Schwingungsgleichung (158) in folgender Form:

$$f(t) = C \cdot e^{\alpha t} \cos(\beta\, t - \vartheta)\,. \tag{165}$$

Der Oszillator führt also schwach gedämpfte Schwingungen aus, deren Periode und Dämpfung durch die Werte von β und α bestimmt sind. Die Amplitude C und die Phasenkonstante ϑ hängen vom Anfangszustand ab.

Bezeichnen wir mit ν_0 die Anzahl der Schwingungen in der Zeiteinheit, so ist

$$\nu_0 = \frac{\beta}{2\,\pi} = \frac{1}{2\,\pi}\sqrt{\frac{K}{L}}\,, \tag{166}$$

also, nach Gleichung (144), bis auf die begangene kleine Vernachlässigung ebensogroß wie im Falle der ungedämpften Schwingungen.

Für das logarithmische Dämpfungsdekrement der Schwingungen, d. h. für den natürlichen Logarithmus des Quotienten zweier um die Zeit $\frac{1}{\nu_0}$ einer Schwingung auseinanderliegenden Werte von $f(t)$ ergibt sich:

(167) $$\log e^{-\frac{\alpha}{\nu_0}} = -\frac{\alpha}{\nu_0} = \frac{K}{3\,c^3 L^2} \cdot 2\,\pi \sqrt{\frac{L}{K}} = \sigma,$$

wodurch die oben in (156) zur Abkürzung eingeführte Konstante σ eine einfache physikalische Bedeutung erhält.

Statt durch die Energiekonstanten K und L kann man die Natur des Oszillators auch durch die Schwingungskonstanten ν_0 und σ charakterisieren, und erhält dann aus (166) und (167):

(168) $$K = \frac{16\,\pi^4\,\nu_0^3}{3\,\sigma\,c^3}, \qquad L = \frac{4\,\pi^2\,\nu_0}{3\,\sigma\,c^3}.$$

Mit den neuen Konstanten lautet die Schwingungsgleichung (165):

(169) $$f(t) = C e^{-\sigma \nu_0 t} \cos(2\,\pi\,\nu_0\,t - \vartheta).$$

§ 111. Nachdem wir bisher nur solche Schwingungen des Oszillators betrachtet hatten, die ohne jede Einwirkung von außen erfolgen und daher lediglich in dem Abklingen einer durch einen beliebig gegebenen Anfangszustand bedingten Erregung bestehen, wollen wir nun auch noch den allgemeineren Fall untersuchen, daß gleichzeitig von außen her auf den Oszillator gewisse Wirkungen ausgeübt werden, oder mit anderen Worten, daß der Oszillator sich in einem von vornherein gegebenen elektromagnetischen Felde befindet. Die elektrische und die magnetische Feldstärke dieses äußeren Feldes bezeichnen wir von jetzt an mit $\mathfrak{E}$ und mit $\mathfrak{H}$. Dann erweitert sich die Energiegleichung (157) dadurch, daß die Energie U des Oszillators jetzt außer durch die Ausstrahlung von Energie auch noch durch die Arbeit geändert wird, welche das äußere elektromagnetische Feld an dem Oszillator leistet, und diese Arbeit wird, da die Achse des elektrischen Dipols mit der z-Achse zusammenfällt, für ein Zeitelement dt ausgedrückt durch die Größe $\mathfrak{E}_z \cdot df = \mathfrak{E}_z \cdot \dot{f} \cdot dt$, wobei $\mathfrak{E}_z$ die z-Komponente der äußeren elektrischen Feldstärke am Orte des Oszillators bezeichnet, d. h. derjenigen elektrischen Feldstärke, welche am Orte des Oszillators bestehen würde, wenn derselbe gar nicht vorhanden wäre. Die übrigen Komponenten des äußeren Feldes haben keinen Einfluß auf die Schwingungen des Oszillators.

Da nun die Energie U des Oszillators im Zeitelement dt um den Betrag der angegebenen äußeren Arbeit vergrößert wird, so ist die während der hinlänglich großen Zeit T vom Oszillator absorbierte Energie:[1]

$$\int_t^{t+T} \mathfrak{E}_z \dot{f}\, dt \tag{170}$$

und die vervollständigte Energiegleichung (157) lautet:

$$\int_t^{t+T} \left(\frac{dU}{dt} - \frac{2}{3c^3} \dot{f} \dddot{f} - \mathfrak{E}_z \dot{f} \right) dt = 0 .$$

Daraus folgt, wenn man wieder den mit dt multiplizierten Ausdruck $= 0$ setzt und den Wert von U aus (142) substituiert, als Schwingungsgleichung des Oszillators:

$$K f + L \ddot{f} - \frac{2}{3c^3} \dddot{f} = \mathfrak{E}_z \tag{171}$$

oder, wenn man nach (168) für die Konstanten K und L die Konstanten ν_0 und σ einführt:

$$16 \pi^4 \nu_0^3 f + 4 \pi^2 \nu_0 \ddot{f} - 2 \sigma \dddot{f} = 3 \sigma c^3 \mathfrak{E}_z . \tag{172}$$

Sobald man aus dieser Gleichung mittels des gegebenen Anfangszustandes und der gegebenen äußeren Feldstärke $\mathfrak{E}_z$ die Schwingungsfunktion $f(t)$ des Oszillators berechnet hat, ist auch die Aufgabe gelöst, die Rückwirkung des Oszillators auf das äußere elektromagnetische Feld zu bestimmen. Denn außerhalb des Oszillators superponieren sich die Komponenten des ursprünglich gegebenen Feldes: die „primäre“ Welle und die Komponenten der vom Oszillator emittierten Kugelwelle (145): die „sekundäre“ Welle, wobei F durch (148) gegeben ist, überall einfach durch Addition, und damit ist der ganze Vorgang für alle Zeiten eindeutig bestimmt.

---

[1] Man sieht hieraus, daß die „absorbierte Energie“ im allgemeinen auch negativ sein kann, d. h. daß unter Umständen durch die auffallende Strahlung dem Oszillator direkt Energie entzogen wird. Beispiele für diesen (in der Wärmestrahlung nicht realisierbaren) Fall werden wir im ersten Kapitel des fünften Abschnitts finden.

Zweites Kapitel. Ein Resonator unter der Einwirkung einer ebenen periodischen Welle.

§ 112. Als erste Anwendung der Schwingungsgleichung (172) betrachten wir den Fall, daß eine ebene einfach periodische Welle, die sich längs der x-Achse fortpflanzen möge und deren elektrische Feldstärke in die Richtung der z-Achse fällt, den im vorigen Kapitel betrachteten Oszillator trifft. Dann haben wir für die primäre, erregende Welle nach den allgemeinen MAXWELLschen Gleichungen (52):

$$\mathfrak{E}_x = 0 \qquad \mathfrak{H}_x = 0$$

$$\mathfrak{E}_y = 0 \qquad \mathfrak{H}_y = - C \cos\left[2\pi\nu\left(t - \frac{x}{c}\right) - \vartheta\right]$$

$$\mathfrak{E}_z = C \cos\left[2\pi\nu\left(t - \frac{x}{c}\right) - \vartheta\right] \qquad \mathfrak{H}_z = 0 .$$

Hierbei ist ν die Schwingungszahl, C (positiv) die Amplitude und ϑ die Phasenkonstante der primären Welle.

Die Schwingungen des Oszillators werden im allgemeinen von seinem Anfangszustand abhängen. Wenn aber die Zeit t hinlänglich groß genommen wird, nämlich wenn $\sigma\nu_0 t$ eine große Zahl ist, so wird nach (169) der Anfangszustand für die Schwingungen des Oszillators gänzlich bedeutungslos werden, und dieselben sind dann durch die primäre Welle allein vollkommen bestimmt. In diesem Falle, den wir im folgenden ausschließlich betrachten werden, spielt der Oszillator die Rolle eines Resonators, seine Schwingungen erfolgen in derselben Periode wie die der ihn erregenden primären Welle. Setzt man in die Schwingungsgleichung (172) den Wert von $\mathfrak{E}_z$ für den Ort des Oszillators $x = 0$:

$$\mathfrak{E}_z = C \cos(2\pi\nu t - \vartheta), \tag{173}$$

so ergibt dieselbe integriert:

$$f(t) = \frac{3c^3 C \sin\gamma}{16\pi^3\nu^3} \cos(2\pi\nu t - \vartheta - \gamma), \tag{174}$$

wobei

$$\operatorname{ctg}\gamma = \frac{\pi\nu_0(\nu_0^2 - \nu^2)}{\sigma\nu^3}. \tag{175}$$

Um den Winkel γ eindeutig zu bestimmen, setzen wir noch fest, daß er zwischen 0 und π liegt. Dann ist $\sin\gamma$ ebenso wie C stets positiv.

§ 113. Wenn das Verhältnis der Schwingungszahl ν der primären Welle zu der Eigenschwingungszahl ν_0 des Resonators mittelgroß und von 1 merklich verschieden ist, so nimmt, da σ eine kleine Zahl ist, $\operatorname{ctg}\gamma$ große Werte an, positiv oder negativ, je nachdem ν kleiner oder größer als ν_0. Dadurch wird der Winkel γ nahezu 0 oder nahezu π, und die dem $\sin\gamma$ proportionale Schwingungsamplitude des Resonators wird klein. Auch für die Grenzfälle $\nu = 0$ und $\nu = \infty$ werden die Schwingungen des Resonators unmerklich, da in beiden Fällen der Quotient $\frac{\sin\gamma}{\nu^3}$ kleine Werte annimmt. Damit also ein merkliches Mitschwingen des Resonators stattfindet, müssen die Schwingungszahlen ν und ν_0 nahezu übereinstimmen. In diesem Falle weicht die Phase der Resonatorschwingung merklich ab von der der primären Welle, da der Phasenunterschied gerade γ beträgt. Für $\nu = \nu_0$ wird die Schwingungsamplitude des Resonators ein Maximum, und der Phasenunterschied γ wird gleich $\frac{\pi}{2}$. Dann geht in demselben Augenblick, wo die äußere Feldstärke $\mathfrak{E}_z$ ihren Maximalwert erreicht, $f(t)$ durch Null hindurch, d. h. der elektrische Dipol ist in diesem Augenblick ungeladen und daher der Ladungsstrom $\dot{f}(t)$ in ihm in der größten Entwickelung begriffen. Überhaupt ist dann $\dot{f}(t)$ einfach proportional $\mathfrak{E}_z$, und damit die vom Resonator absorbierte Energie (170) ein Maximum.

§ 114. Im allgemeinen Falle, bei beliebigem Verhältnis von ν zu ν_0, ist die vom Resonator absorbierte Energie nach (170) pro Zeiteinheit:

$$\frac{1}{T}\int\limits_t^{t+T} \mathfrak{E}_z \dot{f}\, dt = \overline{\mathfrak{E}_z \dot{f}},$$

wo der Wert von $\mathfrak{E}_z$ aus (173) zu entnehmen ist, während $\dot{f}$ sich nach (174) ergibt zu:

$$\dot{f} = -\frac{3\,c^3\,C\sin\gamma}{8\,\pi^2\,\nu^2}\sin(2\pi\,\nu t - \vartheta - \gamma) \tag{176}$$

$$= -\frac{3\,c^3\,C\sin\gamma}{8\,\pi^2\,\nu^2}\left[\sin(2\pi\,\nu t - \vartheta)\cos\gamma - \cos(2\pi\,\nu t - \vartheta)\sin\gamma\right].$$

Bedenkt man nun, daß der zeitliche Mittelwert von $\sin(2\pi\,\nu t - \vartheta)$ $\cos(2\pi\,\nu t - \vartheta)$ gleich Null, und der von $\cos^2(2\pi\,\nu t - \vartheta)$ gleich $\frac{1}{2}$

ist, so folgt für die pro Zeiteinheit vom Resonator absorbierte Energie:

$$\overline{\mathfrak{E}_z \dot{f}} = \frac{3 c^3 C^2 \sin^2 \gamma}{16 \pi^2 \nu^2}, \tag{177}$$

also positiv. Da der ganze Vorgang periodisch ist, so muß diese Energie zugleich auch die emittierte Energie darstellen. In der Tat ergibt sich aus (176):

$$\ddot{f} = - \frac{3 c^3 C \sin \gamma}{4 \pi \nu} \cos(2 \pi \nu t - \vartheta - \gamma)$$

und für die vom Resonator pro Zeiteinheit emittierte Energie nach (151):

$$\frac{2}{3 c^3} \overline{\ddot{f}^2} = \frac{3 c^3 C^2 \sin^2 \gamma}{16 \pi^2 \nu^2}.$$

Für den zeitlichen Mittelwert der Energie des Resonators ergibt sich aus (142) und (168) auf ähnliche Weise:

$$U = \tfrac{1}{2} K \overline{f^2} + \tfrac{1}{2} L \overline{\dot{f}^2}$$

$$= \frac{3 c^3 \nu_0 (\nu_0{}^2 + \nu^2)}{64 \pi^2 \sigma \nu^6} C^2 \sin^2 \gamma. \tag{178}$$

§ 115. Fassen wir nun die Rückwirkung der Resonatorschwingungen auf die primäre Welle kurz ins Auge. Während sich die Komponenten der Feldstärken der primären und der sekundären, vom Resonator emittierten Welle überall einfach durch Addition übereinanderlagern, ist das mit den Energiestrahlungen keineswegs der Fall. Denn schon aus dem Umstand, daß der Resonator vermöge seiner Schwingungen nach allen Richtungen des Raumes Energie durch Strahlung emittiert, folgt nach dem Prinzip der Erhaltung der Energie, daß er durch seine Schwingungen gleichzeitig der primären Welle Energie entziehen muß. In der Tat ergibt sich bei näherer Betrachtung, daß die primäre und die sekundäre Welle da, wo sie sich gemeinsam fortpflanzen, nämlich in denjenigen vom Resonator ausgehenden Richtungen, die mit der positiven x-Achse kleine Winkel bilden, stets in der Weise interferieren, daß sie sich gegenseitig schwächen, und eine direkte Berechnung der Energieströmung nach dem POYNTINGschen Satz, auf die aber hier nicht näher eingegangen zu werden braucht, zeigt, daß die Schwächung der primären Welle im ganzen genommen gerade gleich ist der vom Resonator absorbierten Energie. Der Resonator absorbiert also pro Zeiteinheit aus der primären Welle einen bestimmten, durch (177) ausgedrückten Energiebetrag und zerstreut ihn nach allen Richtungen.

Drittes Kapitel. Ein Resonator unter der Einwirkung stationärer Wärmestrahlung. Entropie und Temperatur des Resonators.

§ 116. Nachdem wir im vorigen Kapitel die Schwingungsgleichung (172) eines linearen Oszillators auf den speziellen Fall angewendet haben, daß eine ebene periodische Welle als Erreger dient, wollen wir uns jetzt den Oszillator einer stationären Wärmestrahlung ausgesetzt denken. Dieser Fall unterscheidet sich insofern wesentlich von dem vorigen, als eine ebene periodische Welle niemals, auch nicht bei entsprechend großer Schwingungszahl, als Wärmestrahlung gedeutet werden kann. Denn zu einer endlichen Wärmestrahlungsintensität gehört nach § 16 immer auch ein endlicher Öffnungswinkel der Strahlen, und nach § 18 immer ein Spektralbezirk von endlicher Breite. Eine absolut ebene und absolut periodische Welle besitzt aber den Öffnungswinkel Null und die Spektralbreite Null. Daher kann auch bei einer ebenen periodischen Welle weder von Entropie noch von Temperatur der Strahlung die Rede sein. Die nähere Erklärung dieses für die elektromagnetische Theorie der Wärmestrahlung prinzipiell wichtigen Umstandes werden wir im nächsten Abschnitt kennen lernen.

Wir stellen uns also nun den Oszillator in einem allseitig durch vollständig reflektierende Wände begrenzten, von beliebiger Wärmestrahlung erfüllten Vakuum befindlich vor. Dann wird sich im Verlauf der Zeit, wie in jedem nach außen abgeschlossenen, mit beliebiger emittierender und absorbierender Substanz besetzten Raum, ein stationärer Zustand herausbilden, in welchem das Vakuum von unpolarisierter, nach allen Richtungen gleichmäßiger Wärmestrahlung durchzogen wird. Von dieser Wärmestrahlung absorbiert und emittiert der Oszillator nur solche Strahlen, deren Schwingungszahl nahe gleich ν_0 ist; folglich übt er auf diese, und nur auf diese, einen Einfluß aus. Für alle übrigen Strahlenarten verhält er sich wie eine diathermane Substanz: sie streichen über ihn hinweg, ohne ihn zu alterieren oder von ihm alteriert zu werden.

§ 117. Wir fragen nun nach dem Zusammenhang, welcher nach Eintritt des stationären Zustandes zwischen den Resonatorschwingungen und der Wärmestrahlung im Vakuum besteht. Die Gleichung (172) verlangt nur die Berücksichtigung der Komponente $\mathfrak{E}_z$ der erregenden Welle am Orte des Oszillators als Funktion der Zeit. Diese Größe setzt sich aus allen den Oszillator treffenden Wärmestrahlen zusammen und läßt sich, wie kompliziert sie auch sein möge, für ein begrenztes Zeitintervall, etwa von $t = 0$ bis $t = \mathfrak{T}$, als Fourierche Reihe schreiben:

$$\mathfrak{E}_z = \sum_{n=0}^{n=\infty} C_n \cos\left(\frac{2\pi n t}{\mathfrak{T}} - \vartheta_n\right), \tag{179}$$

wobei die Summation über alle positiven ganzen Zahlen n zu erstrecken ist, während die Konstanten C_n (positiv) und ϑ_n von Glied zu Glied beliebig variieren können. Das Zeitintervall $\mathfrak{T}$, die Grundperiode der Fourierschen Reihe, wollen wir so groß wählen, daß alle Zeiten t, die wir in der Folge betrachten, in dies Zeitintervall hineinfallen, also $0 < t < \mathfrak{T}$. Daher ist jedenfalls das Produkt $\nu_0 \mathfrak{T}$ eine ungeheuer große Zahl. Außerdem aber wollen wir immer t so groß nehmen, daß der Zustand des Oszillators zur Zeit $t = 0$ auf die Vorgänge zur Zeit t gar nicht mehr von Einfluß ist. Dies erfordert, nach (169), daß das Produkt $\sigma \nu_0 t$, und um so mehr

$$\sigma \nu_0 \mathfrak{T} \text{ eine große Zahl} \tag{180}$$

ist. Diese Bedingungen lassen sich ohne weiteres erfüllen, da ja die Größe von $\mathfrak{T}$ nach oben hin durch nichts begrenzt ist.

§ 118. Wenn wir auch über die Funktion $\mathfrak{E}_z$ im einzelnen nichts wissen, so steht sie doch in einem angebbaren Zusammenhang mit den Eigenschaften der Wärmestrahlung. Zunächst haben wir für die räumliche Strahlungsdichte im Vakuum nach der Maxwellschen Theorie:

$$u = \frac{1}{8\pi} \cdot \left(\overline{\mathfrak{E}_x^2} + \overline{\mathfrak{E}_y^2} + \overline{\mathfrak{E}_z^2} + \overline{\mathfrak{H}_x^2} + \overline{\mathfrak{H}_y^2} + \overline{\mathfrak{H}_z^2}\right).$$

Da nun wegen des stationären, nach allen Richtungen gleichmäßigen Strahlungszustandes die genannten sechs Mittelwerte einander gleich sind, so folgt:

$$u = \frac{3}{4\pi} \overline{\mathfrak{E}_z^2}$$

und aus (179):

$$u = \frac{3}{8\pi} \sum C_n^2, \tag{181}$$

ferner für die spezifische Intensität der in irgendeiner Richtung fortschreitenden Strahlung nach (21):

$$K = \frac{c\,u}{4\pi} = \frac{3c}{32\pi^2} \sum C_n^2. \tag{182}$$

§ 119. Nehmen wir nun auch die spektrale Zerlegung der letzten beiden Gleichungen vor. Zunächst haben wir nach (22):

$$u = \int_0^\infty \mathfrak{u}_\nu \, d\nu = \frac{3}{8\pi} \sum_0^\infty C_n^2. \tag{183}$$

Auf der rechten Seite der Gleichung zerfällt die Summe $\sum$ in die einzelnen den Ordnungszahlen n entsprechenden Glieder, von denen jedes einer einfach periodischen „Partialschwingung" mit der Schwingungszahl $\nu = \frac{n}{\mathfrak{T}}$ entspricht. Genau genommen stellt diese Beziehung keine stetige Aufeinanderfolge von Schwingungszahlen ν dar, da n eine ganze Zahl ist. Aber n ist für die hier in Betracht kommenden Schwingungszahlen so ungeheuer groß, daß die den fortlaufenden Werten von n entsprechenden Schwingungszahlen ν sehr dicht beieinander liegen. Daher umfaßt auch das Intervall $d\nu$, obwohl gegen ν unendlich klein, dennoch eine große Anzahl von Partialschwingungen, etwa n', wobei:

$$d\nu = \frac{n'}{\mathfrak{T}}. \tag{184}$$

Setzen wir nun in der Gleichung (183) die dem Intervall $d\nu$ entsprechenden Energiedichten, die ja von denen der übrigen Spektralbezirke unabhängig sind, auf beiden Seiten einander gleich, so ergibt sich:

$$\mathfrak{u}_\nu \, d\nu = \frac{3}{8\pi} \sum_n^{n+n'} C_n^2$$

oder nach (184):

$$\mathfrak{u}_\nu = \frac{3\mathfrak{T}}{8\pi} \cdot \frac{1}{n'} \sum_n^{n+n'} C_n^2 = \frac{3\mathfrak{T}}{8\pi} \cdot \overline{C_n^2}, \tag{185}$$

wobei wir mit $\overline{C_n^2}$ den Mittelwert von C_n^2 in dem Intervall von n bis $n+n'$ bezeichnen. Daß ein solcher Mittelwert, dessen Größe unabhängig ist von n', falls nur n' klein gegen n genommen wird, überhaupt existiert, ist natürlich nicht von vornherein selbstverständlich, sondern durch eine besondere, der stationären Wärmestrahlung eigentümliche Eigenschaft der Funktion $\mathfrak{E}_z$ bedingt. Dagegen läßt sich, da viele Glieder zu dem Mittelwert beitragen, über die Größe eines einzelnen Gliedes C_n^2 gar nichts aussagen, und ebensowenig etwas über den Zusammenhang zweier aufeinanderfolgender Glieder. Dieselben sind vielmehr als völlig unabhängig voneinander anzusehen.

In ganz ähnlicher Weise ergibt sich mit Benutzung von (24) für die spezifische Intensität eines monochromatischen geradlinig polarisierten, nach irgendeiner Richtung fortschreitenden Strahles:

$$\mathfrak{K}_\nu = \frac{3c\mathfrak{T}}{64\pi^2}\overline{C_n^2}. \tag{186}$$

Man ersieht hieraus u. a., daß nach der elektromagnetischen Strahlungstheorie ein monochromatischer Licht- oder Wärmestrahl keineswegs durch eine einzelne einfach periodische Welle dargestellt wird, sondern stets durch eine Übereinanderlagerung einer großen Anzahl von einfach periodischen Wellen, aus denen die Intensität des Strahles durch Bildung eines Mittelwertes sich zusammensetzt. Dem entspricht auch die aus der Optik bekannte Tatsache, daß zwei Strahlen von derselben Farbe, aber verschiedener Herkunft niemals miteinander interferieren, was notwendig der Fall sein müßte, wenn jeder Strahl einfach periodisch wäre.

§ 120. Nachdem wir den Zusammenhang der den Resonator erregenden Schwingung $\mathfrak{E}_z$ mit der im Vakuum stattfindenden Wärmestrahlung, soweit er sich angeben läßt, festgestellt haben, berechnen wir die entsprechende Schwingung des Resonators, die sich aus (172) und (179) durch einen einfachen Vergleich mit (174) und (175) folgendermaßen ergibt:

$$f(t) = \frac{3c^3}{16\pi^3}\sum_{n=0}^{n=\infty}\frac{C_n \sin\gamma_n}{\nu^3}\cos(2\pi\nu t - \vartheta_n - \gamma_n), \tag{187}$$

wobei gesetzt ist:

$$\nu = \frac{n}{\mathfrak{T}} \quad \text{und} \quad \operatorname{ctg}\gamma_n = \frac{\pi\nu_0(\nu_0^2 - \nu^2)}{\sigma\nu^3}. \tag{188}$$

Zunächst erkennt man hieraus, wie schon in § 113 bemerkt wurde, daß nur diejenigen in $\mathfrak{E}_z$ enthaltenen Partialschwingungen merklichen Einfluß auf die Schwingungen des Resonators haben, für welche $\frac{\nu}{\nu_0}$ nahe gleich 1 ist. Wenn ν von einem Werte, der merklich kleiner ist als ν_0, durch ν_0 hindurch zu einem Werte, der merklich größer ist als ν_0, übergeht, so wächst der Winkel γ von 0 durch $\frac{\pi}{2}$ hindurch bis π. Je kleiner das Dämpfungsdekrement σ des Resonators ist, um so schmaler ist das Gebiet der Schwingungszahlen ν, in welchem γ von 0 oder von π merklich verschieden ist, und um so steiler erfolgt innerhalb dieses Gebietes das Anwachsen des Winkels γ von 0 bis π. Jedoch ist wichtig zu bemerken, daß, wie klein auch σ sei, für zwei benachbarte Glieder der Summe $\sum$, z. B. für die Ordnungszahlen n und $n+1$, γ stets nur sehr wenig verschiedene Werte annimmt. Denn es ist nach (188):

$$\operatorname{ctg}\gamma_{n+1} - \operatorname{ctg}\gamma_n = \frac{\pi \mathfrak{T} \nu_0 \left(\mathfrak{T}^2 \nu_0^2 - (n+1)^2\right)}{\sigma (n+1)^3} - \frac{\pi \mathfrak{T} \nu_0 (\mathfrak{T}^2 \nu_0^2 - n^2)}{\sigma n^3}$$

$$= -\frac{2\pi}{\sigma \nu_0 \mathfrak{T}},$$

also nach (180) klein. Bei dieser Berechnung ist davon Gebrauch gemacht, daß n groß gegen 1 und daß $\frac{\nu}{\nu_0}$ nahe $=1$.

Man kann also sagen, daß der Winkel γ zwar schnell, aber doch merklich stetig von 0 bis π wächst, wenn ν durch ν_0 hindurchgeht. Ganz anders verhalten sich die Größen C_n und ϑ_n, die sich von einem Glied der FOURIERschen Reihe zum andern sprungweise und gänzlich unregelmäßig ändern können.

§ 121. Die vom Resonator in der Zeiteinheit absorbierte Energie ergibt sich nach (170) durch die Bildung des zeitlichen Mittelwertes von $\mathfrak{E}_z f'$, wobei $\mathfrak{E}_z$ aus (179), und f' aus (187) zu entnehmen ist. Die Berechnung ergibt ganz analog dem Ausdruck (177):

$$\overline{\mathfrak{E}_z f'} = \frac{3c^3}{16\pi^2} \sum_{n=0}^{n=\infty} \frac{C_n^2 \sin^2\gamma_n}{\nu^2}. \qquad (189)$$

Man ersieht auch hier wieder, daß nur diejenigen Partialschwingungen vom Resonator merklich absorbiert werden, deren Schwingungszahl ν der Eigenschwingungszahl ν_0 des Resonators nommen wird, überhaupt existiert, ist natürlich nicht von vornherein selbstverständlich, sondern durch eine besondere, der stationären Wärmestrahlung eigentümliche Eigenschaft der Funktion $\mathfrak{E}_z$ bedingt. Dagegen läßt sich, da viele Glieder zu dem Mittelwert beitragen, über die Größe eines einzelnen Gliedes C_n^2 gar nichts aussagen, und ebensowenig etwas über den Zusammenhang zweier aufeinanderfolgender Glieder. Dieselben sind vielmehr als völlig unabhängig voneinander anzusehen.

In ganz ähnlicher Weise ergibt sich mit Benutzung von (24) für die spezifische Intensität eines monochromatischen geradlinig polarisierten, nach irgendeiner Richtung fortschreitenden Strahles:

$$\mathfrak{K}_\nu = \frac{3c\mathfrak{T}}{64\pi^2}\overline{C_n^2}. \tag{186}$$

Man ersieht hieraus u. a., daß nach der elektromagnetischen Strahlungstheorie ein monochromatischer Licht- oder Wärmestrahl keineswegs durch eine einzelne einfach periodische Welle dargestellt wird, sondern stets durch eine Übereinanderlagerung einer großen Anzahl von einfach periodischen Wellen, aus denen die Intensität des Strahles durch Bildung eines Mittelwertes sich zusammensetzt. Dem entspricht auch die aus der Optik bekannte Tatsache, daß zwei Strahlen von derselben Farbe, aber verschiedener Herkunft niemals miteinander interferieren, was notwendig der Fall sein müßte, wenn jeder Strahl einfach periodisch wäre.

§ 120. Nachdem wir den Zusammenhang der den Resonator erregenden Schwingung $\mathfrak{E}_z$ mit der im Vakuum stattfindenden Wärmestrahlung, soweit er sich angeben läßt, festgestellt haben, berechnen wir die entsprechende Schwingung des Resonators, die sich aus (172) und (179) durch einen einfachen Vergleich mit (174) und (175) folgendermaßen ergibt:

$$f(t) = \frac{3c^3}{16\pi^3}\sum_{n=0}^{n=\infty}\frac{C_n \sin\gamma_n}{\nu^3}\cos(2\pi\nu t - \vartheta_n - \gamma_n), \tag{187}$$

wobei gesetzt ist:

$$\nu = \frac{n}{\mathfrak{T}} \quad \text{und} \quad \operatorname{ctg}\gamma_n = \frac{\pi\nu_0(\nu_0^2 - \nu^2)}{\sigma\nu^3}. \tag{188}$$

Zunächst erkennt man hieraus, wie schon in § 113 bemerkt wurde, daß nur diejenigen in $\mathfrak{E}_z$ enthaltenen Partialschwingungen merklichen Einfluß auf die Schwingungen des Resonators haben, für welche $\frac{\nu}{\nu_0}$ nahe gleich 1 ist. Wenn ν von einem Werte, der merklich kleiner ist als ν_0, durch ν_0 hindurch zu einem Werte, der merklich größer ist als ν_0, übergeht, so wächst der Winkel γ von 0 durch $\frac{\pi}{2}$ hindurch bis π. Je kleiner das Dämpfungsdekrement σ des Resonators ist, um so schmaler ist das Gebiet der Schwingungszahlen ν, in welchem γ von 0 oder von π merklich verschieden ist, und um so steiler erfolgt innerhalb dieses Gebietes das Anwachsen des Winkels γ von 0 bis π. Jedoch ist wichtig zu bemerken, daß, wie klein auch σ sei, für zwei benachbarte Glieder der Summe $\sum$, z. B. für die Ordnungszahlen n und $n+1$, γ stets nur sehr wenig verschiedene Werte annimmt. Denn es ist nach (188):

$$\operatorname{ctg}\gamma_{n+1} - \operatorname{ctg}\gamma_n = \frac{\pi \mathfrak{T} \nu_0 \left(\mathfrak{T}^2 \nu_0^2 - (n+1)^2\right)}{\sigma (n+1)^3} - \frac{\pi \mathfrak{T} \nu_0 (\mathfrak{T}^2 \nu_0^2 - n^2)}{\sigma n^3}$$

$$= -\frac{2\pi}{\sigma \nu_0 \mathfrak{T}},$$

also nach (180) klein. Bei dieser Berechnung ist davon Gebrauch gemacht, daß n groß gegen 1 und daß $\frac{\nu}{\nu_0}$ nahe $= 1$.

Man kann also sagen, daß der Winkel γ zwar schnell, aber doch merklich stetig von 0 bis π wächst, wenn ν durch ν_0 hindurchgeht. Ganz anders verhalten sich die Größen C_n und ϑ_n, die sich von einem Glied der FOURIERschen Reihe zum andern sprungweise und gänzlich unregelmäßig ändern können.

§ 121. Die vom Resonator in der Zeiteinheit absorbierte Energie ergibt sich nach (170) durch die Bildung des zeitlichen Mittelwertes von $\mathfrak{E}_z f'$, wobei $\mathfrak{E}_z$ aus (179), und f' aus (187) zu entnehmen ist. Die Berechnung ergibt ganz analog dem Ausdruck (177):

$$\overline{\mathfrak{E}_z f'} = \frac{3c^3}{16\pi^2} \sum_{n=0}^{n=\infty} \frac{C_n^2 \sin^2 \gamma_n}{\nu^2}. \tag{189}$$

Man ersieht auch hier wieder, daß nur diejenigen Partialschwingungen vom Resonator merklich absorbiert werden, deren Schwingungszahl ν der Eigenschwingungszahl ν_0 des Resonators nahe liegt; denn nur für diese weicht $\frac{\sin\gamma_n}{\nu}$ merklich von Null ab.

Dividieren wir die vom Resonator absorbierte Energie durch die spezifische Intensität $\mathfrak{K}_0$ der auf ihn treffenden monochromatischen Strahlung von der Schwingungszahl ν_0, so erhalten wir eine Größe, die wir als ein Maß für das Absorptionsvermögen des Resonators ansehen können.

§ 122. Jetzt führen wir den zwar nicht von vornherein selbstverständlichen, aber durch die elementarsten Erfahrungen auf dem Gebiete der Wärmestrahlung bewährten und von uns schon seit dem § 12 fortwährend benutzten Satz ein, daß das Absorptionsvermögen unabhängig ist von der Intensität der auffallenden Strahlung. Für den vorliegenden Fall ergibt sich dann durch Vergleichung von (189) und (186), daß der Quotient

$$\frac{\sum_0^\infty \frac{C_n^2 \sin^2\gamma_n}{\nu^2}}{\mathfrak{T}\,\overline{C_n^2}} = A \tag{190}$$

unabhängig ist von den Amplituden C_n. Der Wert von A läßt sich leicht aus dem speziellen Fall ableiten, daß alle Amplituden C_n einander gleich sind. Dann wird der Mittelwert $\overline{C_n^2}$ gleich C_n^2 selber, und es folgt:

$$A = \frac{1}{\mathfrak{T}} \sum_0^\infty \frac{\sin^2\gamma_n}{\nu^2}.$$

Der Wert der Summe $\sum$ läßt sich am einfachsten berechnen durch Verwandlung in ein Integral. Wir schreiben zunächst:

$$A = \frac{1}{\mathfrak{T}} \sum_0^\infty \frac{\sin^2\gamma_n}{\nu^2} \cdot \Delta n,$$

wobei Δn, die Differenz zweier aufeinanderfolgenden Ordnungszahlen, gleich 1. Die entsprechenden Schwingungszahlen seien ν und $\nu + d\nu$, dann ist $\frac{\Delta n}{\mathfrak{T}} = d\nu$, und:

$$A = \int_0^\infty \frac{\sin^2\gamma_n}{\nu^2}\, d\nu,$$

wo nun γ_n nach § 120 sich stetig mit ν ändert. Die Substitution von (188) ergibt:

$$A = \int_0^\infty \frac{d\nu}{\nu^2} \cdot \frac{1}{1 + \frac{\pi^2 \nu_0^2 (\nu_0^2 - \nu^2)^2}{\sigma^2 \nu^6}}.$$

Zum Werte dieses Integrals tragen nur diejenigen Glieder merklich bei, für welche ν nahezu gleich ν_0 ist. Daher läßt sich einfacher schreiben:

$$A = \frac{1}{\nu_0^2} \int_0^\infty \frac{d\nu}{1 + \frac{4\pi^2 (\nu - \nu_0)^2}{\sigma^2 \nu_0^2}}$$

und, wenn man statt ν als Integrationsvariable einführt:

$$x = \frac{2\pi(\nu - \nu_0)}{\sigma \nu_0},$$

$$A = \frac{\sigma}{2\pi\nu_0} \int_{-\infty}^{+\infty} \frac{dx}{1 + x^2} = \frac{\sigma}{2\nu_0}.$$

Somit erhalten wir aus (190):

$$\sum_0^\infty \frac{C_n^2 \sin^2 \gamma_n}{\nu^2} = \frac{\sigma}{2\nu_0} \mathfrak{T} \overline{C_n^2} \tag{191}$$

und aus (186) für die spezifische Intensität der monochromatischen Strahlung von der Schwingungszahl ν_0:

$$\mathfrak{K}_0 = \frac{3c\nu_0}{32\pi^2\sigma} \cdot \sum_0^\infty \frac{C_n^2 \sin^2 \gamma_n}{\nu^2}. \tag{192}$$

Ebensogroß wie die vom Resonator in der Zeiteinheit absorbierte Energie (189) ist wegen des stationären Zustandes die von ihm in der Zeiteinheit emittierte Energie, wie man auch direkt durch Berechnung der Größe (151) mit Benutzung von (187) finden kann.

§ 123. Für den zeitlichen Mittelwert der Energie des Resonators endlich ergibt sich aus (142), (168) und (187) durch einen Vergleich mit (178):

$$U = \frac{3c^3 \nu_0}{64\pi^2 \sigma} \sum_0^\infty \frac{\nu_0^2 + \nu^2}{\nu^6} C_n^2 \sin^2 \gamma_n$$

oder auch, da zum Wert der Summe nur diejenigen Glieder

merklich beitragen, deren Schwingungszahl ν der Eigenschwingungszahl ν_0 des Resonators nahe liegt:

$$U = \frac{3c^3}{32\pi^2\sigma\nu_0}\sum_0^\infty \frac{C_n^2 \sin^2\gamma_n}{\nu^2}.$$

Durch Vergleich mit (192) ergibt sich so die einfache Beziehung:

$$U = \frac{c^2}{\nu_0^2}\,\mathfrak{K}_0 \tag{193}$$

zwischen der mittleren Schwingungsenergie des Resonators und der spezifischen Strahlungsintensität eines monochromatischen geradlinig polarisierten Strahles von der Periode des Resonators. Hierbei ist besonders bemerkenswert, daß die Dämpfungskonstante σ des Resonators in diese Relation gar nicht eingeht.

Ferner erhält man mit Rücksicht auf (24):

$$U = \frac{c^3 \mathfrak{u}_0}{8\pi\nu_0^2} \tag{194}$$

als Beziehung zwischen der mittleren Energie des Resonators und der räumlichen Strahlungsdichte der Schwingungszahl ν_0 im stationären Strahlungszustand.

Endlich ergibt sich durch Vergleichung mit (185):

$$U = \frac{3c^3\mathfrak{T}}{64\pi^2\nu_0^2}\cdot\overline{C_n^2}, \tag{195}$$

wodurch die Energie des Resonators direkt in Zusammenhang gebracht wird mit der elektrischen Feldstärke (179) der ihn erregenden Welle. Der Mittelwert von C_n^2 ist, wie in (185), zu bilden aus einer großen Anzahl von Partialschwingungen, deren Schwingungszahlen ν der Eigenzahl ν_0 des Resonators nahe liegen.

§ 124. Wir denken uns nun mit dem betrachteten System, das aus einem gleichmäßig durchstrahlten, von vollständig und diffus reflektierenden Wänden begrenzten Vakuum und dem darin befindlichen ruhenden Resonator besteht, eine unendlich kleine reversible Zustandsänderung vorgenommen, etwa so, daß wir die Strahlung adiabatisch und unendlich langsam komprimieren, wie das im vorigen Abschnitt beschrieben wurde. Dann bleibt nach dem zweiten Hauptsatz der Thermodynamik die Gesamtentropie des Systems ungeändert, dagegen wird durch die Kompression die Intensität $\mathfrak{K}$ jeder einzelnen Schwingungszahl verändert, und infolgedessen auch die Energie U des Resonators. Denn diese ist im stationären Zustand nach (193) proportional der Intensität $\mathfrak{K}_0$ der ihn erregenden monochromatischen Strahlung. Der Resonator wird also einen Teil der durch die Kompression erzeugten Strahlungsenergie absorbieren und diesen Energiebetrag der freien Wärmestrahlung im Vakuum entziehen.

Zur bequemeren Übersicht können wir uns jeden unendlich kleinen Kompressionsvorgang so in zwei Epochen zerlegt denken, daß in der ersten Epoche die Kompression stattfindet und dabei die Strahlung sich verhält, als ob der Resonator gar nicht vorhanden wäre, und daß dann in der zweiten Epoche der Resonator soviel Energie aus der ihn erregenden Strahlung absorbiert, daß die durch die Vorgänge der ersten Epoche gestörte Beziehung (193) wieder in Gültigkeit tritt. Während der ersten Epoche bleibt nach den Ergebnissen des vorigen Abschnitts die Entropie der Wärmestrahlung im Vakuum für sich konstant; während der zweiten Epoche aber ändert sich die Entropie der Wärmestrahlung durch Abgabe von Wärme an den Resonator. Da nun die Gesamtentropie des Systems konstant bleiben muß, so folgt daraus, daß nicht nur der freien Wärmestrahlung, sondern auch dem Resonator eine Entropie zukommt, deren Änderung die Entropieänderung der freien Wärmestrahlung gerade kompensiert. Da der thermodynamische Zustand des Resonators allein von seiner Energie U abhängt, so ist auch die Entropie S des Resonators durch U bestimmt.

§ 125. Es ist leicht, die Beziehung zwischen der Entropie des Resonators und der räumlichen Entropiedichte der monochromatischen Vakuumstrahlung von der Schwingungszahl ν_0 im stationären Zustand aufzustellen. Denn nach dem zweiten Hauptsatz ist der stationäre Zustand unter allen Zuständen dadurch ausgezeichnet, daß in ihm die Gesamtentropie des Systems ein Maximum besitzt. Die Gesamtentropie besteht aber aus der Entropie des Resonators: S, und aus der Entropie (113) der äußeren Strahlung:

$$V \cdot \int_0^\infty \mathfrak{s} \, d\nu ,$$

wobei V das Volumen des gleichmäßig durchstrahlten Vakuums bedeutet. Dann ist also für den absolut stabilen Strahlungszustand, d. h. für einen Resonator in einem von schwarzer Strahlung erfüllten Vakuum von bestimmtem Volumen:

$$\delta S + V \int_0^\infty \delta \mathfrak{s} \, d\nu = 0$$

oder:

$$\frac{dS}{dU} \delta U + V \int_0^\infty \frac{\partial \mathfrak{s}}{\partial \mathfrak{u}} \delta \mathfrak{u} \, d\nu = 0 .$$

Die einzige Bedingung, welcher die Variation δ zu genügen hat, ist die, daß die Gesamtenergie des Systems die nämliche bleibt, also:

$$\delta U + V \int_0^\infty \delta \mathfrak{u} \, d\nu = 0 .$$

Wir wollen nun die räumliche Energiedichte $\mathfrak{u}$ und infolgedessen auch die räumliche Entropiedichte $\mathfrak{s}$ aller Strahlenarten unvariiert lassen bis auf einen schmalen Spektralbezirk in der Umgebung der Schwingungszahl ν_0, von der Breite $\Delta \nu_0$, wobei $\Delta \nu_0$ klein gegen ν_0, im übrigen aber beliebig ist. Dann reduzieren sich die letzten Gleichungen auf:

$$\frac{dS}{dU} \delta U + V \frac{\partial \mathfrak{s}_0}{\partial \mathfrak{u}_0} \delta \mathfrak{u} \, \Delta \nu_0 = 0$$

und

$$\delta U + V \delta \mathfrak{u} \, \Delta \nu_0 = 0 .$$

Hieraus folgt:

$$\frac{dS}{dU} = \frac{\partial \mathfrak{s}_0}{\partial \mathfrak{u}_0} . \tag{196}$$

Die vier Größen S, U, $\mathfrak{s}_0$, $\mathfrak{u}_0$ hängen, bei gegebenem ν_0, von einer einzigen Variabeln ab. Denn S ist eine bestimmte Funktion von U, $\mathfrak{s}_0$ eine solche von $\mathfrak{u}_0$, und U ist mit $\mathfrak{u}_0$ durch die Beziehung (194) verbunden. Man kann daher, bei konstantem ν_0, auch schreiben:

$$\frac{dS}{d\mathfrak{s}_0} = \frac{dU}{d\mathfrak{u}_0} = \frac{c^3}{8 \pi \nu_0^2}$$

und erhält hieraus durch Integration, mit Fortlassung der physikalisch bedeutungslosen Integrationskonstanten:

$$S = \frac{c^3}{8\pi\nu_0^2}\mathfrak{S}_0 \tag{197}$$

als Beziehung zwischen der Entropie des Resonators und der räumlichen Dichte der Strahlungsentropie von der Schwingungszahl ν_0 im stationären Zustand. Ferner nach (133):

$$S = \frac{c^2}{\nu_0^2}\mathfrak{L}_0 \tag{198}$$

als Beziehung zwischen der Entropie des Resonators und der spezifischen Intensität der monochromatischen geradlinig polarisierten Entropiestrahlung von der Schwingungszahl ν_0.

§ 126. Die Gleichung (196) besitzt eine einfache physikalische Bedeutung, sie liefert nämlich mit Rücksicht auf (117):

$$\frac{dS}{dU} = \frac{1}{T}, \tag{199}$$

wobei T die Temperatur der den Resonator erregenden Strahlung bedeutet. Wenn wir also allgemein den reziproken Wert der Größe $\frac{dS}{dU}$, welche nur von der Energie und der natürlichen Beschaffenheit des Resonators abhängt, als „Temperatur des Resonators" definieren, so gilt der Satz, daß im stationären Strahlungszustand die Temperatur des Resonators gleich ist der Temperatur der ihn erregenden monochromatischen Strahlung.

§ 127. Über die Abhängigkeit der Entropie S eines Resonators von seiner Energie U kann man etwas erfahren aus dem WIENschen Verschiebungsgesetz, indem man etwa in die Form (134) desselben $\nu = \nu_0$, und für $\mathfrak{L}_0$ und $\mathfrak{K}_0$ die sich aus (198) und (193) ergebenden Werte setzt. Man erhält dann:

$$S = F\left(\frac{U}{\nu_0}\right), \tag{200}$$

wobei die Funktion F außer ihrem Argument nur universelle Konstante enthält, also namentlich auch keine auf die natürliche Beschaffenheit des Resonators bezügliche Konstante. Dies ist die einfachste unter allen bisher von uns aufgestellten Formen des WIENschen Verschiebungsgesetzes, da die Fortpflanzungsgeschwindigkeit c in ihr überhaupt nicht, und die Schwingungszahl ν_0 nur ein einziges Mal in der ersten Potenz vorkommt. Es ist auch leicht zu verstehen, daß die Einfachheit der mathematischen Relation ihren Grund haben wird in der Einfachheit des durch die Resonatorschwingungen dargestellten physikalischen Vorganges.

Dieser Umstand läßt darauf schließen, daß von dieser einfachen Beziehung aus am ehesten ein Einblick in die Natur der Verschiebungsfunktion F zu gewinnen sein wird. Wenn die analytische Form dieser universellen Funktion gefunden ist, so ergibt sich nach § 92 f. ohne weiteres daraus das Gesetz der Energieverteilung im Normalspektrum. Eine Lösung dieser Aufgabe scheint aber nicht möglich zu sein ohne ein näheres Eingehen auf den Begriff der Entropie, und dieser Begriff wird, vom Standpunkte der elektromagnetischen Strahlungstheorie aus, erst dann vollkommen verständlich, wenn man ihn mit dem der Wahrscheinlichkeit in Zusammenhang bringt, wie im nächsten Abschnitt näher ausgeführt werden soll.

Vierter Abschnitt.

Entropie und Wahrscheinlichkeit.

Erstes Kapitel. Einleitung. Grundlegende Sätze und Definitionen.

§ 128. Da mit der Einführung von Wahrscheinlichkeitsbetrachtungen in die elektromagnetische Strahlungstheorie ein vollkommen neues, den Grundlagen der Elektrodynamik gänzlich fremdes Element in den Bereich der Untersuchungen eintritt, so erhebt sich gleich zu Beginn dieses Abschnitts die prinzipielle Vorfrage nach der Berechtigung und nach der Notwendigkeit solcher Betrachtungen. Man könnte nämlich bei oberflächlicher Überlegung leicht zu der Schlußfolgerung neigen, daß für Wahrscheinlichkeitsrechnungen in einer rein elektrodynamischen Theorie überhaupt kein Platz vorhanden wäre. Denn da die elektromagnetischen Feldgleichungen zusammen mit den Anfangs- und den Grenzbedingungen den zeitlichen Verlauf eines elektrodynamischen Vorganges bekanntlich eindeutig bestimmen, so wären Betrachtungen, die außerhalb der Feldgleichungen stehen, prinzipiell unberechtigt, in jedem Falle aber entbehrlich. Entweder führen sie nämlich zu denselben Ergebnissen wie die elektrodynamischen Grundgleichungen — dann wären sie überflüssig; oder sie führen zu anderen Ergebnissen — dann wären sie unrichtig.

Trotz dieses scheinbar unausweichlichen Dilemmas steckt in jener Überlegung doch eine Lücke. Denn die Ergebnisse, zu denen in der elektromagnetischen Wärmestrahlungstheorie die elektrodynamischen Grundgleichungen allein genommen führen, sind keineswegs eindeutig, sondern sie sind im Gegenteil vieldeutig, und sogar von unendlich hoher Ordnung vieldeutig. Knüpfen wir, um dies einzusehen, an das spezielle im letzten

Kapitel betrachtete Beispiel an, daß ein Resonator von der dort betrachteten elementaren Art sich in einem nach allen Richtungen gleichmäßig durchstrahlten Vakuum befindet. Wir zogen dort den Schluß, daß sich im Laufe der Zeit ein stationärer Schwingungszustand herstellt, in welchem die vom Resonator in der Zeiteinheit absorbierte und emittierte Energie einen konstanten, der Intensität $\mathfrak{K}_0$ der ihn erregenden monochromatischen Strahlung proportionalen Wert besitzt. Aber diesen Schluß konnten wir, wie zu Beginn des § 122 ausdrücklich hervorgehoben wurde, nur thermodynamisch, keineswegs elektrodynamisch begründen, während man doch vom Standpunkt der elektrodynamischen Strahlungstheorie aus verlangen müßte, daß, wie alle Begriffe, so auch alle Sätze der Wärmestrahlung aus rein elektrodynamischen Überlegungen heraus entwickelt werden. Wollte man nun versuchen, den allgemeinen Zusammenhang zwischen der vom Resonator absorbierten Energie und der Intensität der ihn erregenden Strahlung ganz ohne die Einmischung thermodynamischer Erfahrungen auf rein elektrodynamischem Wege abzuleiten, so würde man bald finden, daß es einen solchen allgemeinen Zusammenhang gar nicht gibt, oder mit anderen Worten, daß man über die vom Resonator absorbierte Energie, bei gegebener Intensität der ihn erregenden Strahlung, vom Standpunkt der reinen Elektrodynamik aus überhaupt gar nichts aussagen kann, solange von den Werten der Amplituden C_n und der Phasenkonstanten ϑ_n der einzelnen in der erregenden Strahlung enthaltenen Partialschwingungen nichts Näheres bekannt ist. Denn sowohl die absorbierte Energie als auch die Intensität der erregenden Strahlung werden durch gewisse Mittelwerte dargestellt, die aus den Größen C_n und ϑ_n jedesmal in verschiedener Weise zu bilden sind, und die daher nicht allgemein aus einander berechnet werden können, ebensowenig, wie man etwa den Mittelwert von C_n aus dem Mittelwert von C_n^2 allgemein berechnen kann. Wenn also auch die Intensität der Strahlung, die von allen Seiten auf den Resonator fällt, für alle Spektralbezirke als Funktion der Richtung und eventuell der Zeit vollständig gegeben ist, und auch der Anfangszustand des Resonators bekannt ist, so lassen sich die Schwingungen des Resonators daraus doch noch nicht eindeutig berechnen, auch nicht annähernd, auch nicht für hinreichend lange Zeitepochen. Vielmehr kann der Resonator, falls über die Einzelwerte der C_n und ϑ_n geeignet verfügt wird, durch die nämliche auffallende Strahlungsintensität zu gänzlich verschiedenartigen Schwingungen veranlaßt werden. Ja, wir werden später, im ersten Kapitel des nächsten Abschnitts, einen speziellen, mit allen elektrodynamischen Gesetzen vollkommen verträglichen Vorgang näher besprechen, wo der Resonator, so seltsam das klingt, die von allen Seiten auf ihn fallende Strahlung vollständig und fortwährend absorbiert, ohne überhaupt jemals die geringste Menge Energie auszustrahlen (§ 172); ferner auch einen anderen Vorgang, wo die vom Resonator absorbierte Energie sogar negativ ist,[1] wo also die auffallende Strahlung dem Resonator Energie entzieht, bis seine Energie gleich Null wird! (§ 173)

An einem einzigen solchen Beispiel sehen wir, daß durch die Intensität der erregenden Strahlung die Schwingungen des Resonators noch keineswegs bestimmt sind, und daß daher in einem Falle, wo nach den Gesetzen der Thermodynamik und nach allen Erfahrungen ein eindeutiges Resultat zu erwarten ist, die reine Elektrodynamik vollkommen im Stiche läßt, da für sie die vorliegenden Daten noch lange nicht hinreichen, um die in den elektrodynamischen Differentialgleichungen auftretenden Konstanten eindeutig zu bestimmen.

§ 129. Ehe wir diesen Umstand und die damit für die elektrodynamische Theorie der Wärmestrahlung verbundene Schwierigkeit weiter verfolgen, möge darauf hingewiesen werden, daß bei der mechanischen Wärmetheorie, speziell der kinetischen Gastheorie, genau der gleiche Umstand und die gleiche Schwierigkeit vorliegt. Denn wenn etwa in einem strömenden Gase zur Zeit $t = 0$ an jeder Stelle die Geschwindigkeit, die Dichte und die Temperatur des Gases gegeben ist und außerdem die Grenzbedingungen vollständig bekannt sind, so wird man nach allen Erfahrungen erwarten, daß dadurch der zeitliche Verlauf des Vorganges eindeutig bestimmt ist. Vom rein mechanischen Standpunkt aus ist das aber keineswegs der Fall; denn durch die sichtbare Geschwindigkeit, die Dichte und die Temperatur des Gases sind noch lange nicht die Orte und die Geschwindigkeiten aller einzelnen Moleküle gegeben, und diese müßte man

[1] Vgl. die Anmerkung zu § 111.

9*

genau kennen, wenn man aus den Bewegungsgleichungen den zeitlichen Verlauf des Vorganges vollständig berechnen wollte. Auch hier läßt sich leicht zeigen, daß bei den nämlichen Werten der sichtbaren Geschwindigkeit, der Dichte und der Temperatur unendlich viele gänzlich verschiedenartige Vorgänge mechanisch möglich sind, von denen einige den Grundsätzen der Thermodynamik, namentlich dem zweiten Hauptsatz, direkt widersprechen.

§ 130. Aus diesen Überlegungen sehen wir, daß, wenn es sich um die Berechnung des zeitlichen Verlaufs eines thermodynamischen Vorganges handelt, sowohl die mechanische Wärmetheorie als auch die elektrodynamische Theorie der Wärmestrahlung mit derjenigen Formulierung der Anfangs- und Grenzbedingungen, welche in der Thermodynamik zur eindeutigen Bestimmung des Vorganges vollkommen hinreicht, keineswegs auskommt, sondern daß vom Standpunkt der reinen Mechanik bez. Elektrodynamik betrachtet noch unendlich viele Lösungen des Problems existieren. Infolgedessen bleibt, falls man nicht überhaupt ganz darauf verzichten will, die thermodynamischen Vorgänge mechanisch bez. elektrodynamisch zu begreifen, nur die eine Möglichkeit übrig, durch Einführung von besonderen ergänzenden Hypothesen die Anfangs- und Grenzbedingungen insoweit näher zu präzisieren, daß die mechanischen oder elektrodynamischen Gleichungen auf ein eindeutiges und mit der Erfahrung übereinstimmendes Resultat führen. Wie man eine derartige Hypothese zu formulieren hat, dafür läßt sich aus den Prinzipien der Mechanik oder Elektrodynamik selber natürlicherweise kein Anhaltspunkt gewinnen; denn diese lassen ja gerade den Fall ganz offen. Ebendeswegen ist aber auch von vornherein j e d e mechanische oder elektrodynamische Hypothese zulässig, welche eine nähere, durch direkte Messungen gar nicht kontrollierbare Spezialisierung der gegebenen Anfangs- und Grenzbedingungen enthält. Welcher Hypothese vor den übrigen der Vorzug zu geben ist, darüber kann die Entscheidung nur dadurch gewonnen werden, daß man die Resultate, zu denen die Hypothese führt, hinterher im Lichte der thermodynamischen Erfahrungssätze prüft.

§ 131. Nun ist es sehr bemerkenswert, daß, obwohl hiernach die definitive Prüfung der verschiedenen zulässigen Hypothesen erst a posteriori erfolgen kann, man doch schon durch eine Betrachtung a priori, ohne sich irgendwie auf die Thermodynamik zu stützen, einen festen Anhaltspunkt für den Inhalt der aufzustellenden Hypothese gewinnen kann. Fassen wir nämlich einmal wieder das obige Beispiel (§ 128) ins Auge, daß ein Resonator bei gegebenem Anfangszustand einer Strahlung von gegebener Intensität ausgesetzt ist. Dann ist, wie damals besprochen wurde, der Schwingungsvorgang im Resonator, solange man die unkontrollierbaren Einzelwerte der C_n und ϑ_n in der erregenden Strahlung ganz offen läßt, unendlich vieldeutig. Wenn man nun aber alle die unendlich verschiedenen Fälle, wie sie den verschiedenen bei der gegebenen Strahlungsintensität möglichen Werten der C_n und ϑ_n entsprechen, näher untersucht, und die Resultate, zu denen sie einzeln führen, miteinander vergleicht, so findet man, daß die ungeheure Mehrzahl dieser Fälle in den Mittelwerten zu ganz übereinstimmenden Resultaten führt, während diejenigen Fälle, in denen sich merkliche Abweichungen zeigen, nur in verhältnismäßig verschwindend geringer Anzahl auftreten, nämlich dann, wenn gewisse ganz spezielle weitgehende Bedingungen zwischen den einzelnen Größen C_n und ϑ_n erfüllt sind. Nimmt man also an, daß solche spezielle Bedingungen nicht gelten, so ergibt sich, wie verschieden auch die Konstanten C_n und ϑ_n im übrigen gewählt werden mögen, für den Resonator eine Schwingung, die, wenn auch natürlich nicht in allen Einzelheiten, so doch in bezug auf alle meßbaren Mittelwerte — und diese sind die einzigen, welche kontrolliert werden können — eine ganz bestimmte genannt werden kann. Und, was nun das Bemerkenswerte dabei ist: gerade die auf diese Weise erhaltene Schwingung entspricht den Forderungen des zweiten Hauptsatzes der Thermodynamik, wie im nächsten Abschnitt näher ausgeführt werden wird (vgl. § 182).

In der Mechanik verhält es sich genau ebenso. Wenn man, um auf das vorige Beispiel (§ 129) zurückzukommen, alle nur denkbaren Lagen und Geschwindigkeiten der einzelnen Gasmoleküle ins Auge faßt, die mit den gegebenen Werten der sichtbaren Geschwindigkeit, der Dichte und der Temperatur des Gases verträglich sind, und für jede Kombination derselben den mechanischen Vorgang genau nach den Bewegungsgleichungen berechnet, so findet man ebenfalls, daß in der ungeheuren Mehrzahl der Fälle Vorgänge resultieren, die, wenn auch nicht in den Einzelheiten, so doch in allen meßbaren Mittelwerten miteinander übereinstimmen, und die außerdem dem zweiten Hauptsatz der Thermodynamik Genüge leisten. Nur einige wenige besondere Fälle, in denen zwischen den Koordinaten und den Geschwindigkeiten der Moleküle ganz spezielle Bedingungen bestehen, liefern abweichende Ergebnisse.

§ 132. Nach diesen Betrachtungen ist klar, daß die Hypothesen, deren Einführung oben als notwendig nachgewiesen wurde, ihren Zweck vollständig erfüllen, wenn ihr Inhalt nichts weiter besagt, als daß derartige besondere Fälle, die speziellen Bedingungen zwischen den einzelnen direkt nicht kontrollierbaren Konstanten entsprechen, in der Natur nicht vorkommen. In der Mechanik leistet dies die Hypothese,[1] daß die Wärmebewegungen „molekular-ungeordnet" sind, in der Elektrodynamik leistet das Entsprechende die Hypothese der „natürlichen Strahlung", welche besagt, daß zwischen den zahlreichen verschiedenen Partialschwingungen (179) eines Strahles keine anderen Beziehungen bestehen, als diejenigen, welche durch die meßbaren Mittelwerte bedingt sind (§ 181). Wenn wir zur Abkürzung alle Zustände und alle Vorgänge, für welche eine solche Hypothese gilt, als „elementar ungeordnet" bezeichnen, so liefert der Satz, daß in der Natur alle Zustände und alle Vorgänge, welche zahlreiche unkontrollierbare Bestandteile enthalten, elementar ungeordnet sind, die Vorbedingung, aber auch die sichere Gewähr für die eindeutige Bestimmbarkeit der meßbaren Vorgänge, sowohl in der Mechanik als auch in der Elektrodynamik, und zugleich für die Gültigkeit des zweiten Hauptsatzes der Thermodynamik, womit dann selbstverständlich auch der für den zweiten Hauptsatz charakteristische Begriff der Entropie und der damit unmittelbar verbundene der Temperatur seine mechanische bez. elektrodynamische Erklärung finden muß. Zugleich folgt hieraus, daß die Begriffe der Entropie und der Temperatur ihrem Wesen nach an die Bedingung der elementaren Unordnung geknüpft sind. Eine rein periodische absolut ebene Welle besitzt weder Entropie noch Temperatur, weil sie

---

[1] L. Boltzmann, Vorlesungen über Gastheorie, **1**, p. 21, 1896. Wiener Sitzungsber., **78**, Juni 1878, am Schluß. Vgl. auch S. H. Burbury, Nature, **51**, p. 78, 1894.

gar keine unkontrollierbaren Größen enthält, also auch nicht elementar ungeordnet sein kann, ebensowenig wie das bei der Bewegung eines einzelnen starren Atoms der Fall ist. Erst das unregelmäßige Zusammenwirken sehr vieler Partialschwingungen verschiedener Perioden, die sich unabhängig voneinander nach den verschiedenen Richtungen des Raumes fortpflanzen, oder das ungeregelte Durcheinanderschwirren sehr vieler Atome schafft die Vorbedingung für die Gültigkeit der Hypothese der elementaren Unordnung und somit für die Existenz einer Entropie und einer Temperatur.

§ 133. Welche mechanische bez. elektrodynamische Größe stellt nun aber die Entropie eines Zustandes dar? Offenbar hängt diese Größe irgendwie mit der „Wahrscheinlichkeit" des Zustandes zusammen. Denn da die elementare Unordnung und der Mangel jeglicher Einzelkontrolle zum Wesen der Entropie gehört, so können nur kombinatorische oder Wahrscheinlichkeitsbetrachtungen die nötigen Anhaltspunkte zur Berechnung ihrer Größe liefern. Schon die Hypothese der elementaren Unordnung selbst ist ja ihrem Wesen nach ein Wahrscheinlichkeitssatz, da sie aus einer ungeheuren Anzahl von gleichmöglichen Fällen eine bestimmte Anzahl herausgreift und dieselben als in der Natur nicht existent bezeichnet.

Da nun der Begriff der Entropie, ebenso wie der Inhalt des zweiten Hauptsatzes der Thermodynamik, ein universeller ist, und da andererseits die Wahrscheinlichkeitssätze nicht minder universelle Bedeutung besitzen, so ist zu vermuten, daß der Zusammenhang zwischen Entropie und Wahrscheinlichkeit ein sehr enger sein wird. Wir stellen daher unseren ferneren Ausführungen folgenden Satz an die Spitze: Die Entropie eines physikalischen Systems in einem bestimmten Zustand hängt lediglich ab von der Wahrscheinlichkeit dieses Zustandes. Die Zulässigkeit und Fruchtbarkeit dieses Satzes wird sich später in verschiedenen Fällen zeigen. Einen allgemeinen strengen Beweis desselben zu liefern werden wir aber hier nicht versuchen. Ja, ein derartiger Versuch würde offenbar an dieser Stelle nicht einmal einen Sinn haben. Denn solange die „Wahrscheinlichkeit" eines Zustandes nicht zahlenmäßig definiert ist, läßt sich der obige Satz auch nicht zahlenmäßig auf seine Richtigkeit prüfen. Man könnte sogar vielleicht auf den ersten Blick vermuten, daß er aus diesem Grunde überhaupt keinen bestimmten physikalischen Inhalt besitzt. Indessen läßt sich durch eine einfache Deduktion zeigen, daß man, ohne noch auf den Begriff der Wahrscheinlichkeit eines Zustandes näher einzugehen, auf Grund des obigen Satzes doch schon in der Lage ist, die Art der Abhängigkeit der Entropie von der Wahrscheinlichkeit ganz allgemein zu fixieren.

§ 134. Bezeichnet nämlich S die Entropie, W die Wahrscheinlichkeit eines physikalischen Systems in einem bestimmten Zustand, so besagt der obige Satz, daß

$$S = f(W), \tag{201}$$

wobei $f(W)$ eine universelle Funktion des Arguments W bedeutet. Wie man nun auch W näher definieren möge, soviel läßt sich jedenfalls dem mathematischen Wahrscheinlichkeitsbegriffe als feststehend entnehmen, daß die Wahrscheinlichkeit eines Systems, das aus zwei voneinander ganz unabhängigen Systemen zusammengesetzt ist, gleich dem Produkte der Wahrscheinlichkeiten der beiden Einzelsysteme ist. Denken wir uns z. B. als erstes System irgendeinen Körper auf der Erde, als zweites System einen durchstrahlten Hohlraum auf dem Sirius, so ist die Wahrscheinlichkeit dafür, daß sich der irdische Körper in einem bestimmten Zustand 1, und zugleich die Hohlraumstrahlung in einem bestimmten Zustand 2 befindet:

$$W = W_1 \cdot W_2, \tag{202}$$

wenn W_1 und W_2 die Wahrscheinlichkeiten dafür sind, daß sich das betreffende System in dem betreffenden Zustande befindet. Sind nun S_1 und S_2 die Entropien der Einzelsysteme in den beiden Zuständen, so ist nach (201):

$$S_1 = f(W_1), \quad S_2 = f(W_2).$$

Aber nach dem zweiten Hauptsatz der Thermodynamik ist die Gesamtentropie beider voneinander unabhängigen Systeme: $S = S_1 + S_2$, folglich nach (201) und (202):

$$f(W_1 W_2) = f(W_1) + f(W_2).$$

Aus dieser Funktionalgleichung läßt sich f berechnen. Differentiiert man nämlich auf beiden Seiten nach W_1, bei konstantem W_2, so kommt:

$$W_2 \dot{f}(W_1 W_2) = \dot{f}(W_1).$$

Differentiiert man ferner nach W_2, bei konstantem W_1, so kommt:

$$\dot{f}(W_1 W_2) + W_1 W_2 \ddot{f}(W_1 W_2) = 0$$

oder

$$\dot{f}(W) + W\ddot{f}(W) = 0.$$

Das allgemeine Integral dieser Differentialgleichung zweiter Ordnung ist:

$$f(W) = k \log W + \text{const.}$$ Also nach (201):

$$S = k \log W + \text{const}, \qquad (203)$$

wodurch die Abhängigkeit der Entropie von der Wahrscheinlichkeit allgemein bestimmt ist. Die universelle Integrationskonstante k ist für ein irdisches System dieselbe wie für ein kosmisches, und wenn ihr numerischer Wert für dieses bestimmt ist, gilt er auch für jenes. Die zweite, additive, Integrationskonstante hat, weil die Entropie S eine willkürliche additive Konstante enthält, keine physikalische Bedeutung, und kann nach Belieben fortgelassen werden.

§ 135. Die Beziehung (203) enthält eine allgemeine Methode, um den Ausdruck der Entropie S durch Wahrscheinlichkeitsbetrachtungen zu berechnen. Doch wird dieselbe natürlich erst dann von praktischem Nutzen, wenn die Größe W der Wahrscheinlichkeit eines physikalischen Systems in einem gegebenen Zustand zahlenmäßig angegeben werden kann. Die Aufsuchung der allgemeinsten und präzisesten Definition dieser Größe gehört zu den wichtigsten Aufgaben der mechanischen bez. elektrodynamischen Wärmetheorie. Zunächst erfordert sie ein näheres Eingehen auf den Begriff des „Zustandes" eines physikalischen Systems.

Unter dem „Zustand" eines physikalischen Systems zu einer bestimmten Zeit verstehen wir den Inbegriff aller derjenigen voneinander unabhängigen Größen, durch welche der zeitliche Verlauf der in dem System stattfindenden Vorgänge, soweit sie der Messung zugänglich sind, bei gegebenen Grenzbedingungen eindeutig bestimmt wird; die Kenntnis des Zustandes ist also genau äquivalent der Kenntnis der „Anfangsbedingungen". Daher ist z. B. bei einem aus unveränderlichen Molekülen bestehenden Gase der Zustand bestimmt durch das Gesetz der Raum- und Geschwindigkeitsverteilung, d. h. durch die Angabe der Anzahl der Moleküle, deren Koordinaten und Geschwindigkeitskomponenten innerhalb je eines einzelnen kleinen „Intervalls" oder „Gebietes" liegen. Die auf die verschiedenen Gebiete entfallenden Molekülzahlen sind im allgemeinen ganz unabhängig voneinander, da der Zustand ja kein Gleichgewichts- oder stationärer Zustand zu sein braucht; sie müssen also alle einzeln bekannt sein, wenn der Zustand des Gases als gegeben betrachtet werden soll. Dagegen ist es für die Charakterisierung des Zustandes nicht erforderlich, nähere Details bezüglich der innerhalb eines einzelnen Elementargebiets befindlichen Moleküle anzugeben; denn hier tritt als Ergänzung die Hypothese der molekularen Unordnung ein, welche trotz der mechanischen Unbestimmtheit die Eindeutigkeit des zeitlichen Vorganges verbürgt.

Bei einem Licht- oder Wärmestrahl ist der Zustand bestimmt durch die Richtung, die spektrale Energieverteilung und den Polarisationszustand (§ 17). Näheres über die Amplituden und Phasen der einzelnen periodischen Partialschwingungen des Strahles zu wissen ist nicht nötig, da auch hier die Hypothese der elementaren Unordnung als Ergänzung eingreift.

Man sieht, daß der so definierte Zustandsbegriff, im statistischen Sinne, wohl zu unterscheiden ist von dem Zustandsbegriff im absolut mechanischen oder elektrodynamischen Sinne, nach welchem ein Zustand erst dann als gegeben betrachtet werden darf, wenn die Koordinaten und Geschwindigkeitskomponenten jedes einzelnen Moleküls, bez. die Amplituden und Phasen aller einzelnen Partialschwingungen genau bekannt sind. In einem derartigen Zustand würden gar keine unkontrollierbaren Elemente mehr vorkommen und daher auch keinerlei Wahrscheinlichkeitsbetrachtungen am Platze sein.

§ 136. Wenn wir nun von der Wahrscheinlichkeit W eines bestimmten elementar ungeordneten Zustandes reden, so ist damit ausgedrückt, daß ein solcher Zustand auf verschiedene Arten realisiert werden kann. Denn jeder Zustand, der viele gleichartige unkontrollierbare Bestandteile enthält, entspricht einer gewissen „Verteilung", nämlich im ersten Beispiel der Verteilung der Koordinaten und der Geschwindigkeitskomponenten auf die Gasmoleküle, im zweiten Beispiel der Verteilung der Amplituden und Phasen auf die einzelnen Partialschwingungen. Eine Verteilung ist aber immer eine Zuordnung einer Gruppe von unter sich gleichartigen Elementen (Koordinaten, Geschwindigkeitskomponenten, Amplituden, Phasen) zu einer anderen Gruppe von unter sich gleichartigen Elementen (Molekülen, Partialschwingungen). Solange man nun einen bestimmten Zustand ins Auge faßt, kommt es offenbar nur darauf an, wieviel Elemente der beiden Gruppen einander wechselseitig zugeordnet sind, nicht aber darauf, welche individuellen Elemente der einen Gruppe bestimmten individuellen Elementen der anderen Gruppe zugeordnet sind. Daher kann ein bestimmter Zustand durch eine große Anzahl voneinander verschiedener individueller Zuordnungen zustande kommen. Nennen wir also jede besondere Verteilung, bei der die Elemente der einen Gruppe den Elementen der anderen Gruppe individuell zugeordnet sind, eine „Komplexion", so enthält ein bestimmter Zustand im allgemeinen eine große Anzahl von verschiedenen Komplexionen. Diese Zahl, d. h. die Anzahl der Komplexionen, welche ein gegebener Zustand umfaßt, definieren wir nun als die Wahrscheinlichkeit W des Zustandes, und erhalten dadurch ein Mittel, um in gegebenen Fällen W und dann aus (203) die Entropie S des Zustandes zu berechnen. Nähere Erläuterungen über die Art dieser Berechnung werden in den nächsten beiden Kapiteln ausführlich zur Sprache kommen.

§ 137. Hier sei nur noch auf einen Punkt hingewiesen, in dem sich die hier gebrauchte Definition der Wahrscheinlichkeit von der sonst üblichen der mathematischen Wahrscheinlichkeit eines Ereignisses unterscheidet. Die letztere wird bekanntlich als ein echter Bruch definiert, nämlich als der Quotient aus der Anzahl der dem Ereignis günstigen durch die Anzahl aller gleichmöglichen Fälle. Im Unterschied davon wird hier die Wahrscheinlichkeit W eines physikalischen Zustandes durch eine ganze Zahl, und zwar durch eine große Zahl ausgedrückt. Man könnte versucht sein, den Unterschied der beiden Definitionen dadurch zu beseitigen, daß man die Anzahl der Komplexionen, welche ein Zustand umfaßt, noch dividiert durch die Anzahl „aller möglichen" Komplexionen, und diesen Quotienten als die Wahrscheinlichkeit des Zustandes bezeichnet. Allein es würden hier bei der Frage nach der Anzahl aller möglichen Komplexionen unter Umständen Schwierigkeiten entstehen, die wir lieber vermeiden wollen, indem wir jene Frage gar nicht aufwerfen und bei der oben gegebenen Definition der Wahrscheinlichkeit W

eines Zustandes stehen bleiben. Für die Berechnung der Entropie wird der besprochene Unterschied ohnehin belanglos, da er nach (203) nur auf die Hinzufügung einer additiven negativen Konstanten hinausläuft.

Zweites Kapitel. Entropie eines idealen einatomigen Gases.

§ 138. Im vorigen Kapitel wurde die Berechtigung und die Notwendigkeit der Einführung von Wahrscheinlichkeitsbetrachtungen in die mechanische und in die elektrodynamische Theorie der Wärme nachgewiesen, und aus dem allgemeinen Zusammenhang der Entropie S mit der Wahrscheinlichkeit W, welcher in der Gleichung (203) ausgedrückt ist, eine Methode abgeleitet, um die Entropie eines physikalischen Systems in einem gegebenen Zustand zu berechnen. Bevor nun diese Methode zur Bestimmung der Entropie der strahlenden Wärme angewendet wird, soll sie in diesem Kapitel dazu benutzt werden, um die Entropie eines idealen einatomigen Gases in einem beliebig gegebenen Zustand zu berechnen. Alles Wesentliche dieser Berechnung findet sich zwar schon in den zum Teil noch weiter ausgreifenden Untersuchungen von L. Boltzmann[1] über die mechanische Theorie der Wärme; indessen wird es sich doch empfehlen, hier auf jenen ganz einfachen Fall besonders einzugehen, einmal um die Berechnungsweise und die physikalische Bedeutung der mechanischen Entropie mit derjenigen der Strahlungsentropie bequemer vergleichen zu können, dann aber hauptsächlich deshalb, um die Bedeutung der universellen Konstanten k der Gleichung (203) in der kinetischen Gastheorie klar hervortreten zu lassen; und dazu genügt natürlich die Behandlung eines einzigen speziellen Falles.

§ 139. Wir denken uns ein ideales, aus N gleichartigen einatomigen Molekülen bestehendes Gas in einem gegebenen Zustand und fragen nach der Entropie des Gases in diesem Zustand. Da der Zustand als gegeben vorausgesetzt ist, so ist das Gesetz der Raum- und der Geschwindigkeitsverteilung als bekannt anzunehmen (§ 135). Betrachten wir also das Raum-

---

[1] L. Boltzmann, Sitzungsber. d. Akad. d. Wissensch. zu Wien (II) **76**, p. 373, 1877. Vgl. auch Gastheorie **1**, p. 38, 1896.

gebiet, welches durch die Raumkoordinaten x, y, z und ihre Differentiale dx, dy, dz, und das Geschwindigkeitsgebiet, welches durch die Geschwindigkeitskomponenten ξ, η, ζ und ihre Differentiale $d\xi$, $d\eta$, $d\zeta$ charakterisiert wird, so ist die Anzahl der Moleküle, deren Koordinaten und Geschwindigkeiten zugleich in diesen beiden Gebieten liegen, als gegeben anzusehen. Die Ausdehnung eines solchen „Elementargebietes":

$$dx \cdot dy \cdot dz \cdot d\xi \cdot d\eta \cdot d\zeta = d\sigma$$

ist klein gegen die äußere Begrenzung des Gesamtgebietes, aber doch immerhin so groß zu denken, daß sich viele Moleküle darin befinden; denn sonst könnte der Zustand nicht elementar ungeordnet sein. Wir setzen daher die Anzahl der in dem Elementargebiet $d\sigma$ befindlichen Moleküle gleich:

$$f(x, y, z, \xi, \eta, \zeta) \cdot d\sigma. \tag{204}$$

f ist hier als eine endliche bekannte Funktion der Koordinaten und der Geschwindigkeitskomponenten anzusehen, deren analytischer Ausdruck das gesamte Verteilungsgesetz und damit den Zustand des Gases eindeutig darstellt. Denn auf die speziellere Anordnung der Moleküle innerhalb eines einzelnen Elementargebietes kommt es weiter nicht an. Wir wollen f als stetig und differentiierbar voraussetzen; im übrigen muß f nur die eine Bedingung erfüllen, daß sich durch Integration über alle Elementargebiete die Gesamtzahl der Gasmoleküle ergibt:

$$\int f \, d\sigma = N. \tag{205}$$

§ 140. Es handelt sich jetzt im wesentlichen um die Bestimmung der Wahrscheinlichkeit W für die gegebene Raum- und Geschwindigkeitsverteilung, welche nach § 136 gleich ist der Anzahl von Komplexionen, die dieser Verteilung entsprechen. Zu diesem Zwecke nehmen wir zunächst, was bisher nicht wesentlich war, alle Elementargebiete $d\sigma$ als gleich groß an.

Nun kann man die gegebene Raum- und Geschwindigkeitsverteilung anschaulich illustrieren dadurch, daß man die verschiedenen gleich großen Elementargebiete numeriert, die Nummern nebeneinander schreibt, und unter jede Nummer die Anzahl der Moleküle setzt, welche in dem betreffenden Gebiet liegen. Hätten wir z. B. nur 10 Moleküle und nur 7 Elementargebiete, so würde eine bestimmte Verteilung durch folgendes Ziffernbild dargestellt:

1	2	3	4	5	6	7
1	2	0	0	1	4	2,

welches besagt, daß

1 Molekül im 1. Elementargebiet,
2 Moleküle „ 2. „
0 Molekül „ 3. „
0 „ „ 4. „
1 „ „ 5. „
4 Moleküle „ 6. „
2 „ „ 7. „ liegen.

Diese bestimmte Verteilung kann nun durch viele verschiedene individuelle Zuordnungen oder Komplexionen realisiert werden, je nachdem ein bestimmtes ins Auge gefaßtes Molekül in dieses oder in jenes Elementargebiet zu liegen kommt. Um sich eine einzelne derartige Komplexion zu versinnbildlichen, kann man die Moleküle mit Ziffern versehen, diese nebeneinander schreiben, und unter jede Molekülziffer die Nummer desjenigen Elementargebiets setzen, welchem das betreffende Molekül bei dieser Komplexion angehört. Für die oben angeführte Verteilung erhalten wir so als Ausdruck einer einzelnen dazugehörigen beliebig herausgegriffenen Komplexion das folgende Ziffernbild:

$$\left\{\begin{matrix} 1 & 2 & 3 & 4 & 5 & 6 & 7 & 8 & 9 & 10 \\ 6 & 1 & 7 & 5 & 6 & 2 & 2 & 6 & 6 & 7. \end{matrix}\right. \tag{206}$$

Hierdurch ist ausgedrückt, daß

das 2. Molekül . . . im 1. Elementargebiet,
„ 6. u. 7. Molekül . . „ 2. „
„ 4. Molekül . . . „ 5. „
„ 1., 5., 8. u. 9. Molekül „ 6. „
„ 3. u. 10. Molekül . „ 7. „ liegt.

Wie man durch einen Vergleich mit der vorigen Tabelle unmittelbar erkennt, entspricht diese Komplexion in der Tat in allen Stücken dem oben angegebenen Verteilungsgesetz, und ebenso lassen sich leicht viele andere Komplexionen angeben, welche zu dem nämlichen Verteilungsgesetz gehören. Die gesuchte Zahl aller möglichen Komplexionen ergibt sich nun durch die Betrachtung der unteren der beiden Ziffernreihen (206). Denn da die Anzahl der Moleküle gegeben ist, so enthält die

Ziffernreihe eine bestimmte Anzahl Stellen. Da ferner das Verteilungsgesetz gegeben ist, so kommt jede Ziffer (d. h. jedes Elementargebiet) stets gerade so oft in der Reihe vor, als die Anzahl der Moleküle beträgt, die in dem betreffenden Elementargebiet liegen. Im übrigen bedingt jede Veränderung des Ziffernbildes eine neue individuelle Zuordnung der Moleküle zu den Gebieten, also eine neue Komplexion. Die Anzahl der möglichen Komplexionen, oder die Wahrscheinlichkeit W des gegebenen Zustandes, ist also gleich der Anzahl der unter den genannten Bedingungen möglichen „Permutationen mit Wiederholung". In dem gewählten einfachen Zahlenbeispiel ergibt sich hierfür nach einer bekannten Formel der Ausdruck:

$$\frac{10!}{1!\ 2!\ 0!\ 0!\ 1!\ 4!\ 2!} = 37800.$$

Die Form dieses Ausdrucks ist so gewählt, daß sie leicht auf den hier vorliegenden Fall der Gasmoleküle verallgemeinert werden kann. Der Zähler enthält die Fakultät oder Faktorielle der Gesamtzahl N der betrachteten Moleküle, der Nenner das Produkt der Fakultäten der Molekülzahlen, welche in jedem einzelnen Elementargebiet liegen, und welche in unserem Falle durch den Ausdruck (204) gegeben sind.

Daher erhalten wir für die gesuchte Wahrscheinlichkeit der gegebenen Raum- und Geschwindigkeitsverteilung, und somit des gegebenen Gaszustandes:

$$W = \frac{N!}{\Pi (f\, d\sigma)!}.$$

Das Zeichen Π bedeutet das Produkt, erstreckt über alle Elementargebiete $d\sigma$.

§ 141. Daraus ergibt sich nach (203) für die Entropie des Gases in dem gegebenen Zustand:

$$S = k \log N! - k \sum \log (f\, d\sigma)! + \text{const.}$$

Die Summation $\sum$ ist über alle Elementargebiete $d\sigma$ zu erstrecken.

Da $f\, d\sigma$ eine große Zahl ist, so läßt sich für die Fakultät derselben die STIRLINGsche Formel anwenden, welche für eine große Zahl n abgekürzt lautet:[1]

[1] Vgl. z. B. E. CZUBER, Wahrscheinlichkeitsrechnung (Leipzig, B. G. Teubner), p. 22, 1903.

(207) $$n! = \left(\frac{n}{e}\right)^n \sqrt{2\pi n}.$$

Also, mit Fortlassung unwesentlicher Glieder:

$$\log n! = n(\log n - 1).$$

Daher wird, wenn man $f\,d\sigma$ statt n setzt:

$$S = k \log N! - k \sum f\,d\sigma\,[\log(f\,d\sigma) - 1] + \text{const}.$$

Das Summenzeichen $\sum$ ersetzen wir weiterhin durch das Integralzeichen. Ferner wollen wir alle additiven konstanten Glieder in die const aufgenommen denken. Dazu gehört zunächst das Glied mit $N!$, ferner der Faktor $d\sigma$ hinter dem Logarithmus, weil alle Elementargebiete gleich groß sind, und weil $\sum f\,d\sigma = N$ konstant ist, endlich das Glied mit -1. So bleibt für die Entropie des Gases der Ausdruck übrig:

(208) $$S = \text{const} - k \int f \log f\,d\sigma,$$

gültig für jede beliebige gegebene Raum- und Geschwindigkeitsverteilung der Gasmoleküle, also für jeden Zustand des Gases.

§ 142. Wir wollen nun speziell die Entropie des Gases in einem Gleichgewichtszustand bestimmen, und fragen daher zunächst nach derjenigen Form des Verteilungsgesetzes, welches dem thermodynamischen Gleichgewicht entspricht. Nach dem zweiten Hauptsatz der Thermodynamik ist ein Gleichgewichtszustand durch die Bedingung ausgezeichnet, daß bei gegebenen Werten des Gesamtvolumens V und der Gesamtenergie U die Entropie S ihren Maximalwert annimmt. Setzen wir also das Gesamtvolumen der Gasmoleküle:

$$V = \iiint dx\,dy\,dz$$

und die Gesamtenergie derselben:

(209) $$U = \frac{m}{2} \int (\xi^2 + \eta^2 + \zeta^2) f\,d\sigma$$

als gegeben voraus (m Masse eines Moleküls), so muß für den Gleichgewichtszustand die Bedingung gelten:

$$\delta S = 0$$

oder nach (208):

(210) $$\int (\log f + 1)\,\delta f\,d\sigma = 0,$$

wobei die Variation δf sich auf eine beliebige mit den gegebenen Werten von N, V und U verträgliche Änderung des Verteilungsgesetzes bezieht.

Nun ist wegen der Unveränderlichkeit der Gesamtzahl N der Moleküle nach (205):

$$\int \delta f \, d\sigma = 0$$

und wegen der Unveränderlichkeit der Gesamtenergie U nach (209):

$$\int (\xi^2 + \eta^2 + \zeta^2) \, \delta f \, d\sigma = 0 .$$

Folglich ist zur Erfüllung der Bedingung (210) für alle zulässigen δf hinreichend und notwendig, daß

$$\log f + \beta (\xi^2 + \eta^2 + \zeta^2) = \text{const}$$

oder:

$$f = \alpha \, e^{-\beta(\xi^2 + \eta^2 + \zeta^2)} , \tag{211}$$

wobei α und β konstant. Im Gleichgewichtszustand ist also die Raumverteilung der Moleküle gleichmäßig, d. h. unabhängig von x, y, z, und die Geschwindigkeitsverteilung ist die bekannte MAXWELLsche.

§ 143. Die Werte der Konstanten α und β ergeben sich aus denen von N, V und U. Denn die Substitution des gefundenen Ausdrucks von f in (205) ergibt:

$$N = V \alpha \iiint_{-\infty}^{+\infty} e^{-\beta(\xi^2 + \eta^2 + \zeta^2)} \, d\xi \, d\eta \, d\zeta = V \alpha \left(\frac{\pi}{\beta}\right)^{\frac{3}{2}}$$

und die Substitution von f in (209) ergibt:

$$U = V \cdot \frac{m}{2} \alpha \iiint_{-\infty}^{+\infty} (\xi^2 + \eta^2 + \zeta^2) \cdot e^{-\beta(\xi^2 + \eta^2 + \zeta^2)} \, d\xi \, d\eta \, d\zeta ,$$

$$U = \frac{3}{4} V m \alpha \frac{1}{\beta} \left(\frac{\pi}{\beta}\right)^{\frac{3}{2}} .$$

Daraus folgt:

$$\alpha = \frac{N}{V} \cdot \left(\frac{3 m N}{4 \pi U}\right)^{\frac{3}{2}} , \qquad \beta = \frac{3 m N}{4 U}$$

und daraus endlich nach (208) der Ausdruck der Entropie S des Gases im Gleichgewichtszustand bei gegebenen Werten von N, V und U:

$$S = \text{const} + k N \left(\tfrac{3}{2} \log U + \log V\right) . \tag{212}$$

Hier enthält die additive Konstante Glieder mit N und mit m, nicht aber solche mit U oder mit V.

§ 144. Die hier durchgeführte Bestimmung der Entropie eines einatomigen Gases stützt sich lediglich auf den allgemeinen durch die Gleichung (203) ausgedrückten Zusammenhang zwischen Entropie und Wahrscheinlichkeit; insbesondere haben wir bei unserer Berechnung an keiner Stelle von irgend einem speziellen Satz der Lehre von den Gasen Gebrauch gemacht. Daher ist es von Wichtigkeit, zu sehen, wie nun aus dem gefundenen Ausdruck der Entropie das gesamte thermodynamische Verhalten eines einatomigen Gases, namentlich die Zustandsgleichung und die Werte der spezifischen Wärme, direkt mittels der Hauptsätze der Thermodynamik erschlossen werden kann. Denn aus der allgemeinen thermodynamischen Definitionsgleichung der Entropie:

$$d\,S = \frac{d\,U + p\,d\,V}{T} \tag{213}$$

ergeben sich die partiellen Differentialquotienten von S nach U und nach V:

$$\left(\frac{\partial S}{\partial U}\right)_V = \frac{1}{T}, \qquad \left(\frac{\partial S}{\partial V}\right)_U = \frac{p}{T}.$$

Folglich für unser Gas, mit Benutzung von (212):

$$\left(\frac{\partial S}{\partial U}\right)_V = \frac{3}{2}\frac{k\,N}{U} = \frac{1}{T} \tag{214}$$

und

$$\left(\frac{\partial S}{\partial V}\right)_U = \frac{k\,N}{V} = \frac{p}{T}. \tag{215}$$

Die zweite dieser Gleichungen:

$$p = \frac{k\,N\,T}{V}$$

enthält die Gesetze von Boyle, Gay Lussac und Avogadro, das letztere deshalb, weil der Druck nur von der Anzahl N, nicht von der Beschaffenheit der Moleküle abhängt. Schreibt man sie in der gewöhnlichen Form:

$$p = \frac{R\,n\,T}{V},$$

wo n die Anzahl der G r a m m o l e k ü l e oder der Mole des Gases, bezogen auf $O_2 = 32$ g, und R die absolute Gaskonstante bezeichnet:

$$R = 831 \cdot 10^5 \frac{\text{erg}}{\text{grad}}, \tag{216}$$

so ergibt sich durch Vergleichung:

$$k = \frac{Rn}{N}. \tag{217}$$

Nennen wir also ω das Verhältnis der Molzahl zur Molekülzahl, oder, was dasselbe ist, das Verhältnis der Molekülmasse zur Molmasse, $\omega = \frac{n}{N}$, so kommt:

$$k = \omega R. \tag{218}$$

Hieraus kann man, wenn ω gegeben ist, die universelle Konstante k berechnen, und umgekehrt.

Die Gleichung (214) lautet:

$$U = \tfrac{3}{2} k N T. \tag{219}$$

Da nun andererseits die Energie eines idealen Gases:

$$U = A n c_v T,$$

wo c_v die Wärmekapazität eines Mol bei konstantem Volumen in Kalorien, A das mechanische Wärmeäquivalent bedeutet:

$$A = 419 \cdot 10^5 \frac{\text{erg}}{\text{cal}}, \tag{220}$$

so folgt:

$$c_v = \frac{3}{2} \frac{kN}{An}$$

und mit Berücksichtigung von (217):

$$c_v = \frac{3}{2} \frac{R}{A} = \frac{3}{2} \cdot \frac{831 \cdot 10^5}{419 \cdot 10^5} = 3{,}0 \tag{221}$$

als Molwärme irgend eines einatomigen Gases bei konstantem Volumen in Kalorien.[1]

Für die Molwärme c_p bei konstantem Druck folgt aus dem ersten Hauptsatz der Thermodynamik:

$$c_p - c_v = \frac{R}{A},$$

also mit Rücksicht auf (221):

$$c_p - c_v = \frac{2}{3} c_v, \qquad \frac{c_p}{c_v} = \frac{5}{3},$$

wie für einatomige Gase bekannt.

---

[1] Vgl. F. Richarz, Wied. Ann. **67**, p. 705, 1899.

10*

Die mittlere Energie oder die mittlere lebendige Kraft L eines Moleküls ergibt sich aus (219) zu:

$$\frac{U}{N} = L = \frac{3}{2} k T. \tag{222}$$

Man sieht, daß sich alle diese Beziehungen lediglich aus der Identifizierung des mechanischen Ausdrucks (208) mit dem thermodynamischen Ausdruck (213) der Entropie ergeben.

Drittes Kapitel.
Berechnung der Strahlungsentropie und Folgerungen daraus. Energieverteilungsgesetz. Elementarquanta.

§ 145. Nachdem wir gesehen haben, wie man für ein ideales Gas den Ausdruck der Entropie direkt aus der Wahrscheinlichkeit des Zustandes berechnen kann, und wie sich daraus alle thermodynamischen Eigenschaften des Gases durch eine direkte Anwendung der Hauptsätze der Thermodynamik ableiten lassen, wollen wir jetzt denselben Gedankengang für die strahlende Wärme durchführen. Aus dem WIENschen Verschiebungsgesetz erhielten wir in der Gleichung (119) einen Ausdruck für die räumliche Entropiedichte $\mathfrak{s}$ als Funktion der räumlichen Energiedichte $\mathfrak{u}$, ferner in der Gleichung (134) einen Ausdruck für die Entropie $\mathfrak{L}$ eines einzelnen Strahles als Funktion seiner spezifischen Intensität $\mathfrak{K}$, endlich in der Gleichung (200) einen Ausdruck für die Entropie S eines der Wärmestrahlung ausgesetzten Resonators als Funktion seiner Energie U. Jeder dieser drei Ausdrücke enthält eine bis jetzt noch unbekannt gebliebene universelle Funktion eines einzigen Arguments, und die Berechnung dieser Funktion ist es, worauf es im folgenden ankommt. Wenn diese Aufgabe für eine der drei genannten Ausdrücke gelöst ist, sind damit auch die beiden anderen Ausdrücke gefunden, vermöge der bekannten früher abgeleiteten Beziehungen zwischen den Größen $\mathfrak{s}$, $\mathfrak{L}$ und S untereinander, und der Größen $\mathfrak{u}$, $\mathfrak{K}$ und U untereinander. Wir können daher von vornherein an irgend eine jener drei Gleichungen anknüpfen. Am meisten empfiehlt es sich natürlich, die einfachste unter ihnen auszuwählen, und das ist, wie schon früher hervorgehoben, die Resonatorgleichung (200):

$$S = F\left(\frac{U}{\nu}\right), \tag{223}$$

wenn wir die Schwingungszahl der Eigenperiode des Resonators von jetzt an kurz mit ν ohne Index bezeichnen. Die Funktion F enthält außer ihrem Argument nur universelle Konstante.

§ 146. Bei der näheren Untersuchung der Entropie eines Resonators von gegebener Energie ist die erste Frage die nach der Art der elementaren Unordnung, auf welcher die Entropie beruht und ohne welche sie keine Bedeutung besitzt (§ 132). Die Antwort hierauf gibt ein Blick auf die Gleichungen (187) und (195). Hiernach sind die Schwingungen eines der stationären Wärmestrahlung ausgesetzten Resonators zusammengesetzt aus einer großen Reihe von Partialschwingungen, und seine Energie ist ein Mittelwert aus sehr vielen im einzelnen nicht kontrollierbaren Größen. Diese zahlreichen voneinander unabhängigen Partialschwingungen sind es also, die bei dem Resonator in bezug auf die elementare Unordnung dieselbe Rolle spielen, die bei einem Gase den zahlreichen durcheinanderfliegenden Molekülen zukommt. So wenig man bei einem Gase von einer endlichen Entropie sprechen kann, wenn alle Moleküle gleiche und gleichgerichtete, oder auch nur in irgend einer Weise geordnete Geschwindigkeiten besitzen, ebensowenig kommt einem Resonator eine endliche Entropie zu, wenn seine Schwingungen etwa einfach periodisch sind oder wenn sie überhaupt nach irgend einem bestimmten Gesetz erfolgen, das alles bis ins einzelne regelt. Denn dann ist der Schwingungsvorgang nicht mehr elementar ungeordnet. Daher besitzt z. B. ein Resonator, der von außen überhaupt nicht erregt wird, dessen Schwingungen also einfach mit konstanter Dämpfung nach Gleichung (169) abklingen, keine endliche Entropie und keine endliche Temperatur, obwohl er eine endliche Energie besitzen kann.

Ob nun die Resonatorschwingungen elementar ungeordnet sind oder nicht, kann man offenbar gar nicht beurteilen, wenn man den Zustand des Resonators nur zu einem bestimmten Zeitpunkt berücksichtigt. Denn dann bleibt es noch ganz unentschieden, ob der Zustand sich mit der Zeit regelmäßig oder regellos ändert. Damit stimmt auch ganz überein, daß wir die

Energie U eines der stationären Wärmestrahlung ausgesetzten Resonators nur als einen zeitlichen Mittelwert definieren können, wie in § 123 geschehen ist. Aus diesem Grund besitzt auch die Entropie S eines Resonators nicht für einen Zeitpunkt, sondern nur für ein Zeitintervall Bedeutung, das viele Resonatorschwingungen umfaßt, und wir können nur von einem zeitlichen Mittelwert der Entropie reden.[1] Kurz gesagt: bei den Wärmeschwingungen eines Resonators ist die Unordnung eine zeitliche, während sie bei den Molekularbewegungen eines Gases eine räumliche ist. Doch fällt dieser Unterschied für die Berechnung der Entropie des Resonators nicht so schwer ins Gewicht, als es vielleicht auf den ersten Anblick scheinen möchte; denn er läßt sich durch eine einfache Betrachtung beseitigen, was im Interesse einer gleichförmigen Behandlung von Vorteil ist.

Der zeitliche Mittelwert U der Energie eines einzelnen in einem stationär durchstrahlten Vakuum befindlichen Resonators ist nämlich offenbar identisch mit dem in einem bestimmten Zeitpunkt genommenen Mittelwert der Energien einer sehr großen Anzahl N von genau gleichbeschaffenen Resonatoren, die sich in dem nämlichen stationären Strahlungsfelde befinden, aber so weit voneinander entfernt, daß ihre Schwingungen sich nicht direkt merklich beeinflussen. Natürlich muß zu diesem Zweck das Feld von hinreichender räumlicher Ausdehnung genommen werden. Damit ist die Frage nach der Verteilung der Energie unter die einzelnen Partialschwingungen eines einzigen Resonators zurückgeführt auf die räumliche Verteilung der Energie auf die N Resonatoren, wie es dem bei den Gasmolekülen behandelten Fall besser entspricht.

§ 147. Um nun die Entropie dieses Systems von N in einem stationären Strahlungsfelde befindlichen gleichbeschaffenen Resonatoren in einem gegebenen Zustand zu berechnen, müssen wir nach den Ausführungen des § 135 zunächst nach denjenigen Größen fragen, welche den physikalischen Zustand des Systems bestimmen. Das ist hier einzig und allein die mittlere Energie U

[1] Bei der Anwendung auf nichtstationäre Felder muß das dem Mittelwert zugrunde gelegte Zeitintervall so klein genommen werden, daß das Feld als stationär betrachtet werden kann. Vgl. § 3.

eines einzelnen Resonators, bez. die Gesamtenergie U_N des ganzen Systems von Resonatoren, welche mit U durch die Gleichung:

$$N \cdot U = U_N \tag{224}$$

zusammenhängt. Denn da das Strahlungsfeld stationär ist, so ist durch die Energie der physikalische Zustand des ganzen Systems bestimmt.[1] In diesem Punkt liegt der wesentlichste Unterschied zwischen dem hier behandelten Fall und dem früheren eines Gases. Denn dort war der Zustand bedingt durch die Art der Raum- und Geschwindigkeitsverteilung unter den Molekülen, die von vornherein ganz beliebig angenommen werden konnte. Erst wenn das Verteilungsgesetz gegeben war, konnte der Zustand als bekannt angenommen werden. Hier dagegen genügt die Angabe der Gesamtenergie U_N der N Resonatoren für die Bestimmung des Zustandes; die speziellere Art der Verteilung der Energie U_N unter die einzelnen Resonatoren unterliegt nicht mehr der Kontrolle, sie ist ganz dem Zufall, der elementaren Unordnung, anheimgegeben. Denn die Bedingung, daß das Strahlungsfeld stationär ist, bedeutet hier nicht etwa einen speziellen Fall unter vielen anderen, sondern sie gehört hier mit zu den notwendigen Voraussetzungen; sonst könnte man den Quotienten $\frac{U_N}{N}$ nicht mehr, wie wir es getan haben, mit dem zeitlichen Mittelwert der Energie eines einzelnen Resonators identifizieren.

§ 148. Es handelt sich nun weiter um die Wahrscheinlichkeit W des durch die Energie U_N bestimmten Zustandes der N Resonatoren, d. h. um die Anzahl der individuellen Zuordnungen oder Komplexionen, welche der Verteilung der Energie U_N auf die N Resonatoren entsprechen (§ 136). Wir könnten hier ganz analog wie bei den Gasmolekülen verfahren, indem wir nur berücksichtigen, daß der gegebene Zustand des Resonatorensystems nicht, wie dort, eine einzige, sondern eine große Anzahl von verschiedenen Verteilungsgesetzen zuläßt, da die Anzahl der Resonatoren, welche eine bestimmte Größe der Energie besitzen (besser: welche in ein bestimmtes „Energiegebiet" hineinfallen), keine vorgeschriebene ist, sondern variieren

[1] Das „System" umfaßt natürlich nur die N Resonatoren selber; das Strahlungsfeld gehört nicht mit dazu.

kann. Betrachten wir nun alle mögliche Arten von Energieverteilungsgesetzen und berechnen für jedes derselben die ihm entsprechende Anzahl von Komplexionen, genau wie bei den Gasmolekülen, so erhalten wir durch Addition sämtlicher so erhaltener Komplexionszahlen die gesuchte Wahrscheinlichkeit W des gegebenen physikalischen Zustandes,

Schneller und bequemer als auf dem angegebenen Wege kommen wir folgendermaßen zu demselben Ziele. Wir teilen die gegebene Gesamtenergie U_N in eine große Anzahl P gleiche Teile von der Größe ε, deren jeden wir ein Energieelement nennen. Dann ist:

$$P = \frac{U_N}{\varepsilon}. \tag{225}$$

Diese P Energieelemente sind nun auf alle mögliche Weise unter die N Resonatoren zu verteilen, wobei es aber nicht darauf ankommt, welche Energieelemente, sondern nur wieviel Energieelemente auf einen bestimmten Resonator entfallen. Denken wir uns also die N Resonatoren numeriert und die Ziffern nebeneinander in eine Reihe geschrieben, und zwar jede Ziffer so oft, als die Zahl der Energieelemente beträgt, die auf den betreffenden Resonator entfallen, so erhalten wir durch eine solche Ziffernreihe das Bild einer bestimmten Komplexion, in welcher jedem individuellen Resonator eine bestimmte Energie zukommt. Die Anordnung der Ziffern in der Reihe ist für die Komplexion gleichgültig, da eine bloße Umstellung der Ziffern an der Energie eines bestimmten Resonators nichts ändert. Besitzt in der Komplexion ein Resonator gar keine Energie, so kommt seine Ziffer in der Reihe gar nicht vor. Die Gesamtzahl der Zifferstellen ist notwendig P, d. h. die Zahl der zu verteilenden Energieelemente. Somit ist die Anzahl aller möglichen verschiedenen Komplexionen gleich der Anzahl der möglichen „Kombinationen mit Wiederholung von N Elementen zur P. Klasse“:

$$W = \frac{(N + P - 1)!}{(N - 1)!\, P!}$$

und dies ist zugleich die gesuchte Wahrscheinlichkeit des gegebenen Zustandes der N Resonatoren. Wenn beispielsweise $N = 3$, $P = 4$, so sind die Bilder aller möglichen Komplexionen:

1	1	1	1		1	1	3	3		2	2	2	2
1	1	1	2		1	2	2	2		2	2	2	3
1	1	1	3		1	2	2	3		2	2	3	3
1	1	2	2		1	2	3	3		2	3	3	3
1	1	2	3		1	3	3	3		3	3	3	3

Die Anzahl aller möglichen Komplexionen ist hier $W = 15$, wie es der Formel entspricht.

Für die Entropie S_N des Resonatorsystems erhalten wir daher nach Gleichung (203), da N und P große Zahlen sind, mit Weglassung der additiven Konstanten:

$$S_N = k \log \frac{(N+P)!}{N!\,P!}$$

und mit Benutzung der STIRLINGschen Formel (207):

$$S_N = k\{(N+P)\log(N+P) - N\log N - P\log P\}.$$

Ersetzt man nun nach (225) P durch U_N und nach (224) U_N durch U, so ergibt sich nach leichter Umformung für die Entropie der N gleichbeschaffenen Resonatoren:

$$S_N = kN\left\{\left(1+\frac{U}{\varepsilon}\right)\log\left(1+\frac{U}{\varepsilon}\right) - \frac{U}{\varepsilon}\log\frac{U}{\varepsilon}\right\}$$

und für die Entropie eines einzelnen Resonators:

$$S = \frac{S_N}{N} = k\left\{\left(1+\frac{U}{\varepsilon}\right)\log\left(1+\frac{U}{\varepsilon}\right) - \frac{U}{\varepsilon}\log\frac{U}{\varepsilon}\right\}.$$

Ein Vergleich dieses Ausdrucks mit (223) zeigt, daß das Energieelement ε proportional der Schwingungszahl ν der Eigenperiode des Resonators sein muß. Wir setzen daher:

$$\varepsilon = h\nu, \tag{226}$$

wobei h konstant, und erhalten dadurch:

$$S = k\left\{\left(1+\frac{U}{h\nu}\right)\log\left(1+\frac{U}{h\nu}\right) - \frac{U}{h\nu}\log\frac{U}{h\nu}\right\} \tag{227}$$

als Lösung des behandelten Problems.

§ 149. Auffallend an diesem Resultat ist zunächst das Auftreten einer neuen universellen Konstante h von der Dimension eines Produkts aus Energie und Zeit. Hierin liegt ein wesentlicher Unterschied gegenüber dem Ausdruck der Entropie eines Gases, wo die Größe eines Elementargebiets, die wir $d\sigma$ nannten, aus dem Schlußresultat ganz verschwindet, da sie sich nur in der physikalisch bedeutungslosen additiven Konstanten geltend macht. Es kann wohl keinem Zweifel unterliegen, daß die Konstante h bei den elementaren Schwingungsvorgängen in einem Emissionszentrum eine gewisse Rolle spielt, zu deren Ergründung von elektrodynamischer Seite her unsere bisherige Theorie jedoch keine näheren Anhaltspunkte liefert.[1] Und doch wird die Thermodynamik der Strahlung erst dann zum vollständig befriedigenden Abschluß gelangt sein, wenn die Konstante h in ihrer vollen universellen Bedeutung erkannt ist. Ich möchte dieselbe als „elementares Wirkungsquantum" oder als „Wirkungselement" bezeichnen, weil sie von derselben Dimension ist wie diejenige Größe, welcher das Prinzip der kleinsten Wirkung seinen Namen verdankt.

§ 150. Es ist von Interesse sich besonders zu vergewissern, daß man zu dem nämlichen Ausdruck der Entropie wie oben gelangt, wenn man bei der Berechnung der Anzahl von Komplexionen, die einem gegebenen Zustand entsprechen, nicht von vornherein auf die Energie, die ja immerhin eine zusammengesetzte Größe ist, Bezug nimmt, sondern direkt auf den elektromagnetischen Zustand der einzelnen Resonatoren zurückgeht, was für die Berechnung nicht ganz so einfach, aber allgemeiner und daher rationeller ist. Es handelt sich hierbei im wesentlichen um die richtige Ausmessung der „Elementargebiete" des Zustandsbereichs, da deren Größe ja der Berechnung der Komplexionen zugrunde gelegt wird und somit in letzter Linie den Maßstab für die Vergleichung der Wahrscheinlichkeiten verschiedener Zustände liefert. Der elektromagnetische Zustand eines Resonators ist nach § 104 bestimmt durch die Werte von f und $\dot{f}$. Trägt man also in einer Koordinatenebene f als Abszisse, $\dot{f}$ als Ordinate auf, so entspricht jeder Punkt der Ebene einem bestimmten Zustand des Resonators, und umgekehrt. Doch ist die Größe eines Flächenelements in dieser Ebene keineswegs im allgemeinen ein Maß der Wahrscheinlichkeit dafür, daß der Zustand des Resonators durch einen Punkt innerhalb des Flächenelements dargestellt wird. Dieser einfache Satz gilt vielmehr nur dann, wenn man als Ordinate statt $\dot{f}$ den der

[1] Vgl. die Anmerkung in § 109.

Koordinate f entsprechenden „Impuls" (oder das „Moment" von f), nämlich die Größe:

$$\frac{\partial U}{\partial \dot{f}} = g,$$

d. h. nach (142):

$$g = L\dot{f} \tag{228}$$

nimmt.[1] Wir denken uns also f und g als die Koordinaten eines Punktes der Zustandsebene, und fragen zunächst nach der Größe der Wahrscheinlichkeit dafür, daß die Energie des Resonators zwischen den Werten U und $U + \triangle U$ liegt. Diese Wahrscheinlichkeit wird gemessen durch die Größe desjenigen Flächenstücks in der Ebene der Zustandsvariabeln f und g, welches von den Kurven $U =$ const und $U + \triangle U =$ const begrenzt wird.

Nun ist die Energie des Resonators in dem Zustandspunkt (f, g) nach (142) und (228) gegeben durch:

$$U = \frac{1}{2} K f^2 + \frac{1}{2} \frac{g^2}{L},$$

folglich ist die Kurve $U =$ const eine Ellipse mit den Halbachsen:

$$\sqrt{\frac{2U}{K}} \quad \text{und} \quad \sqrt{2UL}.$$

Ihr Flächeninhalt beträgt mithin:

$$\pi \sqrt{\frac{2U}{K}} \cdot \sqrt{2UL} = 2\pi U \sqrt{\frac{L}{K}} = \frac{U}{\nu},$$

wenn man nach Gleichung (166) die Schwingungszahl ν der Eigenperiode des Resonators einführt. Ebenso ergibt sich der Flächeninhalt der Ellipse $U + \triangle U =$ const als:

$$\frac{U + \triangle U}{\nu}.$$

Die Differenz der beiden Flächenräume, das Maß der gesuchten Wahrscheinlichkeit, beträgt mithin $\frac{\triangle U}{\nu}$. Denken wir uns nun die ganze Zustandsebene durch eine große Anzahl derartiger Ellipsen so in einzelne Abschnitte geteilt, daß die von je zwei aufeinanderfolgenden Ellipsen begrenzten ringförmigen Flächenstücke einander gleich sind, d. h. so, daß

---

[1] Vgl. z. B. L. Boltzmann, Gastheorie II, p. 62 ff., 1898, oder J. W. Gibbs, Elementary Principles in Statistical Mechanics, Chapter I, 1902.

$$\frac{\triangle U}{\nu} = \text{const},$$

so erhalten wir dadurch diejenigen Abschnitte $\triangle U$ der Energie, welche gleichen Wahrscheinlichkeiten entsprechen und welche daher als die Energieelemente zu bezeichnen sind. Setzen wir die Größe eines Energieelements $\triangle U = \varepsilon$ und die const der letzten Gleichung gleich h, so kommen wir genau zu der früheren Gleichung (226) zurück, ohne daß wir das WIENsche Verschiebungsgesetz herangezogen haben. Zugleich zeigt sich uns hier das elementare Wirkungsquantum h in einer neuen Bedeutung, nämlich als die Größe eines Elementargebiets in der Zustandsebene eines Resonators, gültig für Resonatoren von ganz beliebiger Schwingungsperiode. Der Umstand, daß die Konstante h als eine bestimmte endliche Größe eingeführt wird, ist charakteristisch für die ganze hier entwickelte Theorie. Würde man h unendlich klein annehmen, so käme man zu einem Strahlungsgesetz, welches als ein spezieller Fall aus dem allgemeinen hervorgeht (das RAYLEIGHsche Gesetz, vgl. § 154 und namentlich § 166).

§ 151. Die Gleichung (227) führt zunächst mit Rücksicht auf die Beziehungen (198) und (193) zu dem Ausdruck der Entropiestrahlung $\mathfrak{L}$ eines monochromatischen geradlinig polarisierten Strahles von der spezifischen Strahlungsintensität $\mathfrak{K}$ und der Schwingungszahl ν:

$$\mathfrak{L}_\nu = \frac{k\nu^2}{c^2}\left\{\left(1+\frac{c^2\mathfrak{K}}{h\nu^3}\right)\log\left(1+\frac{c^2\mathfrak{K}}{h\nu^3}\right) - \frac{c^2\mathfrak{K}}{h\nu^3}\log\frac{c^2\mathfrak{K}}{h\nu^3}\right\} \tag{229}$$

als bestimmtere Fassung der Gleichung (134) des WIENschen Verschiebungsgesetzes.

Ferner folgt mit Rücksicht auf (197) und (194) die räumliche Entropiedichte $\mathfrak{s}$ einer gleichmäßigen monochromatischen unpolarisierten Strahlung in ihrer Abhängigkeit von der räumlichen Energiedichte $\mathfrak{u}$:

$$\mathfrak{s}_\nu = \frac{8\pi k\nu^2}{c^3}\left\{\left(1+\frac{c^3\mathfrak{u}_\nu}{8\pi h\nu^3}\right)\log\left(1+\frac{c^3\mathfrak{u}_\nu}{8\pi h\nu^3}\right) - \frac{c^3\mathfrak{u}_\nu}{8\pi h\nu^3}\log\frac{c^3\mathfrak{u}_\nu}{8\pi h\nu^3}\right\} \tag{230}$$

als bestimmtere Fassung der Gleichung (119).

§ 152. Nun wollen wir in jede der drei Gleichungen (227), (229), (230) die Temperatur T des Resonators bez. der monochromatischen Strahlung einführen und die Energiegrößen U, $\mathfrak{K}$

und $\mathfrak{u}$ durch die Temperatur T ausdrücken. Wir benutzen dazu je eine der Gleichungen (199), (135) und (117), und erhalten dann:
Für die Energie des Resonators:

$$U = \frac{h\nu}{e^{\frac{h\nu}{kT}} - 1}. \tag{231}$$

Für die spezifische Intensität eines monochromatischen geradlinig polarisierten Strahles von der Schwingungszahl ν:

$$\mathfrak{K} = \frac{h\nu^3}{c^2} \cdot \frac{1}{e^{\frac{h\nu}{kT}} - 1}. \tag{232}$$

Für die räumliche Energiedichte der gleichmäßigen monochromatischen unpolarisierten Strahlung von der Schwingungszahl ν:

$$\mathfrak{u} = \frac{8\pi h\nu^3}{c^3} \cdot \frac{1}{e^{\frac{h\nu}{kT}} - 1}. \tag{233}$$

Unter allen verschiedenartig zusammengesetzten Strahlungen ist die schwarze Strahlung dadurch ausgezeichnet, daß alle darin enthaltenen monochromatischen Strahlen die nämliche Temperatur besitzen (§ 93). Daher liefern diese Gleichungen auch das Gesetz der Energieverteilung im Normalspektrum, d. h. im Emissionsspektrum eines in bezug auf das Vakuum schwarzen Körpers.

Bezieht man die spezifische Intensität eines monochromatischen Strahles nicht auf die Schwingungszahl ν, sondern, wie es in der Experimentalphysik gewöhnlich geschieht, auf die Wellenlänge λ, so erhält man, mit Benutzung von (15) und (16), den Ausdruck:

$$E_\lambda = \frac{c^2 h}{\lambda^5} \cdot \frac{1}{e^{\frac{ch}{k\lambda T}} - 1} \tag{234}$$

als die Intensität eines monochromatischen geradlinig polarisierten Strahles von der Wellenlänge λ, der von einem auf der Temperatur T befindlichen schwarzen Körper senkrecht zur Oberfläche in das Vakuum emittiert wird. Die dazu gehörige räumliche Strahlungsdichte der unpolarisierten Strahlung ergibt sich durch Multiplikation von E_λ mit $\frac{8\pi}{c}$.

Zur Geschichte der Gleichung (234) vgl. weiter unten § 189. Ihre experimentelle Prüfung hat bisher eine gute Übereinstimmung mit der Erfahrung ergeben.[1] Doch sind nach O. LUMMER und E. PRINGSHEIM[2] die bisherigen Messungen noch nicht ausreichend, um vom rein experimentellen Standpunkt aus die Allgemeingültigkeit jener Formel als gesichert hinstellen zu können.

§ 153. Für kleine Werte von λT (d. h. klein gegen die Konstante $\frac{ch}{k}$) geht (234) über in die Gleichung:

$$E_\lambda = \frac{c^2 h}{\lambda^5} \cdot e^{-\frac{ch}{k\lambda T}}, \tag{235}$$

welche das „WIENsche Energieverteilungsgesetz" ausspricht.[3]

Die spezifische Strahlungsintensität $\mathfrak{K}$ wird dann nach (232):

$$\mathfrak{K} = \frac{h\nu^3}{c^2} \cdot e^{-\frac{h\nu}{kT}} \tag{236}$$

und die räumliche Energiedichte $\mathfrak{u}$ nach (233):

$$\mathfrak{u} = \frac{8\pi h\nu^3}{c^3} \cdot e^{-\frac{h\nu}{kT}}. \tag{237}$$

Für die Energie eines Resonators von der Schwingungszahl ν erhält man dann aus (231):

$$U = h\nu e^{-\frac{h\nu}{kT}}. \tag{238}$$

Die Entropie S des Resonators als Funktion der Energie U wird dann nach (227), da der Quotient $\frac{U}{h\nu}$ kleine Werte annimmt:

$$S = -\frac{kU}{h\nu} \log \frac{U}{eh\nu}. \tag{239}$$

Diese Beziehungen gelten also für jede Wellenlänge bei hinreichend tiefen Temperaturen, und für jede Temperatur bei hinreichend kurzen Wellen.

---

[1] Vgl. namentlich H. RUBENS und F. KURLBAUM, Sitzungsber. d. Akad. d. Wissensch. zu Berlin, vom 25. Okt. 1900, p. 929. DRUDES Ann. **4**, p. 649, 1901; und F. PASCHEN, DRUDES Ann. **4**, p. 277, 1901.

[2] O. LUMMER und E. PRINGSHEIM, DRUDES Ann. **6**, p. 210, 1901.

[3] W. WIEN, WIED. Ann. **58**, p. 662, 1896.

§ 154. Für große Werte von λT hingegen wird aus (234):

$$E_\lambda = \frac{c k T}{\lambda^4}, \tag{240}$$

eine Beziehung, die zuerst von Lord RAYLEIGH[1] aufgestellt worden ist, und die wir daher als „RAYLEIGHsches Strahlungsgesetz" bezeichnen können.

Für die spezifische Strahlungsintensität $\mathfrak{K}$ ergibt sich dann aus (232):

$$\mathfrak{K} = \frac{k \nu^2 T}{c^2} \tag{241}$$

und für die räumliche Energiedichte $\mathfrak{u}$ der monochromatischen Strahlung aus (233):

$$\mathfrak{u} = \frac{8 \pi k \nu^2 T}{c^3}. \tag{242}$$

Die Energie eines Resonators wird dann nach (231):

$$U = k T, \tag{243}$$

also einfach proportional der absoluten Temperatur und ganz unabhängig von der Schwingungszahl ν der Eigenperiode, wie überhaupt von der natürlichen Beschaffenheit des Resonators.

Für die Entropie S des Resonators als Funktion seiner Energie U endlich ergibt sich unter derselben Voraussetzung, da $\frac{U}{h\nu}$ dann große Werte annimmt:

$$S = k \log U + \text{const}. \tag{244}$$

Es ist von Interesse, den einfachen für lange Wellen oder hohe Temperaturen gültigen Wert (243) der Schwingungsenergie eines Resonators zu vergleichen mit der früher in (222) berechneten mittleren lebendigen Kraft L der Bewegung eines einatomigen Moleküls bei der nämlichen Temperatur. Der Vergleich ergibt:

$$U = \tfrac{2}{3} L. \tag{245}$$

Diese Beziehung, und damit auch die Identität der Konstanten k für die Molekularbewegungen und für die Strahlungsvorgänge, wird von einer ganz anderen Seite her in sehr bemerkenswerter Weise bestätigt durch eine Folgerung aus der Elektronentheorie. Nach den Anschauungen dieser Theorie hat man sich nämlich

[1] Lord RAYLEIGH, Phil. Mag. **49**, p. 539, 1900.

die von uns betrachteten linearen Schwingungen eines elementaren Oszillators vorzustellen als geradlinige Bewegungen eines Elektrons. Dann muß nach einem Satz der statistischen Mechanik in einem von Wärmestrahlung erfüllten Gase beim thermodynamischen Gleichgewichtszustand die mittlere lebendige Kraft dieser geradlinigen Elektronenbewegung gleich sein dem dritten Teil der mittleren lebendigen Kraft der fortschreitenden Bewegung eines Moleküls; denn die letztere Bewegung wird durch drei voneinander unabhängige Komponenten bestimmt, besitzt also drei Grade von Bewegungsfreiheit, während dagegen den Elektronenschwingungen in unserem Oszillator nur ein einziger Grad von Bewegungsfreiheit zukommt. Nun ist einerseits die mittlere lebendige Kraft der Elektronenschwingungen gleich der Hälfte der ganzen Schwingungsenergie, also $\frac{1}{2}U$, andererseits ist der dritte Teil der mittleren lebendigen Kraft der fortschreitenden Bewegung eines Moleküls gleich $\frac{1}{3}L$, also folgt daraus die Beziehung (245). Sind verschiedenartige Resonatoren mit verschiedenen Eigenschwingungen im Gase vorhanden, so müssen sie alle die nämliche mittlere Schwingungsenergie besitzen, ebenso wie die mittlere lebendige Kraft der fortschreitenden Bewegung verschiedener Molekülarten die gleiche ist. In der Tat ist nach (243) U von ν unabhängig.[1]

§ 155. Für die räumliche Gesamtdichte u der schwarzen Strahlung bei irgend einer Temperatur T ergibt sich aus (233):

$$u = \int_0^\infty \mathfrak{u}\, d\nu = \frac{8\pi h}{c^3} \int_0^\infty \frac{\nu^3\, d\nu}{e^{\frac{h\nu}{kT}} - 1}$$

oder:
$$u = \frac{8\pi h}{c^3} \int_0^\infty \left(e^{-\frac{h\nu}{kT}} + e^{-\frac{2h\nu}{kT}} + e^{-\frac{3h\nu}{kT}} + \ldots\right) \nu^3\, d\nu$$

---

[1] Vgl. hierzu A. Einstein, Drudes Ann. **17**, p. 132, 1905. Die dort hervorgehobene der Strahlungstheorie entgegenstehende Schwierigkeit rührt daher, daß die Beziehung (245) dort von vornherein als allgemeingültig vorausgesetzt wird, während nach der hier dargestellten Theorie der erwähnte Satz der statistischen Mechanik nur für hinreichend große Werte des Produkts λT Gültigkeit beanspruchen kann. Näheres über diesen prinzipiell wichtigen Punkt s. § 166.

und durch gliedweise Integration:

$$u = \frac{48\pi h}{c^3}\left(\frac{kT}{h}\right)^4 \alpha, \tag{246}$$

wobei zur Abkürzung gesetzt ist:

$$\alpha = 1 + \frac{1}{2^4} + \frac{1}{3^4} + \frac{1}{4^4} + \ldots = 1{,}0823. \tag{247}$$

Hierdurch ist das STEFAN-BOLTZMANNsche Gesetz (75) ausgedrückt, mit der näheren Maßgabe, daß die Konstante dieses Gesetzes:

$$a = \frac{48\pi\alpha k^4}{c^3 h^3}. \tag{248}$$

§ 156. Für diejenige Wellenlänge λ_m, welcher im Spektrum der schwarzen Strahlung das Maximum der Strahlungsintensität E_λ entspricht, ergibt sich aus der Gleichung (234):

$$\left(\frac{dE_\lambda}{d\lambda}\right)_{\lambda=\lambda_m} = 0.$$

Die Ausführung der Differentiation liefert, wenn man zur Abkürzung setzt:

$$\frac{ch}{k\lambda_m T} = \beta,$$

$$e^{-\beta} + \frac{\beta}{5} - 1 = 0.$$

Die Wurzel dieser transzendenten Gleichung ist:

$$\beta = 4{,}9651, \tag{249}$$

mithin ist $\lambda_m T = \frac{ch}{\beta k}$, also konstant, wie es das WIENsche Verschiebungsgesetz verlangt. Durch Vergleichung mit (109) erhält man die Bedeutung der Konstanten b:

$$b = \frac{ch}{\beta k}. \tag{250}$$

§ 157. Zahlenwerte. Mit Hilfe der gemessenen Werte von a und b lassen sich die universellen Konstanten h und k leicht berechnen. Es folgt nämlich aus den Gleichungen (248) und (250):

$$h = \frac{a\beta^4 b^4}{48\pi\alpha c}, \qquad k = \frac{a\beta^3 b^3}{48\pi\alpha}, \tag{251}$$

Dies ergibt mit den angegebenen Werten der Konstanten a, b, α, β, c, in (79), (110), (247), (249) und (51):

$$h = 6{,}548 \cdot 10^{-27}\,\mathrm{erg} \cdot \mathrm{sec}, \qquad k = 1{,}346 \cdot 10^{-16} \frac{\mathrm{erg}}{\mathrm{grad}}. \tag{252}$$

§ 158. Zur Enthüllung der vollen physikalischen Bedeutung des elementaren Wirkungsquantums h wird es noch mannigfacher Forschungsarbeit bedürfen. Dagegen gestattet der gefundene Wert von k leicht, den allgemeinen Zusammenhang zwischen Entropie S und Wahrscheinlichkeit W, wie er durch die universelle Gleichung (203) ausgedrückt ist, nunmehr auch numerisch im C.G.S.-System anzugeben. Es ist nämlich danach ganz allgemein die Entropie eines physikalischen Systems:

$$S = 1{,}346 \cdot 10^{-16} \cdot \log W \frac{\mathrm{erg}}{\mathrm{grad}} \tag{253}$$

zuzüglich einer willkürlichen additiven Konstanten. Diese Gleichung kann als die allgemeinste bisher existierende Definition der Entropie angesehen werden.

In der Anwendung auf die kinetische Gastheorie erhalten wir aus Gleichung (218) für das Verhältnis der Molekülmasse zur Molmasse:

$$\omega = \frac{k}{R} = \frac{1{,}346 \cdot 10^{-16}}{831 \cdot 10^5} = 1{,}62 \cdot 10^{-24}, \tag{254}$$

d. h. auf ein Mol gehen

$$\frac{1}{\omega} = 6{,}175 \cdot 10^{23}$$

Moleküle, wobei immer das Sauerstoffmol $O_2 = 32\,\mathrm{g}$ vorausgesetzt ist. Daher ist z. B. die absolute Masse eines Wasserstoffatoms ($\frac{1}{2} H_2 = 1{,}008\,\mathrm{g}$) gleich $1{,}63 \cdot 10^{-24}\,\mathrm{g}$. Damit wird die Anzahl der bei 0^0 C. und Atmosphärendruck in 1 ccm enthaltenen Moleküle eines idealen Gases:

$$\mathfrak{N} = \frac{76 \cdot 13{,}6 \cdot 981}{831 \cdot 10^5 \cdot 273 \cdot \omega} = 2{,}76 \cdot 10^{19}. \tag{255}$$

Die mittlere lebendige Kraft der fortschreitenden Bewegung eines Moleküls bei der absoluten Temperatur $T = 1$ ist nach (222) im absoluten C.G.S.-System:

$$\tfrac{3}{2} k = 2{,}02 \cdot 10^{-16}. \tag{256}$$

Allgemein wird die mittlere lebendige Kraft der fortschreitenden

Bewegung eines Moleküls durch das Produkt dieser Zahl und der absoluten Temperatur T ausgedrückt.

Das Elementarquantum der Elektrizität, oder die freie elektrische Ladung eines einwertigen Ions oder Elektrons ist im elektrostatischen Maße:

$$e = \omega \cdot 9658 \cdot 3 \cdot 10^{10} = 4{,}69 \cdot 10^{-10}. \qquad (257)$$

Da die hier benutzten Formeln mit absoluter Genauigkeit gelten so dürfen diese Zahlen so lange als die exaktesten Bestimmungen der genannten physikalischen Größen angesehen werden, bis die zur Berechnung der universellen Konstante k benutzten Werte der Strahlungskonstanten a und b durch neuere Messungen verbessert werden.

§ 159. Natürliche Maßeinheiten. Alle bisher in Gebrauch genommenen physikalischen Maßsysteme, auch das sogenannte absolute C.G.S.-System, verdanken ihren Ursprung insofern dem Zusammentreffen zufälliger Umstände, als die Wahl der jedem System zugrunde liegenden Einheiten nicht nach allgemeinen, notwendig für alle Orte und Zeiten bedeutungsvollen Gesichtspunkten, sondern wesentlich mit Rücksicht auf die speziellen Bedürfnisse unserer irdischen Kultur getroffen ist. So sind die Einheiten der Länge und der Zeit aus den gegenwärtigen Dimensionen und der gegenwärtigen Bewegung unseres Planeten hergeleitet worden, ferner die Einheit der Maße und der Temperatur aus der Dichte und den Fundamentalpunkten des Wassers, als derjenigen Flüssigkeit, die an der Erdoberfläche die wichtigste Rolle spielt, genommen bei einem Druck, der der mittleren Beschaffenheit der uns umgebenden Atmosphäre entspricht. An dieser Willkür würde prinzipiell auch nichts Wesentliches geändert werden, wenn etwa zur Längeneinheit die unveränderliche Wellenlänge des Na-Lichtes genommen würde. Denn die Auswahl gerade des Na unter den vielen chemischen Elementen könnte wiederum nur etwa durch sein häufiges Vorkommen auf der Erde oder etwa durch seine für unser Auge glänzende Doppellinie, die keineswegs einzig in ihrer Art dasteht, gerechtfertigt werden. Es wäre daher sehr wohl denkbar, daß zu einer anderen Zeit, unter veränderten äußeren Bedingungen, jedes der bisher in Gebrauch genommenen Maßsysteme seine ursprüngliche natürliche Bedeutung teilweise oder gänzlich verlieren würde.

11*

Dem gegenüber dürfte es nicht ohne Interesse sein, zu bemerken, daß mit Zuhilfenahme der beiden in dem Ausdrucke (227) der Strahlungsentropie auftretenden Konstanten h und k die Möglichkeit gegeben ist, Einheiten für Länge, Masse, Zeit und Temperatur aufzustellen, welche, unabhängig von speziellen Körpern oder Substanzen, ihre Bedeutung für alle Zeiten und für alle, auch außerirdische und außermenschliche Kulturen notwendig behalten und welche daher als „natürliche Maßeinheiten" bezeichnet werden können.

Die Mittel zur Festsetzung der vier Einheiten für Länge, Masse, Zeit und Temperatur werden gegeben durch die beiden erwähnten Konstanten h und k, ferner durch die Größe der Lichtfortpflanzungsgeschwindigkeit c im Vakuum und durch die der Gravitationskonstante f. Bezogen auf Zentimenter, Gramm, Sekunde und Celsiusgrad sind die Zahlenwerte dieser vier Konstanten die folgenden:

$$h = 6{,}548 \cdot 10^{-27} \, \frac{\mathrm{g\,cm^2}}{\mathrm{sec}},$$

$$k = 1{,}346 \cdot 10^{-16} \, \frac{\mathrm{g\,cm^2}}{\mathrm{sec^2\,grad}},$$

$$c = 3 \cdot 10^{10} \, \frac{\mathrm{cm}}{\mathrm{sec}},$$

$$f = 6{,}685 \cdot 10^{-8} \, \frac{\mathrm{cm^3}}{\mathrm{g\,sec^2}}.$$ [1]

Wählt man nun die „natürlichen Einheiten" so, daß im neuen Maßsystem jede der vorstehenden vier Konstanten den Wert 1 annimmt, so erhält man als Einheit der Länge die Größe:

$$\sqrt{\frac{f h}{c^3}} = 4{,}03 \cdot 10^{-33}\,\mathrm{cm},$$

als Einheit der Masse:

$$\sqrt{\frac{c h}{f}} = 5{,}42 \cdot 10^{-5}\,\mathrm{g},$$

als Einheit der Zeit:

$$\sqrt{\frac{f h}{c^5}} = 1{,}34 \cdot 10^{-43}\,\mathrm{sec},$$

als Einheit der Temperatur:

[1] F. Richarz und O. Krigar-Menzel, Wied. Ann. **66**, p. 190, 1898.

$$\frac{1}{k}\sqrt{\frac{c^5 h}{f}} = 3{,}63 \cdot 10^{32}\,\text{grad Cels}\,.$$

Diese Größen behalten ihre natürliche Bedeutung so lange bei, als die Gesetze der Gravitation, der Lichtfortpflanzung im Vakuum und die beiden Hauptsätze der Thermodynamik in Gültigkeit bleiben, sie müssen also, von den verschiedensten Intelligenzen nach den verschiedensten Methoden gemessen, sich immer wieder als die nämlichen ergeben.

§ 160. Man bezeichnet häufig das Normalspektrum der Licht- und Wärmestrahlung als zusammengesetzt aus einer großen Anzahl von regelmäßigen periodischen Schwingungen. Diese Ausdrucksweise ist insofern vollkommen berechtigt, als sie an die Zerlegung der Gesamtschwingung in eine Fourier sche Reihe, nach Gleichung (179), anknüpft, und eignet sich häufig in hervorragendem Maße dazu, die Betrachtungen bequem und übersichtlich zu gestalten; sie darf aber nicht zu der Auffassung verleiten, als ob jene „Regelmäßigkeit“ auf einer besonderen physikalischen Eigenschaft der elementaren Schwingungsvorgänge im Spektrum beruhe; denn die Zerlegbarkeit in eine Fourier sche Reihe ist mathematisch selbstverständlich und lehrt daher in physikalischer Beziehung nichts Neues. Man könnte im Gegenteil mit vollem Rechte behaupten, daß es in der ganzen Natur keinen unregelmäßigeren Vorgang gibt als die Schwingungen in den Strahlen eines Normalspektrums. Insbesondere hängen diese Schwingungen in keiner irgendwie charakteristischen Weise zusammen mit den speziellen Vorgängen in den Emissionszentren der Strahlen, etwa mit der Periode oder mit der Dämpfung der emittierenden Oszillatoren; denn gerade das Normalspektrum ist ja dadurch vor allen anderen Spektren ausgezeichnet, daß alle von der speziellen Natur der emittierenden Substanz herrührenden individuellen Verschiedenheiten vollkommen ausgeglichen und verwischt sind. Es wäre daher auch ein gänzlich aussichtsloses Unternehmen, wenn man etwa versuchen wollte, aus den Elementarschwingungen in den Strahlen des Normalspektrums Schlüsse zu ziehen auf die speziellen Eigenschaften der die Strahlen emittierenden Oszillatoren.

Die schwarze Strahlung läßt sich in der Tat ebensowohl wie aus regelmäßig periodischen Schwingungen, so auch aus gänzlich unregelmäßigen Einzelimpulsen zusammengesetzt ansehen. Die besonderen Regelmäßigkeiten, die wir an spektral zerlegtem monochromatischen Licht beobachten, rühren lediglich her von den besonderen Eigenschaften der benutzten Spektralapparate: des dispergierenden Prismas (Eigenperioden der Moleküle), des Beugungsgitters (Spaltbreite). Daher ist es unzutreffend, einen Unterschied zwischen Lichtstrahlen und Röntgenstrahlen, letztere als elektromagnetischen Vorgang im Vakuum angenommen, in dem Umstand zu erblicken, daß in ersteren die elementaren Schwingungen mit größerer Regelmäßigkeit erfolgen. Die Zerlegbarkeit in eine FOURIERsche Reihe von Partialschwingungen mit konstanten Amplituden und Phasen gilt für beide Arten von Strahlen in ganz gleicher Weise. Was aber die Lichtschwingungen vor den Röntgenschwingungen auszeichnet, ist wesentlich die viel kleinere Schwingungszahl ihrer Partialschwingungen, welche die Möglichkeit ihrer spektralen Zerlegung bedingt, und außerdem wahrscheinlich auch die viel größere zeitliche Gleichmäßigkeit der Strahlungsintensität in jedem Gebiete des Spektrums, die aber keineswegs auf einer besonderen Eigenschaft der elementaren Schwingungsvorgänge, sondern lediglich auf der Konstanz der Mittelwerte beruht.

§ 161. Die im § 152 ausgedrückten Beziehungen zwischen Strahlungsintensität und Temperatur gelten für die Strahlung im reinen Vakuum. Befindet sich die Strahlung in einem Medium vom Brechungsexponenten n, so wird die Abhängigkeit der Strahlungsintensität von der Schwingungszahl und der Temperatur durch den Satz des § 39 geregelt, daß das Produkt der spezifischen Strahlungsintensität $\mathfrak{K}_\nu$ und des Quadrats der Fortpflanzungsgeschwindigkeit der Strahlung für alle Substanzen den nämlichen Wert hat. Die Form dieser universellen Funktion (42) ergibt sich ohne weiteres aus (232) zu:

$$\mathfrak{K}\, q^2 = \frac{\varepsilon_\nu}{\alpha_\nu}\, q^2 = \frac{h\,\nu^3}{e^{\frac{h\,\nu}{k\,T}} - 1}. \tag{258}$$

Da nun der Brechungsexponent n der Fortpflanzungsgeschwindigkeit umgekehrt proportional ist, so tritt für ein Medium mit dem Brechungsexponent n an die Stelle von (232) die allgemeinere Beziehung:

$$\mathfrak{K}_\nu = \frac{h\nu^3 n^2}{c^2} \cdot \frac{1}{e^{\frac{h\nu}{kT}} - 1} \tag{259}$$

und ebenso an die Stelle von (233) die allgemeinere Beziehung:

$$\mathfrak{u} = \frac{8\pi h \nu^3 n^3}{c^3} \cdot \frac{1}{e^{\frac{h\nu}{kT}} - 1}. \tag{260}$$

Diese Ausdrücke gelten natürlich zugleich für die Emission eines in bezug auf das Medium mit dem Brechungsexponenten n schwarzen Körpers.

§ 162. Wir wollen die gefundenen Strahlungsgesetze nun dazu benutzen, um die Temperatur einer monochromatischen unpolarisierten Strahlung von gegebener Intensität zu berechnen, welche von einer kleinen Fläche (Spalt) in senkrechter Richtung emittiert und durch ein beliebiges System diathermaner, durch zentrierte brechende (oder spiegelnde) Kugelflächen voneinander getrennter Medien nahe der Achse hindurchgegangen ist. Eine solche Strahlung besteht aus homozentrischen Bündeln und entwirft daher hinter jeder brechenden Fläche ein reelles oder virtuelles Bild der emittierenden Fläche, wiederum senkrecht zur Achse. Das letzte Medium nehmen wir zunächst, wie das erste, als reines Vakuum an. Dann handelt es sich für die Bestimmung der Strahlungstemperatur nach Gleichung (232) nur um die Berechnung der spezifischen Strahlungsintensität $\mathfrak{K}_\nu$ im letzten Medium, und diese ist gegeben durch die Gesamtintensität der monochromatischen Strahlung I_ν, die Größe der Bildfläche F, und den räumlichen Öffnungswinkel Ω des durch einen Punkt des Bildes hindurchgehenden Strahlenkegels. Denn die spezifische Strahlungsintensität $\mathfrak{K}_\nu$ ist nach (13) dadurch bestimmt, daß durch ein Flächenelement $d\sigma$ in senkrechter Richtung innerhalb des Elementarkegels $d\Omega$ in der Zeit dt die dem Schwingungsintervall von ν bis $\nu + d\nu$ entsprechende Energiemenge:

$$2\,\mathfrak{K}_\nu \, d\sigma \, d\Omega \, d\nu \, dt$$

unpolarisierten Lichtes hindurchgestrahlt wird. Bedeutet nun $d\sigma$ ein Element der Bildfläche im letzten Medium, so besitzt hiernach die gesamte betrachtete auf das Bild fallende monochromatische Strahlung die Intensität:

$$I_\nu = 2\,\mathfrak{K}_\nu \int d\sigma \int d\Omega .$$

I_ν ist von der Dimension einer Energiemenge, da das Produkt $d\nu \cdot dt$ eine reine Zahl ist. Das erste Integral ist die ganze Fläche F des Bildes, das zweite ist der räumliche Öffnungswinkel Ω des durch einen Punkt der Bildfläche hindurchgehenden Strahlenkegels. Daher erhält man:

$$I_\nu = 2\,\mathfrak{K}_\nu F \Omega , \tag{261}$$

und daraus mit Benutzung von (232) als Temperatur der Strahlung:

$$T = \frac{h\nu}{k} \cdot \frac{1}{\log\left(\frac{2 h \nu^3 F \Omega}{c^2 I_\nu} + 1\right)} . \tag{262}$$

Wenn das betrachtete diathermane Medium nicht das Vakuum ist, sondern den Brechungsexponenten n besitzt, so tritt an die Stelle von (232) die allgemeinere Beziehung (259), und man erhält statt der letzten Gleichung:

$$T = \frac{h\nu}{k} \cdot \frac{1}{\log\left(\frac{2 h \nu^3 F \Omega n^2}{c^2 I_\nu} + 1\right)} \tag{263}$$

oder, mit Substitution der Zahlenwerte von c, h und k:

$$T = \frac{0{,}487 \cdot 10^{-10} \cdot \nu}{\log\left(\frac{1{,}46 \cdot 10^{-47} \nu^3 F \Omega n^2}{I_\nu} + 1\right)} \text{ grad Cels.}$$

Hierbei ist der natürliche Logarithmus zu nehmen, und I_ν ist in erg, ν in reziproken Sekunden, F in Quadratzentimetern auszudrücken. Bei sichtbaren Strahlen wird man den Summanden 1 im Nenner meistens weglassen können.

Die so berechnete Temperatur bleibt der betrachteten Strahlung so lange erhalten, als sie sich in dem diathermanen Medium ungestört fortpflanzt, auch wenn sie sich bis in beliebige Entfernungen und in beliebig große Räume ausbreitet. Denn wenn auch in größeren Entfernungen eine immer kleinere Energiemenge durch ein Flächenelement von bestimmter Größe hindurchstrahlt, so verteilt sich dieselbe dafür auf einen um so schmaleren, von dem Elemente ausgehenden Strahlenkegel, so daß der Wert von $\mathfrak{K}$ ganz ungeändert bleibt. Daher ist die freie

Ausbreitung der Strahlung ein vollkommen reversibler Vorgang. Die Umkehrung desselben läßt sich etwa mit Hilfe eines passenden Hohlspiegels oder einer Sammellinse realisieren.

Fragen wir nun weiter nach der Temperatur der Strahlung in den übrigen Medien, die zwischen den einzelnen brechenden oder spiegelnden Kugelflächen liegen. In jedem dieser Medien besitzt die Strahlung eine bestimmte Temperatur, die durch die letzte Formel gegeben ist, wenn man sie auf das von der Strahlung in diesem Medium erzeugte reelle oder virtuelle Bild bezieht.

Die Schwingungszahl ν der monochromatischen Strahlung ist selbstverständlich in allen Medien dieselbe; ferner ist nach den Gesetzen der geometrischen Optik das Produkt $n^2 F \Omega$ in allen Medien gleich. Wenn daher auch noch die Gesamtintensität der Strahlung I_ν bei der Brechung (oder Reflexion) an einer Fläche konstant bleibt, so bleibt auch T konstant, oder mit anderen Worten: Die Temperatur eines homozentrischen Strahlenbündels wird durch regelmäßige Brechung oder Reflexion nicht geändert, falls dabei kein Energieverlust der Strahlung eintritt. Jede Schwächung der Gesamtintensität I_ν aber, durch Spaltung der Strahlung, sei es in zwei oder in viele verschiedene Richtungen, wie bei der diffusen Reflexion, führt zu einer Erniedrigung der Temperatur T des Strahlenbündels. Tatsächlich findet ja im allgemeinen bei jeder Brechung oder Reflexion ein bestimmter Energieverlust durch Reflexion oder Brechung, und mithin auch eine Temperaturerniedrigung statt. Hier kommt also der prinzipielle Unterschied scharf zur Geltung, den es macht, ob eine Strahlung lediglich durch freie Ausbreitung, oder ob sie durch Spaltung bez. Absorption geschwächt wird. Im ersten Fall bleibt die Temperatur konstant, im zweiten wird sie erniedrigt.[1]

§ 163. Nachdem die Gesetze der Emission eines schwarzen Körpers festgestellt sind, läßt sich mit Hilfe des Kirchhoffschen Gesetzes (48) das Emissionsvermögen E eines beliebigen Körpers berechnen, wenn sein Absorptionsvermögen A, bez. sein

---

[1] Deshalb ist aber doch die reguläre Brechung und Reflexion kein irreversibler Vorgang. Denn die Entropie zweier kohärenter Strahlen ist nicht gleich der Summe der Entropien der Einzelstrahlen, weil dieselben nicht unabhängig voneinander sind im Sinne der Wahrscheinlichkeitsrechnung. Vgl. § 158 und § 134, sowie einen demnächst in Drudes Annalen erscheinenden Aufsatz von M. Laue (Anm. bei der Korrektur).

Reflexionsvermögen $1 - A$ bekannt ist. Bei Metallen gestaltet sich diese Berechnung besonders einfach für lange Wellen, nachdem E. HAGEN und H. RUBENS[1] experimentell gezeigt haben, daß das Reflexionsvermögen und überhaupt das ganze optische Verhalten der Metalle in dem genannten Spektralgebiete durch die einfachen MAXWELLschen Gleichungen des elektromagnetischen Feldes für homogene Leiter dargestellt wird und somit nur von der spezifischen Leitungsfähigkeit für stationäre galvanische Ströme abhängt. Daher läßt sich das Emissionsvermögen eines Metalls für lange Wellen vollständig ausdrücken durch seine galvanische Leitungsfähigkeit in Verbindung mit den Formeln für die schwarze Strahlung.[2]

§ 164. Es gibt aber auch einen Weg zur direkten theoretischen Bestimmung sowohl der galvanischen Leitungsfähigkeit und damit des Absorptionsvermögens A als auch des Emissionsvermögens E von Metallen gerade für lange Wellen, wenn man von den Anschauungen der Elektronentheorie ausgeht, wie sie von E. RIECKE[3] und namentlich von P. DRUDE[4] für die thermischen und elektrischen Vorgänge in den Metallen ausgebildet worden sind. Hiernach beruhen alle diese Vorgänge auf den schnellen unregelmäßigen Bewegungen der Elektronen, die zwischen den entgegengesetzt elektrisch geladenen ponderablen Metallmolekülen hin und her fliegen und von ihnen, wie auch untereinander, beim Zusammenstoß abprallen, ähnlich wie Gasmoleküle, wenn sie ein festes Hindernis oder sich gegenseitig treffen. Die Geschwindigkeit der Wärmebewegungen der ponderablen Moleküle ist nämlich gegen die der Elektronen zu vernachlässigen, weil im stationären Zustand die mittlere lebendige Kraft der Bewegung eines ponderablen Moleküles gleich derjenigen eines Elektrons ist (vgl. oben § 154), und weil die träge Masse eines ponderablen Moleküles die eines Elektrons um mehr als das Tausendfache übertrifft. Besteht nun im Innern des Metalls ein elektrisches Feld, so werden die entgegengesetzt geladenen Partikel nach entgegengesetzten Seiten getrieben, mit durchschnittlichen Geschwindigkeiten, die wesentlich mit von der mittleren freien

---

[1] E. HAGEN und H. RUBENS, DRUDES Ann. **11**, p. 873, 1903.
[2] E. ASCHKINASS, DRUDES Ann. **17**, p. 960, 1905.
[3] E. RIECKE, WIED. Ann. **66**, p. 353, 1898.
[4] P. DRUDE, DRUDES Ann. **1**, p. 566, 1900.

Weglänge abhängen, und hieraus resultiert die Leitungsfähigkeit des Metalls für den galvanischen Strom. Andererseits ergibt sich das Emissionsvermögen des Metalls für strahlende Wärme aus der Berechnung der Stöße der Elektronen. Denn solange ein Elektron mit konstanter Geschwindigkeit in konstanter Richtung fliegt, bleibt seine kinetische Energie konstant und es findet keine Ausstrahlung von Energie statt; sobald es aber durch einen Stoß eine Änderung seiner Geschwindigkeitskomponenten erleidet, wird ein gewisser aus der Elektrodynamik zu berechnender Betrag von Energie, der sich stets in der Form einer FOURIERschen Reihe darstellen läßt, in den umgebenden Raum ausgestrahlt, ganz ebenso, wie man sich die Röntgenstrahlung entstanden denkt durch den Anprall der von der Kathode fortgeschleuderten Elektronen gegen die Antikathode. Durchführen läßt sich diese Berechnung bis jetzt allerdings nur unter der Voraussetzung, daß während der Zeit einer Partialschwingung der FOURIERschen Reihe eine große Anzahl von Elektronenstößen stattfinden, d. h. für verhältnismäßig lange Wellen.

Diese Methode kann man nun offenbar dazu benutzen, um auf einem neuen, von dem früher eingeschlagenen ganz unabhängigen Wege die Gesetze der schwarzen Strahlung für lange Wellen abzuleiten. Denn wenn man das so berechnete Emissionsvermögen E des Metalls dividiert durch das mittels der galvanischen Leitungsfähigkeit bestimmte Absorptionsvermögen A desselben Metalls, so muß nach dem KIRCHHOFFschen Gesetz (48) das Emissionsvermögen des schwarzen Körpers, unabhängig von jeder speziellen Substanz, resultieren. Auf diese Weise hat H. A. LORENTZ[1] in einer tiefgehenden Untersuchung das Strahlungsgesetz eines schwarzen Körpers abgeleitet und ist dabei zu einem Resultat gekommen, das inhaltlich genau mit der Gleichung (240) übereinstimmt, wobei auch die Konstante k mit der Gaskonstante R wieder durch die Gleichung (217) zusammenhängt. Diese Art der Begründung der Strahlungsgesetze ist zwar auf das Gebiet langer Wellen beschränkt, aber dafür gewährt sie einen tieferen, höchst bedeutungsvollen Einblick in den Mechanismus der Elektronenbewegungen und der durch sie bedingten Strahlungsvorgänge in Metallen. Zugleich wird dadurch die oben in § 160 geschilderte

[1] H. A. LORENTZ, Proc. Kon. Akad. v. Wet. Amsterdam 1903, p. 666.

Auffassung ausdrücklich bestätigt, wonach das Normalspektrum als zusammengesetzt aus einer großen Schar gänzlich unregelmäßiger Elementarvorgänge betrachtet werden kann.

§ 165. Eine weitere interessante Bestätigung des Strahlungsgesetzes schwarzer Körper für lange Wellen und des Zusammenhanges der Strahlungskonstanten k mit der absoluten Masse der ponderablen Moleküle hat vor kurzem J. H. JEANS[1] aufgefunden, auf einem schon vorher von Lord RAYLEIGH[2] beschrittenen Wege, welcher sich dadurch wesentlich von dem hier eingeschlagenen unterscheidet, daß er die Heranziehung von speziellen Wechselwirkungen zwischen der Materie (Moleküle, Oszillatoren) und dem Äther ganz vermeidet und im wesentlichen nur auf die Vorgänge im durchstrahlten Vakuum eingeht. Den Ausgangspunkt dieser Betrachtungsweise liefert folgender Satz der statistischen Mechanik (vgl. oben § 154). Wenn in einem den HAMILTONschen Bewegungsgleichungen gehorchenden System, dessen Zustand durch die Werte einer großen Anzahl von unabhängigen Variabeln bestimmt ist, und dessen Gesamtenergie sich additiv aus verschiedenen von den einzelnen Zustandsvariablen quadratisch abhängigen Teilen zusammensetzt, irreversible Prozesse stattfinden, so vollziehen sich diese durchschnittlich immer in der Richtung, daß die auf die einzelnen unabhängigen Zustandsvariablen entfallenden Teilenergien sich gegenseitig ausgleichen, so daß schließlich, bei Erreichung des statistischen Gleichgewichts, alle im Mittel einander gleich geworden sind. Nach diesem Satze läßt sich also in einem solchen System die stationäre Energieverteilung angeben, sobald man nur die unabhängigen Variabeln kennt, durch welche der Zustand bestimmt ist.

Wir denken uns nun ein reines Vakuum in der Form eines Würfels von der Kantenlänge l, mit metallisch spiegelnden Seitenflächen. Legt man den Koordinatenanfangspunkt in eine Ecke des Würfels und die Koordinatenachsen in die anstoßenden Kanten, so wird ein in diesem Hohlraum möglicher elektromagnetischer Vorgang dargestellt durch das folgende System von Gleichungen:

---

[1] J. H. JEANS, Phil. Mag., **10**, p. 91, 1905.
[2] Lord RAYLEIGH, Nature **72**, p. 54 und p. 243, 1905.

$$\left.\begin{aligned}
\mathfrak{E}_x &= \cos\frac{\mathfrak{a}\pi x}{l}\cdot\sin\frac{\mathfrak{b}\pi y}{l}\cdot\sin\frac{\mathfrak{c}\pi z}{l}\cdot(e_1\cos 2\pi\nu t + e_1'\sin 2\pi\nu t),\\
\mathfrak{E}_y &= \sin\frac{\mathfrak{a}\pi x}{l}\cdot\cos\frac{\mathfrak{b}\pi y}{l}\cdot\sin\frac{\mathfrak{c}\pi z}{l}\cdot(e_2\cos 2\pi\nu t + e_2'\sin 2\pi\nu t),\\
\mathfrak{E}_z &= \sin\frac{\mathfrak{a}\pi x}{l}\cdot\sin\frac{\mathfrak{b}\pi y}{l}\cdot\cos\frac{\mathfrak{c}\pi z}{l}\cdot(e_3\cos 2\pi\nu t + e_3'\sin 2\pi\nu t),\\
\mathfrak{H}_x &= \sin\frac{\mathfrak{a}\pi x}{l}\cdot\cos\frac{\mathfrak{b}\pi y}{l}\cdot\cos\frac{\mathfrak{c}\pi z}{l}\cdot(h_1\sin 2\pi\nu t - h_1'\cos 2\pi\nu t),\\
\mathfrak{H}_y &= \cos\frac{\mathfrak{a}\pi x}{l}\cdot\sin\frac{\mathfrak{b}\pi y}{l}\cdot\cos\frac{\mathfrak{c}\pi z}{l}\cdot(h_2\sin 2\pi\nu t - h_2'\cos 2\pi\nu t),\\
\mathfrak{H}_z &= \cos\frac{\mathfrak{a}\pi x}{l}\cdot\cos\frac{\mathfrak{b}\pi y}{l}\cdot\sin\frac{\mathfrak{c}\pi z}{l}\cdot(h_3\sin 2\pi\nu t - h_3'\cos 2\pi\nu t),
\end{aligned}\right\}\quad(264)$$

wobei $\mathfrak{a}$, $\mathfrak{b}$, $\mathfrak{c}$ irgend drei positive ganze Zahlen bedeuten. Die Grenzbedingungen sind in diesen Ausdrücken dadurch befriedigt, daß für die sechs Grenzflächen $x = 0$, $x = l$, $y = 0$, $y = l$, $z = 0$, $z = l$ die tangentiellen Komponenten der elektrischen Feldstärke $\mathfrak{E}$ verschwinden. Die MAXWELLschen Feldgleichungen (52) werden ebenfalls, wie man durch Substitution erkennen kann, sämtlich befriedigt, wenn zwischen den Konstanten gewisse Beziehungen bestehen, welche sich alle in einen einzigen Satz zusammenfassen lassen: Bezeichnet man mit a eine gewisse positive Konstante, so bestehen zwischen den neun quadratisch angeordneten Größen:

$$\begin{matrix}
\frac{\mathfrak{a}c}{2l\nu} & \frac{\mathfrak{b}c}{2l\nu} & \frac{\mathfrak{c}c}{2l\nu}\\[1ex]
\frac{h_1}{a} & \frac{h_2}{a} & \frac{h_3}{a}\\[1ex]
\frac{e_1}{a} & \frac{e_2}{a} & \frac{e_3}{a}
\end{matrix}$$

alle diejenigen Relationen, welche die neun sogenannten „Richtungskosinus“ zweier orthogonaler rechtshändiger Koordinatensysteme, d. h. die Kosinus der Winkel je zweier Achsen der beiden Systeme, erfüllen.

Daher ist die Summe der Quadrate der Glieder jeder Horizontalreihe oder jeder Vertikalreihe $= 1$, also z. B.

$$\frac{c^2}{4l^2\nu^2}(\mathfrak{a}^2 + \mathfrak{b}^2 + \mathfrak{c}^2) = 1 \qquad (265)$$

$$h_1^2 + h_2^2 + h_3^2 = a^2 = e_1^2 + e_2^2 + e_3^2\,,$$

Ferner ist die Summe der Produkte entsprechender Glieder in je zwei Parallelreihen gleich null. Also z. B.

$$\mathfrak{a} e_1 + \mathfrak{b} e_2 + \mathfrak{c} e_3 = 0, \tag{266}$$

$$\mathfrak{a} h_1 + \mathfrak{b} h_2 + \mathfrak{c} h_3 = 0.$$

Ferner gelten Relationen von der Form:

$$\frac{h_1}{a} = \frac{e_2}{a} \cdot \frac{\mathfrak{c}\, c}{2 l \nu} - \frac{e_3}{a} \cdot \frac{\mathfrak{b}\, c}{2 l \nu},$$

also:

$$h_1 = \frac{c}{2 l \nu} (\mathfrak{c} e_2 - \mathfrak{b} e_3), \text{ usw.} \tag{267}$$

Wenn die ganzen Zahlen $\mathfrak{a}$, $\mathfrak{b}$, $\mathfrak{c}$ gegeben sind, so ist nach (265) die Schwingungszahl ν unmittelbar dadurch bestimmt. Dann lassen sich von den sechs Größen e_1, e_2, e_3, h_1, h_2, h_3 nur noch zwei beliebig wählen, die übrigen sind dann eindeutig, und zwar linear und homogen, durch sie bestimmt. Nimmt man z. B. e_1 und e_2 willkürlich an, so berechnet sich e_3 aus (266) und dann folgen die Werte von h_1, h_2, h_3 aus den Relationen von der Form (267). Zwischen den gestrichenen Konstanten e_1', e_2', e_3', h_1', h_2', h_3' bestehen genau dieselben Beziehungen wie zwischen den ungestrichenen, von denen sie ihrerseits ganz unabhängig sind. Daher lassen sich auch von ihnen noch zwei beliebig wählen, etwa h_1' und h_2', so daß von allen in den obigen Gleichungen vorkommenden Konstanten bei gegebenen $\mathfrak{a}$, $\mathfrak{b}$, $\mathfrak{c}$ noch vier Konstante unbestimmt bleiben. Bildet man nun für jedes beliebige Zahlensystem $\mathfrak{a}$, $\mathfrak{b}$, $\mathfrak{c}$ Ausdrücke von der Form (264) und summiert die entsprechenden Feldkomponenten, so erhält man wiederum eine Lösung der MAXWELLschen Feldgleichungen und Grenzbedingungen, welche aber nun so allgemein ist, daß sie jeden beliebigen in dem betrachteten Hohlwürfel möglichen elektromagnetischen Vorgang darzustellen vermag. Denn über die in den einzelnen partikulären Lösungen unbestimmt gebliebenen Konstanten e_1, e_2, h_1', h_2' kann immer so verfügt werden, daß der Vorgang jedem beliebigen Anfangszustand ($t = 0$) angepaßt werden kann.

Ist nun, wie wir bisher angenommen haben, der Hohlraum ganz von Materie entblößt, so ist der Strahlungsvorgang bei gegebenem Anfangszustand in allen seinen Einzelheiten eindeutig bestimmt. Er zerfällt in eine Reihe von stehenden Schwingungen, deren jede durch eine der betrachteten partikulären Lösungen dargestellt wird, und die vollständig unabhängig voneinander verlaufen. Von Irreversibilität kann also hierbei keine Rede sein, und daher auch nicht von der Tendenz zu einem Ausgleich der auf die einzelnen Partialschwingungen entfallenden Teilenergien. Sobald man aber auch nur eine Spur von Materie in dem Hohlraum befindlich annimmt, welche die elektrodynamischen Schwingungen beeinflussen können, z. B. einige Gasmoleküle, welche Strahlung emittieren und absorbieren, so wird der Vorgang ungeordnet, und es wird sich, wenn auch langsam, ein Übergang von weniger wahrscheinlichen zu wahrscheinlicheren Zuständen vollziehen. Ohne nun auf irgendeine nähere Einzelheit in der elektromagnetischen Konstitution der Moleküle einzugehen, kann man aus dem oben angeführten Satz der statistischen Mechanik den Schluß ziehen, daß unter allen möglichen Vorgängen derjenige einen stationären Charakter besitzt, bei welchem die Energie sich gleichmäßig auf alle unabhängigen Variabeln des Zustandes verteilt hat.

Bestimmen wir also diese unabhängigen Variabeln. Zunächst sind es die Geschwindigkeitskomponenten der Gasmoleküle. Jeder der drei voneinander unabhängigen Geschwindigkeitskomponenten eines Moleküls entspricht mithin im stationären Zustand durchschnittlich die Energie $\frac{1}{3}L$, wo L die mittlere Energie eines Moleküls darstellt und durch (222) gegeben ist (vgl. § 154). Ebenso groß ist also die Teilenergie, welche im stationären Zustand durchschnittlich auf jede der unabhängigen Variabeln des elektromagnetischen Systems entfällt.

Nun ist nach den obigen Ausführungen der elektromagnetische Zustand des ganzen Hohlraums für jede einzelne, je einem bestimmten Wertensystem der Zahlen $\mathfrak{a}$, $\mathfrak{b}$, $\mathfrak{c}$ entsprechende stehende Schwingung in irgendeinem Zeitpunkt durch vier voneinander unabhängige Größen bestimmt. Folglich ist für die Strahlungsvorgänge die Anzahl der unabhängigen Zustandsvariablen 4mal so groß als die Anzahl der möglichen Wertensysteme der positiven ganzen Zahlen $\mathfrak{a}$, $\mathfrak{b}$, $\mathfrak{c}$.

Wir wollen nun die Anzahl der möglichen Wertensysteme $\mathfrak{a}$, $\mathfrak{b}$, $\mathfrak{c}$ berechnen, welche den Schwingungen innerhalb eines bestimmten kleinen Spektralbezirks, etwa zwischen den Schwingungszahlen ν und $\nu + d\nu$, entsprechen. Nach (265) genügen diese Wertensysteme den Ungleichungen:

$$(268) \qquad \left(\frac{2\,l\,\nu}{c}\right)^2 < \mathfrak{a}^2 + \mathfrak{b}^2 + \mathfrak{c}^2 < \left(\frac{2\,l(\nu + d\nu)}{c}\right)^2,$$

wobei nicht nur $\frac{2\,l\,\nu}{c}$, sondern auch $\frac{2\,l\,d\nu}{c}$ als große Zahl zu denken ist. Versinnlichen wir uns jedes Wertensystem der $\mathfrak{a}$, $\mathfrak{b}$, $\mathfrak{c}$ durch einen Punkt, indem wir die Werte der positiven ganzen Zahlen $\mathfrak{a}$, $\mathfrak{b}$, $\mathfrak{c}$ als Koordinaten in einem rechtwinkligen Koordinatensystem auffassen, so erfüllen die so erhaltenen Punkte einen Oktanten des unendlichen Raumes, und die Bedingung (268) ist gleichbedeutend mit der, daß die Entfernung eines dieser Punkte vom Anfangspunkt der Koordinaten zwischen den Werten $\frac{2\,l\,\nu}{c}$ und $\frac{2\,l(\nu + d\nu)}{c}$ liegt. Die gesuchte Zahl ist daher gleich der Zahl der Punkte, welche zwischen den beiden Kugelflächenoktanten gelegen sind, die den Radien $\frac{2\,l\,\nu}{c}$ und $\frac{2\,l(\nu + d\nu)}{c}$ entsprechen. Da nun jedem Punkt ein Würfel vom Volumen 1 entspricht, und umgekehrt, so ist jene Zahl einfach gleich dem Volumen der genannten Kugelschicht, also gleich:

$$\frac{1}{8} \cdot 4\pi \left(\frac{2\,l\,\nu}{c}\right)^2 \cdot \frac{2\,l\,d\nu}{c},$$

und die Anzahl der entsprechenden unabhängigen Zustandsvariabeln gleich dem Vierfachen davon:

$$\frac{16\pi\, l^3 \nu^2\, d\nu}{c^3}.$$

Da nun auf jede unabhängige Zustandsvariable beim stationären Vorgang durchschnittlich die Teilenergie $\frac{L}{3}$ entfällt, so kommt auf das Intervall der Schwingungszahlen von ν bis $\nu + d\nu$ im ganzen die Energie:

$$\frac{16\pi\, l^3 \nu^2\, d\nu}{3\,c^3} \cdot L.$$

Dies ergibt, da das Volumen des Hohlraums l^3 ist, für die räumliche Energiedichte der Schwingungszahl ν:

$$\mathfrak{u}\, d\nu = \frac{16\pi\, \nu^2\, d\nu}{3\,c^3} \cdot L,$$

und mit Substitution des Wertes von L aus (222):

$$\mathfrak{u} = \frac{8\pi\nu^2 k T}{c^3}. \tag{269}$$

§ 166. Ein Vergleich der letzten Formel mit (242) zeigt, daß wir durch die statistische Mechanik zu genau demselben Zusammenhang zwischen Strahlungsdichte, Temperatur und Schwingungszahl geführt werden, wie durch das aus den Resonatorschwingungen abgeleitete Strahlungsgesetz, allerdings nur für hinreichend lange Wellen, bez. hohe Temperaturen. Denn nur unter dieser Bedingung ist die Gleichung (242) gültig. Aus dieser Beschränkung erwächst für die Anwendung der statistischen Mechanik auf die Strahlungsvorgänge eine gewisse Schwierigkeit. Denn würde man den Satz von der gleichmäßigen Energieverteilung auf alle unabhängige Zustandsvariabeln vollständig unbeschränkt anwenden, so müßte jene Beziehung ganz allgemein für alle Temperaturen und Schwingungszahlen gelten, und das würde, wie man leicht sieht, die Unmöglichkeit einer stationären Energieverteilung zur Folge haben, da die Energiedichte zugleich mit der Schwingungszahl unbegrenzt zunehmen würde.

Diese Schwierigkeit sucht J. H. JEANS[1] durch die Annahme zu heben, daß in einem mit emittierender und absorbierender Substanz versehenen durchstrahlten Hohlraum gar kein wirklich stabiler Strahlungszustand existiert, sondern daß die gesamte vorhandene Energie im Laufe der Zeit in Wärmestrahlung von immer höheren Schwingungszahlen übergeht, bis schließlich die Geschwindigkeit der Molekularbewegung unmerklich klein und die absolute Temperatur derselben daher gleich Null geworden ist.

Einer solchen Annahme kann ich mich aber nicht anschließen. Denn wenn je ein aus der alltäglichen Erfahrung genommener Satz dadurch an Zuverlässigkeit gewinnt, daß die verschiedensten aus ihm gezogenen Folgerungen sich als mit den feinsten Messungen in Übereinstimmung erweisen, so trifft dies bei dem Satze zu, daß die Strahlung in einem mit Materie versehenen Hohlraum einem Endzustand mit bestimmter endlicher Energieverteilung zwischen Materie und Äther zustrebt.

[1] J. H. JEANS, Proc. Roy. Soc. Vol. 76 A, p. 296, 545, 1905.

Alle bisher in der Theorie der Wärmestrahlung gezogenen, zum Teil auf den ersten Blick sehr kühn erscheinenden thermodynamischen Konsequenzen, von dem KIRCHHOFFschen Satze der Proportionalität des Emissionsvermögens und des Absorptionsvermögens angefangen, beruhen auf der Annahme der Existenz eines im thermodynamischen Sinn absoluten Gleichgewichtszustandes, und ihnen allen würde der Boden entzogen, wenn man jene Annahme fallen ließe; dagegen ist noch niemals eine Folgerung jenes Satzes im Widerspruch mit der Erfahrung befunden worden. Andererseits hat sich bisher nicht die Spur einer Andeutung dafür gezeigt, die auf die Vermutung führen könnte, daß wir es bei der schwarzen Strahlung nicht mit einem wirklich stabilen Zustand zu tun haben, im Gegenteil: schon die einfache Tatsache, daß ein Körper durch Wärmestrahlung erwärmt werden kann, daß also strahlende Energie ohne Kompensation in Energie der Molekularbewegung übergehen kann, ließe sich von jenem Standpunkt aus wohl nur schwer mit dem zweiten Hauptsatz der Thermodynamik in Einklang bringen.

Ich bin daher der Meinung, daß die besprochene Schwierigkeit nur durch eine unberechtigte Anwendung des Satzes von der Gleichmäßigkeit der Energieverteilung auf alle unabhängigen Zustandsvariabeln hervorgerufen ist. In der Tat ist für die Gültigkeit dieses Satzes die Voraussetzung wesentlich, daß die Zustandsverteilung unter allen bei gegebener Gesamtenergie von vornherein möglichen Systemen eine „ergodische" ist,[1] oder kurz ausgedrückt, daß die Wahrscheinlichkeit dafür, daß der Zustand des Systems in einem bestimmten kleinen „Elementargebiet" (§ 150) liegt, einfach proportional ist der Größe dieses Gebiets, wenn dasselbe auch noch so klein genommen wird. Diese Voraussetzung ist aber bei der stationären Energiestrahlung nicht erfüllt; denn die Elementargebiete dürfen nicht beliebig klein genommen werden, sondern ihre Größe ist eine endliche, durch den Wert des elementaren Wirkungsquantums h bestimmte. Nur wenn man das Wirkungselement h unendlich klein annehmen dürfte, würde man zu dem Gesetz der gleichmäßigen Energieverteilung gelangen. In der Tat geht für unendlich kleines h,

[1] L. BOLTZMANN, Gastheorie II, p. 92, 101, 1898.

wie man aus der Formel (233) ersieht, die allgemeine Energieverteilung in die spezielle hier abgeleitete (269) über, und es gelten dann überhaupt alle Beziehungen des § 154, entsprechend dem RAYLEIGHschen Strahlungsgesetz. Dann werden, entsprechend dem Satz von der gleichmäßigen Energieverteilung, auch die Energien aller Resonatoren einander gleich, was im allgemeinen nicht der Fall ist.

Natürlich muß dem Wirkungselement h auch eine direkte elektrodynamische Bedeutung zukommen; aber welcher Art diese ist, bleibt zunächst noch eine offene Frage.

---

12 *

Fünfter Abschnitt.

Irreversible Strahlungsvorgänge.

Erstes Kapitel.
Einleitung. Direkte Umkehrung eines Strahlungsvorgangs.

§ 167. Nach den Entwickelungen im vorigen Abschnitt kann man die Natur der Wärmestrahlung innerhalb eines im stabilen thermodynamischen Gleichgewicht befindlichen isotropen Mediums in allen Stücken als bekannt ansehen. Die Intensität der nach allen Richtungen gleichmäßigen Strahlung hängt für alle Wellenlängen ausschließlich von der Temperatur und von der Fortpflanzungsgeschwindigkeit ab, nach der Gleichung (259) für die schwarze Strahlung in einem beliebigen Medium. Aber damit ist die Aufgabe der Theorie noch nicht erledigt; dieselbe hat vielmehr auch davon noch Rechenschaft abzulegen, in welcher Weise und durch welche Vorgänge eine ursprünglich in dem Medium vorhandene ganz beliebig gegebene Strahlung, wenn das Medium nach außen durch eine undurchlässige Hülle abgeschlossen ist, allmählich in den stabilen, dem Maximum der Entropie entsprechenden Zustand der schwarzen Strahlung übergeht, ähnlich wie ein in ein festes Gefäß eingeschlossenes Gas, in dem ursprünglich beliebig gegebene Strömungen und Temperaturdifferenzen vorhanden waren, allmählich in den Zustand der Ruhe und der gleichmäßigen Temperaturverteilung übergeht.

Diese sehr viel schwierigere Frage läßt sich bis jetzt nur in beschränktem Umfang beantworten. Zunächst ist nach den ausführlichen Erörterungen im ersten Kapitel des vorigen Abschnitts klar, daß man, da es sich hier um irreversible Prozesse handelt, mit den Prinzipien der reinen Elektrodynamik allein nicht auskommen wird. Denn der reinen Elektrodynamik ist, ebenso wie der reinen Mechanik, der zweite Hauptsatz der Thermodynamik oder das Prinzip der Vermehrung der Entropie inhaltlich fremd. Dies zeigt sich am unmittelbarsten in dem Umstand, daß die Grundgleichungen sowohl der Mechanik als auch der Elektrodynamik die direkte zeitliche Umkehrung eines jeden Vorgangs gestatten, in geradem Widerspruch mit dem Prinzip der Vermehrung der Entropie. Selbstverständlich müssen hier alle Arten von Reibung und galvanischer Stromleitung ausgeschlossen gedacht werden; denn diese Vorgänge gehören, da sie stets mit Wärmeerzeugung verbunden sind, nicht mehr der reinen Mechanik oder Elektrodynamik an.

Setzt man dies voraus, so kommt in den Grundgleichungen der Mechanik die Zeit t nur in den Beschleunigungskomponenten, also in der Form des Quadrats ihres Differentials vor. Wenn man also in den Bewegungsgleichungen als Zeitvariable die Größe $-t$ statt t einführt, so behalten dieselben ihre Form unverändert bei, und daraus folgt, daß, wenn man bei einer Bewegung eines Systems materieller Punkte in irgend einem Augenblick die Geschwindigkeitskomponenten sämtlicher Punkte plötzlich umkehrt, die Bewegung genau rückwärts verlaufen muß. Für die elektrodynamischen Vorgänge in einem homogenen nichtleitenden Medium gilt ganz Ähnliches. Wenn man in den MAXWELLschen Gleichungen des elektrodynamischen Feldes überall $-t$ statt t schreibt, und außerdem das Vorzeichen der magnetischen Feldstärke $\mathfrak{H}$ umkehrt, so bleiben die Gleichungen, wie man leicht sehen kann, unverändert, und daraus folgt, daß, wenn bei irgend einem elektrodynamischen Vorgang in einem gewissen Zeitpunkt die magnetische Feldstärke überall plötzlich umgekehrt wird, während die elektrische Feldstärke ihren Wert behält, der ganze Vorgang in umgekehrter Richtung verlaufen muß.

§ 168. So unmittelbar dieser Satz, daß alle rein elektrodynamischen Vorgänge auch in umgekehrter Richtung verlaufen können, für die Ausbreitung elektrodynamischer Wellen im reinen Vakuum zu erweisen ist, so scheint er auf den ersten Blick seine Gültigkeit zu verlieren bei den im dritten Abschnitt von uns betrachteten Oszillatorschwingungen. Denn die Gleichung (171) für die Schwingungen eines solchen Oszillators lautet:

$$Kf + L\ddot{f} - \frac{2}{3c^3}\dddot{f} = \mathfrak{E}_z, \tag{270}$$

wobei $\mathfrak{E}_z$ die z-Komponente der elektrischen Feldstärke der erregenden primären Welle am Orte des Oszillators bezeichnet. Führt man nun in dieser Gleichung die Differentiation von f, statt nach t, nach $-t$ aus, so ändert das Dämpfungsglied sein Vorzeichen, und man könnte daher zu glauben versucht sein, daß ein Schwingungsvorgang nicht in umgekehrter Richtung verlaufen kann. Das wäre aber ein Fehlschluß. Denn zu dem ganzen Schwingungsvorgang gehört nicht nur die Schwingung des Oszillators selber, sondern auch die ihn erregende primäre Welle, und wenn von einer Umkehrung des Vorgangs gesprochen wird, so muß nicht nur der Oszillator, sondern auch das äußere Feld in Betracht gezogen werden. Nun sendet der schwingende Oszillator nach dem, was wir früher (§ 107) gesehen haben, eine bestimmte Kugelwelle in das umgebende Vakuum hinaus, folglich wird er bei der Umkehrung des ganzen Vorgangs nicht nur von der umgekehrten ursprünglich primären Welle $\mathfrak{E}_z$, sondern zugleich auch von der umgekehrten Kugelwelle, d. h. von einer nach innen zu auf den Oszillator als Zentrum fortschreitenden Kugelwelle erregt, und es frägt sich, wie seine Schwingungen sich nun, unter diesem doppelten Einfluß, gestalten. Diese Frage soll im folgenden allgemein untersucht werden.

§ 169. Zuerst betrachten wir die Eigenschaften der von außen nach innen auf den Oszillator als Zentrum zu fortschreitenden Kugelwelle, welche die Umkehrung der direkten vom Oszillator ausgesandten Kugelwelle darstellt. Die direkte Welle ist gegeben durch die Gleichungen (145), wenn man darin für die Funktion F den in (148) angegebenen Ausdruck setzt. Deshalb wird die umgekehrte Welle, die entsteht, wenn zur Zeit $t = 0$ alle magnetischen Feldstärken der direkten Welle plötzlich umgekehrt werden, dargestellt durch die nämlichen Gleichungen (145), wenn man darin für die Funktion F setzt:

$$F = \frac{1}{r} \cdot f\left(-t - \frac{r}{c}\right). \tag{271}$$

Denn durch Substitution in (145) ergeben sich daraus die nämlichen Ausdrücke für die Komponenten der Feldstärke wie für die direkte Welle, nur daß überall $-t$ statt t steht, und daß das Vorzeichen der magnetischen Feldstärke umgekehrt ist. In der Tat bezeichnet die Wellenfunktion (271) eine nach innen fortschreitende Kugelwelle.

Es handelt sich jetzt um die Größe der elektrischen Kraft, mit der diese sich auf den Oszillator zusammenziehende Kugelwelle den Oszillator erregt. Das ist nach den allgemeinen Sätzen des § 111 die z-Komponente der elektrischen Feldstärke, welche diese Welle am Orte des Oszillators besitzen würde, wenn derselbe gar nicht vorhanden wäre. Lassen wir also einmal den Oszillator ganz weg und betrachten den Verlauf der von außen nach innen fortschreitenden Kugelwelle (271). Dieselbe wird im Kugelmittelpunkt, also im Anfangspunkt der Koordinaten, durch sich selber hindurchgehen, und zwar so, daß das elektromagnetische Feld der Welle endlich und stetig bleibt; denn es ist unmöglich, daß eine ursprünglich endliche Welle im freien Vakuum irgendwo und irgendwann eine unendlich große Feldstärke erzeugt. Nehmen wir also den Ausdruck der durch sich selbst hindurchgegangenen nach außen fortschreitenden Kugelwelle mit in die Wellenfunktion F auf, so erhalten wir vollständiger als in (271):

$$F = \frac{1}{r} f\left(-t - \frac{r}{c}\right) + \frac{1}{r} g\left(-t + \frac{r}{c}\right),$$

wobei nun f die nach innen fortschreitende, g die durch sich selbst hindurchgegangene nach außen fortschreitende Welle bedeutet. Da für $r = 0$ F für alle Zeiten endlich bleiben muß, so ergibt sich:

$$f(-t) + g(-t) = 0.$$

Folglich:
$$F = \frac{1}{r} f\left(-t - \frac{r}{c}\right) - \frac{1}{r} f\left(-t + \frac{r}{c}\right). \tag{272}$$

Mit diesem Werte von F stellen die Gleichungen (145) die direkte Umkehrung der vom schwingenden Oszillator emittierten Welle auch in ihrem späteren Verlauf vor. Sie schreitet nach innen fort und streicht über den Oszillator hinweg, ohne dortselbst unendlich oder unstetig zu werden, wie das überhaupt bei jeder den Oszillator erregenden Welle der Fall ist. Berechnen wir jetzt die z-Komponente der elektrischen Feldstärke dieser Welle am Orte des Oszillators. Für hinreichend kleine Werte von r wird:

$$f\left(-t-\frac{r}{c}\right)=f(-t)-\frac{r}{c}\dot{f}(-t)+\frac{1}{2}\frac{r^2}{c^2}\ddot{f}(-t)-\frac{1}{1\cdot 2\cdot 3}\frac{r^3}{c^3}\dddot{f}(-t),$$

$$f\left(-t+\frac{r}{c}\right)=f(-t)+\frac{r}{c}\dot{f}(-t)+\frac{1}{2}\frac{r^2}{c^2}\ddot{f}(-t)+\frac{1}{1\cdot 2\cdot 3}\frac{r^3}{c^3}\dddot{f}(-t),$$

wobei die Ableitungen $\dot{f}, \ddot{f}, \dddot{f}$ von f immer, auch im folgenden, nach dem Argument, nicht etwa nach t, genommen zu denken sind.

Daraus folgt für die Wellenfunktion (272):

$$F=-\frac{2}{c}\dot{f}(-t)-\frac{1}{3}\frac{r^2}{c^3}\dddot{f}(-t).$$

Dies ergibt nach (145) für die z-Komponente der elektrischen Feldstärke am Orte $r=0$, da $r^2=x^2+y^2+z^2$:

$$\mathfrak{E}_z=\frac{\partial^2 F}{\partial z^2}-\frac{1}{c^2}\frac{\partial^2 F}{\partial t^2}=\frac{4}{3}\frac{1}{c^3}\dddot{f}(-t). \tag{273}$$

§ 170. Nachdem wir nun die elektrische Kraft festgestellt haben, mit welcher die sich nach innen fortpflanzende, durch die Umkehrung der ursprünglich emittierten Kugelwelle entstandene Kugelwelle den Oszillator erregt, wollen wir die Schwingungen berechnen, welche der Oszillator ausführt, wenn außer der umgekehrten primären Welle auch noch die umgekehrte Kugelwelle auf ihn einwirkt. Bezeichnen wir die gesuchte Schwingung mit $f'(t)$, so muß nach der allgemeinen Schwingungsgleichung (270) gelten:

$$Kf'(t)+L\ddot{f}'(t)-\frac{2}{3c^3}\dddot{f}'(t)=\mathfrak{E}_z(-t)+\frac{4}{3}\frac{1}{c^3}\dddot{f}(-t). \tag{274}$$

Denn als erregende Schwingung haben wir hier einzusetzen einmal die der umgekehrten primären Welle, und dann die der umgekehrten Kugelwelle (273).

Die Gleichung (274) wird befriedigt, wenn man setzt:

$$f'(t)=f(-t), \tag{275}$$

also $\quad \dot{f}'(t)=-\dot{f}(-t),\quad \ddot{f}'(t)=\ddot{f}(-t),\quad \dddot{f}'(t)=-\dddot{f}(-t).$

Denn dann ergibt sich aus (274):

$$Kf(-t)+L\ddot{f}(-t)-\frac{2}{3c^3}\dddot{f}(-t)=\mathfrak{E}_z(-t)$$

und diese Gleichung ist genau erfüllt für alle positiven Werte von t, wenn die Gleichung (270) für alle negativen Werte von t, d. h. für alle Zeiten, welche dem Augenblick der Umkehrung, $t = 0$, vorausgehen, erfüllt ist. Hierdurch ist die Schwingung (275) des Oszillators für alle positiven Zeiten gegeben, und sie stellt, wie man sieht, in allen Stücken die Umkehrung des direkten Schwingungsvorgangs dar.

§ 171. Fassen wir nun weiter die bei dem umgekehrten Vorgang vom Oszillator emittierte Welle ins Auge, so wird dieselbe nach (148) dargestellt durch eine nach außen fortschreitende Kugelwelle mit der Wellenfunktion:

$$\frac{1}{r} f'\left(t - \frac{r}{c}\right). \tag{276}$$

Dieser Welle superponiert sich die oben betrachtete durch das Kugelzentrum hindurchgegangene und nun ebenfalls nach außen fortschreitende Kugelwelle, deren Wellenfunktion nach (272) den Wert besitzt:

$$-\frac{1}{r} f\left(-t + \frac{r}{c}\right). \tag{277}$$

Nun ist gemäß der Beziehung (275) die Summe der beiden Wellenfunktionen (276) und (277) gleich Null, die beiden nach außen fortschreitenden Kugelwellen vernichten sich also gegenseitig, und wir erhalten das Resultat, daß bei dem betrachteten umgekehrten Vorgang überhaupt keine Kugelwelle vom Oszillator nach außen geht.

Wenn man alles zusammenfaßt, so besteht mithin der umgekehrte Vorgang einfach darin, daß die ursprüngliche primäre Welle wieder rückwärts geht und daß der Oszillator die ursprünglich emittierte Welle wieder in sich aufnimmt, ohne eine neue Welle nach außen zu entsenden, wobei er seine Schwingungen genau in umgekehrter Reihenfolge wiederholt.[1]

§ 172. Nehmen wir als Beispiel den speziellen Fall, daß überhaupt keine primäre Welle vorhanden ist, die den Oszillator zu Schwingungen anregt. Dann erfolgt einfaches Abklingen der Oszillatorschwingungen, mit konstantem Dekrement, nach Gleichung (169), indem die Schwingungsenergie in Form von Kugelwellen allmählich in den Raum hinausgestrahlt wird. Kehrt

---

[1] Vgl. L. Boltzmann, Sitzungsber. d. Berliner Akad. d. Wissensch. vom 3. März 1898, p. 182.

man nun zu irgend einer Zeit die magnetischen Feldstärken der emittierten Kugelwelle überall plötzlich um und ebenso den Strom $\dot{f}$ im Oszillator, so geht der Vorgang genau umgekehrt vor sich: die Kugelwelle kehrt zum Oszillator zurück, seine Schwingungen werden mit konstantem Inkrement verstärkt, und er saugt die früher ausgestrahlte Energie vollständig wieder ein, ohne dabei irgendwelche Energiebeträge auszustrahlen.

§ 173. Ein anderes drastisches Beispiel erhalten wir, wenn ein Oszillator, der ursprünglich gar keine Energie besitzt ($f = 0$, $\dot{f} = 0$) von einer ebenen periodischen Welle getroffen wird. Der Oszillator wird allmählich immer stärker ins Mitschwingen geraten, bis der in § 112 ff. ausführlich untersuchte stationäre Vorgang entsteht, wobei der Oszillator in der Periode der erregenden primären Welle schwingt, mit einem größeren oder kleineren Phasenunterschied gegen dieselbe, und zugleich eine gewisse Menge Energie nach allen Richtungen des Raumes ausstrahlt, die er der primären Welle entzieht (§ 115). Kehren wir nun den ganzen Vorgang plötzlich um, so wird die primäre Welle den umgekehrten Weg gehen; beim Passieren des Oszillators nimmt sie die früher an ihn abgegebene Energie wieder von ihm zurück, und dadurch wird, obwohl der Oszillator gleichzeitig von der zu ihm zurückkehrenden Kugelwelle gespeist wird, und obwohl er jetzt gar keine Energie mehr durch Emission verliert, dennoch seine Energie beständig verkleinert, bis sie zuletzt Null wird, indem f und $\dot{f}$ beide verschwinden. Wir haben hier also den Fall, daß eine primäre Welle einem Oszillator, auf den sie fällt, die gesamte Schwingungsenergie vollständig entzieht.

§ 174. Die letzten Beispiele umgekehrt verlaufender elektrodynamischer Strahlungsvorgänge machen es besonders deutlich, daß hier von einer Gültigkeit der Gesetze der Wärmestrahlung, insbesondere von einer steten Vermehrung der Entropie, nicht die Rede sein kann, und wir kommen daher wieder auf den schon im vorigen Abschnitt gezogenen Schluß zurück, daß eine Ableitung jener Gesetze aus der reinen Elektrodynamik nicht erfolgen kann ohne besondere Hypothesen, deren Inhalt auf eine Einschränkung der elektrodynamischen Möglichkeiten hinausläuft. Man erkennt aber auch sogleich, daß die zuletzt beschriebenen, vom Standpunkt der Thermodynamik ganz unverständlichen Vorgänge nur durch eine besondere Eigentümlichkeit der den Oszillator erregenden Wellen zustande kommen. Die bei der Umkehrung eines Strahlungsvorganges von verschiedenen Seiten auf den Oszillator fallenden Strahlen sind nämlich in ganz spezieller Weise abhängig von der Vorgeschichte des Oszillators und dadurch auch abhängig voneinander. Würde man nur solche Strahlungsvorgänge als möglich zulassen, bei denen die von verschiedenen Seiten auf den Oszillator fallenden Strahlen keinerlei gesetzmäßige Beziehungen zueinander aufweisen, so könnte ein derartiger umgekehrter Vorgang gar nicht zustande kommen, und die Irreversibilität der Strahlungsvorgänge erschiene wenigstens nicht von vornherein ausgeschlossen. Eine solche Beschränkung in den Eigenschaften der auffallenden Strahlung ist nun enthalten in der Hypothese der „natürlichen Strahlung", von der in den nächsten Kapiteln gezeigt werden soll, daß sie in der Tat die Gültigkeit des zweiten Hauptsatzes der Thermodynamik auch für strahlende Energie gewährleistet.

Zweites Kapitel. Ein Oszillator in beliebigem Strahlungsfelde. Hypothese der natürlichen Strahlung.

§ 175. Wir behandeln jetzt genau die nämliche Aufgabe wie im 3. Kapitel des dritten Abschnitts, nur unter der allgemeineren Voraussetzung, daß das Vakuum, in dem sich der Oszillator befindet, nicht von gleichmäßiger stationärer Strahlung, sondern von örtlich und zeitlich beliebig veränderlichen Strahlen erfüllt ist, die nach allen möglichen Richtungen den Raum durchkreuzen.

Sei wieder $f(t)$ das Moment des von dem linearen Oszillator zur Zeit t dargestellten elektrischen Dipols, $\mathfrak{E}_z(t)$ die in die Richtung der Oszillatorachse fallende Komponente der Feldstärke des elektromagnetischen Feldes, welches von den im Vakuum sich fortpflanzenden Wellen am Orte des Oszillators gebildet wird, so ist die Schwingung des Oszillators bestimmt durch seinen Anfangszustand (für $t = 0$) und durch die Differentialgleichung (172).

Wir beschränken von vornherein die ganze Betrachtung auf ein begrenztes, wenn auch sehr großes, nötigenfalls nach Jahren zählendes Zeitintervall, etwa von $t=0$ bis $t=T$. Dann läßt sich die Funktion $\mathfrak{E}_z(t)$ für dieses Zeitintervall jedenfalls folgendermaßen schreiben:

$$\mathfrak{E}_z = \int_0^\infty d\nu \cdot C_\nu \cos(2\pi\nu t - \vartheta_\nu), \tag{278}$$

wobei C_ν (positiv) und ϑ_ν gewisse Funktionen der positiven Integrationsvariabeln ν bedeuten, deren Werte übrigens durch das Verhalten der Größe $\mathfrak{E}_z$ in dem genannten Zeitintervall bekanntlich noch nicht bestimmt sind, sondern außerdem noch von der Art abhängen, wie die Zeitfunktion $\mathfrak{E}_z$ über jenes Intervall hinaus nach beiden Seiten fortgesetzt wird. Daher besitzen die Größen C_ν und ϑ_ν einzeln gar keine bestimmte physikalische Bedeutung, und es wäre auch ganz unrichtig, wenn man die Schwingung $\mathfrak{E}_z$ sich etwa als ein kontinuierliches Spektrum von periodischen Schwingungen mit den konstanten Amplituden C_ν vorstellen würde, wie man übrigens auch schon daraus erkennt, daß der Charakter der Schwingung $\mathfrak{E}_z$ sich ja im Laufe der Zeit beliebig ändern kann. Wie die spektrale Zerlegung der Schwingung $\mathfrak{E}_z$ vorzunehmen ist, und zu welchen Resultaten sie führt, wird unten im § 180 gezeigt werden.

Wir wollen das Zeitintervall T so groß wählen, daß nicht nur $\nu_0 T$, sondern auch $\sigma\nu_0 T$ durch eine große Zahl ausgedrückt wird, und wollen im folgenden immer nur solche zwischen 0 und T gelegene Zeiten t betrachten, für welche $\sigma\nu_0 t$, und um so mehr $\nu_0 t$, große Werte hat. Diese Festsetzung gewährt nämlich den Vorteil, daß wir dann von dem Anfangszustand des Oszillators (für $t=0$) ganz absehen können, weil derselbe sich zur Zeit t nur mit einem Gliede von der Größenordnung $e^{-\sigma\nu_0 t}$ geltend macht und daher dann keinen merklichen Einfluß auf den Zustand mehr ausübt.

Unter den gemachten Voraussetzungen ergibt sich für irgendeine erregende Schwingung $\mathfrak{E}_z$ als allgemeine Lösung der Schwingungsgleichung (172), wie leicht durch Vergleich mit (174) zu verifizieren:

$$f(t) = \frac{3c^3}{16\pi^3}\int d\nu \cdot \frac{C_\nu}{\nu^3} \cdot \sin\gamma_\nu \cdot \cos(2\pi\nu t - \vartheta_\nu - \gamma_\nu), \tag{279}$$

wobei zur Abkürzung gesetzt ist:

$$\operatorname{ctg}\gamma_\nu = \frac{\pi\nu_0(\nu_0^2 - \nu^2)}{\sigma\nu^3}. \tag{280}$$

Um γ_ν eindeutig zu machen, wollen wir noch festsetzen, daß γ_ν zwischen 0 und π liegt.

Da σ klein ist, so weicht $\frac{\sin \gamma_\nu}{\nu^3}$ nur dann merklich von Null ab, wenn ν / ν_0 nahezu $= 1$, d. h. es tragen nur diejenigen Glieder des FOURIERschen Integrals merklich zur Resonanzerregung bei, deren Index ν der Eigenschwingung ν_0 des Oszillators nahe liegt. Man kann daher in vielen Fällen hinter dem Integralzeichen ν durch ν_0 ersetzen, wovon wir im folgenden öfters Gebrauch machen werden.

§ 176. Die „Intensität der erregenden Schwingung“[1] J als Funktion der Zeit t definieren wir als den Mittelwert von $\mathfrak{E}_z^2$ in dem Zeitintervall von t bis $t + \tau$, wobei τ möglichst klein genommen ist gegen die Zeit T, aber immer noch groß gegen die Zeit $1 / \nu_0$, d. h. gegen die Zeitdauer einer Eigenschwingung des Oszillators. In dieser Festsetzung liegt eine gewisse Unbestimmtheit, welche bewirkt, daß im allgemeinen J nicht nur von t, sondern auch von τ abhängig bleiben wird. Wenn dies der Fall ist, kann man von einer Intensität der erregenden Schwingung überhaupt nicht reden; denn es gehört mit zum Begriff der Schwingungsintensität, daß ihr Betrag sich innerhalb der Zeitdauer einer einzelnen Schwingung nur unmerklich ändert (vgl. oben § 3). Daher wollen wir künftig nur solche Vorgänge in Betracht ziehen, bei denen unter den angegebenen Bedingungen ein nur von t abhängiger Mittelwert von $\mathfrak{E}_z^2$ existiert. Die später vorzunehmende weitere Beschränkung auf den Fall der „natürlichen Strahlung“ wird zugleich auch die Erfüllung der hier als notwendig erkannten Bedingung enthalten. Um ihr in mathematischer Hinsicht zu genügen, wollen wir zunächst annehmen, daß die Größen C_ν in (278) für alle diejenigen Werte von ν unmerklich klein sind, welche gegen ν_0 verschwinden, oder, anders ausgedrückt, daß in der erregenden Schwingung $\mathfrak{E}_z$ keine ganz langsamen Perioden von merklicher Amplitude enthalten sind.

Zur Berechnung von J bilden wir nun aus (278) den Wert von $\mathfrak{E}_z^2$ und bestimmen den Mittelwert $\overline{\mathfrak{E}_z^2}$ dieser Größe durch Integration nach t von t bis $t + \tau$, Division durch τ und Übergang

---

[1] nicht zu verwechseln mit der „Feldintensität“ (Feldstärke) $\mathfrak{E}_z$ der erregenden Schwingung.

zur Grenze durch gehörige Verkleinerung von τ. Es ergibt sich so zunächst:

$$\mathfrak{E}_z^2 = \int_0^\infty\int_0^\infty d\nu' \, d\nu \, C_{\nu'} C_\nu \cos(2\pi\nu' t - \vartheta_{\nu'}) \cos(2\pi\nu t - \vartheta_\nu).$$

Vertauscht man die Werte von ν und ν', so ändert sich die Funktion unter dem Integralzeichen nicht; daher setzen wir fest:

$$\nu' > \nu$$

und schreiben:

$$\mathfrak{E}_z^2 = 2\int\int d\nu' \, d\nu \, C_{\nu'} C_\nu \cos(2\pi\nu' t - \vartheta_{\nu'}) \cos(2\pi\nu t - \vartheta_\nu)$$

oder:

$$\mathfrak{E}_z^2 = \int\int d\nu' \, d\nu \, C_{\nu'} C_\nu \{\cos[2\pi(\nu' - \nu)t - \vartheta_{\nu'} + \vartheta_\nu] + \cos[2\pi(\nu' + \nu)t - \vartheta_{\nu'} - \vartheta_\nu]\}.$$

Folglich:

$$J = \overline{\mathfrak{E}_z^2} = \frac{1}{\tau}\int_t^{t+\tau} \mathfrak{E}_z^2 dt$$

$$= \int\int d\nu' \, d\nu \, C_{\nu'} C_\nu \left\{ \frac{\sin\pi(\nu' - \nu)\tau \cdot \cos[\pi(\nu' - \nu)(2t + \tau) - \vartheta_{\nu'} + \vartheta_\nu]}{\pi(\nu' - \nu)\tau} + \frac{\sin\pi(\nu' + \nu)\tau \cdot \cos[\pi(\nu' + \nu)(2t + \tau) - \vartheta_{\nu'} - \vartheta_\nu]}{\pi(\nu' + \nu)\tau} \right\}.$$

Da nach der oben gemachten Voraussetzung alle diejenigen C_ν unmerklich klein sind, für welche ν gegen ν_0 verschwindet, so kann man in dem vorstehenden Ausdruck ν, und um so mehr ν', als von gleicher oder höherer Größenordnung wie ν_0 annehmen. Lassen wir nun τ immer kleiner werden, so ist vermöge der Bedingung, daß $\nu_0 \tau$ groß bleibt, der Nenner $(\nu' + \nu)\tau$ des zweiten Bruches jedenfalls groß, während der des ersten Bruches, $(\nu' - \nu)\tau$, mit abnehmendem τ unter jeden endlichen Betrag herabsinken kann. Daher reduziert sich das Integral für genügend kleine Werte von $\nu' - \nu$ auf:

$$\int\int d\nu' \, d\nu \, C_{\nu'} C_\nu \cos[2\pi(\nu' - \nu)t - \vartheta_{\nu'} + \vartheta_\nu],$$

also in der Tat unabhängig von τ. Die übrigen Glieder des Doppelintegrals, welche größeren Werten von $\nu' - \nu$, d. h. schnelleren Änderungen mit der Zeit entsprechen, hängen im allgemeinen von τ ab und müssen daher verschwinden, wenn die Intensität J nicht von τ abhängen soll. Daher ist in unserem Falle, wenn man noch

$$\mu = \nu' - \nu (> 0)$$

als zweite Integrationsvariable statt ν' einführt:

$$J = \iint d\mu\, d\nu\, C_{\nu+\mu} C_\nu \cos(2\pi\mu t - \vartheta_{\nu+\mu} + \vartheta_\nu) \tag{281}$$

oder:

$$\left.\begin{aligned} J &= \int d\mu (A_\mu \sin 2\pi\mu t + B_\mu \cos 2\pi\mu t), \\ \text{wobei:} \quad A_\mu &= \int d\nu\, C_{\nu+\mu} C_\nu \sin(\vartheta_{\nu+\mu} - \vartheta_\nu) \\ B_\mu &= \int d\nu\, C_{\nu+\mu} C_\nu \cos(\vartheta_{\nu+\mu} - \vartheta_\nu). \end{aligned}\right\} \tag{282}$$

Hierdurch ist die Intensität J der erregenden Schwingung, falls sie überhaupt existiert, als Funktion der Zeit t in der Form eines FOURIERschen Integrals dargestellt.

§ 177. Schon in dem Begriff der Schwingungsintensität J liegt die Voraussetzung enthalten, daß diese Größe mit der Zeit t viel langsamer variiert als die Schwingung $\mathfrak{E}_z$ selber. Dasselbe folgt aus der Berechnung von J im vorigen Paragraphen. Denn dort ist für alle in Betracht kommenden Wertenpaare von C_ν und $C_{\nu'}$ $\nu\tau$ und $\nu'\tau$ groß, dagegen $(\nu' - \nu)\tau$ klein; folglich a fortiori

$$\frac{\nu' - \nu}{\nu} = \frac{\mu}{\nu} \text{ klein}, \tag{283}$$

und demgemäß sind die FOURIERschen Integrale $\mathfrak{E}_z$ in (278) und J in (282) in ganz verschiedener Weise mit der Zeit veränderlich. Wir werden daher im folgenden in bezug auf die Abhängigkeit von der Zeit zwei verschiedenartig veränderliche Arten von Größen zu unterscheiden haben: schnell veränderliche Größen, wie $\mathfrak{E}_z$ und das mit $\mathfrak{E}_z$ durch die Differentialgleichung (172) verbundene f, und langsam veränderliche Größen, wie J und ebenso auch U, die Energie des Oszillators, deren Wert wir im nächsten Paragraphen berechnen wollen. Doch ist dieser Unterschied in der zeitlichen Veränderlichkeit der genannten Größen nur ein relativer, da der absolute Wert des Differentialquotienten von J nach der Zeit von der Größe der Zeiteinheit abhängt und durch geeignete Wahl derselben beliebig groß gemacht werden kann. Man ist daher nicht berechtigt, $J(t)$ oder $U(t)$ schlechthin als langsam veränderliche Funktionen von t zu bezeichnen. Wenn wir diese Ausdrucksweise der Kürze halber in der Folge dennoch anwenden, so geschieht das stets im relativen Sinne, nämlich mit Bezug auf das abweichende Verhalten der Funktionen $\mathfrak{E}_z(t)$ oder $f(t)$.

Was nun aber die Abhängigkeit der Phasenkonstante ϑ_ν von ihrem Index ν anbetrifft, so besitzt diese notwendig die Eigenschaft der schnellen Veränderlichkeit im absoluten Sinne. Denn obwohl μ klein ist gegen ν, ist doch die Differenz $\vartheta_{\nu+\mu} - \vartheta_\nu$ im allgemeinen nicht klein, weil sonst die Größen A_μ und B_μ in (282) zu spezielle Werte erhalten würden, und daraus folgt, daß $(\partial \vartheta_\nu / \partial \nu) \cdot \nu$ durch eine große Zahl dargestellt wird. Hieran ändert auch ein Wechsel der Zeiteinheit oder eine Verlegung des Anfangspunktes der Zeit nichts Wesentliches.

Die schnelle Veränderlichkeit der Größen ϑ_ν und ebenso C_ν mit ν ist also eine im absoluten Sinne notwendige Bedingung für die Existenz einer bestimmten Schwingungsintensität J, oder mit anderen Worten: für die Möglichkeit der Einteilung der von der Zeit abhängigen Größen in schnell veränderliche und in langsam veränderliche — einer Einteilung, die auch in anderen physikalischen Theorien häufig gemacht wird und auf welche sich alle folgenden Untersuchungen gründen.

§ 178. Die im vorstehenden eingeführte Unterscheidung zwischen schnell veränderlichen und langsam veränderlichen Größen ist in physikalischer Beziehung hier deshalb wichtig, weil wir im folgenden nur die langsame Abhängigkeit von der Zeit als direkt meßbar annehmen wollen. Damit nähern wir uns eben den in der Optik und in der Wärmestrahlung tatsächlich stattfindenden Verhältnissen. Unsere Aufgabe wird dann darin bestehen, Beziehungen ausschließlich zwischen langsam veränderlichen Größen aufzustellen; denn diese allein sind es, welche mit den Ergebnissen der Erfahrung verglichen werden können. Wir bestimmen daher nun zunächst die Werte der wichtigsten hier in Betracht kommenden langsam veränderlichen Größen, nämlich die Energie des Oszillators und den Betrag der vom Oszillator emittierten und absorbierten Energie.

Die Energie des Oszillators, die in (142) ausgedrückt ist, besteht aus zwei Teilen: der potentiellen Energie und der kinetischen Energie. Da wegen der kleinen Dämpfung der Mittelwert dieser beiden Energiearten jedenfalls der nämliche ist, d. h.

$$K\overline{f^2} = L\overline{\left(\frac{df}{dt}\right)^2}, \tag{284}$$

wie sich übrigens auch direkt aus (168) und (279) ableiten läßt, so können wir auch schreiben:

$$U = K\overline{f^2}, \tag{285}$$

indem wir mit $\overline{f^2}$ den Mittelwert von f^2 in dem Zeitintervall von t bis $t + \tau$ bezeichnen. Dieser Mittelwert berechnet sich nach (279) genau in der nämlichen Weise wie der von $\mathfrak{E}_z^2$ in § 176, nur daß hier $\frac{3\, c^3\, C_\nu \sin \gamma_\nu}{16\, \pi^3\, \nu^3}$ statt C_ν, und $\vartheta_\nu + \gamma_\nu$ statt ϑ_ν zu setzen ist. Wir erhalten daher analog (281) mit Rücksicht auf den Wert von K in (168):

$$\left.\begin{array}{l} U = \dfrac{3\, c^3}{16\, \pi^2\, \sigma} \displaystyle\iint d\mu\, d\nu\, C_{\nu+\mu}\, C_\nu \frac{\sin \gamma_{\nu+\mu} \sin \gamma_\nu}{\nu^3} \times \\ \cos(2\, \pi\, \mu\, t - \vartheta_{\nu+\mu} + \vartheta_\nu - \gamma_{\nu+\mu} + \gamma_\nu). \end{array}\right\} \tag{286}$$

Hierbei ist davon Gebrauch gemacht, daß μ kein ist gegen ν, und daß in dem Integral nur diejenigen Glieder merklich in Betracht kommen, für welche ν nahe gleich ν_0.

Statt dessen kann man schreiben:

$$\left.\begin{array}{l} U = \displaystyle\int d\mu\, (a_\mu \sin 2\, \pi\, \mu\, t + b_\mu \cos 2\, \pi\, \mu\, t), \\ \text{wobei:} \\ a_\mu = \dfrac{3\, c^3}{16\, \pi^2\, \sigma} \displaystyle\int d\nu\, C_{\nu+\mu}\, C_\nu \frac{\sin \gamma_{\nu+\mu} \sin \gamma_\nu}{\nu^3} \sin(\vartheta_{\nu+\mu} - \vartheta_\nu + \gamma_{\nu+\mu} - \gamma_\nu), \\ b_\mu = \dfrac{3\, c^3}{16\, \pi^2\, \sigma} \displaystyle\int d\nu\, C_{\nu+\mu}\, C_\nu \frac{\sin \gamma_{\nu+\mu} \sin \gamma_\nu}{\nu^3} \cos(\vartheta_{\nu+\mu} - \vartheta_\nu + \gamma_{\nu+\mu} - \gamma_\nu). \end{array}\right\} \tag{287}$$

Ebenso wie C_ν und ϑ_ν, so ist auch γ_ν, wie man aus (280) erkennt, im absoluten Sinne schnell veränderlich mit ν. Man darf daher, obwohl μ klein ist gegen ν, den Winkel $\gamma_{\nu+\mu}$ nicht etwa annähernd gleich γ_ν setzen, nämlich dann nicht, wenn μ von gleicher oder sogar höherer Größenordnung ist, wie $\sigma\, \nu_0$.

§ 179. Der Betrag der vom Oszillator in der Zeit dt emittierten Energie, als einer „langsam veränderlichen" Größe, ergibt sich aus der Gleichung (151) zu:

$$\frac{2}{3c^3}\overline{\ddot{f}^2(t)}\cdot dt$$

oder nach (279), durch Bildung des Mittelwertes in derselben Weise wie oben:

$$= \frac{3c^3 dt}{8\pi^2}\iint d\nu\, d\mu\, C_{\nu+\mu}\, C_\nu \frac{\sin\gamma_{\nu+\mu}\sin\gamma_\nu}{\nu^2}$$

$$\cos(2\pi\mu t - \vartheta_{\nu+\mu} + \vartheta_\nu - \gamma_{\nu+\mu} + \gamma_\nu),$$

also durch Vergleich mit (286):

$$= 2\sigma\nu_0 U dt. \tag{288}$$

Die in einem Zeitelement vom Oszillator emittierte Energie ist proportional der Energie des Oszillators, ferner seiner Schwingungszahl und seinem logarithmischen Dekrement.

Der Betrag der vom Oszillator in der Zeit dt absorbierten Energie, als einer „langsam veränderlichen" Größe, läßt sich entweder aus (170) berechnen durch die Bildung des Mittelwertes von $\mathfrak{E}_z\left(\frac{df}{dt}\right)$ mit Hilfe der bekannten Ausdrücke für $\mathfrak{E}_z$ und f, oder auch direkt aus einer Anwendung des Prinzips der Erhaltung der Energie, welche besagt, daß die im Zeitelement dt vom Oszillator absorbierte Energie gleich ist der Summe der in dem Zeitelement erfolgten Energiezunahme und der emittierten Energie:

$$\overline{\mathfrak{E}_z \dot{f}}\, dt = dU + 2\sigma\nu_0 U dt. \tag{289}$$

Setzt man hierin für U den in (287) gegebenen Wert, so ergibt sich für die in der Zeit dt vom Oszillator absorbierte Energie der Wert:

$$\left\{\begin{aligned} & dt\cdot\int d\mu\,(a_\mu'\sin 2\pi\mu t + b_\mu'\cos 2\pi\mu t), \\ & \text{wobei:} \\ & \quad a_\mu' = 2\sigma\nu_0 a_\mu - 2\pi\mu b_\mu, \\ & \quad b_\mu' = 2\sigma\nu_0 b_\mu + 2\pi\mu a_\mu. \end{aligned}\right. \tag{290}$$

Diese Größen wollen wir nun mit der Intensität der erregenden Schwingung in eine allgemeine Beziehung bringen, wobei immer festzuhalten ist, daß das Verhältnis $\mu : \sigma \nu_0$ beliebig große und kleine Werte annehmen kann.

§ 180. Von den bisher in unseren Gleichungen auftretenden Energiegrößen dürfen wir als direkt meßbar ansehen nur die Intensität J der erregenden Schwingung und die Energie U des Oszillators. Dieselben stehen aber im allgemeinen in keinem einfachen Zusammenhang miteinander, da die Energie des Oszillators nicht allein von der Gesamtintensität J der erregenden Schwingung $\mathfrak{E}_z$, sondern noch von spezielleren Eigentümlichkeiten, nämlich von den spektralen Eigenschaften dieser Schwingung abhängt. Man kann nun offenbar die Eigenschaften einer bestimmten erregenden Schwingung dadurch weiter verfolgen, daß man die zu untersuchende Schwingung auf verschiedene Resonatoren wirken läßt und die Energie mißt, welche ein jeder Resonator einzeln unter dem Einfluß derselben erregenden Schwingung annimmt. Es ist dies ganz die nämliche Methode, welche in der Akustik zur Analyse eines Klanges angewendet wird.

Hierauf gründen wir unsere Definition der in der Gesamtintensität J enthaltenen Intensität $\mathfrak{J}_\nu$ einer bestimmten Schwingungszahl ν. Wir setzen nämlich:

$$J = \int_0^\infty \mathfrak{J}_\nu \, d\nu \tag{291}$$

und definieren $\mathfrak{J}_\nu$, eine „langsam veränderliche" Funktion der beiden Variabeln ν und t, durch die Energie, welche ein Resonator mit der Schwingungszahl ν unter dem Einfluß der erregenden Schwingung $\mathfrak{E}_z$ annimmt. Den Resonator nehmen wir der Einfachheit halber von der gleichen Beschaffenheit an wie den bis jetzt betrachteten Oszillator.

Doch ist hier noch ein wichtiger Punkt zu erledigen. Da nämlich die Energie eines von der Schwingung $\mathfrak{E}_z$ erregten Resonators nicht allein von seiner Eigenschwingung, sondern außerdem auch von seiner Dämpfung abhängt, so ist noch auf eine geeignete Wahl der Dämpfungskonstanten des zur Messung der Intensität $\mathfrak{J}_\nu$ benutzten Resonators Rücksicht zu nehmen. Damit der Resonator auf eine bestimmte Schwingungszahl und

13*

nicht etwa auf ein endliches Intervall von Schwingungszahlen merklich reagiert, muß sein Dämpfungsdekrement klein sein. Es darf aber auch andererseits nicht allzu klein genommen werden; denn ein Resonator mit sehr kleiner Dämpfung braucht sehr lange Zeit zum Abklingen, und ein solcher Resonator würde den Zweck, durch sein Mitschwingen jederzeit eine gleichzeitige Eigenschaft der ihn erregenden, im allgemeinen mit der Zeit veränderlichen Schwingung anzugeben, nicht erfüllen, da seine Energie nicht von der gleichzeitigen Beschaffenheit, sondern zugleich auch von der Vorgeschichte der erregenden Schwingung abhängen würde. Die Energie des Resonators würde also nicht die Intensität $\mathfrak{J}_\nu$ selber, sondern einen gewissen, über einen größeren Zeitraum erstreckten Mittelwert dieser Größe zum Ausdruck bringen.

Um diesen Umstand zu berücksichtigen, wählen wir das logarithmische Dekrement ϱ aller zur Analyse der erregenden Schwingung $\mathfrak{E}_z$ benutzten Resonatoren zwar klein gegen 1, machen aber doch $\varrho\,\nu$ groß gegen alle μ, was stets möglich ist, da nach (283) μ klein ist gegen ν. Dann ist der Zustand eines analysierenden Resonators, z. B. desjenigen mit der Schwingungszahl ν_0, vollständig bestimmt durch die gleichzeitige Beschaffenheit der erregenden Schwingung, und man kann sagen, daß der Resonator alle Intensitätsschwankungen der erregenden Schwingung momentan anzeigt. In der Tat ersieht man z. B. leicht aus (290), wenn man darin ϱ statt σ setzt, daß die Glieder mit dem Faktor μ gegen die Glieder mit dem Faktor $\varrho\,\nu_0$ verschwinden und daß dadurch die vom Resonator absorbierte Energie proportional wird seiner augenblicklichen Energie U, was nur dann möglich ist, wenn der Zustand des Resonators nur von der gleichzeitigen Beschaffenheit der erregenden Schwingung abhängt.

Unter den gemachten Voraussetzungen ist die in der Gesamtintensität J der erregenden Schwingung enthaltene Intensität der Schwingungszahl ν_0, die wir kurz mit $\mathfrak{J}_0$ bezeichnen wollen, nach (286) als Funktion der Zeit gegeben durch:

$$\mathfrak{J}_0 = \varkappa_0 \cdot \frac{3\,c^3}{16\,\pi^2\,\varrho} \iint d\mu\; d\nu\; C_{\nu+\mu}\; C_\nu \frac{\sin^2\delta_\nu}{\nu^3} \times$$

$$\cos(2\,\pi\,\mu\,t - \vartheta_{\nu+\mu} + \vartheta_\nu).$$

Hier ist $\varkappa_0$ ein von ν_0 abhängiger, sogleich zu bestimmender Proportionalitätsfaktor; der Winkel δ_ν geht aus γ_ν in (280) hervor, wenn man darin ϱ statt σ setzt, also:

$$\operatorname{ctg} \delta_\nu = \frac{\pi \nu_0 (\nu_0^2 - \nu^2)}{\varrho \nu^3} \tag{292}$$

und $\delta_{\nu+\mu}$ ist $= \delta_\nu$ gesetzt, da μ klein ist gegen $\varrho \nu_0$. Der Proportionalitätsfaktor $\varkappa_0$ bestimmt sich aus der Bedingung (291). Schreibt man nämlich diese Bedingung nach (281) in der Form:

$$\iint d\mu \, d\nu \, C_{\nu+\mu} C_\nu \cos(2\pi\mu t - \vartheta_{\nu+\mu} + \vartheta_\nu) = \int_0^\infty \mathfrak{J}_0 \, d\nu_0,$$

so folgt aus dem soeben für $\mathfrak{J}_0$ gefundenen Ausdruck, da μ und ν nicht von ν_0 abhängen:

$$1 = \int_0^\infty d\nu_0 \cdot \frac{3 c^3 \varkappa_0}{16 \pi^2 \varrho \nu^3} \cdot \sin^2 \delta_\nu,$$

oder nach (292):

$$\frac{16 \pi^2 \nu^3}{3 c^3} = \int_0^\infty d\nu_0 \cdot \frac{\varkappa_0}{\varrho} \cdot \frac{1}{1 + \pi^2 \dfrac{\nu_0^2 (\nu_0^2 - \nu^2)^2}{\varrho^2 \nu^6}}.$$

Da nun ϱ klein ist gegen 1, so braucht man nur diejenigen Werte der Funktion unter dem Integralzeichen zu berücksichtigen, für welche ν_0 nahe $= \nu$ ist, und erhält so ganz ähnlich wie in § 122:

$$\frac{16 \pi^2 \nu^3}{3 c^3} = \int_0^\infty d\nu_0 \cdot \frac{\varkappa_0}{\varrho} \cdot \frac{1}{1 + \dfrac{4 \pi^2 (\nu_0 - \nu)^2}{\varrho^2 \nu^2}} = \frac{\varkappa \nu}{2},$$

wenn $\varkappa$ den Wert von $\varkappa_0$ für $\nu_0 = \nu$ bedeutet. Daraus ergibt sich:

$$\varkappa_0 = \frac{32 \pi^2 \nu_0^2}{3 c^3}.$$

Daher ist die Intensität $\mathfrak{J}_0$ der Schwingungszahl ν_0:

$$\left.\begin{aligned}
\mathfrak{J}_0 &= \int d\mu \, (\mathfrak{A}_\mu^0 \sin 2\pi\mu t + \mathfrak{B}_\mu^0 \cos 2\pi\mu t), \\
&\text{wobei:} \\
\mathfrak{A}_\mu^0 &= \frac{2 \nu_0^2}{\varrho} \int d\nu \, C_{\nu+\mu} C_\nu \frac{\sin^2 \delta_\nu}{\nu^3} \sin(\vartheta_{\nu+\mu} - \vartheta_\nu), \\
\mathfrak{B}_\mu^0 &= \frac{2 \nu_0^2}{\varrho} \int d\nu \, C_{\nu+\mu} C_\nu \frac{\sin^2 \delta_\nu}{\nu^3} \cos(\vartheta_{\nu+\mu} - \vartheta_\nu).
\end{aligned}\right\} \tag{293}$$

Im allgemeinen werden die Werte von $\mathfrak{A}_\mu^0$ und $\mathfrak{B}_\mu^0$ noch von ϱ abhängig sein. In diesem Falle kann man von einer Intensität der Schwingungszahl ν_0 in bestimmtem Sinne gar nicht reden. Wir wollen nun für das Folgende die Voraussetzung machen, daß eine jede Schwingungszahl ν eine ganz bestimmte, mit der Zeit „langsam veränderliche" Schwingungsintensität $\mathfrak{J}_\nu$ besitzt, unabhängig von der zu ihrer Messung dienenden Größe ϱ. Dann ist zugleich auch die schon in § 176 eingeführte Bedingung erfüllt, daß eine Gesamtintensität

$$J = \int_0^\infty \mathfrak{J}_\nu \, d\nu$$

der erregenden Schwingung $\mathfrak{E}_z$ existiert. Auf die Frage, weshalb und inwieweit diese Annahme, welche übrigens in der Wärme- und Lichtstrahlung bisher tatsächlich stets gemacht wurde, in der Natur gerechtfertigt ist, soll hier nicht näher eingegangen werden.

§ 181. Wir haben jetzt die erregende Schwingung $\mathfrak{E}_z$, die zu den „schnell veränderlichen" und daher nicht direkt meßbaren Größen gehört, so weit analysiert, daß wir ihre Gesamtintensität J zu jeder Zeit in eine Reihe von meßbaren Größen zerlegt haben: den Intensitäten $\mathfrak{J}_\nu$ der verschiedenen Schwingungszahlen ν. Weitere Mittel, um „langsam veränderliche" Eigenschaften von $\mathfrak{E}_z$ abzuleiten, besitzen wir nicht; die Methoden der Analyse sind also hiermit erschöpft Was wir durch sie von der schnell veränderlichen Schwingung $\mathfrak{E}_z$ kennen gelernt haben, ist aber im Vergleich zu der in ihr noch enthaltenen Mannigfaltigkeit von Eigenschaften nur äußerst wenig. Die Funktionen C_ν und ϑ_ν selber, in ihrer Abhängigkeit von ν, sind und bleiben uns innerhalb eines breiten Spielraumes gänzlich unbekannt.

Stellen wir nun zunächst dasjenige zusammen, was wir durch Messung der Intensität $\mathfrak{J}_0$ der Schwingungszahl ν_0, als einer langsam veränderlichen Funktion der Zeit t, über die schnell veränderlichen Größen C_ν und ϑ_ν erfahren können. Als meßbar haben wir in (293) die Größen $\mathfrak{A}_\mu^0$ und $\mathfrak{B}_\mu^0$ zu betrachten, für alle Werte von μ. Setzen wir nun:

$$(294) \qquad \begin{cases} C_{\nu+\mu}\, C_\nu \sin(\vartheta_{\nu+\mu} - \vartheta_\nu) = \mathfrak{A}_\mu^0 + \xi, \\ C_{\nu+\mu}\, C_\nu \cos(\vartheta_{\nu+\mu} - \vartheta_\nu) = \mathfrak{B}_\mu^0 + \eta, \end{cases}$$

wobei ξ und η schnell veränderliche Funktionen von ν und μ sind, so folgt aus (293):

$$\mathfrak{A}_\mu^0 = \mathfrak{A}_\mu^0 \cdot \frac{2\,\nu_0^2}{\varrho} \cdot \int d\nu \frac{\sin^2 \delta_\nu}{\nu^3} + \frac{2\,\nu_0^2}{\varrho} \int \xi \frac{\sin^2 \delta_\nu}{\nu^3}\, d\nu .$$

Nun ist mit Rücksicht auf (292):

$$\frac{2\,\nu_0^2}{\varrho} \int \frac{\sin^2 \delta_\nu}{\nu^3}\, d\nu = 1 .$$

Folglich:

$$\int \xi \frac{\sin^2 \delta_\nu}{\nu^3}\, d\nu = 0 .$$

Ebenso:

$$\int \eta \frac{\sin^2 \delta_\nu}{\nu^3}\, d\nu = 0 .$$

Da $\frac{\sin \delta_\nu}{\nu}$ für alle Werte von ν verschwindet, deren Verhältnis zu ν_0 nicht nahe gleich 1 ist, so stellt die Größe $\mathfrak{A}_\mu^0$ in (294) den langsam veränderlichen Mittelwert der schnell veränderlichen Größe $C_{\nu+\mu}\, C_\nu \sin(\vartheta_{\nu+\mu} - \vartheta_\nu)$ für ν nahe gleich ν_0 vor, und ebenso $\mathfrak{B}_\mu^0$ den entsprechenden Mittelwert der schnell veränderlichen Größe $C_{\nu+\mu}\, C_\nu \cos(\vartheta_{\nu+\mu} - \vartheta_\nu)$.[1]

Kehren wir nun zu der Untersuchung des Oszillators mit der Schwingungszahl ν_0 und dem Dämpfungsdekrement σ zurück, so ist zunächst von vornherein einleuchtend, daß zur Berechnung des Einflusses, welchen die erregende Schwingung $\mathfrak{E}_z$ auf den Oszillator ausübt, die Kenntnis der Mittelwerte $\mathfrak{A}_\mu^0$ und $\mathfrak{B}_\mu^0$ im allgemeinen noch nicht genügt, sondern daß dazu die Größen C_ν und ϑ_ν selber bekannt sein müssen. In der Tat ersieht man aus dem in (287) abgeleiteten Ausdruck der Energie U des Oszillators, daß diese erst dann genau berechnet werden kann, wenn man die Werte von $C_{\nu+\mu}\, C_\nu \sin(\vartheta_{\nu+\mu} - \vartheta_\nu)$ und von $C_{\nu+\mu}\, C_\nu \cos(\vartheta_{\nu+\mu} - \vartheta_\nu)$ für jeden Wert von ν anzugeben vermag, für den $\nu : \nu_0$ nahe gleich 1 ist. Mit anderen Worten: die

---

[1] Man könnte auch sehr viel einfacher die Intensität $\mathfrak{J}_\nu$ einer bestimmten Schwingungszahl ν durch die genannten Mittelwerte definieren, indem man das für die Gesamtintensität J aufgestellte Integral (281) einfach in der Form der Gleichung (291) schreibt und daraus die Werte $\mathfrak{A}_\mu$ und $\mathfrak{B}_\mu$ ableitet. Dann geht aber die hier benutzte physikalische Bedeutung der Definition verloren.

in der erregenden Schwingung enthaltene Intensität $\mathfrak{J}_0$ der Schwingungszahl ν_0, auch wenn sie für alle Zeiten bekannt ist, bestimmt im allgemeinen noch nicht die Energie U des von der Schwingung getroffenen Oszillators.

Somit bleibt nichts anderes übrig, als entweder auf die Konstatierung eines allgemeinen Zusammenhangs der Größen U und $\mathfrak{J}_0$ überhaupt zu verzichten, was aber den Ergebnissen aller Erfahrung zuwiderlaufen würde, oder mittels einer neu einzuführenden Hypothese die vorhandene Kluft zu überbrücken. Die physikalischen Tatsachen entscheiden für die zweite Alternative.

Die Hypothese, welche wir jetzt als die nächstliegende und wohl einzig mögliche einführen und für alles folgende beibehalten wollen, besteht in der Annahme, daß bei der Berechnung von U aus der Gleichung (287) in den Integralen, welche die Werte der Koeffizienten a_μ und b_μ angeben, für die schnell veränderlichen Größen $C_{\nu+\mu} C_\nu \sin(\vartheta_{\nu+\mu} - \vartheta_\nu)$ und $C_{\nu+\mu} C_\nu \cos(\vartheta_{\nu+\mu} - \vartheta_\nu)$ — die einzigen von C_ν und ϑ_ν abhängigen Größen, die in diesen Integralen vorkommen — ohne merklichen Fehler ihre langsam veränderlichen Mittelwerte $\mathfrak{A}_\mu^0$ und $\mathfrak{B}_\mu^0$ gesetzt werden können. Damit erhält dann die Aufgabe, U aus $\mathfrak{J}_0$ zu berechnen, eine ganz bestimmte, durch Messungen zu verifizierende Lösung. Um aber auszudrücken, daß die hier abzuleitenden Gesetze nicht für jede Art Schwingungen, sondern nur mit Ausschließung gewisser besonderer Einzelfälle gelten, wollen wir jede Art Strahlung, auf welche die hier eingeführte Hypothese paßt, als „natürliche" Strahlung bezeichnen. Dieser Name empfiehlt sich deshalb, weil, wie sich im nächsten Kapitel zeigen wird, der so charakterisierten Strahlung gerade die Eigenschaften der Wärmestrahlung zukommen.

Man kann den Begriff der natürlichen Strahlung noch anschaulicher, aber weniger direkt, als oben geschehen, auch dahin fassen, daß bei ihr die Abweichungen der unmeßbaren schnell veränderlichen Größen $C_{\nu+\mu} C_\nu \sin(\vartheta_{\nu+\mu} - \vartheta_\nu)$ usw. von ihren meßbaren langsam veränderlichen Mittelwerten $\mathfrak{A}_\mu^0$ usw. gänzlich unregelmäßig sind, entsprechend der „elementaren Unordnung" (§ 132).

§ 182. Gemäß der im vorigen Paragraphen eingeführten Hypothese ergibt sich aus der Gleichung (287):

$$a_\mu = \frac{3\,c^3}{16\,\pi^2\,\sigma}\int d\nu\,\frac{\sin\gamma_{\nu+\mu}\sin\gamma_\nu}{\nu^3}\cdot$$
$$\left(\mathfrak{A}_\mu^0 \cos(\gamma_{\nu+\mu}-\gamma_\nu) + \mathfrak{B}_\mu^0 \sin(\gamma_{\nu+\mu}-\gamma_\nu)\right),$$

$$b_\mu = \frac{3\,c^3}{16\,\pi^2\,\sigma}\int d\nu\,\frac{\sin\gamma_{\nu+\mu}\sin\gamma_\nu}{\nu^3}\cdot$$
$$\left(\mathfrak{B}_\mu^0 \cos(\gamma_{\nu+\mu}-\gamma_\nu) - \mathfrak{A}_\mu^0 \sin(\gamma_{\nu+\mu}-\gamma_\nu)\right),$$

oder:

$$a_\mu = \frac{3\,c^3}{16\,\pi^2\,\sigma}(\mathfrak{A}_\mu^0\,\alpha + \mathfrak{B}_\mu^0\,\beta),$$

$$b_\mu = \frac{3\,c^3}{16\,\pi^2\,\sigma}(\mathfrak{B}_\mu^0\,\alpha - \mathfrak{A}_\mu^0\,\beta),$$

wobei:

$$\alpha = \int_0^\infty d\nu\,\frac{\sin\gamma_{\nu+\mu}\sin\gamma_\nu}{\nu^3}\cos(\gamma_{\nu+\mu}-\gamma_\nu),$$

$$\beta = \int_0^\infty d\nu\,\frac{\sin\gamma_{\nu+\mu}\sin\gamma_\nu}{\nu^3}\sin(\gamma_{\nu+\mu}-\gamma_\nu).$$

Nun ergibt sich mit Berücksichtigung der in (280) gegebenen Werte von $\operatorname{ctg}\gamma_\nu$ und $\operatorname{ctg}\gamma_{\nu+\mu}$ durch elementare Rechnungen, wobei besonders zu beachten ist, daß σ klein ist und daß μ im allgemeinen von derselben Größenordnung ist wie $\sigma\nu_0$:

$$\alpha = \frac{\sigma}{2\,\nu_0^2}\cdot\frac{1}{1+\dfrac{\pi^2\,\mu^2}{\sigma^2\,\nu_0^2}},$$

$$\beta = \frac{\pi\,\mu}{2\,\nu_0^3}\cdot\frac{1}{1+\dfrac{\pi^2\,\mu^2}{\sigma^2\,\nu_0^2}}.$$

Folglich, wenn man daraus a_μ und b_μ berechnet und die so erhaltenen Werte in (290) einsetzt:

$$a_\mu' = \frac{3\,c^3\,\sigma}{16\,\pi^2\,\nu_0}\,\mathfrak{A}_\mu^0,$$

$$b_\mu' = \frac{3\,c^3\,\sigma}{16\,\pi^2\,\nu_0}\,\mathfrak{B}_\mu^0.$$

Die in der Zeit dt vom Oszillator absorbierte Energie ist also nach (290):

$$dt \cdot \frac{3\,c^3\,\sigma}{16\pi^2\nu_0} \cdot \int d\,\mu\,(\mathfrak{A}_\mu^0 \sin 2\,\pi\,\mu\,t + \mathfrak{B}_\mu^0 \cos 2\,\pi\,\mu\,t),$$

oder nach (293):

$$(295) \qquad = dt \cdot \frac{3\,c^3\,\sigma}{16\,\pi^2\,\nu_0} \cdot \mathfrak{J}_0\,.$$

Die in einem Zeitelement vom Oszillator absorbierte Energie ist proportional der in der erregenden Schwingung enthaltenen Intensität seiner Eigenperiode, ferner seinem logarithmischen Dekrement und dem Kubus der Lichtgeschwindigkeit, und umgekehrt proportional der Schwingungszahl.

Bei der natürlichen Strahlung wird also stets positive Energie absorbiert, was im allgemeinen, wie schon in der § 111 gemachten Bemerkung betont wurde, nicht der Fall zu sein braucht.

Durch Substitution des Wertes der absorbierten Energie in (289) erhält man schließlich die Fundamentalgleichung der entwickelten Theorie:

$$dt \cdot \frac{3\,c^3\,\sigma}{16\,\pi^2\,\nu_0} \cdot \mathfrak{J}_0 = d\,U + 2\,\sigma\,\nu_0\,U dt$$

oder:

$$(296) \qquad \frac{d\,U}{d\,t} + 2\sigma\,\nu_0\,U = \frac{3\,c^3\,\sigma}{16\,\pi^2\,\nu_0}\,\mathfrak{J}_0\,.$$

Diese Differentialgleichung kann zur Berechnung der Energie U des Oszillators benutzt werden, wenn die seiner Schwingungszahl ν_0 entsprechende Intensität $\mathfrak{J}_0$ der erregenden Schwingung als Funktion der Zeit gegeben ist. Da die Funktionen $U(t)$ und $\mathfrak{J}_0(t)$ hier nicht mehr durch FOURIERsche Integrale dargestellt zu werden brauchen, so können wir von jetzt ab auch die früher in § 175 eingeführte Beschränkung in bezug auf das betrachtete Zeitintervall wieder aufheben, und diese und die folgenden Gleichungen als für alle positiven und negativen Zeiten gültig ansehen.

Die allgemeine Lösung der Differentialgleichung (296) ist:

$$U = \frac{3\,c^3\,\sigma}{16\pi^2\,\nu_0} \int_{-\infty}^{t} \mathfrak{J}_0\,(x)\,e^{2\,\sigma\,\nu_0\,(x-t)}\,d\,x\,.$$

Für konstantes $\mathfrak{J}_0$ hat man:

$$U = \frac{3\,c^3}{16\pi^2\,\nu_0{}^2}\,\mathfrak{J}_0\,.$$

Bei konstanter Bestrahlung ist die Energie des Oszillators proportional der in der erregenden Schwingung enthaltenen Intensität seiner Schwingungszahl, ferner dem Kubus der Lichtgeschwindigkeit, und umgekehrt proportional dem Quadrat der Schwingungszahl, aber unabhängig von der Dämpfung.

Nachdem wir so die Abhängigkeit der Energie des Oszillators von der Intensität der erregenden Schwingung festgestellt haben, wird es unsere nächste Aufgabe sein, die letztere Größe in Zusammenhang zu bringen mit der im umgebenden Felde stattfindenden Energiestrahlung. Dies geschieht nach bekannten Methoden im nächsten Abschnitt und führt zur Formulierung der Gesetze der Energie und der Entropie.

Drittes Kapitel.
Erhaltung der Energie und Vermehrung der Entropie.

§ 183. Indem wir jetzt zur Untersuchung der Vorgänge in dem den Oszillator umgebenden elektromagnetischen Felde übergehen, wollen wir überall im folgenden von den im vorigen Kapitel abgeleiteten Resultaten Gebrauch machen, selbstverständlich unter der Voraussetzung, daß dabei überall und zu allen Zeiten die Bedingungen der natürlichen Strahlung erfüllt sind. Dementsprechend brauchen wir künftig nie mehr mit Amplituden und Phasen zu rechnen, sondern stets nur mit Intensitäten und Energien, d. h. mit „langsam veränderlichen" (im Sinne des § 177) Größen. In diesem Sinne ist auch die Bedeutung der unten benutzten Raumelemente und Zeitelemente zu verstehen, nämlich als Größen, welche unendlich klein sind gegen die Dimensionen der betrachteten Räume und Zeiten, aber immer noch groß gegen die betrachteten Wellenlängen und Schwingungsdauern. Die Wände des durchstrahlten Vakuums denken wir uns als ruhende, absolut spiegelnde Flächen, deren Krümmungsradien groß sind gegen alle in Betracht kommenden Wellenlängen (§ 2).

Dann ist die totale Energie U_t des durchstrahlten Vakuums und einer darin in gehörigen Abständen voneinander befindlichen Anzahl von Oszillatoren der betrachteten Art von der Form:

$$U_t = \Sigma U + \int u \, d\tau, \tag{297}$$

wobei U die Energie eines einzelnen Oszillators, Σ die Summation über alle Oszillatoren, und u die Dichte der strahlenden Energie im Raumelement $d\tau$ des Vakuums bezeichnet. Da die Oszillatoren verschwindend kleine Räume einnehmen, so ist es gleichgültig, ob in dem Integral die Integration auch über die von den Oszillatoren erfüllten Räume erstreckt wird oder nicht.

Die räumliche Dichte u der elektromagnetischen Energie in einem Punkte des Vakuums ist:

$$u = \frac{1}{8\pi}\left(\overline{\mathfrak{E}_x^2} + \overline{\mathfrak{E}_y^2} + \overline{\mathfrak{E}_z^2} + \overline{\mathfrak{H}_x^2} + \overline{\mathfrak{H}_y^2} + \overline{\mathfrak{H}_z^2}\right),$$

wo $\mathfrak{E}_x^2$, $\mathfrak{E}_y^2$, $\mathfrak{E}_z^2$, $\mathfrak{H}_x^2$, $\mathfrak{H}_y^2$, $\mathfrak{H}_z^2$ die Quadrate der Feldstärken bedeuten, als „langsam veränderliche" Größen (§ 177) betrachtet, und daher mit dem auf den Mittelwert deutenden Querstrich versehen. Da für jeden einzelnen Strahl die mittlere elektrische und magnetische Energie gleich sind, kann man immer schreiben:

$$u = \frac{1}{4\pi}\left(\overline{\mathfrak{E}_x^2} + \overline{\mathfrak{E}_y^2} + \overline{\mathfrak{E}_z^2}\right). \tag{298}$$

Nun wollen wir die Intensität $J = \overline{\mathfrak{E}_z^2}$ der einen Oszillator erregenden Schwingung (§ 176) aus den Intensitäten der den Oszillator von allen Seiten treffenden Wärmestrahlen berechnen.

Zu diesem Zweck müssen wir auch auf die Polarisation der den Oszillator treffenden monochromatischen Strahlen Rücksicht nehmen. Fassen wir also zunächst ein Strahlenbündel ins Auge, welches den Oszillator innerhalb eines Elementarkegels trifft, dessen Spitze im Oszillator liegt, und dessen Öffnung $d\,\Omega$ durch (5) gegeben ist, wobei die Polarwinkel ϑ und φ die Richtung der Fortpflanzung der Strahlen bezeichnen, so zerfällt das ganze Strahlenbündel in eine Reihe monochromatischer Bündel, von denen eines die Hauptwerte der Intensität $\mathfrak{K}$ und $\mathfrak{K}'$ (§ 17) besitzen möge. Bezeichnen wir nun den Winkel, welchen die zur Hauptintensität $\mathfrak{K}$ gehörige Polarisationsebene mit der durch die Richtung des Strahles und die Z-Achse (die Oszillatorachse) gelegten Ebene bildet, mit ω, einerlei in welchem Quadranten, so läßt sich nach (8) die spezifische Intensität des monochromatischen Bündels zerlegen in die beiden geradlinig und senkrecht aufeinander polarisierten Komponenten:

$$\mathfrak{K}\cos^2\omega + \mathfrak{K}'\sin^2\omega,$$
$$\mathfrak{K}\sin^2\omega + \mathfrak{K}'\cos^2\omega,$$

von denen die erste in der durch die Z-Achse gehenden Ebene polarisiert ist, da sie für $\omega = 0$ gleich $\mathfrak{K}$ wird. Diese Komponente liefert k e i n e n Beitrag zu dem Werte von $\overline{\mathfrak{E}_z^2}$ im Oszillator, weil die elektrische Feldstärke eines geradlinig polarisierten Strahles senkrecht steht auf der Polarisationsebene. Es bleibt also nur übrig die zweite Komponente, deren elektrische Feldstärke den Winkel $\frac{\pi}{2} - \vartheta$ mit der Z-Achse bildet. Nun ist nach dem POYNTINGschen Satze die Intensität eines geradlinig polarisierten Strahles im Vakuum gleich $\frac{c}{4\pi}$ mal dem mittleren Quadrat der elektrischen Feldstärke. Folglich ist das mittlere Quadrat der elektrischen Feldstärke des hier betrachteten Strahlenbündels:

$$\frac{4\pi}{c}(\mathfrak{K}\sin^2\omega + \mathfrak{K}'\cos^2\omega)\,d\nu\,d\Omega$$

und das mittlere Quadrat der Komponente davon in der Richtung der Z-Achse:

$$\frac{4\pi}{c}(\mathfrak{K}\sin^2\omega + \mathfrak{K}'\cos^2\omega)\sin^2\vartheta\,d\nu\,d\Omega\,. \tag{299}$$

Durch Integration über alle Schwingungszahlen und alle Öffnungswinkel erhalten wir mithin den gesuchten Wert:

$$\overline{\mathfrak{E}_z^2} = \frac{4\pi}{c}\int\sin^2\vartheta\,d\Omega\int d\nu\,(\mathfrak{K}_\nu\sin^2\omega_\nu + \mathfrak{K}_\nu'\cos^2\omega_\nu) = J. \tag{300}$$

Sind speziell alle Strahlen unpolarisiert und die Strahlungsintensität nach allen Richtungen konstant, so ist $\mathfrak{K}_\nu = \mathfrak{K}_\nu'$ und, da:

$$\int\sin^2\vartheta\,d\Omega = \iint\sin^3\vartheta\,d\vartheta\,d\varphi = \frac{8\pi}{3},$$

$$\overline{\mathfrak{E}_z^2} = \frac{32\pi^2}{3c}\int\mathfrak{K}_\nu\,d\nu = \overline{\mathfrak{E}_x^2} = \overline{\mathfrak{E}_y^2}$$

und durch Substitution in (298):

$$u = \frac{8\pi}{c}\int\mathfrak{K}_\nu\,d\nu,$$

übereinstimmend mit (22) und (24).

Nehmen wir nun nach § 180 die spektrale Zerlegung der Intensität J vor:

$$J = \int\mathfrak{J}_\nu\,d\nu,$$

so ergibt sich durch Vergleichung mit (300) für die in der

erregenden Schwingung enthaltene Intensität einer bestimmten Schwingungszahl ν der Wert:

$$(301) \qquad \mathfrak{J}_\nu = \frac{4\pi}{c}\int \sin^2\vartheta\, d\Omega\, (\mathfrak{K}_\nu \sin^2\omega_\nu + \mathfrak{K}_\nu' \cos^2\omega_\nu).$$

Da nun $\mathfrak{J}$ mit der Energie U des Oszillators durch die Gleichung (296) zusammenhängt, so ist hiermit die Möglichkeit gegeben, die Schwingung des Oszillators zu berechnen, wenn die Intensitäten und Polarisationen aller den Oszillator treffenden Strahlen für alle Zeiten bekannt sind. Insbesondere ergibt sich für unpolarisierte und nach allen Richtungen gleichmäßige Strahlung:

$$\mathfrak{J} = \frac{32\pi^2}{3c}\mathfrak{K}$$

und nach (296):

$$\frac{dU}{dt} + 2\sigma\nu U = \frac{2c^2\sigma}{\nu}\mathfrak{K}.$$

Der Index 0 kann von jetzt an weggelassen werden. Ist die Strahlung auch noch unabhängig von der Zeit, oder der Strahlungszustand „stationär", so ist auch U von der Zeit unabhängig und:

$$302) \qquad U = \frac{c^2}{\nu^2}\mathfrak{K},$$

übereinstimmend mit Gleichung (193).

§ 184. Die ganze in der Zeit dt von dem Oszillator absorbierte Energie beträgt nach (295):

$$dt \cdot \frac{3c^3\sigma}{16\pi^2\nu} \cdot \mathfrak{J}$$

oder nach (301):

$$dt \cdot \frac{3c^2\sigma}{4\pi\nu}\int \sin^2\vartheta\, d\Omega (\mathfrak{K}\sin^2\omega + \mathfrak{K}'\cos^2\omega).$$

Daher wird von der in der Richtung (ϑ, φ) auf den Oszillator fallenden Strahlung in der Zeit dt der Energiebetrag:

$$dt \cdot \frac{3c^2\sigma}{4\pi\nu}(\mathfrak{K}\sin^2\omega + \mathfrak{K}'\cos^2\omega)\sin^2\vartheta\, d\Omega$$

absorbiert.

Nun beträgt die spezifische Intensität der in der Richtung (ϑ, φ) auf den Oszillator fallenden Strahlung, soweit sie „absorbierbar" ist, d. h. die dem Oszillator entsprechende Schwingungszahl und Polarisation besitzt, nach (299), da der Faktor $\frac{4\pi}{c}\,d\nu\,d\Omega$ hier wegzulassen ist:

$$(\mathfrak{K}\sin^2\omega + \mathfrak{K}'\cos^2\omega)\sin^2\vartheta. \tag{303}$$

Daraus ergibt sich der Satz: der absolute Betrag der vom Oszillator in der Zeit dt absorbierten Energie wird erhalten, wenn man die spezifische Intensität der in irgendeiner Richtung (ϑ, φ) auf ihn fallenden absorbierbaren Strahlung mit

$$dt\cdot\frac{3c^2\sigma}{4\pi\nu}\cdot d\Omega \tag{304}$$

multipliziert und diesen Ausdruck über alle Richtungen (ϑ, φ) integriert. Der Faktor $\frac{3c^2\sigma}{4\pi\nu}$ bestimmt also die Breite des vom Oszillator aufgefangenen Strahlenbündels, indem er ein Maß liefert für das Produkt aus dem Querschnitt dieses Bündels am Orte des Oszillators und seiner Spektralbreite.

Auf der anderen Seite beträgt die vom Oszillator in der Zeit dt nach allen Richtungen emittierte Energie nach (288):

$$dt\cdot 2\sigma\nu U$$

oder, was dasselbe ist:

$$dt\cdot\frac{3\sigma\nu}{4\pi}\,U\cdot\int\sin^2\vartheta\,d\Omega.$$

Da nun die Intensität der vom Oszillator in der Richtung (ϑ, φ) emittierten Strahlung nach (150) unabhängig ist von φ und proportional $\sin^2\vartheta$, so beträgt die in der Zeit dt in dieser Richtung emittierte Energie:

$$dt\cdot\frac{3\sigma\nu}{4\pi}\,U\sin^2\vartheta\,d\Omega$$

und die spezifische Intensität der vom Oszillator in derselben Richtung emittierten Strahlung, durch Division mit (304):

$$\frac{\nu^2 U\sin^2\vartheta}{c^2}. \tag{305}$$

Für den am Schluß des vorigen Paragraphen betrachteten „stationären" Strahlungszustand ist

$$\mathfrak{K} = \mathfrak{K}' \quad \text{und} \quad U = \frac{c^2}{\nu^2}\,\mathfrak{K}.$$

Man sieht also, daß im stationären Strahlungszustand die spezifische Intensität (303) der in irgendeiner Richtung auf den

Oszillator fallenden absorbierbaren Strahlung gleich ist der spezifischen Intensität (305) der in derselben Richtung vom Oszillator emittierten Strahlung, wie es sein muß.

§ 185. Wir wollen nun, als Vorbereitung für die folgenden Deduktionen, die Eigenschaften der verschiedenen den Oszillator passierenden Strahlenbündel noch näher ins Auge fassen. Von allen Seiten treffen Strahlen auf den Oszillator; betrachten wir wieder diejenigen von ihnen, welche in der Richtung (ϑ, φ), innerhalb des Elementarkegels $d\Omega$, dessen Spitze im Oszillator liegt, auf ihn zulaufen, so können wir sie uns zunächst zerlegt denken in ihre monochromatischen Bestandteile, und brauchen uns nur mit demjenigen dieser Bestandteile weiter zu beschäftigen, welcher der Schwingungszahl ν des Oszillators entspricht; denn alle übrigen Strahlen streichen über den Oszillator einfach hinweg, ohne ihn zu beeinflussen oder von ihm beeinflußt zu werden. Die spezifische Intensität des monochromatischen Strahles von der Schwingungszahl ν ist:

$$\mathfrak{K} + \mathfrak{K}',$$

wenn $\mathfrak{K}$ und $\mathfrak{K}'$ die Hauptintensitäten vorstellen. Dieser Strahl wird nun je nach den Richtungen seiner Hauptpolarisationsebenen in zwei Komponenten (8) zerlegt.

Die eine Komponente:

$$\mathfrak{K} \cos^2 \omega + \mathfrak{K}' \sin^2 \omega$$

geht direkt über den Oszillator hinweg und tritt völlig ungeändert auf der anderen Seite wieder aus; sie liefert also einen in der Richtung (ϑ, φ) innerhalb der Kegelöffnung $d\Omega$ vom Oszillator ausgehenden geradlinig polarisierten Strahl, dessen Polarisationsebene durch die Achse des Oszillators hindurchgeht, und dessen Intensität beträgt:

$$\mathfrak{K} \cos^2 \omega + \mathfrak{K}' \sin^2 \omega = \mathfrak{K}''. \tag{306}$$

Die andere, senkrecht auf der vorigen polarisierte Komponente:

$$\mathfrak{K} \sin^2 \omega + \mathfrak{K}' \cos^2 \omega$$

zerfällt wiederum in zwei Teile:

$$(\mathfrak{K} \sin^2 \omega + \mathfrak{K}' \cos^2 \omega) \cos^2 \vartheta$$

und:

$$(\mathfrak{K} \sin^2 \omega + \mathfrak{K}' \cos^2 \omega) \sin^2 \vartheta,$$

von denen der erste ungeändert durch den Oszillator hindurchpassiert, der zweite dagegen absorbiert wird. Statt des letzteren erscheint aber in der vom Oszillator ausgehenden Strahlung die Intensität des emittierten Strahles (305):

$$\frac{\nu^2 U \sin^2 \vartheta}{c^2}.$$

Diese liefert zusammen mit dem ersten, unverändert gebliebenen Teile die gesamte Intensität des vom Oszillator in der Richtung (ϑ, φ) innerhalb des Öffnungswinkels $d\Omega$ ausgehenden, senkrecht auf (306) polarisierten Strahles:

$$(\mathfrak{K} \sin^2 \omega + \mathfrak{K}' \cos^2 \omega) \cos^2 \vartheta + \frac{\nu^2 U}{c^2} \sin^2 \vartheta = \mathfrak{K}'''. \qquad (307)$$

Im ganzen haben wir also schließlich in der Richtung (ϑ, φ) innerhalb $d\Omega$ vom Oszillator ausgehend einen aus zwei senkrecht zueinander polarisierten Komponenten zusammengesetzten Strahl, dessen eine Polarisationsebene durch die Achse des Oszillators geht und dessen Hauptintensitäten die Werte $\mathfrak{K}''$ und $\mathfrak{K}'''$ besitzen.

§ 186. Es ist nun leicht, sich Rechenschaft zu geben von der Erhaltung der Gesamtenergie des Systems auf Grund der lokalen darin stattfindenden Energieänderungen.

Wenn gar kein Oszillator im Felde vorhanden ist, so behält ein jedes der unendlich vielen elementaren Strahlenbündel beim geradlinigen Fortschreiten mit seiner spezifischen Intensität auch seine Energie unverändert bei, auch bei der Reflexion an einer als eben und absolut spiegelnd vorausgesetzten Grenzfläche des Feldes.

Jeder Oszillator dagegen bewirkt im allgemeinen eine Änderung der ihn treffenden Strahlenbündel. Berechnen wir die ganze Energieänderung, die der oben betrachtete Oszillator in der Zeit dt in dem ihn umgebenden Felde hervorruft. Dabei brauchen wir nur diejenigen monochromatischen Strahlen zu berücksichtigen, welche der Schwingungszahl ν des Oszillators entsprechen, da die übrigen durch ihn gar nicht alteriert werden.

In der Richtung (ϑ, φ) innerhalb des Elementarkegels $d\Omega$ wird der Oszillator von einem irgendwie polarisierten Strahlenbündel getroffen, dessen Intensität durch die Summe der beiden Hauptintensitäten $\mathfrak{K}$ und $\mathfrak{K}'$ gegeben ist. Dieses Strahlenbündel läßt, der Bedeutung des Ausdrucks (304) gemäß, in der Zeit dt die Energie:

$$(\mathfrak{K} + \mathfrak{K}')\,dt \cdot \frac{3\,c^2\,\sigma}{4\,\pi\,\nu} \cdot d\,\Omega$$

auf den Oszillator fallen, und dadurch wird auf der Seite der ankommenden Strahlen der nämliche Energiebetrag dem Felde entzogen. Auf der anderen Seite geht dafür vom Oszillator in derselben Richtung (ϑ, φ) ein in bestimmter Weise polarisiertes Strahlenbündel aus, dessen Intensität durch die Summe der beiden Hauptintensitäten $\mathfrak{K}''$ und $\mathfrak{K}'''$ gegeben ist. Dadurch wird dem umgebenden Felde in der Zeit dt der Energiebetrag

$$(\mathfrak{K}'' + \mathfrak{K}''')\,dt \cdot \frac{3\,c^2\,\sigma}{4\,\pi\,\nu}\,d\,\Omega$$

zugeführt.

Im ganzen beträgt also die in der Zeit dt eingetretene Energieänderung des den Oszillator umgebenden Feldes, durch Subtraktion des vorletzten Ausdruckes vom letzten und Integration über $d\,\Omega$:

$$dt \cdot \frac{3\,c^2\,\sigma}{4\,\pi\,\nu} \int d\,\Omega\,(\mathfrak{K}'' + \mathfrak{K}''' - \mathfrak{K} - \mathfrak{K}')\,.$$

Nimmt man dazu die in derselben Zeit eingetretene Energieänderung des Oszillators:

$$dt \cdot \frac{d\,U}{d\,t}\,,$$

so verlangt das Prinzip der Erhaltung der Energie, daß die Summe der letzten beiden Ausdrücke verschwindet, d. h. daß

$$\frac{d\,U}{d\,t} + \frac{3\,c^2\,\sigma}{4\,\pi\,\nu} \int d\,\Omega\,(\mathfrak{K}'' + \mathfrak{K}''' - \mathfrak{K} - \mathfrak{K}') = 0\,, \tag{308}$$

und das ist in der Tat der Inhalt der beiden Gleichungen (296) und (301), wenn man berücksichtigt, daß nach (306) und (307):

$$\mathfrak{K}'' + \mathfrak{K}''' - \mathfrak{K} - \mathfrak{K}' = \left(\frac{\nu^2\,U}{c^2} - \mathfrak{K}\sin^2\omega - \mathfrak{K}'\cos^2\omega\right)\sin^2\vartheta\,.$$

§ 187. Wir bilden jetzt, entsprechend der totalen Energie U_t in (297), die totale Entropie des betrachteten Systems:

$$S_t = \sum S + \int s\,d\,\tau\,. \tag{309}$$

Die Summation Σ ist wieder über alle Oszillatoren, die Integration über alle Raumelemente $d\tau$ des durchstrahlten Feldes zu erstrecken. S, die Entropie eines einzelnen Oszillators, ist als Funktion von U gegeben durch (227), und s, die räumliche Entropiedichte in einem Punkte des Feldes, ist als Funktion aller Hauptstrahlungsintensitäten $\mathfrak{K}$ und $\mathfrak{K}'$ gegeben durch (131), in Verbindung mit (129) und (229).

Nun wollen wir die Änderung berechnen, welche die totale Entropie S_t unseres Systems im Zeitelement dt erleidet. Wir halten uns dabei genau an die analoge im vorigen Paragraphen für die Energie des Systems durchgeführte Rechnung.

Wenn gar kein Oszillator vorhanden ist, so behält ein jedes der im Vakuum vorhandenen unendlich vielen Strahlenbündel beim geradlinigen Fortschreiten zugleich mit seiner spezifischen Intensität seine Entropie unverändert bei, auch bei der Reflexion an einer als eben und absolut spiegelnd vorausgesetzten Grenzfläche des Feldes. Durch die Strahlungsvorgänge im freien Felde kann also keine Entropieänderung des Systems hervorgerufen werden (vgl. § 162). Dagegen bewirkt jeder Oszillator im allgemeinen eine Entropieänderung der ihn treffenden Strahlenbündel. Berechnen wir die ganze Entropieänderung, welche der oben betrachtete Oszillator in der Zeit dt in dem ihn umgebenden Felde hervorruft. Dabei brauchen wir nur diejenigen monochromatischen Strahlen zu berücksichtigen, welche der Schwingungszahl ν des Oszillators entsprechen, da die übrigen durch ihn gar nicht alteriert werden.

In der Richtung (ϑ, φ), innerhalb des Elementarkegels $d\Omega$, wird der Oszillator von einem irgendwie polarisierten Strahlenbündel getroffen, dessen Energiestrahlung die Hauptintensitäten $\mathfrak{K}$ und $\mathfrak{K}'$, und dessen Entropiestrahlung daher die Intensität $\mathfrak{L}+\mathfrak{L}'$ (§ 98) besitzt. Dieses Strahlenbündel läßt, der Bedeutung des Ausdruckes (304) gemäß, in der Zeit dt die Entropie:

$$(\mathfrak{L} + \mathfrak{L}') \cdot dt \cdot \frac{3 c^2 \sigma}{4 \pi \nu} \cdot d\Omega$$

auf den Oszillator fallen, und dadurch wird auf der Seite der ankommenden Strahlen der nämliche Entropiebetrag dem Felde entzogen. Auf der anderen Seite geht vom Oszillator in derselben Richtung (ϑ, φ) ein in bestimmter Weise polarisiertes Strahlenbündel aus, dessen Energiestrahlung die Hauptintensitäten $\mathfrak{K}''$ und $\mathfrak{K}'''$, und dessen Entropiestrahlung daher die

14*

entsprechende Intensität $\mathfrak{L}'' + \mathfrak{L}'''$ besitzt. Dadurch wird dem umgebenden Felde in der Zeit dt die Entropie:

$$(\mathfrak{L}'' + \mathfrak{L}''')\,dt \cdot \frac{3c^2\sigma}{4\pi\nu} \cdot d\Omega$$

zugeführt. Im ganzen beträgt also die in der Zeit dt eingetretene Entropieänderung des den Oszillator umgebenden Feldes, durch Subtraktion des vorletzten Ausdruckes vom letzten und Integration über $d\Omega$:

$$dt \cdot \frac{3c^2\sigma}{4\pi\nu} \cdot \int d\Omega\,(\mathfrak{L}'' + \mathfrak{L}''' - \mathfrak{L} - \mathfrak{L}'). \tag{310}$$

Nimmt man dazu die in derselben Zeit erfolgte Entropieänderung des Oszillators:

$$\frac{dS}{dt} \cdot dt = \frac{dS}{dU} \cdot \frac{dU}{dt} \cdot dt,$$

so ergibt sich durch Addition zu (310) und Summation über alle Oszillatoren die gesuchte Änderung der totalen Entropie des Systems:

$$\frac{dS_t}{dt} \cdot dt = dt \cdot \sum \left[\frac{3c^2\sigma}{4\pi\nu} \int d\Omega(\mathfrak{L}'' + \mathfrak{L}''' - \mathfrak{L} - \mathfrak{L}') + \frac{dS}{dU} \cdot \frac{dU}{dt}\right]. \tag{311}$$

Wir wollen nun weiter den Nachweis führen, daß der Ausdruck hinter dem $\sum$-Zeichen stets positiv ist, inbegriffen den Grenzfall Null. Zu diesem Zwecke setzen wir für $\frac{dU}{dt}$ den in (308) gegebenen Wert und erhalten dadurch:

$$\frac{dS_t}{dt} = \sum \frac{3c^2\sigma}{4\pi\nu} \int d\Omega\Big(\mathfrak{L}'' - \mathfrak{K}''\frac{dS}{dU} + \mathfrak{L}''' - \mathfrak{K}'''\frac{dS}{dU} - \mathfrak{L} + \mathfrak{K}\frac{dS}{dU} - \mathfrak{L}' + \mathfrak{K}'\frac{dS}{dU}\Big).$$

Es erübrigt jetzt nur noch zu zeigen, daß der eingeklammerte Ausdruck für alle beliebigen Werte der positiven Größen U, $\mathfrak{K}$, $\mathfrak{K}'$, ϑ, ω positiv ist, wobei nach Gleichung (306):

$$\mathfrak{K}'' = \mathfrak{K}\cos^2\omega + \mathfrak{K}'\sin^2\omega \tag{312}$$

und nach Gleichung (307):

$$\mathfrak{K}''' = (\mathfrak{K}\sin^2\omega + \mathfrak{K}'\cos^2\omega)\cos^2\vartheta + \frac{\nu^2 U}{c^2}\sin^2\vartheta.$$

Setzen wir zur Abkürzung die positive Größe:

$$\mathfrak{K} \sin^2 \omega + \mathfrak{K}' \cos^2 \omega = \mathfrak{K} + \mathfrak{K}' - \mathfrak{K}'' = \mathfrak{K}_0''' , \qquad (313)$$

so ist hiernach:

$$\mathfrak{K}''' = \mathfrak{K}_0''' \cos^2 \vartheta + \frac{\nu^2 U}{c^2} \sin^2 \vartheta . \qquad (314)$$

Wir wollen zunächst das Glied:

$$\mathfrak{L}''' - \mathfrak{K}''' \frac{dS}{dU} = f(\mathfrak{K}''')$$

ins Auge fassen, indem wir darin U und folglich auch $\frac{dS}{dU}$ als konstant, dagegen $\mathfrak{K}'''$ und folglich auch $\mathfrak{L}'''$ als variabel betrachten. Mit Berücksichtigung von (229) und (227) ergibt sich dann:

$$\frac{df}{d\mathfrak{K}'''} = \frac{d\mathfrak{L}'''}{d\mathfrak{K}'''} - \frac{dS}{dU}$$

$$= \frac{k}{h\nu} \log\left(\frac{h\nu^3}{c^2 \mathfrak{K}'''} + 1\right) - \frac{k}{h\nu} \log\left(\frac{h\nu}{U} + 1\right),$$

$$\frac{d^2 f}{d\mathfrak{K}'''^2} = - \frac{k}{h\nu \mathfrak{K}'''} \cdot \frac{1}{1 + \frac{c^2 \mathfrak{K}'''}{h\nu^3}} < 0 .$$

Daraus folgt, daß die Funktion $f(\mathfrak{K}''')$ ein einziges Maximum besitzt, und zwar für $\mathfrak{K}''' = \frac{\nu^2}{c^2} U$.

Da nun nach (314) $\mathfrak{K}'''$ zwischen $\mathfrak{K}_0'''$ und $\frac{\nu^2 U}{c^2}$ liegt, so ist jedenfalls:

$$f(\mathfrak{K}''') > f(\mathfrak{K}_0'''),$$

d. h.

$$\mathfrak{L}''' - \mathfrak{K}''' \frac{dS}{dU} > \mathfrak{L}_0''' - \mathfrak{K}_0''' \frac{dS}{dU},$$

und um den Beweis durchzuführen, genügt es, den Ausdruck:

$$\mathfrak{L}'' - \mathfrak{K}'' \frac{dS}{dU} + \mathfrak{L}_0''' - \mathfrak{K}_0''' \frac{dS}{dU} - \mathfrak{L} + \mathfrak{K} \frac{dS}{dU} - \mathfrak{L}' + \mathfrak{K}' \frac{dS}{dU}$$

oder, was nach (313) dasselbe ist, den Ausdruck:

$$(\mathfrak{L}'' + \mathfrak{L}_0''') - (\mathfrak{L} + \mathfrak{L}')$$

als positiv zu erweisen. Hierzu wollen wir setzen:

$$\mathfrak{K} + \mathfrak{K}' = \mathfrak{K}'' + \mathfrak{K}_0''' = \mathfrak{S} .$$

$\mathfrak{K}''$ und $\mathfrak{K}_0'''$ liegen nach (312) und (313) zwischen $\mathfrak{K}$ und $\mathfrak{K}'$.

Betrachten wir jetzt die Größe:

$$\mathfrak{L} + \mathfrak{L}' = F(\mathfrak{K})$$

als Funktion von $\mathfrak{K}$ allein, indem wir $\mathfrak{S}$ konstant nehmen und daher $\mathfrak{K}'$ als von $\mathfrak{K}$ abhängig ansehen, so handelt es sich nur noch um das Vorzeichen des Ausdruckes:

$$F(\mathfrak{K}'') - F(\mathfrak{K}).$$

Nun ergibt sich nach (229) durch Differentiation:

$$\frac{dF}{d\mathfrak{K}} = \frac{k}{h\nu}\log\left(\frac{h\nu^3}{c^2\mathfrak{K}} + 1\right) - \frac{k}{h\nu}\log\left(\frac{h\nu^3}{c^2\mathfrak{K}'} + 1\right)$$

$$\frac{d^2F}{d\mathfrak{K}^2} = -\frac{k}{h\nu\mathfrak{K}}\cdot\frac{1}{1+\frac{c^2\mathfrak{K}}{h\nu^3}} - \frac{k}{h\nu\mathfrak{K}'}\cdot\frac{1}{1+\frac{c^2\mathfrak{K}'}{h\nu^3}} < 0.$$

Daraus folgt, daß die Funktion $F(\mathfrak{K})$ ein einziges Maximum besitzt, und zwar für $\mathfrak{K} = \mathfrak{K}' = \frac{\mathfrak{S}}{2}$, und daß sie zu beiden Seiten dieses Maximums symmetrisch abfällt. Je näher also das Argument $\mathfrak{K}$ dem Werte $\frac{\mathfrak{S}}{2}$ kommt, einerlei von welcher Seite her, desto größer wird der Wert von F.

Nun liegt $\mathfrak{K}''$ dem Werte $\frac{\mathfrak{S}}{2}$, welcher das arithmetische Mittel sowohl von $\mathfrak{K}$ und $\mathfrak{K}'$, als auch von $\mathfrak{K}''$ und $\mathfrak{K}_0'''$ bildet, jedenfalls näher als $\mathfrak{K}$, weil $\mathfrak{K}''$ und $\mathfrak{K}_0'''$ zwischen $\mathfrak{K}$ und $\mathfrak{K}'$ liegen. Folglich ist $F(\mathfrak{K}'') > F(\mathfrak{K})$, und damit ist der Beweis für die Vermehrung der Entropie geliefert.

Jeder der betrachteten Strahlungsvorgänge verläuft also einseitig in dem Sinne wachsender Entropie, bis mit dem Maximum der Entropie auch der stationäre Strahlungszustand erreicht wird, welcher durch die Beziehungen charakterisiert ist:

$$\mathfrak{K} = \mathfrak{K}' = \mathfrak{K}'' = \mathfrak{K}_0''' = \mathfrak{K}''' = \frac{\nu^2 U}{c^2}.$$

Viertes Kapitel. Anwendung auf einen speziellen Fall. Schluß.

§ 188. Wir wollen noch zur Betrachtung des speziellen Falles übergehen, daß das Feld, in welchem der Oszillator liegt, sich in einem Zustand stationärer Strahlung befindet, während die Schwingungsenergie des Oszillators anfangs eine ganz beliebige sein möge. Eine Folge der stationären Strahlung im Vakuum ist, daß die auf den Oszillator fallenden Strahlen unpolarisiert und nach allen Richtungen gleichmäßig intensiv sind, d. h.

$$\mathfrak{K} = \mathfrak{K}' = \mathfrak{K}_0 = \text{const}$$

unabhängig von der Zeit und von der Richtung der Strahlung. Dann folgt aus (306) und (307):

$$\mathfrak{K}'' = \mathfrak{K}_0, \qquad \mathfrak{K}''' = \mathfrak{K}_0 \cos^2 \vartheta + \frac{\nu^2 U}{c^2} \sin^2 \vartheta. \tag{315}$$

Ferner aus (308), da

$$\int \sin^2 \vartheta \, d\Omega = \frac{8\pi}{3},$$

$$\frac{dU}{dt} + 2\sigma\nu U - \frac{2c^2\sigma}{\nu} \mathfrak{K}_0 = 0, \tag{316}$$

woraus sich die Schwingungsenergie U des Oszillators bei gegebenem Anfangswert als Funktion der Zeit t ergibt.

Die totale Entropieänderung des Systems in der Zeit dt folgt aus (311), da nur ein einziger Oszillator vorhanden ist, und da $\mathfrak{L} = \mathfrak{L}' = \mathfrak{L}'' = \mathfrak{L}_0$, zu:

$$dS_t = dS + \frac{3c^2\sigma}{4\pi\nu} dt \cdot \int d\Omega \, (\mathfrak{L}''' - \mathfrak{L}_0). \tag{317}$$

Hier ist $\mathfrak{L}'''$ nach (229) von $\mathfrak{K}'''$ und dadurch auch vom Winkel ϑ abhängig.

Wenn nicht nur das den Oszillator umgebende Feld, sondern das ganze System sich im stationären Strahlungszustand befände, so wäre auch $\mathfrak{K}''' = \mathfrak{K}_0$, und die Energie U des Oszillators würde speziell:

$$U_0 = \frac{c^2}{\nu^2} \mathfrak{K}_0, \tag{318}$$

wie auch aus (315) zu ersehen. Wir wollen daher diesen Wert U_0 den stationären Wert der Energie des Oszillators nennen; ihm nähert sich U nach (316) für wachsende Zeiten asymptotisch an.

Nun besitze die Energie des Oszillators einen Wert, der nur noch wenig von dem stationären Wert U_0 verschieden ist, d. h.

$$U = U_0 + \Delta U. \tag{319}$$

Die kleine, positive oder negative, Größe ΔU können wir die Abweichung der Oszillatorenergie von ihrem stationären Wert nennen, sie ist zugleich ein Maß für die Gleichgewichtsstörung des ganzen betrachteten Systems. Dann wird aus (316) und (318):

$$\frac{dU}{dt} + 2\sigma\nu\,\Delta U = 0 \tag{320}$$

und aus (315):

$$\mathfrak{K}''' = \mathfrak{K}_0 + \frac{\nu^2}{c^2}\sin^2\vartheta \cdot \Delta U, \tag{321}$$

folglich, durch Entwicklung in eine TAYLORsche Reihe und mit Vernachlässigung höherer Potenzen von ΔU:

$$\mathfrak{L}''' = \mathfrak{L}_0 + \left(\frac{d\mathfrak{L}}{d\mathfrak{K}}\right)_0 \frac{\nu^2}{c^2}\sin^2\vartheta\,\Delta U + \frac{1}{2}\left(\frac{d^2\mathfrak{L}}{d\mathfrak{K}^2}\right)_0 \frac{\nu^4}{c^4}\sin^4\vartheta\,(\Delta U)^2.$$

Führt man dies in (317) ein, so ergibt sich, da:

$$\int \sin^4\vartheta\, d\Omega = \frac{32\pi}{15},$$

$$dS_t = dS + 2\sigma\nu\,dt\cdot\left(\frac{d\mathfrak{L}}{d\mathfrak{K}}\right)_0 \Delta U + \frac{4\sigma\nu^3}{5c^2}dt\cdot\left(\frac{d^2\mathfrak{L}}{d\mathfrak{K}^2}\right)_0 (\Delta U)^2,$$

oder nach (320), mit Elimination des Zeitelements dt:

$$dS_t = dS - dU\left\{\left(\frac{d\mathfrak{L}}{d\mathfrak{K}}\right)_0 + \frac{2\nu^2}{5c^2}\left(\frac{d^2\mathfrak{L}}{d\mathfrak{K}^2}\right)_0 \Delta U\right\}.$$

Andererseits ist:

$$dS = \frac{dS}{dU}\cdot dU$$

und nach (319):

$$\frac{dS}{dU} = \left(\frac{dS}{dU}\right)_0 + \left(\frac{d^2S}{dU^2}\right)_0 \Delta U + \dots.$$

Folglich durch Substitution die Entropieänderung des ganzen Systems, unter Vernachlässigung höherer Potenzen von ΔU:

$$dS_t = dU\left[\left(\frac{dS}{dU}\right)_0 - \left(\frac{d\mathfrak{L}}{d\mathfrak{K}}\right)_0 + \left\{\left(\frac{d^2S}{dU^2}\right)_0 - \frac{2\nu^2}{5c^2}\left(\frac{d^2\mathfrak{L}}{d\mathfrak{K}^2}\right)_0\right\}\Delta U\right]. \tag{322}$$

Da beim thermodynamischen Gleichgewicht die Temperatur der freien Strahlung gleich derjenigen des Oszillators ist, so folgt aus (135) und (199):

$$\left(\frac{d\mathfrak{L}}{d\mathfrak{K}}\right)_0 = \left(\frac{dS}{dU}\right)_0$$

und durch Differentiation dieser Gleichung, mit Rücksicht auf (318):

$$\left(\frac{d^2 \mathfrak{L}}{d \mathfrak{K}^2}\right)_0 = \frac{c^2}{\nu^2} \left(\frac{d^2 S}{d U^2}\right)_0.$$

Dadurch geht der Ausdruck (322) über in:

$$d S_t = \frac{3}{5} d U \cdot \Delta U \cdot \frac{d^2 S}{d U^2} = - \frac{3}{5} k \cdot \frac{d U \cdot \Delta U}{U(U + h \nu)}, \qquad (323)$$

indem der Wert von S aus (227) eingesetzt und der Index 0 bei U jetzt als überflüssig fortgelassen ist.

Dieser Ausdruck stellt also die Entropievermehrung dar, welche in der Natur eintritt, wenn ein in einem stationären Strahlungsfeld befindlicher Oszillator, dessen Energie eine kleine, positive oder negative, Abweichung ΔU von ihrem stationären Wert U aufweist, die unendlich kleine Energieänderung $d U$ erleidet, natürlich auf Kosten bez. zugunsten der Energie des Strahlungsfeldes. Die Entropievermehrung hängt also außer von der Schwingungszahl ν nur ab von $d U$, ΔU und U, nicht von dem Dämpfungsdekrement des Oszillators, und ist überdies, wie auch von vornherein einleuchtet, den Werten von $d U$ und von ΔU proportional. Da sie stets positiv ist, so besitzen $d U$ und ΔU entgegengesetzte Vorzeichen, wie natürlich.

§ 189. Denken wir uns nun, daß in dem betrachteten stationären Strahlungsfeld statt eines einzigen Oszillators eine beliebige Anzahl n mit dem bisher betrachteten ganz gleichbeschaffene Oszillatoren vorhanden sind, in denen sich während des Zeitelementes $d t$ genau die nämlichen Vorgänge[1] abspielen. Dann ist die Energie aller Oszillatoren zusammengenommen die Summe der Einzelenergien: $n U = U_n$, ihre Abweichung von dem stationären Wert: $\Delta U_n = n \cdot \Delta U$, ihre Änderung im Zeitelement $d t$: $d U_n = n\, d U$, endlich ihre Entropie, als Summe aller Einzelentropien: $S_n = n S$.

Die gesamte Entropievermehrung in diesem System, die wir mit $d \sum_t$ bezeichnen wollen, ist gleich dem n fachen des Ausdrucks (323), da sich n einander ganz gleiche Vorgänge gleichzeitig und unabhängig voneinander abspielen, also:

[1] natürlich nur bezüglich der Energien, nicht etwa in dem Sinne, daß die Schwingungen kohärent sind.

(324) $$d\sum_t = n \cdot \frac{3}{5} dU \cdot \Delta U \cdot \frac{d^2 S}{dU^2} = -n \cdot \frac{3}{5} k \cdot \frac{dU \cdot \Delta U}{U(U + h\nu)}$$

oder mit Einführung von U_n, ΔU_n, dU_n und S_n:

(325) $$d\sum_t = \frac{3}{5} dU_n \cdot \Delta U_n \cdot \frac{d^2 S_n}{dU_n^2} = -\frac{3}{5} k \cdot \frac{dU_n \cdot \Delta U_n}{U_n\left(\frac{U_n}{n} + h\nu\right)}.$$

Vergleicht man diesen Ausdruck mit dem ganz analog gebauten in (323), so erhellt, daß die Entropievermehrung des Systems außer von der Schwingungszahl ν nicht allein von U_n, ΔU_n und dU_n abhängt, sondern auch von der Zahl n explizite.

Diese Folgerung, zu der der Ausdruck (227) der Entropie S mit Notwendigkeit führt, erschien mir, da die Gleichungen (323) und (325) sich sonst in ihrem Bau durch nichts unterscheiden, anfangs als auffällig und nicht wohl annehmbar. Daher setzte ich zu einer Zeit, als noch kein direkter Weg zur Berechnung der Strahlungsentropie bekannt war, statt der Gleichung (325) die folgende:

(326) $$d\sum_t = \frac{3}{5} dU_n \cdot \Delta U_n \cdot \frac{d^2 S_n}{dU_n^2} = -dU_n \cdot \Delta U_n \cdot f(U_n)$$

mit der Annahme, daß die positive Funktion f nur von U_n, nicht aber von n abhängt. Dadurch geht (323) über in:

(327) $$dS_t = \frac{3}{5} dU \cdot \Delta U \cdot \frac{d^2 S}{dU^2} = -dU \cdot \Delta U \cdot f(U),$$

und da $d\sum_t = n \cdot dS_t$, wie wir schon oben zur Ableitung von (324) benutzten, so folgt:

$$dU_n \cdot \Delta U_n \cdot f(U_n) = n\, dU \cdot \Delta U \cdot f(U)$$

oder, mit Einführung von U statt U_n:

$$n \cdot f(nU) = f(U).$$

Die allgemeine Lösung dieser Funktionalgleichung ist:

$$f(U) = \frac{\text{const}}{U}$$

und daraus folgt nach (327):

(328) $$\frac{d^2 S}{dU^2} = -\frac{5}{3} f(U) = -\frac{\text{const}}{U}.$$

Die Integration dieser Gleichung ergibt für S gerade diejenige Funktion von U, welche in der Beziehung (239) ausgedrückt ist, und die zum WIENschen Energieverteilungsgesetz führt. Daher hielt ich eine Zeitlang jene Beziehung für den allgemeinen Ausdruck der Entropie eines Oszillators der betrachteten Art, und dementsprechend das WIENsche Energieverteilungsgesetz für das allgemeine Spektralgesetz, was auch die Messungen von F. PASCHEN zu bestätigen schienen.[1]

Erst die Versuche von O. LUMMER und E. PRINGSHEIM[2] haben gezeigt, daß das WIENsche Energieverteilungsgesetz nur bedingungsweise gilt, nämlich dann, wenn die Strahlungsintensität, und daher auch die Energie U, einen verhältnismäßig kleinen Wert besitzt. Für größere Werte von U nähert sich dagegen, wie besonders deutlich aus den Messungen von H. RUBENS und F. KURLBAUM hervorgeht,[3] die Energiestrahlung merklich dem RAYLEIGHschen Gesetz (§ 154), nach welchem statt der Beziehung (328) die folgende gilt:

$$\frac{d^2 S}{d U^2} = - \frac{\text{const}}{U^2}, \tag{329}$$

wie man unmittelbar aus der Gleichung (244) findet.

Versucht man, die beiden in speziellen Gebieten, für kleine U und für große U, gültigen Formeln (328) und (329) in eine einzige allgemeinere zu vereinigen, so bietet sich als die einfachste Fassung die folgende dar:

$$\frac{d^2 S}{d U^2} = - \frac{\text{const}}{U(U + \text{const})},$$

welche mit (323) genau übereinstimmt und durch zweimalige Integration nach U zu der Gleichung (227) führt; denn die Abhängigkeit von der Schwingungszahl ν ist ja durch das WIENsche Verschiebungsgesetz (223) festgelegt.

---

[1] F. PASCHEN, Sitzungsber. d. k. preuß. Akad. d. Wissensch. 1899, p. 405, 893. WIED. Ann. **60**, p. 662, 1897.

[2] O. LUMMER und E. PRINGSHEIM, Verhandlungen der Deutschen Physikalischen Gesellschaft **2**, p. 163, 1900. Vgl. aber auch H. BECKMANN, Inaugural-Dissertation, Tübingen 1898. H. RUBENS, WIED. Ann. **69**, p. 582, 1899.

[3] H. RUBENS und F. KURLBAUM, Sitzungsber. d. k. preuß. Akad. d. Wissensch. vom 25. Oktober 1900, p. 929.

Dies ist der Weg, auf welchem die Beziehung (227) und das dadurch bedingte Strahlungsgesetz (234) ursprünglich gefunden wurde.

§ 190. Schluß. Die hier entwickelte Theorie der irreversibeln Strahlungsvorgänge gibt eine Erklärung dafür, wie in einem durchstrahlten, von Oszillatoren aller möglichen Eigenschwingungen erfüllten Hohlraum bei beliebig angenommenem Anfangszustand sich mit der Zeit ein stationärer Zustand herstellt, indem die Intensitäten und Polarisationen aller Strahlen sich nach Größe und Richtung gegenseitig ausgleichen. Aber die Theorie läßt noch eine wesentliche Lücke. Denn sie behandelt nur die Wechselwirkungen zwischen Strahlen und Oszillatorschwingungen der nämlichen Periode. Für eine bestimmte Schwingungszahl ist die vom zweiten Hauptsatz der Thermodynamik geforderte Vermehrung der Entropie bis zur Erreichung des Maximalwertes in jedem Zeitelement auf rein elektrodynamischem Wege nachgewiesen. Aber für alle Schwingungszahlen zusammengenommen bedeutet das so erreichte Maximum noch nicht das absolute Maximum der Entropie des Systems, und der entsprechende Strahlungszustand bezeichnet im allgemeinen nicht das absolut stabile Gleichgewicht (vgl. § 27). Denn darüber, wie sich Strahlungsintensitäten, die verschiedenen Schwingungszahlen entsprechen, gegenseitig ausgleichen, wie sich also aus einer anfangs vorhandenen beliebigen spektralen Energieverteilung mit der Zeit die normale, der schwarzen Strahlung entsprechende Energieverteilung entwickelt, erteilt diese Theorie keinerlei Aufschluß. Die hier der Betrachtung zugrunde gelegten Oszillatoren beeinflussen eben nur die Intensitäten der Strahlen, die ihrer Eigenschwingung entsprechen, sie vermögen aber nicht, deren Schwingungszahlen zu verändern, solange sie keine anderen Wirkungen ausüben und erleiden als daß sie strahlende Energie emittieren und absorbieren.[1]

Um einen Einblick in diejenigen Vorgänge zu erhalten, durch welche sich in der Natur der Austausch von Energie zwischen Strahlen verschiedener Schwingungszahlen vollzieht, bedürfte es jedenfalls auch der Untersuchung des Einflusses,

---

[1] Vgl. P. Ehrenfest, Wien. Ber. **114** [2a], p. 1301, 1905.

welchen eine Bewegung der Oszillatoren auf die Strahlungsvorgänge ausübt. Denn sobald die Oszillatoren sich bewegen, kommt es zu Zusammenstößen zwischen ihnen, und bei jedem Zusammenstoß müssen Wirkungen ins Spiel treten, welche die Schwingungsenergie der Oszillatoren noch in ganz anderer und in viel radikalerer Weise beeinflussen, als die einfache Emission und Absorption strahlender Energie. Das Endresultat aller derartiger Stoßwirkungen läßt sich allerdings mit Hilfe der im vierten Abschnitt angestellten Wahrscheinlichkeitsbetrachtungen voraussehen; wie im einzelnen und in welchen Zeiträumen aber dies Resultat zustande kommt, dies zu lehren wird erst die Aufgabe einer künftigen Theorie sein. Von einer solchen Theorie sind dann sicherlich auch weitergehende Aufschlüsse über die Konstitution der in der Natur vorhandenen Oszillatoren zu erwarten, schon deshalb, weil sie jedenfalls auch eine nähere Erklärung für die physikalische Bedeutung des universellen Wirkungselements h (§ 149) bringen muß, welche der des elektrischen Elementarquantums sicherlich nicht nachsteht.

Verzeichnis

der vom Verfasser bisher veröffentlichten Schriften aus dem Gebiete der Wärmestrahlung, mit Angabe derjenigen Paragraphen dieses Buches, in denen der nämliche Gegenstand behandelt ist.

Absorption und Emission elektrischer Wellen durch Resonanz. Sitzungsber. d. k. preuß. Akad. d. Wissensch. vom 21. März 1895, p. 289—301. Wied. Ann. **57**, p. 1—14, 1896. (§§ 112—115.)

Über elektrische Schwingungen, welche durch Resonanz erregt und durch Strahlung gedämpft werden. Sitzungsber. d. k. preuß. Akad. d. Wissensch. vom 20. Februar 1896, p. 151—170. Wied. Ann. **60**, p. 577—599, 1897. (§§ 104—115.)

Über irreversible Strahlungsvorgänge. (Erste Mitteilung.) Sitzungsber. d. k. preuß. Akad. d. Wissensch. vom 4. Februar 1897, p. 57—68. (§ 104 ff.)

Über irreversible Strahlungsvorgänge. (Zweite Mitteilung.) Sitzungsber. d. k. preuß. Akad. d. Wissensch. vom 8. Juli 1897, p. 715—717. (§ 168.)

Über irreversible Strahlungsvorgänge. (Dritte Mitteilung.) Sitzungsber. d. k. preuß. Akad. d. Wissensch. vom 16. Dezember 1897, p. 1122—1145. (§§ 169—171.)

Über irreversible Strahlungsvorgänge. (Vierte Mitteilung.) Sitzungsber. d. k. preuß. Akad. d. Wissensch. vom 7. Juli 1898, p. 449—476. (§§ 169—171.)

Über irreversible Strahlungsvorgänge. (Fünfte Mitteilung.) Sitzungsber. d. k. preuß. Akad. d. Wissensch. vom 18. Mai 1899, p. 440—480. (§§ 175—187. § 159.)

Über irreversible Strahlungsvorgänge. Drudes Ann. **1**, p. 69—122, 1900. (§§ 128—132. §§ 175—187. § 159.)

Entropie und Temperatur strahlender Wärme. Drudes Ann. **1**, p. 719—737, 1900. § 101. §§ 188 f. § 162.

Über eine Verbesserung der Wienschen Spektralgleichung. Verhandlungen der Deutschen Physikalischen Gesellschaft **2**, p. 202—204, 1900. § 189.

Ein vermeintlicher Widerspruch des magneto-optischen Faraday-Effektes mit der Thermodynamik. Verhandlungen der Deutschen Physikalischen Gesellschaft **2**, p. 206—210, 1900.

Kritik zweier Sätze des Herrn W. Wien. Drudes Ann. **3**, p. 764—766, 1900.

Zur Theorie des Gesetzes der Energieverteilung im Normalspektrum. Verhandlungen der Deutschen Physikalischen Gesellschaft **2**, p. 237—245, 1900. (§§ 146—152. § 158.)

Über das Gesetz der Energieverteilung im Normalspektrum. Drudes Ann. p. 553—563, 1901. (§§ 145—157.)

Über die Elementarquanta der Materie und der Elektrizität. Drudes Ann. p. 564—566, 1901. (§ 158.)

Über irreversible Strahlungsvorgänge (Nachtrag). Sitzungsber. d. k. preuß. Akad. d. Wissensch. vom 9. Mai 1901, p. 544—555. Drudes Ann. **6**, p. 818—831, 1901. (§ 187.)

Vereinfachte Ableitung der Schwingungsgesetze eines linearen Resonators im stationär durchstrahlten Felde. Physikalische Zeitschrift **2**, p. 530 bis p. 534, 1901. (§§ 116—123.)

Über die Natur des weißen Lichtes. Drudes Ann. **7**, p. 390—400, 1902. (§ 160. § 176. § 180.)

Über die von einem elliptisch schwingenden Ion emittierte und absorbierte Energie. Archives Néerlandaises, Jubelband für H. A. Lorentz, 1900, p. 164—174. Drudes Ann. **9**, p. 619—628, 1902.

Über die Verteilung der Energie zwischen Äther und Materie. Archives Néerlandaises, Jubelband für J. Bosscha, 1901, p. 55—66. Drudes Ann. **9**, p. 629—641, 1902. (§§ 138—144.)